Conservation of Freshwater Fishes: Options for the Future

EDITED BY

M.J. COLLARES-PEREIRA

Centro de Biologia Ambiental, Faculdade de Ciências, Universidade de Lisboa, Lisboa, Portugal

I.G. COWX

Hull International Fisheries Institute, University of Hull, UK

and

M.M. COELHO

Centro de Biologia Ambiental, Faculdade de Ciências, Universidade de Lisboa, Lisboa, Portugal

Fishing News Books
An imprint of Blackwell Science

Blackwell Science

Fishing News Books
A Division of Blackwell Science Ltd
Editorial Offices:
Osney Mead, Oxford OX2 0EL
25 John Street, London WC1N 2BS
23 Ainslie Place, Edinburgh EH3 6AJ
350 Main Street, Malden
MA 02148 5018, USA
54 University Street, Carlton
Victoria 3053, Australia
10, rue Casimir Delavigne
75006 Paris, France

Other Editorial Offices:

Blackwell Wissenschafts-Verlag GmbH
Kurfürstendamm 57
10707 Berlin, Germany

Blackwell Science KK
MG Kodenmacho Building
7-10 Kodenmacho Nihombashi
Chuo-ku, Tokyo 104, Japan

Iowa State University Press
A Blackwell Science Company
2121 S. State Avenue
Ames, Iowa 50014-8300, USA

First published 2002

Set in Times and produced by
Gray Publishing, Tunbridge Wells, Kent
Printed and bound in Great Britain by
MPG Books Ltd, Bodmin, Cornwall

DISTRIBUTORS
Marston Book Services Ltd
PO Box 269
Abingdon
Oxon OX14 4YN
(*Orders:* Tel: 01865 206206
Fax: 01865 721205
Telex: 83355 MEDBOK G)

USA and Canada
Iowa State University Press
A Blackwell Science Company
2121 S. State Avenue
Ames, Iowa 50014-8300
(*Orders:* Tel: 800-862-6657
Fax: 515-292-3348
Web: www.isupress.com
email: orders@isupress.com)

Australia
Blackwell Science Pty Ltd
54 University Street
Carlton, Victoria 3053
(*Orders:* Tel: 03 9347 0300
Fax: 03 9347 5001)

A catalogue record for this title
is available from the British Library

ISBN 0-85238-286-3

Library of Congress
Cataloging-in-Publication Data
is available

For further information on
Fishing News Books, visit our website:
www.blacksci.co.uk/fnb/

Contents

Foreword

From my experiences with the Habitats Directive and the numerous nature conservation projects which the European Commission is supporting through its LIFE-NATURE financial instrument, I am aware that the conservation of aquatic biodiversity, and especially the conservation of fish diversity, remains one of the most difficult challenges facing the European Union (EU) in preserving our natural biological diversity.

While site-oriented approaches often prove to be very effective in providing a favourable conservation status for terrestrial species and habitats, the same does not generally hold for most fish species and other aquatic organisms. Threats such as the introduction of non-native species, water abstraction, dam building and pollution can have far-reaching consequences on a large geographical scale, often with irreversible consequences. For many migratory fish species, over-exploitation at sea or along their river migration routes also have serious consequences, even if the spawning grounds may still be in optimal conditions. Think about what happened to the once thriving (and economically important) salmon and sturgeon populations of many European waters.

One has also to consider the fact that many fish populations are strongly differentiated taxonomically, on a hydrological river basin or even sub-basin level. Endemism and local differentiation are much more important for aquatic species or lakes and running waters than it is for most terrestrial taxa. As was shown on many occasions during the conference, this is widely recognised now amongst fish conservationists, as is the need to conserve local stocks and breeds. This in turn requires special conservation approaches and good scientific knowledge about the actual taxonomic and ecological situation.

One of the most exciting parts of the conference was the presentations of practical fish conservation projects: from the last wild European sturgeon in the Garonne Basin and the large predatory huchen in Austria to the smaller saramugo project in the Guadiana Basin and the *Ladigesocypris* project on Rhodes Island, the projects dealing with cyprinondonts in coastal lagunes of Spain and the apron project of the River Rhône. Even though these projects are covering a wide range of species living under very different conditions, the need to carry out proper public awareness raising and to search for the consensus of all affected stakeholders and user-groups appears to be a recurring and outstanding requirement to their success.

Innovative approaches, and notably those developed through LIFE-NATURE projects, are thus necessary. This was underlined by many of the participants, coming from EU countries as well as from further afield.

Even though the improvement of scientific knowledge on the requirements of fish species and their habitats remains a major objective of all conservation actions, what we mostly need to learn is how to establish the necessary partnerships and structures, how to search for consensus and how to change people's minds if we are to achieve sustainable solutions for the long-term survival of European freshwater fishes.

By organising such a productive symposium, the whole team of the 'Saramugo' project gave a strong impetus to an efficient collaboration among 'fishwatchers'. The LIFE award they have received from the European Commission for their networking efforts is fully deserved.

Let us use these proceedings as a tool to increasingly focus our attention on our invaluable 'freshwater fish' capital.

Bertrand Delpeuch
Deputy Head of Unit 'Nature and Biodiversity'
European Commission – Directorate General Environment

Preface

Freshwater ecosystems have suffered the most intense intervention of all ecosystems over the past 100 years of human history, with severe consequences with respect to fish biodiversity. Therefore scientists, managers and policy makers must face the challenge of developing integrated assessment and decision frameworks.

In October/November 2000, an international symposium on *Freshwater Fish Conservation: Options for the Future* was promoted under the auspices of the EU LIFE project 'A conservation strategy for *Anaecypris hispanica*' in Algarve (Portugal). The objectives of the symposium were:

(1) to bring together experts in fish conservation from temperate and tropical regions, in order to exchange knowledge among countries and to present specific appraisals; to stimulate and encourage those working in the field by highlighting successes and failures; and to identify current and future challenges for the conservation of freshwater fish;
(2) to use the opportunity to assemble the experiences of the current EU LIFE-NATURE projects on fish, following a recommendation of the workshop on 'Mediterranean Stream Fish Ecology and Conservation' held in November 1999 in Rhodes (Greece).

The conference and these proceedings are organised around five main themes:

(1) geographical overview, including a review of the present status of threatened species, especially in Europe;
(2) biology of rare and endangered species, including case studies from the EU LIFE-NATURE projects;
(3) analysis of constraints and opportunities, including the analysis of threats to fish populations and problems related with ecosystem use and sustainability;
(4) aspects of conservation genetics;
(5) methods of managing freshwater fish conservation, including breeding endangered species for stock enhancement, the role of legislation, nature reserves, institutions and policy making in fish conservation.

These proceedings contain selected papers from the symposium and highlight many of the key issues. We hope they will stimulate fish scientists and people involved in freshwater ecosystem conservation to collaborate in future research on a worldwide perspective, aiming to promote wider knowledge. In this sense, it is expected that integrated approaches to land and water use, especially in semi-arid and arid regions, which presently face severe conservation problems due to major shifts in hydrological regimes, will be implemented under the auspices of sustainable regional development.

The production of these proceedings involved considerable effort by a number of colleagues, who critically refereed the papers and made useful suggestions for their improvement, whom we would like to thank: Z. Adamek (Czech Republic), P.R. Almeida (Portugal), J.P. Andrade (Portugal), M. Aprahamian (UK), P. Banarescu (Roumania), A. Brabrand (Norway), F.W.H. Beamish (Canada), P. Berrebi (France), P.G. Bianco (Italy), V.J. Birstein (USA), N. Bogutskaya (Russia), J. Bohlen (Germany), D. Buth (USA), J. Cambray (South Africa), L. Chapman (USA), R. Cortes (Portugal), A. Crivelli (France), A. de Sostoa (Spain), I. Doadrio (Spain), P. Economidis (Greece), B. Elvira (Spain), A. Ferguson (Ireland), J. Freyhof (Germany), P. Gaudin (France), L. Gordo (Portugal), M. Goren (Israel), B. Haenfling (UK), G. Helfman (USA), D. Hefti (Switzerland), K. Hensel (Slovak Republic), P. Hickley (UK), J. Holcik (Slovak Republic), C. Hubbs (USA), S. Hughes (UK), L. Kaufman (USA), A. Kirchhofer (Switzerland), M. Kottelat (Switzerland), A. Lelek (Germany), J. Lobon-Cervia (Spain), S. Lusk (Czech Republic), J. Marques (Portugal), R. Mann (UK), G. Meffe (USA), C.K. Minns (Canada), R. Muller (Switzerland), A. Palau (Spain), E. Pattee (France), M. Penaz (Czech Republic), T. Penzcak (Poland), H. Persat (France), P. Poncin (Belgium), M. Povz (Slovenia), P. Rab (Czech Republic), T. Saat (Estonia), S.K. Saltveit (Norway), I. Schlosser (USA), S. Schumtz (Austria), P. Shaw (UK), P. Skelton (South Africa), D. Skibinski (UK), M.Th. Stoumboudi (Greece), J. Wanzenbock (Austria), S. Weiss (Austria), R. Welcomme (UK), I. Winfield (UK), R. Wootton (UK), E. Vasili'eva (Russia), V. Vasiliev (Russia), J. Velasquez-Cano (Spain).

Finally, we would like to thank Luís Moreira da Costa for his considerable assistance in the preparation and running of this symposium and all the remaining 'freshwater fish team' of the Faculty of Sciences of Lisbon for the continuous help and enthusiasm. Also we are grateful to the following national (FCT, INAG, DGF, UL, FCG, FLAD, FFCUL, FCUL, MB and IPAMB) and international (DG-Environment, EC) institutions for sponsoring this conference, thus ensuring the international coverage of the issues and the overall success achieved.

Maria João Collares-Pereira
University of Lisbon, Portugal

Ian Cowx
University of Hull, UK

Manuela Coelho
University of Lisbon, Portugal

Section I
Geographical overview

Chapter 1
Freshwater fish diversity in Germany, threats and species extinction

J. FREYHOF*
Institute of Freshwater Ecology and Inland Fisheries, Müggelseedamm, Berlin, Germany

Abstract

The fresh waters of Germany have been heavily polluted, all rivers and most streams have been modified, and most lakes suffer eutrophication, and all water bodies have been stocked with both alien and native fish. Applying the IUCN Classification, 46.5% of fish species are threatened or extinct. Diversity of salmoniform fishes has not been studied in recent times, consequently the exact number of species in Germany is unknown. Therefore, it can be only estimated that the German ichthyofauna comprises some 100 freshwater species (including the anadromous and catadromous species), and 14 non-native species have become more or less established. During the 19th and 20th centuries, 12 species disappeared or became extinct. *Romanogobio uranoscopus* and *Cottus poecilopus* disappeared from Germany. *Salvelinus profundus* and *Coregonus gutturosus* are globally extinct as a result of eutrophication in Lake Constance. There are no data on *Coregonus bavaricus*, which is globally threatened. *Coregonus lucinensis* still exists in one small lake. Because of massive stocking with fish of various origins, most native stocks and endemic species of *Coregonus*, *Salmo* and *Salvelinus* might be extinct. *Acipenser ruthenus*, *Coregonus oxyrinchus*, *Salmo salar* and *Rutilus meidingeri* disappeared from the territory of Germany, although they have been recently restored from foreign sources. However, to date recovery programmes for these species have not been successful. Furthermore the chances of restoration of *Huso huso*, *Acipenser stellatus*, *Acipenser gueldenstaedti* and *Acipenser sturio* are very low. By constrast, *Alosa alosa* and *Pelecus cultratus* have returned after they disappeared, and many populations of rheophilous species are increasing in distribution and abundance. Thus, the national Red List tends to show a positive trend.

Keywords: biodiversity, conservation, endemic species, fish communities, stocking.

1.1 Introduction

The loss of biodiversity, and especially of genetic and species diversity, represents a loss to present and future generations. Therefore, protection of biodiversity, and to buffer the negative effects of increasing economic development, must be high priority within society. In the post-glacial era, only very restricted areas of natural habitats in Central Europe remain unaffected (directly or indirectly) by human activities (Lang 1994). Deforestation, pollution, building of weirs for watermills and intensive fishing,

---

*Correspondence: Institute of Freshwater Ecology and Inland Fisheries, Müggelseedamm 310, D-12561 Berlin, Germany (email: freyhof@igb-berlin.de).

have influenced fish populations for more than 5000 years. All rivers and most streams have been modified by dams and bank regulations, and for more than 100 years no river section remains in Germany where new backwaters or oxbows can be created by natural river processes. Since the late 19th century, industrialisation of Germany (covering 357 000 km^2; population – 83.5 million; 234 inhabitants km^{-2}), has progressed tremendously and heavy pollution has affected fish populations in nearly all rivers and streams (Bless 1978). During the last 200 years, most lakes have become eutrophic and all waterbodies have been stocked with both alien and native fish.

For more than 25-five years, the status of fish species has been assessed on a local and national scale to highlight taxa threatened with extinction, and to promote their conservation (Lelek 1976; Blab, Nowak, Trautmann, Sukopp 1977; Bless 1978; Lelek 1980; Bless, Lelek & Waterstraat 1994). The Red Data Book (Bless *et al.* 1994) lists 70 species of lampreys and freshwater fishes. Of these species, 72% are considered endangered, and four have disappeared from Germany. However, Bless *et al.* (1994) noted that taxonomic problems existed in some groups of salmoniform fishes, and grouped nearly all species into one taxon. Kottelat (1997) emphasised that there is no reason to underestimate biodiversity and, applying modern concepts for evaluation of European ichthyofauna, made some improvements in this situation. Furthermore, chaotic taxonomy of the salmoniform fishes was abused by fishery biologists to promote their business by stocking more productive forms into most lakes in Germany. This might have lead to the biggest post-glacial extinction crisis of vertebrates in Central Europe. Since we are still in a very politically, economically, socially and ecologically diverse world, it remains difficult to obtain objective, scientifically-based information on the current status of globally threatened biodiversity. As species are bearers of genetic diversity and building blocks of ecosystems, information on their conservation status and distribution is fundamental to making competent decisions about protection of biodiversity from the local to global scales. Efficient conservation of biological diversity requires well-prepared programmes, which should involve proper studies, protection measures, recovery and wise management of species and their habitats. However, before these programmes can be conducted successfully, it is necessary to resolve taxonomic problems of the species concerned.

Modern European concepts of nature conservation (Fauna–Flora-Habitat, NATURA (MAISC) 2000) and IUCN Guidelines highlight the responsibility of any government, including Germany, to study and conserve endemic and endangered species of plants and animals. This chapter aims to give an overview of threatened German fish, reviewing information's often hidden in regional literature, and it also describes the development of the main factors threatening the fish communities in Germany in the last two centuries.

1.2 Remarks to threatened fish species in Germany

In the following, a brief review of the status of fish species in Germany is provided. It involves extinct species, species that have vanished from Germany, recently recolonised German rivers, endemic to Germany, likely to been overseen and non-native species. A complete checklist of freshwater fishes of Germany, with IUCN (2000) Classifications, is given in Table 1.1. Threatened endemic or nearly endemic species not listed by

Table 1.1 Checklist of freshwater fishes of Germany

Species	National criteria	National status	Global status
Petromyzontidae			
Eudontomyzon vladykovi Oliva & Zanandrea	Few locations	EN	LR
Lampetra fluviatilis (L.)		nt	LR
Lampetra planeri (Bloch)		nt	LR
Petromyzon marinus L.		nt	
Acipenseridae			
Acipenser gueldenstaedtii Brandt & Ratzeburg		EX	EN
Acipenser ruthenus L.		EX	VU
Acipenser stellatus Pallas		EX	EN
Acipenser sturio L.		EX	CR
Huso huso (L.)		EX	EN
Anguillidae			
Anguilla anguilla (L.)	Stocking dependent	LR/cd	
Clupeidae			
Alosa alosa (L.)	Few locations	EN	DD
Alosa fallax Lacepede	Few locations	VU	DD
Cyprinidae			
Abramis ballerus (L.)	Decline	VU	nt
Abramis bjoerkna (L.)		nt	
Abramis brama (L.)		nt	
Abramis sapa (Pallas)		nt	
Alburnoides bipunctatus (Bloch)	Few locations	VU	nt
Alburnus alburnus (L.)		nt	
Aspius aspius (L.)		nt	DD
Barbus barbus (L.)		nt	
Carassius auratus (L.)		Non-native	
Carassius carassius (L.)	Strong decline	EN	LR
Carassius gibelio (Bloch)		nt	
Chalchalburnus chalcoides Gueldenstaedt		nt	DD
Chondrostoma nasus (L.)		nt	
Ctenopharyngodon idella (Val.)		Non-native/ stocked	
Cyprinus carpio L.		nt	
Gobio gobio (L.)		nt	
Hypophthalmichthys molitrix (Val.)		Non-native/ stocked	
Hypophthalmichthys nobilis (Rich.)		Non-native/ stocked	
Leucaspius delineatus (Heckel)		nt	
Leuciscus cephalus (L.)		nt	
Leuciscus idus (L.)		nt	
Leuciscus leuciscus (L.)		nt	
Pelecus cultratus (L.)	Few locations	VU	DD
Phoxinus phoxinus (L.)		nt	
Pimephales promelas Rafinesque		Non-native	

Table 1.1 continued

Species	National criteria	National status	Global status
Pseudorasbora parva (Schlegel)		Non-native	
Rhodeus amarus (Bloch)	Few locations	VU	nt
Romanogobio albipinnatus (Lukasch)		nt	DD
Romanogobio uranoscopus (Agassiz)		EX	DD
Rutilus meidingeri (Heckel)		EX	EN
Rutilus pigus (Lacepède)		DD	DD
Rutilus rutilus (L.)		nt	
Scardinius erythrophthalmus (L.)		nt	
Telestes souffia (Risso)	Few locations	EN	nt
Tinca tinca (L.)		nt	
Vimba vimba (L.)	Few locations	VU	nt
Balitoridae			
Barbatula barbatula (L.)		nt	
Cobitidae			
Cobitis taenia L.		nt	
Misgurnus angullicaudatus (Cantor)		Non-native	
Misgurnus fossilis (L.)	Strong decline	VU	LR
Siluridae			
Silurus glanis L.		nt	
Ictaluridae			
Ameiurus nebulosus (Lesueur)		Non-native	
Ameiurus melas (Rafinesque)		Non-native	
Salmonidae			
Hucho hucho (L.)	Few locations	EN	EN
Oncorhynchus mykiss (Walbaum)		Non-native	
Salmo salar L.		EX	nt
Salmo trutta L.		nt	
Salvelinus fontinalis (Mitchell)		Non-native/ stocked	
Salvelinus monostichus Heckel, 1851	One location	CR	CR (new)
Salvelinus profundus Schillinger		EX	EX (new)
Salvelinus umbla (L.)	Few locations	VU	VU (new)
Thymallidae			
Thymallus thymallus (L.)		nt	
Coregonidae			
Coregonus albula (L.)		nt	DD
Coregonus arenicolus Kottelat	One location	EN	EN (new)
Coregonus bavaricus Hofer	One location	CR	CR (new)
Coregonus gutturosus (Gmelin)		EX	EX (new)
Coregonus hoferi Berg	Data deficient	VU	VU (new)
Coregonus lucinensis Thienemann	One location	CR	CR (new)
Coregonus macrophthalmus (Nüsslin)	One location	EN	EN (new)
Coregonus maraena (Bloch)	Few locations	EN	nt
Coregonus oxyrinchus (L.)		EX	CR (new)
Coregonus peled (Gmelin)		Non-native	
Coregonus renke Schrank	Data deficient	VU	VU (new)
Coregonus wartmanni (Bloch)	One location	EN	EN (new)
Coregonus widegreni Malmgren	Distribution unknown	DD	DD (new)

Table 1.1 continued

Species	National criteria	National status	Global status
Coregonus sp. 'Stechlin'	One location	CR	CR (new)
Coregonus sp. 'Rhine'	Distribution unknown	DD	DD (new)
Osmeridae			
Osmerus eperlanus (L.)		nt	DD
Esocidae			
Esox lucius L.		nt	
Umbridae			
Umbra pygmaea (DeKay)		Non-native	
Lotidae			
Lota lota (L.)		nt	
Gasterosteidae			
Gasterosteus aculeatus L.		nt	
Pungitius pungitius (L.)		nt	
Cottidae			
Cottus gobio L.		nt	
Cottus poecilopus Heckel		EX	nt
Percidae			
Gymnocephalus cernuus (L.)		nt	
Gymnocephalus baloni Holcík & Hensel		DD	DD
Gymnocephalus schraetser (L.)		nt	VU
Zingel streber (Siebold)	Few locations	VU	VU
Zingel zingel (L.)	Few locations	VU	VU
Perca fluviatilis L.		nt	
Sander lucioperca (L.)		nt	
Centrachidae			
Lepomis gibbosus (L.)		Non-native	
Gobiidae			
Neogobius kessleri (Günther)	Invader		DD
Pomatoschistus minutus (Pallas)		nt	
Pomatoschistus microps (Kroyer)		nt	
Proterorhinus marmoratus (Pallas)		nt	
Pleuronectidae			
Pleuronectes flesus L.		nt	

The global status is adapted from IUCN (2000). Endemic or nearly endemic species not listed by IUCN (2000) are classified following the criteria given by IUCN (1995). National status following Bless *et al.* (1994) and own estimations. EX, extinct; CR, critically endangered; EN, endangered; VU, vulnerable; LR, lower risk; DD, data deficient; cd, conservation dependant; nt, not threatened; new, not listed by IUCN (2000), first classified.

IUCN (2000) are classified following the criteria given by IUCN (1995). Nationally threatened categories are based on the recommendations for the use at the regional level given by IUCN (1995). Species status follows Bless *et al.* (1994) and reassessments based on best available data.

1.2.1 *Extinct species*

Lake Constance deepwater charr Salvelinus profundus *Schillinger;*
Lake Constance Kilch Coregonus gutturosus *(Gmelin)*

Eutrophication and decreased oxygen concentration in the deepwater layers (Nümann 1972) of Lake Constance may be responsible for the extinction of these two species after 1950. *Salvelinus profundus* and *Coregonus gutturosus* are considered globally extinct (Kottelat 1997). The taxonomic problems relating to these species are discussed by Kottelat (1997). The late summer (September–October) spawning and deepwater dwelling species *C. gutturosus* has not been found since late 1960 (Ruhle & Kindle 1992). Today, unidentified small *Coregonus* and *Salvelinus* specimens live in deep layers of Lake Constance. However, these specimens do not correspond with diagnostic characters of *C. gutturosus* (Ruhle & Kindle 1992) and *S. profundus* (Kottelat 1997).

1.2.2 *Species vanished from Germany*

Russian sturgeon Acipenser gueldenstaedtii *Brandt & Ratzeburg; stellate sturgeon* Acipenser stellatus *Pallas and beluga* Huso huso *(L.)*

Four species of sturgeons historically occurred in the German part of the River Danube. The only resident species was the sterlet *Acipenser ruthenus* L., which is supported by stocking. Hensel and Holcík (1997) mentioned *Acipenser gueldenstaedti*, *Acipenser stellatus* and *Huso huso* to have occurred occasionally in Germany, but it is not known if these species regularly spawned so far up river. However, from the Austrian stretch of the middle River Danube, intensive catches of *H. huso* and other sturgeons by fishing fences have been documented since 3500 BC (Balon 1968a, b; Weber 1989), and since the late 16th century a massive decline of Danubian sturgeons is documented (Hensel & Holcík 1997). It is likely that these species, reported as occasional visitors to the German part of the River Danube, were only remnants of an early devastated population. In German waters, the last spawning probably occurred during the 19th century (Siebold 1863). The last *H. huso* was caught at the German city of Passau at the beginning of the 20th century (Mohr 1962). At the end of a long period of overfishing, construction of dams in Austria in the 20th century, and the Iron Gates Dam (Romania) built in 1969, completely blocked the migrations of anadromous Danubian sturgeons to Central Europe.

Common sturgeon Acipenser sturio *L.*

Anadromous stocks of sturgeons were known from all rivers draining into the North Sea and the Baltic Sea. Their abundance started to decline during the middle ages and long-term overexploitation can be blamed for the downward trend of sturgeon populations for more than 1000 years (Benecke 1986). As a result of massive overfishing, damming, river regulation and pollution, this critically endangered species has vanished from Germany (Debus 1995). After 1950, the species was caught only accidentally. The last adult specimen from a German freshwater habitat was caught in 1967 in the River Eider (Spratte & Hartmann 1997). Taxonomic problems and similarities between American Atlantic sturgeon, *Acipenser oxyrinchus* Mitchell, and European Baltic sturgeon (Artyukhin & Vecsei 1999), and the high demand by fisheries societies to stock sturgeons will undoubtedly lead to the introduction of this alien species into the Baltic Sea. This is a good example, of how taxonomic confusion can be misused for economic purposes and why it is not just an academic problem.

Stone gudgeon Romanogobio uranoscopus *(Agassiz)*

This species was described from the River Isar at the German city of Munich in 1828. There are no confirmed subsequent records of *Romanogobio uranoscopus* in German waters after 1828. There are occasional statements of *R. uranoscopus* (Kussmaul, Hoffmann & Gessler 1991) being referred to as other species, especially as *Romanogobio albipinnatus* (Lukasch). The reasons why this species vanished from Germany are not known. It is found in the Austrian section of the River Danube (Mikschi & Wolfram-Wais 1999; G. Zauner, personal communication), and therefore, it might be present but overlooked in the German part of this river.

Siberian sculpin Cottus poecilopus *Heckel*

Occurrence of *Cottus poecilopus* was always restricted to several oligotrophic lakes (Großer Plöner, Selenter, Schöh, Breiter Luzin, Schmaler Luzin, Zansen) in northern Germany, (Dunker 1960; A. Waterstraat, personal communication). The reasons for its disappearance from Germany are not known. All known populations disappeared during the 20th century, and the last record dates back to 1978, from the Lake Breiter Luzin (Spieß & Waterstraat 1989; Winkler, Hamann & Waterstraat 1991).

1.2.3 *Species vanished from Germany involved in restitution projects*

Sterlet Acipenser ruthenus *L.*

This small potamodromous sturgeon was regularly found in the German River Danube and a large, resident population was known between Regensburg and Passau (Hensel & Holcík 1997). *Acipenser ruthenus* has been extirpated from the German part of the River Danube since the late 19th century (Bless *et al.* 1994). Multiple reasons, such as construction of weirs, river regulation and water pollution, contributed to the loss of the population. *A. ruthenus* was stocked into the impoundments of the German stretch of the River Danube (E. Bohl, personal communication). This project did not result in an established population. Currently, attempts have stopped and *A. ruthenus* has not been stocked in recent years (J. Gessner, personal communication). Recently, numbers of *A. ruthenus* have increased in the River Rhine basin, where this species might be introduced.

Atlantic salmon Salmo salar *L.*

The Atlantic salmon has a high economic value in the North Sea and the Baltic Sea basins. Its abundance has declined since the middle ages for multiple reasons

(Hoffmann 1996), but particularly during the late 19th century. Between 1940 and 1960, overfishing, building of weirs and water pollution eliminated all German populations (VDSF 2000). Various methods of stocking have been applied since about 1850 but it did not prevent the species' extinction (Schmidt 1996). Since 1978, an increasing number of restitution programmes has been initiated all over the historic range of the species (VDSF 2000). However, none of these projects met the IUCN guidelines for re-introductions, and many of them have not been intended to establish viable populations but just to have salmon for angling. To date, no restitution project has resulted in a self-sustainable population. However, *Salmo salar* is an important 'flagship' species, and many improvements concerning water quality and aquatic habitats for *S. salar*, have improved conditions for other species.

Houting Coregonus oxyrinchus *(L.)*

Coregonus oxyrinchus was a frequent anadromous species in all German rivers draining to the North Sea. Populations of this species started to decline during the late 19th century, and collapsed in the first three decades of the 20th century. The last specimens were caught during World War II (Bauch 1958; Lelek & Buhse 1992). Reasons why this species vanished from Germany are unknown, although overfishing, river regulation and pollution are generally considered as important. A small population has survived only in the River Vida (Denmark), where the houting was rediscovered in 1982. A restitution programme, based on transferred fishes from the River Vida to the German River Treene has been running since 1987. Spawners from this river were also used for artificial reproduction, and offspring obtained were stocked to the River Elbe basin and to the lower River Rhine (Jäger 1999). However, most of these stocking projects have not been aimed at establishing viable populations, and no restitution project has resulted in a self-sustainable population independent of stocking, thus *C. oxyrinchus* should be still considered a species vanished from Germany. Another riverine *Coregonus* species, which exists in the River Rhine is apparently not conspecific with *C. oxyrinchus* (Freyhof & Weibel 1999). Taxonomy and/or any other parameters of these fishes have not been studied.

Pearl roach Rutilus meidingeri *(Heckel)*

In Germany, the only habitat of pearl roach was the prealpine Lake Chiem belonging to the River Danube basin. *Rutilus meidingeri* was of no commercial value. The populations collapsed during the past 50 years, and the subsequent disappearance of the species has been ignored; and the last specimen was caught in 1993 (A. Hartl, personal communication). Since 1995, 40 000–70 000, 0+ *R. meidingeri* from the Austrian Lake Atter have been stocked each year into Lake Chiem (Fuchs, Schlee, Rottmann & Stein 1999). However, no surviving juveniles from the stocked specimens have been recorded. Recently, a juvenile specimen was recorded in the River Alz, 30 km below the Lake. *Rutilus meidingeri* was known to spawn in the affluent of Lake Chiem, the River Alz. During the 1960s, a weir was built 5 km below Lake Chiem and this has interrupted the migration of juvenile *R. meidingeri* which bred in the river from entering the lake. *Rutilis meidingeri* occurs in the Austrian section of the River Danube

(Mikschi & Wolfram-Wais 1999) and was also found in the Slovakian River Danube (Hensel 1979). Therefore it may be also present in the German stretch. However, these specimens could disperse from Austrian lakes, and there is no evidence, that pearl roach spawns in the River Danube. This species is now endemic to Austria, and only three populations remain, in the lakes Atter, Mond and Wolfgang.

1.2.4 *Species that recently recolonised German rivers*

Allis shad Alosa alosa *(L.)*

Prior to the beginning of the 20th century, *Alosa alosa* was an economically-important species, especially in the basin of the River Rhine. However, the last specimen from the River Rhine was caught in 1963 (Bartl & Troschel 1997). The species disappeared from other German rivers draining to the Baltic Sea and the North Sea basins even earlier. Since 1978, the number of recorded allis shad has increased, although it is still very low (less than 30 individuals per year). Despite recent investigations on 0+ fish communities, little is known about the reproductive aspects of this species in the River Rhine. It is possible that allis shad found in the River Rhine are only vagrants from the large French populations. Recently some allis shad were recorded from the coastal lakes of the German part of the Baltic Sea (H.M. Winkler, personal communication). Massive overfishing during the beginning, and heavy pollution in the middle of the 20th century, may be the reasons why allis shad vanished from the River Rhine (Bartl & Troschel 1997).

Razorfish Pelecus cultratus *(L.)*

Pelecus cultratus were irregular visitors in the German section of the River Danube; their massive occurrence is said to be associated with war, pest and hunger in the past (Heckel & Kner 1858). This species was considered to have vanished from freshwater habitats in Germany (Paepke 1981; Bless & Lelek 1984), being just a rare visitor in brackish lakes at the German Baltic coast, penetrating from Poland (Spiess & Waterstraat 1989; Winkler 1989). The reasons for the decline of its population remain unknown, although damming throughout the River Danube might have stopped its migration. However, in the 1980s, *P. cultratus* recolonised the lower part of the German section of the River Danube, and has established in some reservoirs (Harsanyi 1986).

1.2.5 *Globally threatened species endemic or nearly endemic to German waters*

Currently it is impossible to answer the question what is the exact number of native, endangered or extinct species or stocks of salmoniform fishes in Germany. In many lakes, *Coregonus* spp., *Salvelinus* spp. and lacustrine trout *Salmo trutta* s.l. have been or are in a process of speciation and it is difficult to estimate the level of this differentiation using traditional methods. As mentioned above, this has resulted in a chaotic taxonomy of salmoniform fishes. The management of salmoniform fishes is characterised by absolute ignorance of the conservation principles for local natural resources. No serious attempt to stop translocations or study the impact of stocking on local populations has been carried out, and it is not known how many of the historically described species have gone extinct. However, all local, lacustrine, endemic species that still survive are also in acute danger of extinction, because their generally small global range makes them vulnerable to any human activity, especially to eutrophication and stocking of allochtonous fishes.

Prealpine charr Salvelinus umbla *s.l. (L.)*

Lacustrine prealpine charr occur in many German lakes. Up to three 'forms' have been reported from a single lake. It is not known whether all these sympatric 'forms' represent different species, or just individual life history traits or various age classes. Based on Behnke (1972, 1980), Kottelat (1997) grouped all prealpine charrs with *Salvelinus umbla*. However, the morphological data available are of very limited use, and most prealpine stocks cannot be diagnosed and linked with names. Brunner, Douglas & Bernatchez (1998) studied Central European charr populations and demonstrated that they are related to each other following the river basins to which the lakes belong. Furthermore, Englbrecht (2000) demonstrated that it is possible to distinguish charrs from different German lakes in the River Danube basin by genetic markers. Both studies suggested that stocking may not have had an impact on all local populations, but in some cases it is likely. Englbrecht (2000) was able to distinguish an endemic charr from the Danubian Lake König from charrs living in other Danubian lakes, thus the name *Salvelinus monostichus* Heckel is valid. Further studies could lead to the description of some new, local endemic species.

Whitefish Coregonus *species*

Coregonids have been stocked from many sources and even hybridised in aquaculture and subsequently stocked. The stocking material came from Prealpine lakes, Baltic lakes and Scandinavia. It is not known if the stocked fishes reproduced successfully in the lakes or if local populations mixed with them. Currently, mixing of stocks is not as intensive as it used to be just a few decades ago and now, most lakes are stocked with native material coming from the recipient lake. Douglas, Brunner & Bernatchez (1999) found *Coregonus* populations in Swiss lakes were not affected by stocking. Therefore, there is still a chance that some native populations and local endemic species have resisted intensive stocking and eutrophication in their respective lakes.

Felchen Coregonus *spp. of Lake Constance*

Four different endemic *Coregonus* species have been known from Lake Constance (a part of this lake is situated in Austria and Switzerland). The endemic *C. gutturosus* is extinct. *Coregonus arenicolus* Kottelat (a littoral spawner) is rare in Major Lake

Constance (Obersee) but abundant in Minor Lake Constance (Untersee; Ruhle & Kindle 1992). After 1950, the population of this species declined dramatically, due to eutrophication in Lake Constance, and it became endangered. *Coregonus arenicolus* is currently stripped and stocked separately from the other coregonids. The pelagic spawner *Coregonus wartmanni* (Bloch) is restricted almost entirely to Major Lake Constance; the littoral spawner *Coregonus macrophthalmus* Nüsslin occurs in both parts of the lake. The last two species are commercially important, surviving in large numbers. In Lake Constance, these species are separated by their spawning grounds, and therefore rarely mixed by fishermen if stripped for artificial reproduction. *Coregonus wartmanni* has been used extensively for stocking in many Central European Lakes. A species of *Coregonus*, which spawns in the River Rhine (this river enters Lake Constance), was studied by Ruhle and Kindle (1992), however, its taxonomic status is not clear and needs to be examined. Douglas *et al.* (1999) presented data which support the occurrence of an endemic coregonid species flock in Lake Constance.

Renken Coregonus *spp. of prealpine lakes in the River Danube basin*

Kottelat (1997) reported only three species of *Coregonus* from prealpine lakes of Bavaria. However, this number may be underestimated. Vogt and Hofer (1909), Wagler (1937, 1941) and Dottrens (1959) recorded different numbers of *Coregonus* species, although their reviews appear to be rather confused. Among these species, *Coregonus bavaricus* Hofer is the most distinct. It is a summer-spawning (late June to early July), deepwater-dwelling, fish endemic to Lake Ammer (Vogt & Hofer 1909). There is no information, on whether this species still exists. *Coregonus hoferi* Berg described from Lake Chiem, and *Coregonus renke* (Schank) described from Lake Starnberg are two generalised species reported by Kottelat (1997). Unfortunately, several coregonids have been introduced in these and other lakes, and it has never been determined if current species still correspond to the species described in the past. Thus, the prealpine whitefishes may be victims of the intense translocations and cross-breedings that took place during the last two centuries. Douglas *et al.* (1999) found local species flocks of *Coregonus* in Swiss Lakes, and therefore, *Coregonus* might have formed endemic species flocks in the Danubian lakes, as well. There is an urgent need to review the prealpine Coregonids, and to assess if the historically known species still exist in their respective lakes, how many of them are threatened or extinct, and/or how many native populations have been wiped out by stocking.

Baltic houting Coregonus maraena *(Bloch) and Schaalseemaräne* Coregonus widegreni *Malmgren*

Kottelat (1997) associated the species migrating to the Baltic Sea in Germany with *Coregonus maraena* from Lake Madü in Poland. In Germany, anadromous stocks of *C. maraena* are very close to disappearing. They are known to migrate only into the rivers Odra and Peene, and they seem to have vanished from other rivers where the species was known to spawn. In the south-west part of the Baltic Sea, coregonids from north German lakes, and *C. wartmanni* from Lake Constance, were introduced unsuccessfully into the River Schlei (Nellen & Dehus 1985). Since 1992, a restitution programme has been underway using *C. maraena* from the River Odra and Peene. Fish are stocked into the river, and the returning spawners used for artificial reproduction. Kottelat (1997) considered native coregonid populations from two lakes of North Germany (Lake Schaal, Lake Selenter; Dunker 1960) to be conspecific with *Coregonus widegreni*. Although coregonids still exist in both of these lakes, material of various origins has been stocked into the lakes in northern Germany for more than 130 years (Spratte & Hartmann 1997). In Lake Schaal, two 'forms of large whitefish' live in sympatry today (Spieß, Waterstraat, Gallander, Lux & Sedlacek 1993). Nothing is known about the status of *C. widegreni* in Germany, and the native populations might be wiped out by stocking. However, large populations of unstudied coregonids are reported from the coastal waters of the Baltic Sea (H.M. Winkler, personal communication). These fishes may belong to *C. widegreni,* which is known to spawn in the Baltic Sea (Lehtonen & Himberg 1992).

Deep-water cisco Coregonus lucinensis *Thienemann and* Coregonus *sp. 'Stechlin'*

Baltic cisco, *Coregonus albula* L. is native to northern Germany but was also stocked into many natural and artificial lakes in northern and central Germany. In two lakes, endemic spring-spawning, deepwater cisco species exist in sympatry with *C. albula*. These are *Coregonus lucinensis* in Lake Breiter Luzin, and an undescribed species from Lake Stechlin. No special attention has been paid to either of these endemics. Their existence is ignored even in the German Red Data Book, and they have not been involved in any nature conservation activities. Some ecological aspects of *C. lucinensis* were studied in 1984–1986 (Waterstraat 1990) and it is known that *C. lucinensis* still exists (personal observation). Lake Breiter Luzin was formerly oligothrophic but is now mesotrophic (Koschel 1985). Recently, some plans to re-examine *C. lucinensis* and its environment emerged, and the species is expected to be included into the Red Data Book. Management of Lake Breiter Luzin is also expected to improve, especially with respect to this endemic species. The ecology of the spring-spawning, deepwater *Coregonus* from Lake Stechlin has also been studied (Anwand 1998). The post-glacial oligotrophic Lake Stechlin and the adjacent area have been protected as a nature reserve since 1938. This, however, did not prevent engineers from building a nuclear power plant at the lake, which used its water for cooling and then recycled the warmed water to the lake between 1966 and 1989. After German reunification the power plant was closed.

1.2.6 *Species expected to occur in German waters*

Some fish species are known from neighbouring areas and are very likely to be found in Germany in the future. Two of them, *Eudontomyzon mariae* (Berg) and *Sabanejewia balcanica* (Karaman), have been reported in Polish tributaries of the River Odra (M. Przybylski, personal communication). In Germany, no attention is paid to the possibility that *E. mariae* may be mis-identified as *Lampetra planeri*. Problems of correct identification may well be the reason why two more species – *Barbus petenyii* Heckel and *Romanogobio kessleri* (Dybowski) – have never been reported from the German part of the River Danube basin. Both species were recently recorded in the Austrian part of the River Danube – at Engelhartszell, which is very close to the German border (Zauner 1998). Thus, the occurrence of these species in the German section of the River Danube is very probable. The same may also be true for *Cobitis elongatoides* Bacescu & Maier. In Germany, only a few populations of *Cobitis* have been identified carefully. According to preliminary results it appears that diploid *Cobitis taenia* L. occur together with different types of gynogenetic hybrids of *C. taenia* and *C. elongatoides* (Rab, Rabova, Bohlen & Lusk 2000; Kotusz 2000). Diploid *C. elongatoides* were recorded in the Polish part of the River Odra basin (Kotusz 2000), and in the Czech River Elbe and River Danube basins (Rab, Rabova, Bohlen & Lusk 2000). Although there have been no records of 'pure' *C. elongatoides* from Germany, the few Danubian populations might belong to this species.

1.2.7 *Non-native species occurring in German waters*

In Germany, the pioneer of importing fish species for aquaculture was Max von dem Borne (1826–1894). Starting in 1882, he introduced *Oncorhynchus mykiss* (Walbaum), *Salvelinus fontinalis* (Mitchell), *Ameiurus nebulosus* (Lesueur), *Lepomis gibbosus* (L.), *Micropterus dolomieu* Lacépède and *Micropterus salmoides* Lacépède from North America into German waters. Other fishery biologists experimented with *Oncorhynchus gorbuscha* (Walbaum), *Salvelinus namaycush* (Walbaum), *Coregonus peled* (Gmelin) and later with the Chinese cyprinids *Ctenopharyngodon idella* (Valenciennes), *Hypophthalmichthys molitrix* (Valenciennes), *Hypophthalmichthys nobilis* (Richardson) and *Mylopharyngodon piceus* (Richardson). Additional species have escaped from aquaria (*Umbra pygmaea* (DeKay), *Pimephales promelas* Rafinesque and *Misgurnus anguillicaudatus* (Cantor)), or were introduced accidentally (*Pseudorasbora parva* (Schlegel)). In German waters, the introduction of *Micropterus* spp., *O. gorbuscha*, *S. namaycush* failed completely and Chinese cyprinids were not aimed to form sustainable populations and there is no natural reproduction of these species recorded. Stocking non-native species is prohibited today. Because anglers have argued that *O. mykiss* and *S. fontinalis* have been living in Germany for more than 100 years, they are considered as locally native and are still stocked. In northern and central Germany, very few introductions of American salmonids resulted self-sustainable populations. However, in recent years *O. mykiss* have been found to reproduce in many alpine and prealpine streams and seem to influence the stocks of *Salmo trutta* L. and *Thymallus thymallus* (L.) negatively. Nothing has been done to study this problem.

Of the species introduced, only *L. gibbosus* is widespread and *U. pygmaea*, *A. nebulosus* and *P. parva* are locally established throughout Germany (Arnold 1990). *Coregonus peled* is established in some lakes of northern Germany (Spratte & Hartmann, 1997) and others (included in Table 1.1) are established only at a few locations.

Altogether 14 non-native species are more or less abundant. Negative effects to native species by alien species are only very locally known (Arnold 1990). In some hot springs and small streams fed by cooling water of power plants, a variety of poecilids, goodeids and cichlids occur. These faunas are usually not documented and the species were not included into Table 1.1. In adddition to the introduction of non-native species, many translocations are known. The phylogeographic structure of most commercial species has also been destroyed. *Cyprinus carpio* L., *Stizostedion (Sander) lucioperca* (L.), *Aspius aspius* (L.), *Leucaspius delineatus* (Heckel) and *Vimba vimba* (L.), native to the German River Elbe and River Danube basin, have been introduced into the River Rhine basin (Freyhof 1998; 1999). *Abramis sapa* (Pallas) was translocated from the River Danube to the River Rhine (Bischoff, Freyhof & Staas 1998) and *Protherorinus marmoratus* (Pallas) recently invaded the River Rhine basin via a canal from the River Danube (Reinartz & Hilbrich 2000; Schadt 2000). *Neogobius kessleri* (Günther), an invasive species from the lower River Danube, has finally reached the German part of this river and was first recorded in 1999 (Seifert & Hartmann 2000).

1.3 Discussion

About 100 species of freshwater fish and lampreys (including anadromous and catadromous forms) have been recorded from German waters (Table 1.1). Five species will very likely be found in the near future, eight non-native species are established and six non-native species depend on stocking. During the 19th and 20th centuries, 12 species vanished from the territory of Germany. Applying the criteria published by IUCN (1995) to the 86 native species, two (2.3%) are globally extinct, 10 are extirpated from Germany (11.5%), 12 are endangered (14.0%) and 16 are vulnerable (18.6%). Therefore, nearly half of the native fish species (46.5%) is actually threatened. Many more species have stabilised their populations at a very low level for many years (Bless *et al.* 1994). *Alosa alosa* and *P. cultratus* have returned after they had disappeared, and many stocks of rheophilous species show an important increase in distribution and abundance. This is especially true for the River Rhine (Freyhof & Weibel 1999), where heavy pollution lead to the disappearance or massive decrease of many rheophilic species during the middle of the 20th century (Lelek & Buhse 1992). Since the early 1980s, water quality has improved considerably, and the problems of low oxygen concentrations became unimportant in almost all rivers and streams. Therefore, populations of *Alburnoides bipunctatus* (Bloch), *Aspius aspius* (L.), *Barbus barbus* (L.), *Chondrostoma nasus* (L.), *Petromyzon marinus* L., *Lampetra fluviatilis* (L.), *Pleuronectes flesus* L. and anadromous *S. trutta* recovered (Freyhof & Weibel 1999; Korte 1999). Similar to 100 years ago (Bürger 1926), *B. barbus* is again the most abundant fish species in the River Rhine (Staas 2000). A similar situation was found in thousands of hill-streams in central and southern Germany, where the resident populations of *S. trutta*, *Cottus gobio* L., *Barbatula barbatula* (L.) and *L. planeri* recovered, and can no longer be considered endangered (Freyhof & Weibel 1999). Thus, the local Red lists tend to show a positive trend (Bless *et al.* 1994). However, the situation in the former West Germany is remarkably different from that in the former East Germany.

In West Germany, improvement in water quality has been more apparent than in the eastern part of the country. This is especially obvious when the status of rheophilous fish species is compared. For example, *B. barbus* or hill-stream species are no longer endangered in West Germany but still rare in East Germany. Beside the different levels of water pollution, river regulation had different effects in the River Rhine and River Elbe. In the River Rhine, regulation resulted in a fast flowing river with decreased sedimentation and extensive clean gravel substrate usable by lithophilous spawners. By contrast, in the River Elbe, river regulation resulted in a massive loss of gravel substrate and very monotonous sand shores. While in the River Rhine, high water velocities lead to fast washing out of polluted sediments, in the River Elbe most of pollution remained in the sediment and have had, and will have, negative effects for fish feeding and benthic invertebrate faunas.

Analysing the factors responsible for species vanishing in Germany, it has to be noted, that seven (50%) of the lost species were at the margin of their distribution area (Danubian sturgeons, *C. poecilopus*, *R. uranoscopus*, *P. cultratus*). These marginal populations might be more sensitive to environmental changes or they might be less likely to be recolonised than populations in the centre of their distribution area.

Most species vanished because of multiple reasons. The disappearance of *A. sturio*, *A. alosa*, *C. oxyrinchus* and *S. salar* follows the same pattern. Heavy fishing for these species took place during the 19th and early 20th centuries. Populations were no longer resistant to this fishing pressure because their reproductive success was diminished by river regulations, dam constructions and water pollution since the late 19th century. From the River Elbe, it is well documented, that the construction of groynes led to the loss of many spawning habitats of *A. sturio* and *C. oxyrinchus* and consequently to the decrease of their stocks (Wünstemann 2000). Although habitat loss is often reported to be the primary cause of local fish extinction, commercial fishing was involved until the middle of the 20th century (eight of 14 species; 60%). None of the catadromous, but six (43%) of the anadromous species vanished. Habitat loss due to damming the rivers played an important role in vanishing of anadromous species, and damming is the second most important reason for species extinction (seven of 14 species; 50%). Intensive fishing and damming are the main reasons for Danubian sturgeons vanishing, and these fishes have a very low chance of recovery. Intensive fishing and water pollution are responsible for the disappearance of *A. ruthenus*, *A. alosa*, *C. oxyrinchus*, *P. cultratus*, *S. salar* and the stock of *Alosa fallax* known to perform long distance migrations. Habitat loss due to river engineering along the banks could be responsible for disappearance of *R. uranoscopus*, although the reasons why this species disappeared are not fully understood. Heavy changes in river and stream structures all over Germany might have reduced the abundance of some species (for *C. nasus* see Freyhof 1997) but none has vanished from Germany exclusively for this reason.

The case of *R. meidingeri* is typical of the German approach to handling threatened fish species. For decades, nobody cared about the species decline and no study was conducted to analyse the mortality of this species. Only 3 years after the last specimen was caught, *R. meidingeri* were stocked from a foreign source. The philosophy of many fisheries biologists is that stocking is the only way to help endangered fish species.

Although it is often believed that the ichthyofauna of Central Europe is well known, several groups of species characterised with a high degree of endemism, such as salmoniform fishes, have been ignored generally (Kottelat 1997). Almost all of these endemic salmoniform fishes are restricted to lakes and are therefore very sensitive to changes in the lake's ecosystem. Unfortunately, lakes are like islands and they suffer high rates of extinction when habitat modifications begin or when non-native stocks or species are introduced. Eutrophication in many lakes has had detrimental effects on spawning conditions for coregonids and charrs. To keep these economically-important stocks, these species are subject of artificial propagation and stocking. Species with-out economic importance are not supported by stocking. At least two, *C. gutturosus* and *S. profundus*, became extinct. The same may have happened to many other species, which were never recognised as distinct species. The German populations of *C. poecilopus* and *R. meidingeri* were also restricted to lacustrine habitats, and provide good example that lacustrine endemics are more vulnerable to habitat alternations than riverine species with large distribution areas. For more than 150 years, different coregonids have been translocated throughout Central Europe to improve the quality of local stocks (Vogt & Hofer 1909). Eutrophication of many lakes provided another reason for the stocking policy and the improved technology of the late 20th century also lead to massive mixture and translocation of local populations and species. This might have lead to the biggest post-glacial extinction crisis of vertebrate species in Central Europe. How many species and local stocks have become victims of this development? However, German law allows fishermen to translocate and stock more or less as they wish. To overcome this problem, environmentalists have to be involved into breeding programmes of salmoniform fishes and translocations have to be punished by law.

Salmoniform fishes can no longer be ignored and taxonomic problems should be solved with modern tools. It has to be concluded, that environmentalists, ichthyologists and fishery biologists should wake up and realise their responsibility in the management and survival of the unique composition of the German ichthyofauna as a part of our global heritage.

Acknowledgements

It is a pleasure to thank J. Gessner, A. Hartl, A. Lelek, M. Przybylski, I. Steinmann, A. Waterstraat, H.M. Winkler, C. Wolter, and G. Zauner, for contributing with their unpublished information's to this chapter and the two anonymous reviewers for their helpful comments.

References

Anwand K. (1998) Comparisons of annual gonad cycle and fecundity between nominate and deep-water forms of vendace (*Coregonus albula* L.) in Lake Stechlin (State of Brandenburg, Germany). *Journal of Applied Ichthyology* **14**, 97–100.

Arnold A. (1990) Eingebürgerte Fischarten. Die Neue Brehm-Bücherei, A. Ziemsen, Wittenberg Lutherstadt, 144 pp. (in German).

Artyukhin E. & Vecsei P. (1999) On the status of Atlantic sturgeon: conspecificity of European *Acipenser sturio* and North American *Acipenser oxyrinchus*. *Journal of Applied Ichthyology* **15**, 35–37.

Balon E.K. (1968a) Urgeschichte der Donauichthyofauna (vor dem Einfluß des Menschen). *Archiv für Hydrobiologie* Suppl. **34**, 204–227 (in German).

Balon E.K. (1968b) Einfluß des Fischfangs auf die Fischgemeinschaft der Donau. *Archiv für Hydrobiologie* Suppl. **34**, 228–249 (in German).

Bartl G. & Troschel H.J. (1997) Bestandesentwicklung und aktuelle Situation von *Alosa alosa* und *Alosa fallax* im Rheingebiet. *Zeitschrift für Fischkunde* **4**, 119–162 (in German).

Bauch G. (1958) Untersuchungen über die Gründe für den Ertragsrückgang der Elbfischerei zwischen Elbsandsteingebirge und Boitzenburg. *Zeitschrift für Fischerei und deren Hilfswissenschaften* **7**, 161–437.

Behnke R.J. (1972) The systematics of salmonid fishes of recently glaciated lakes. *Journal of the Fisheries Research Board of Canada* **29**, 639–671.

Behnke R.J. (1980) A systematic review of the genus *Salvelinus*. In E. Balon (ed.) *Charrs, Salmonid Fishes of the Genus* Salvelinus. The Hague: Junk, pp. 441–481.

Benecke N. (1986) Some remarks on sturgeon fishing in the southern Baltic region in medieval times. In D.C. Brinkhuizen & A.T. Clason (eds) *Fish and Archaeology*. Oxford: British Archaeological Reports, International Series **294**, pp. 9–17.

Bischoff A., Freyhof J. & Staas S. (1998) Nachweise des Zobels *Abramis sapa* (Pallas, 1811), (Teleostei: Cyprinidae) im Rhein. *Lauterbornia* **33**, 5–8 (in German).

Blab J., Nowak, E., Trautmann, W. & Sukopp, H. (1977) *Rote Liste der gefährdeten Tiere und Pflanzen in der Bundesrepublik Deutschland*. Kilda, Greven, 270 pp. (in German).

Bless R. (1978) *Bestandsänderung der Fischfauna in der Bundesrepublik Deutschland*. Kilda: Greven, 66 pp. (in German).

Bless R. & Lelek A. (1984) Rote Liste der Fische (Pisces) und Rundmäuler (Cyclostomata). In J. Blab & E. Nowak (eds) *Rote Liste der gefährdeten Tiere und Pflanzen in der Bundesrepublik Deutschland*. Kilda: Greven, pp. 30–32 (in German).

Bless R., Lelek A. & Waterstraat A. (1994) Rote Liste und Artenverzeichnis der in Deutschland in Binnengewässern vorkommenden Rundmäuler und Fische, Cyclostomata und Pisces. In E. Nowak, E.J. Blab & R. Bless (eds) *Rote Liste der gefährdeten Wirbeltiere in Deutschland*. Kilda: Greven, pp. 137–156 (in German).

Brunner P.C., Douglas M.R. & Bernatchez L. (1998) Microsatellite and mitochondrial DNA assessment of population structure and stocking effects on Arctic charr *Salvelinus alpinus* (Teleostei: Salmonidae) from central alpine lakes. *Molecular Ecology* **7**, 209–223.

Bürger F. (1926) Die Fischereiverhältnisse im Rhein im Bereich der preußischen Rheinprovinz. *Zeitschrift für Fischerei* **24**, 217–398 (in German).

Debus L. (1995) Historic and recent distribution of *Acipenser sturio* in the North Sea and Baltic Sea. Proceedings of the International Sturgeon Symposium, VNIRO, Moscow, pp. 189–203.

Dottrens E. (1959) Systématique des coregones de l'Europe occidental, basée sur une étude biometrique. *Revue suisse de Zoologie* **66**, 1–66 (in French).

Douglas M.R., Brunner P.C. & Bernatchez L. (1999) Do assemblages of *Coregonus* (Teleostei: Salmoniformes) in the central alpine region of Europe represent species flocks? *Molecular Ecology* **8**, 589–603.

Dunker G. (1960) Die Fische der Nordmark. *Abhandlungen und Verhandlungen des Naturwissenschaftlichen Vereins in Hamburg* **3**, Suppl., 1–432 (in German).

Englbrecht C. (2000) *Untersuchungen zur Phylogeographie und Populationsstruktur zweier paläarktischer Fischarten, der Mühlkoppe* (Cottus gobio L.) *und des alpinen Seesaiblings* (Salvelinus alpinus L.). PhD Dissertation, University of München, 123 pp. (in German).

Freyhof J. (1997) Remarks on the status of *Chondrostoma nasus* (L.) in the system of the river Rhine. *Folia Zoologica* **46**(Suppl. 1), 61–66.

Freyhof J. (1998) *Strukturierende Faktoren für die Fischgemeinschaft der Sieg*. Göttingen: Cuvillier, 164 pp. (in German).

Freyhof J. (1999) Records of *Vimba vimba* (L.) (Teleostei: Cyprinidae) in the River Rhine and its tributaries. *Folia Zoologica* **48**, 84–89.

Freyhof J. & Weibel U. (1999) Vorschlag einer Neufassung der Rote Liste der bestandesgefährdeten Neunaugen und Fische (Petromyzontiformes; Teleostei) in Rheinland-Pfalz sowie ein Verzeichnis aller vorkommender Arten. *Fauna und Flora in Rheinland-Pfalz* **9**, 195–206 (in German).

Fuchs H., Schlee P., Rottmann O. & Stein H. (1999) Untersuchung von Perlfischen (*Rutilus frisii meidingeri*, Heckel) aus dem Wolfgangsee und dem Attersee auf genetische Unterschiede mit molekulargenetischen Markern. *Österreichs Fischerei* **52**, 57–62 (in German).

Harsanyi A. (1986) Sichling, *Pelecus cultratus* in der Donau. *Fischer und Teichwirt* **11**, 330–331 (in German).

Heckel J.J. & Kner R. (1858) *Die Süßwasserfische der Österreichischen Monarchie mit Rücksicht auf die angrenzenden Länder*. Leipzig: Engelmann, 388 pp. (in German).

Hensel K. (1979) *Rutilus (Pararutilus) frisii meidingeri* in the Czechoslovak stretch of the Danube River. *Vestnak Czechoslovenske Spolecnosti Zoologicke* **18**, 250–252.

Hensel K. & Holcík J. (1997) Past and current status of sturgeons in the upper and middle Danube River. *Environmental Biology of Fishes* **46**, 185–200.

Hoffmann R.C. (1996) Economic development and aquatic ecosystems in medieval Europe. *American Historical Review* **101**, 631–669.

IUCN (1995) The IUCN guidelines for reintroductions. *www.rbgkew.org.uk/conservation/RSGguidelines.html*

IUCN (2000) The 2000 IUCN Red List of threatened species. *www.redlist.cymbiont.ca/resultslist.asp*

Jäger T. (1999) Die Wiedereinbürgerung des Nordseeschnäpels. In Verband Deutscher Sportfischer (ed.) *Der Nordseeschnäpel*. Offenbach: Selbstverlag Verband Deutscher Sportfischer, pp. 3–11 (in German).

Korte E. (1999) Bestandsentwicklung der Fischarten der hessischen Rheinaue 1994–1997 – Reproduktionsstrategien, Jungfischaufkommen, Gefährdung, Entwicklungstendenzen. *Umweltplanung, Arbeits-und Umweltschutz* **268**, 1–186 (in German).

Koschel R. (1985) Das Feldberger Seengebiet. *Natur und Umwelt im Bezirk Neubrandenburg* **3**, 1–96 (in German).

Kottelat M. (1997) European Freshwater Fishes. An heuristic checklist of the freshwater fishes of Europe (exclusive of former USSR), with an introduction for non-systematists and comments on nomenclature and conservation. *Biologia* **52**(Suppl. 5), 1–271.

Kotusz J. (2000) Intra- and interpopulation morphological variability in diploid and varied-ploid *Cobitis* from Poland. *Folia Zoologica* **49**(Suppl. 1), 219–226.

Kussmaul R., Hoffmann R. & Gessler M. (1991) Bedrohte Fischarten in Bayern. *Berichte der Bayrischen Landesanstalt für Wasserforschung* **19**, 1–159 (in German).

Lang G. (1994) *Quartäre Vegetationsgeschichte Europas*. Jena: Fischer, 420 pp. (in German).

Lehtonen H. & Himberg M.K-J. (1992) Baltic sea migration patterns of anadromous, *Coregonus lavaretus* (L.) s. str., and sea-spawning european whitefish, *C. l. wildegreni* Malmgren. *Polskie Archivum Hydrobiologii* **39**, 463–472.

Lelek A. (1976) Veränderungen der Fischfauna in einigen Flüssen Zentraleuropas (Donau, Elbe und Rhein). *Schriftenreihe für Vegetationskunde* **10**, 245–308 (in German).

Lelek A. (1980) Fische als Indikatoren der Umweltverhältnisse. *Courier Forschungsinstitut Senkenberg* **41**, 111–129 (in German).

Lelek A. & Buhse G. (1992) *Fische des Rheins*. Berlin: Springer, 214 pp. (in German).

Mikschi E. & Wolfram-Wais A. (1999) *Rote Liste ausgewählter Tiergruppen Niederösterreichs – Fische und Neunauge (Pisces, Cyclostomata)*. Amt der NÖ Landesregierung, St. Pölten, 135 pp. (in German).

Mohr E. (1962) Ganoidei, Störe. In R. Demoll & H.H. Maier (eds) *Handbuch der Binnenfischerei Mitteleuropas. Vol. IIIB*, Stuttgart: Schweizer, pp. 235–261 (in German).

Nellen W. & Dehus P. (1985) Jüngere Ergebnisse zur Situation der Süßwasserfischfauna in Schleswig-Holstein. *Schriften des Naturwissenschaftlichen Vereins Schleswig-Holstein* **55**, 63–80 (in German).

Nümann W. (1972) The Bodensee: effects of exploitation and eutrophication on the salmonid community. *Journal of the Fisheries Research Board of Canada* **29**, 833–847.

Paepke H.-J. (1981) Die gegenwärtige Situation der Süßwasserfischfauna in der DDR. *Archiv für Naturschutz und Landschaftsforschung, Berlin* **21**, 113–130 (in German).

Rab P., Rabova M., Bohlen J. & Lusk S. (2000) Genetic differentiation of the two diploid-polyploid complexes of loaches, genus *Cobitis* (Cobitidae) involving *C. taenia*, *C. elongatoides* and *C.* spp. in the Czech Republic: Karyotypes and cytogenetic diversity. *Folia Zoologica* **49**(Suppl. 1), 55–66.

Reinartz R. & Hilbrich T. (2000) Nachweis der Marmorierten Grundel (*Proterorhinus marmoratus* Palls, 1811) im unterfränkischen Main bei Eltmann (Rheineinzugsgebiet). *Österreichs Fischerei* **53**, 192–194 (in German).

Ruhle C. & Kindle T. (1992) Morphological comparison of river-spawning whitefish of the alpine Rhine with the whitefish of Lake Constance. *Polskie Archiwum Hydrobiologii* **39**, 403–408.

Schadt J. (2000) Neue Fischart im Main entdeckt: Marmorierte Grundel (*Proterorhinos marmoratus*). *Fischer & Teichwirt* **6**, 217–218 (in German).

Schmidt G. (1996) Wiedereinbürgerung des Lachses *Salmo salar* L. in Nordrhein-Westfalen – Allgemeine Biologie des Lachses sowie Konzeption und Stand des Wiedereinbürgerungsprogramms unter besonderer Berücksichtigung der Sieg. *Schriftenreihe der Landesanstalt für Ökologie, Bodenordnung und Forsten/Landesamt für Agrarordnung Nordrhein-Westfalen* **11**, 1–194 (in German).

Seifert K. & Hartmann F. (2000) Die Kesslergrundel (*Neogobius kessleri* Günther 1861), eine neue Fischart in der deutschen Donau. *Lauterbornia* **38**, 105–108 (in German).

Siebold C.T.E.v. (1863) *Die Süßwasserfische von Mitteleuropa*. Leipzig: Engelmann, 430 pp. (in German).

Spieß H.J. & Waterstraat A. (1989) Ergebnisse der Kartierung der Rundmäuler und Fische der DDR (einheimische Arten des Süßwassers). II. Zentrale Tagung des zentralen Arbeitskreises Ichthyofaunistik der Gesellschaft für Natur und Umwelt, Feldberg, pp. 11–31 (in German).

Spieß H.J., Waterstraat A., Gallander G., Lux E. & Sedlacek J. (1993) Die Fischfauna des Schaalsees und einiger Nebengewässer. *Natur und Naturschutz in Mecklenburg-Vorpommern* **30**, 3–29 (in German).

Spratte S. & Hartmann U. (1997) Fischartenkataster Süßwasserfische und Neunaugen in Schleswig-Holstein. Ministerium für Ländliche Räume, Landwirtschaft, Ernährung und Tourismus des Landes Schleswig-Holstein, Kiel, 193 pp. (in German).

Staas S. (2000) Jungfisch-Artengemeinschaften als Bioindikatoren für die ökologische Qualität des nordrhein-westfälischen Rheinabschnitts. Ministerium für Umwelt, Raumordnung und Landwirtschaft, Düsseldorf, 114 pp. (in German).

VDSF (Verband Deutscher Sportfischer) (2000) *Fisch des Jahres 2000 – Der Lachs*. Verband Deutscher Sportfischer, Selbstverlag, Offenbach, 199 pp. (in German).

Vogt C. & Hofer B. (1909) *Die Süßwasserfische von Mitteleuropa. Teil I: Text*. Frankfurt, 558 pp. (in German).

Wagler E. (1937) Die Coregonen in den Seen des Voralpengebietes. IX. Die Systematik der Voralpencoregonen. *Internationale Review der Gesellschaft für Hydrobiologie und Hydrographie* **35**, 345–446 (in German).

Wagler E. (1941) Die Lachsartigen (Salmonidae) II. Teil. Coregonen. In R. Demoll & H.H. Maier (eds) *Handbuch der Binnenfischerei Mitteleuropas. Vol. IIIA*. Stuttgart: Schweizer, pp. 371–501 (in German).

Waterstraat A. (1990) Ökologische Untersuchungen an Populationen der Kleinen Maräne (*Coregonus albula* L.) im Breiten Luzin (Bezirk Neubrandenburg). *Fortschritte der Fischereiwissenschaft* **9**, 93–104 (in German).

Weber E. (1989) Entwicklung der Donaufischerei von Wien bis zur Marchmündung vor und nach der Donauregulierung von 1880–1900. Studie der Internationalen Arbeitsgemeinschaft Donauforschung, Wien, 123 pp. (in German).

Winkler H.M. (1989) Zur Fischfauna der brackigen Küstengewässer der DDR, Kenntnistand und Gefährdung. II. Zentrale Tagung des zentralen Arbeitskreises Ichthyofaunistik der Gesellschaft für Natur und Umwelt, Feldberg, pp. 32–41 (in German).

Winkler H.M., Hamann N. & Waterstraat A. (1991) *Rote Liste der gefährdeten Rundmäuler, Süßwasser- und Wanderfischarten Mecklenburg-Vorpommerns*. Umweltministerium des Landes Mecklenburg-Vorpommern, Schwerin, 28 pp. (in German).

Wünstemann P. (2000) Die Fischfauna von Sachsen-Anhalt. Ministerium für Raumordnung, Landwirtschaft und Umwelt, Magdeburg, 179 pp. (in German).

Zauner G. (1998) Der Semling – eine verschollene Fischart wurde wiederentdeckt. *Österreichs Fischerei* **51**, 218 (in German).

Chapter 2
Status and protection of species and intraspecific diversity of the ichthyofauna in the Czech Republic

S. LUSK,* V. LUSKOVÁ, K. HALACKA
Institute of Vertebrate Biology, AS CR, Department of Ichthyology, Brno, Czech Republic

V. ŠLECHTA and V. ŠLECHTOVÁ
Institute of Animal Physiology and Genetics, Libechov, Czech Republic

Abstract

The ichthyofauna indigenous to the Czech Republic consists of four lamprey and 55 fish species. Of these, two lamprey and eight fish species are now considered absent. In recent years some of the previously lost species have reappeared either through natural migration (*Acipenser ruthenus*, *Pelecus cultratus*, *Gymnocephalus schraetser*) or reintroduction (*Salmo salar*). As a result of natural immigration from the Danube drainage basin, the indigenous ichthyofauna has been enriched by species new to the Czech Republic: i.e. *Proterorhinus marmoratus*, *Stizostedion* (*Sander*) *volgensis*, and *Gymnocephalus baloni*. According to the IUCN (1999) classification, 19 fish species, one wild form, *Cyprinus carpio*, and two lamprey species are endangered to different degrees. Recent investigations on intraspecific diversity have yielded data on the genetic variability in populations of *Chondrostoma nasus* (average variability, low interpopulation diversity), *Vimba vimba* (low genetic variability), *Barbus barbus* (high variability, interpopulation difference), *Phoxinus phoxinus* (high intra- and interpopulation variability), *Alburnoides bipunctatus* (low variability), *Cobitis elongatoides* (distinct differences connected with the level of ploidy and hybridisation in different populations) and *Salmo trutta* m. *fario* (genetic characteristics resulting from human activities).

Keywords: genetic variability, native ichthyofauna, population diversity, Red List.

2.1 Introduction

Information on the biology and distribution of species is a prerequisite to effective protection of biodiversity. The application of new, more progressive methods has deepened and amended information on ichthyofaunal diversity at different levels. Until recently species or subspecies were detected by morphometric analaysis (e.g. Lelek 1987; Maitland & Lyle 1991), but this has been superceded by studies targetting

*Correspondence: Institute of Vertebrate Biology, AS CR, Department of Ichthyology, Kvetná 8, 603 65 Brno, Czech Republic (email: lusk@brno.cas.cz).

the level of genetic diversity (e.g. Meffe 1986; Nelson & Soule 1987). Consequently conservation activities now focus on the population level because genetic analysis allows discrimination between populations (Carvalho & Hauser 1994; IUCN 1999). A similar development in viewing fish biodiversity is also taking place in Czech Republic, where attention to genetic diversity at the population level has received attention in recent years.

Evaluating the degree of threat to particular species in the form of the so-called Red List, using a unified classification, creates a suitable starting point for their subsequent protection. In the Czech Republic the 'Red List of Fishes' underwent a reevaluation based on improved understanding of the genetic diversity, which increased objectiveness of the evaluation of risk to the species (Lusk 1988; 1996a, b; Lusk & Hanel 1996; Lusk, Luskova, Halacka & Lojkasek 2000b). Since new data are continuously being added and changes in the fauna are occurring it was agreed that the Red List for the Czech Republic would be updated at 5-year intervals. This chapter reviews the changes between the 1995 and 2000 versions of the Red List.

In the Czech Republic, investigations on intraspecific or population genetic diversity in fishes began to be developed in the past decade (e.g. Šlechtová, Šlechta & Berrebi 1993; Lusková, Šlechtová, Šlechta & Lusk 1995). The importance of genetic investigations is demonstrated, for instance, by the most recently changed specific pertinence of populations classified for a long time as *Cobitis taenia* and now classified as *Cobitis elongatoides* (Ráb, Rábová, Bohlen & Lusk 2000) (see Bohlen, Ráb, Šlechtová, Rábová, Ritterbusch & Freyhof, Chapter 28).

2.2 Assessment methodology

The hydrographic network of the Czech Republic (7864 km^2) belongs to three drainage basins, viz., the Black Sea – Morava River basin (25.4%), the North Sea – Elbe River basin (65.2%), and the Baltic Sea – Odra River basin (9.4%) (Fig. 2.1). This results in differences in the species composition of fish fauna as well as differences among populations in different parts of the territory.

The latest working version of the IUCN classification system (1999) was used to evaluate the threatened status of each fish species, although modifications suggested by the alternative categories of the IUCN Red List at a regional level (Gärdenfors, Rodríguez, Hylton-Taylor, Hyslop, Mace, Molur & Poss 1999) were not adopted. This is because a species endangered at the regional level is considered of a similar value for a given region as a species endangered on the same level within its whole range. Basic data for a qualified assessment of the degree of threat and status of a species were obtained from ichthyological studies (e.g. Lusk 1995; 1996b). In estimating the threat of extinction, the intensity and extent of population reduction, decrease in numbers in the particular populations, and the extent of range in general and in the Czech Republic in particular, were considered. In addition to data from fisheries surveys, data were obtained from long-term angler catch statistics (e.g. Lusk & Halacka 1995; Lusk, Lusková, Halacka, Šlechta & Šlechtová 1998).

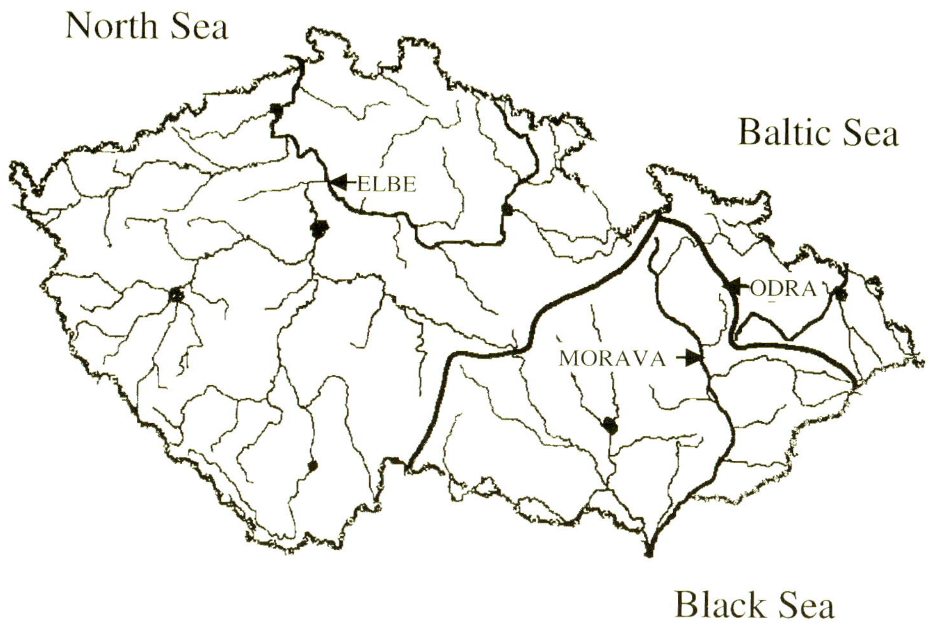

Figure 2.1 Division of the territory of the Czech Republic according to the hydrographic network of the drainage areas of the North, Baltic and Black Seas

2.3 Conservation status

Four lamprey species and 55 fish species are considered native to the waters of the Czech Republic. The 'Red List of Lampreys and Fishes of the Czech Republic, Version 2000' categorises the fauna as follows.

I. Extinct (EX): *Petromyzon marinus* L.; *Lampetra fluviatilis* (L.); *Huso huso* (L.); *Acipenser sturio* L.; *Alosa alosa* (L); *Salmo trutta trutta* L.; *Salmo trutta 'labrax'* Pallas; *Coregonus oxyrinchus* (L.); *Rutilus pigus* (Lacepéde); *Sabanejewia balcanica* (Karaman); *Zingel streber* (Siebold); *Pleuronectes flesus* L.

Of the taxa included in this category, most species occurred in the Czech Republic only occasionally. Only *S. trutta trutta* migrated there for the purpose of reproduction and two of the species occurred permanently (*Z. streber* and *S. balcanica*). *Zingel streber* vanished from the drainage area of the Morava River as early as the beginning of the 20th century as a result of the construction of migration barriers (weirs), subsequent stream fragmentation and water pollution. The population of *S. balcanica* vanished from the Becva River drainage area around 1960 because of poor water quality. There are intentions to repatriate this species.

II. Extinct in the Wild (EW): *Salmo salar* L.; *Hucho hucho* (L.).

At present, these two species occur only if stocked. In the past, *S. salar* was fairly frequent in the Labe (Elbe) River drainage as a result of its reproduction migrations. The species vanished after construction of barriers to migration in the Labe. At present, attempts are being made to repatriate *S. salar* in the Labe drainage. *Hucho hucho*, a species indigenous in the Morava River drainage area, occurs locally where stocking has occurred. There are no suitable conditions for creating an independent and permanent population of *H. hucho*.

III. Threatened – Critically Endangered (CR): *Eudontomyzon mariae* (Berg); *Abramis sapa* (Pallas); *Gobio kessleri* Dybowski; *Pelecus cultratus* (L.); wild form of *Cyprinus carpio* L.; *Gymnocephalus baloni* Holcík & Hensel; *Gymnocephalus schraetser* (L.); *Zingel zingel* (L.).

The occurrence of all species classified in category CR is limited to the Morava River drainage area (Danube River basin). *Eudontomyzon mariae* occurs in a single small stream (Rací creek). The occurrence of *G. kessleri* is limited to the Becva River. The remaining species are only found around the confluence of the Morava and Dyje rivers and their presence depends on the possible migration of these species from the lower reaches of the Morava and from the Danube (Lusk & Holcík 1998). As for the wild form of *C. carpio*, consideration is being given for artificial propagation from wild broodstock and stocking (Lusková, Lusk, Holcík & Halacka 2000).

III. Threatened – Endangered (EN): *Lampetra planeri* (Bloch); *Chondrostoma nasus* (L.); *Misgurnus fossilis* (L.); *Cobitis elongatoides* Bacescu & Maier.

The species included in this subcategory show a considerably wide occurrence in the hydrological network of the Czech Republic but in the past decade some of their populations have vanished from a number of localities and a heavy drop in numbers has been recorded in others. Originally, *C. nasus* occurred in the Morava and Odra drainage areas. Populations of this species, have appeared in the Labe drainage area during the past two decades, but are not included in this evaluation. In the course of the past 50 years an enormous drop in the catches of *C. nasus* has been recorded (Lusk 1995; Lusk & Halacka 1995) (Fig. 2.2). Populations of *C. elongatoides* have vanished from a number of localities (Lusk, Lusková & Halacka 2000a). In the Czech Republic this species occurs in populations consisting of diploid individuals, and in hybrid polyploid complexes which, besides genomes of *C. elongatoides*, also contain one or two genomes of additional species of the genus *Cobitis* (Ráb *et al.* 2000; Šlechtová, Lusková, Šlechta, Lusk, Halacka & Bohlen 2000).

III. Threatened – Vulnerable (VU): *Leuciscus idus* (L.); *Phoxinus phoxinus* (L.); *Gobio albipinnatus* Lukasch; *Alburnoides bipunctatus* (Bloch); *Abramis ballerus* (L.); *Vimba vimba* (L.); *Carassius carassius* (L.); *Lota lota* (L.); *Cottus gobio* L.; *Cottus poecilopus* Heckel.

Species included in this subcategory either show discrete regional occurrence (*L. idus*, *A. ballerus*, *G. albipinnatus*) or their numbers have fallen markedly (*C. carassius*, *G. gobio*, *C. poecilopus*, *V. vimba*, *A. bipunctatus*). While the status of *L. lota* has improved in recent years as a result of mass stocking, extinction of the original population diversity of this species must be considered likely.

IV. Near Threatened (NT): *Thymallus thymallus* (L.); *Leucaspius delineatus* (Heckel); *Barbus barbus* (L.); *Rhodeus amarus* (Bloch); *Anguilla anguilla* (L.); *Stizostedion (Sander) volgensis* (Gmelin).

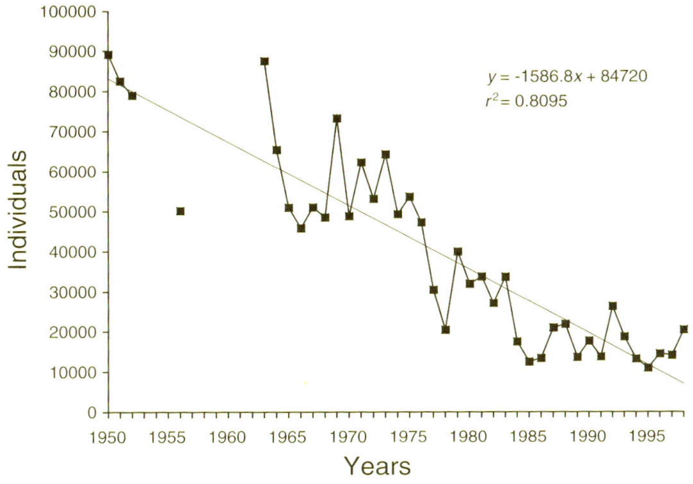

Figure 2.2 Trends in anglers' catches of *C. nasus* in the Czech Republic in 1950–1998

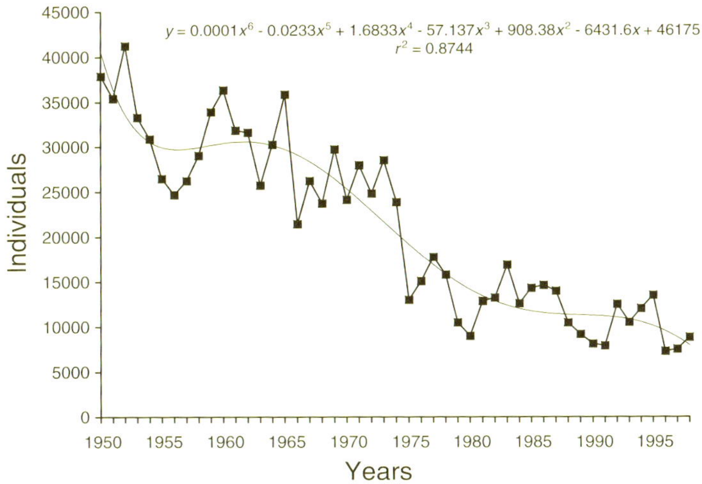

Figure 2.3 Trends in anglers' catches of *B. barbus* in the Czech Republic in 1950–1998

The occurrence and numbers of most populations of *T. thymallus* and all *A. anguilla* depend on stocking of artificially reared or imported stocking material. The numbers of *B. barbus* (Fig. 2.3) and *R. amarus* have markedly dropped during recent years on a nationwide scale. For the time being, the occurrence of *S. volgensis* is limited to the confluence of the Morava and Dyje rivers. *Stizostedion volgensis* was found there as late as after 1990 (Lusk, Lusková & Halacka 1996).

V. Least Concern (LC): *Salmo trutta* m. *fario* (L.); *Esox lucius* L.; *Rutilus rutilus* (L.); *Leuciscus leuciscus* (L.); *Leuciscus cephalus* (L.); *Scardinius erythrophthalmus* (L.);

Aspius aspius (L.); *Tinca tinca* (L.); *Gobio gobio* (L.); *Alburnus alburnus* (L.); *Abramis bjoerkna* (L.); *Abramis brama* (L.); *Barbatula barbatula* (L.); *Silurus glanis* L.; *Perca fluviatilis* L.; *Gymnocephalus cernuus* (L.); *Stizostedion (Sander) lucioperca* (L.); *Proterorhinus marmoratus* (Pallas).

The development and status of populations as well as the nationwide distribution of most species of the LC category give no cause for concern. Some of the species are objects of production cultures (*T. tinca*, *E. lucius*, *S. glanis*, *S. lucioperca*). Some other species are reared to strengthen populations inhabiting natural waters and subject to sport angling.

VI. Not Evaluated (NE): *Acipenser ruthenus* L.

Occasional individuals are known to occur in the lower reaches of the River Dyje.

2.4 Protection of species diversity

In the Czech Republic, legal regulations (Law no. 114/1992/Sb. On Nature Protection, and Intimation no. 395/1992/Sb.) directly protect some of the species mentioned above, according to their risk of extinction or decline. The category of 'critically endangered' species includes *L. planeri*, *E. mariae*, *Z. zingel* (vanished at present), *Z. streber*, *G. kessleri*, *S. balcanica* (vanished at present); the category of 'critically endangered' species includes *P. cultratus*, *A. bipunctatus* and *C. taenia* (*C. elongatoides* – see Introduction); and the category of 'endangered' species includes *R. pigus* (vanished), the wild form of *C. carpio*, *P. phoxinus*, *M. fossilis*, *L. lota*, *C. gobio* and *C. poecilopus*. As a part of the preparations for the Czech Republic entering the European Union (EU), it will be necessary to include, in national legislation, regulations that are valid in the EU (Council Directive 92/43/EEC).

On the basis of current knowledge, the Czech Republic will need to improve the necessary legislative tools to protect fish species diversity. The greatest risk is seen in ill-considered introductions or possible penetration of exotic fish species (e.g. *Perccottus glenii*). In addition, special attention must be paid to the cultivated forms of *C. carpio*, the subject of large scale production in fishponds. The cultivated forms of carp have also become a major, preferred target of sport anglers and are being stocked in most open water bodies inhabited by cyprinid fish fauna (such as larger rivers, various types of impoundment, valley dam lakes), but they are not subject to legislation.

2.5 Intraspecific diversity

Investigations on intra- and interpopulation variability were implemented on the basis of molecular analysis. Attention has been paid to species which are the target of aquaculture production (reared forms of *C. carpio*, *T. tinca*, *C. lavaretus*, *C. peled*, *S. glanis*, *Hypophthalmichthys molitrix*, *Aristichthys nobilis*) and are stocked to natural water bodies for sport angling, plus a few species that occur only in natural populations (*S. trutta* m. *fario*, *C. nasus*, *B. barbus*, *V. vimba*). These species are subject to artificial reproduction and the stocking material serves to strengthen the natural populations. A more detailed consideration of the latter species is given below.

2.5.1 Chondrostoma nasus

In the territory of the Czech Republic the natural distribution of this species is limited to the Black Sea and the Baltic Sea river systems. In the past two decades, *C. nasus* has been successfully introduced into the Labe River system (Lusk 1996a). There is a low interpopulation diversity among populations inhabiting the drainage areas of the Morava and Odra rivers in the Czech Republic (Lusková *et al.* 1995; Lusková, Šlechtová, Povz, Šlechta & Lusk 1997). This absence of interpopulation diversity detectable by biochemical genetic methods is in common with studies on other European populations (Gollmann 1995; Gollmann, Bouvet, Karakousis & Triantaphyllidis 1997).

2.5.2 Barbus barbus

This is the characteristic species of the so-called barbel zone. In the streams of the Czech Republic, this zone occupies approximately one third to one half of the length of streams with water discharges of $0.5–1.0\,m^3 s^{-1}$. A total of 264 individuals from 10 populations in streams of all three sea basins were examined. The observed genetic distances (according to Nei (1972) and Rogers (1972)) between the populations examined were very small, indicating low genetic differentiation (Šlechtová, Šlechta, Lusková, Lusk & Berrebi 1998). Compared with data for *B. barbus* from western Europe, however, the genetic variability of Czech populations was much higher (Šlechtová *et al.* 1993).

2.5.3 Vimba vimba

The status of the populations of this species varies in the Czech Republic. In the waters of the Black Sea basin (the drainage areas of the Morava and Dyje rivers) and in those of the Baltic Sea basin (the Odra River drainage area) the species is critically endangered; in the Labe River (North Sea basin) it is vulnerable. A total of 195 individuals from three populations in the Czech Republic were analysed genetically and found to have low variable in terms of allele frequency. The populations under study were similar on the basis of genetical similarity (Nei – 98.3–99.9%, Rogers – 95.7–98.9%).

2.5.4 Alburnoides bipunctatus

The species occurs in fairly abundant populations in the Morava River drainage area (Black Sea basin). It has vanished from other drainage areas or is rare (the Labe and

Odra rivers). A total of 67 individuals from three localities in the Morava River drainage area were analysed and found to exhibit high genetic similarity, even though the populations are fragmented, forming relatively isolated local populations (Lusk *et al.*, unpublished).

2.5.5 Phoxinus phoxinus

The species inhabits waters of the so-called trout zone. Analysis of 187 individuals from seven populations indicated geographic differentiation of populations. In recent years the occurrence of this species and the population numbers have declined. It is necessary to enlarge the study area and to ascertain the population structure of this species.

The results described above were obtained by analysing products of protein (enzyme) coding loci. In many cases the information value of these markers is insufficient to characterise and possibly even differentiate between populations. Consequently, investigations are being launched on intraspecific diversity by means of molecular methods (above all, studies of mtDNA variability).

2.5.6 *Species important in aquaculture*

Native species *E. lucius*, *C. carpio*, *T. tinca* and *S. lucioperca* are utilised for culture purposes. The populations of these species, inhabiting free waters, result from stocking and, essentially, they are not native wild populations. Much the same holds true of additional species, such as *S. trutta* m. *fario*, *T. thymallus*, *A. aspius* and *L. lota*. While these species are not reared for aquaculture production, they are artificially bred and the stocking material is released into open waters. Because the originality of their maternal population was not observed, the particular populations have become mixed to the extent that inter-population differences were obliterated and one 'large population' was created, showing similar intrapopulation characteristics. The intraspecific variability between 30 populations of *S. trutta* m. *fario*, a species that has been stocked for over a century (Fric 1875), was very low regardless of their locality (unpublished results). Notwithstanding this practise that degrades the original genetic integrity, the status of natural populations of trout is good. In the Czech Republic (unlike many other countries where parental fishes from farmed stocks are used for fry production), maternal fish for artificial reproduction are obtained from natural populations and, after being stripped, they are released back to the same populations. The stocking material is then reared in streams under natural conditions. Therefore, applying such stocking material also to other populations is very effective and shows minimal aspects which would reduce the survival of the original stocks. A similar enhancement process has been on-going with species, such as *L. cephalus*, *C. nasus*, *B. barbus* and, to a lesser extent, *V. vimba*. In this way, alleles from other populations are introduced into the genetic characteristics of local populations, thereby changing their genetic characteristics, e.g. in *C. nasus* (Lusková *et al.* 1997).

Recently, protection activities aimed at utilising artificial reproduction and to use the stocking material to strengthen natural populations of certain other species

(*C. carassius*, *A. bipunctatus*, *P. phoxinus*, *M. fossilis*) have been performed. In view of the hitherto experience with stocking that does not observe the geographic originality of the parental fish, such practice presents a serious threat to the genetic integrity of the natural populations of the above-mentioned species. In this way they could become endangered before their original population structures are known and utilised for their protection.

2.5.7 *Possible protection of intraspecific diversity*

At present, the Czech Republic lacks the legislation to effectively protect the genetic variability of wild populations of native fish species. Protection pertains only to a species *per se*, whereas lower units such as populations are not included. Legal regulations pertaining to introduction also mention only exotic species and not alien populations of native ones. For the time being, an appeal has been made to the fishery public to maintain the genetic integrity of populations by stocking only material reared from broodstock from the recipient water body, or at least the same basic hydrological regions (river basins, drainage areas of larger rivers).

Acknowledgements

The results presented here have been obtained within the research grants reg. nr A6087704 and reg. nr A307865 from the Grant Agency of the Academy of Sciences of the Czech Republic, and grants reg. nr 206/99/1519 and reg. nr 206/00/0668 from the Grant Agency of the Czech Republic.

References

Carvalho G.R. & Hauser L. (1994) Molecular genetics and the stock concept in fisheries. *Reviews in Fish Biology Fisheries* **4**, 326–350.

Council of Ministries of the European Communities (1992) Council Directive 92/43/EEC of 21 May 1992 on the conservation of natural habitats and of wild fauna and flora. *Official Journal of the European Communities* **L 206/7**, 1–44.

Fric A. (1875) Artificial breeding of fishes in Bohemia. *Prag*, 36 pp. (in Czech).

Gärdenfors U., Rodríguez J.P., Hylton-Taylor C., Hyslop C., Mace G., Molur S. & Poss S. (1999) Draft guidelines for the application of IUCN Red List criteria at national and regional levels. *Species* **1999**, 58–70.

Gollmann G. (1995) Biochemical markers in the population genetics of *Chondrostoma nasus*. *Folia Zoologica* **44**(Suppl. 1), 83–89.

Gollmann G., Bouvet I., Karakousis Y. & Triantaphyllidis C. (1997) Genetic variability in Chondrostoma from Austrian, French and Greek rivers (Teleostei, Cyprinidae). *Journal of Zoology and Systematic Evolutionary Research* **35**, 165–169.

IUCN (1999) IUCN Red List criteria review provisional report: draft of the proposed changes and recommendations. *Species* **1999**, 43–57.

Lelek A. (1987) *The Freshwater Fishes of Europe. Vol. 9. Threatened Fishes of Europe*. Wiesbadden: AULA-Verlag, 343 pp.

Lusk S. (1988) Gene pool of fishes in Czechoslovakia: present state and conservation efforts. *Práce VÚRH Vodnany* **18**, 11–26.

Lusk S. (1995) The status of *Chondrostoma nasus* in the waters of the Czech Republic. *Folia Zoologica* **44**(Suppl. 1), 45–56.

Lusk S. (1996a) The status of the fish fauna in the Czech Republic. In A. Kirchhofer & D. Hefti (eds) *Conservation of Endangered Freshwater Fish in Europe*. Basel/Switzerland: Birkhauser Verlag, pp. 89–98.

Lusk S. (1996b) Development and status of populations of *Barbus barbus* in the waters of the Czech Republic. *Folia Zoologica* **44**(Suppl. 1), 39–46.

Lusk S. & Halacka K. (1995) Anglers' catches as an indicator of population size of the nase, *Chondrostoma nasus*. *Folia Zoologica* **44**, 185–192.

Lusk S. & Hanel L. (1996) The Red List of lampreys and fishes in the Czech Republic – Version 1995. *Biodiversity of Fishes in the Czech Republic* **I**, 16–25.

Lusk S. & Holcík J. (1998) The importance of an unhindered connection of the Morava and Dyje river systems in the territory of the Czech Republic with the Danube. *Biodiversity of Fishes in the Czech Republic CR* **II**, 69–84.

Lusk S., Lusková V. & Halacka K. (1996) Distribution of species of the family Percidae in the waters of the Czech Republic. *Acta Universitatis Carolinae Biologica* **40**, 139–145.

Lusk S., Lusková V. & Halacka K. (2000a) On the occurrence of populations of the genera *Cobitis* and *Sabanejewia* (Pisces, Cobitidae) in the Czech Republic. *Folia Zoologica* **49**(Suppl. 1), 97–106.

Lusk S., Lusková V., Halacka K. & Lojkásek B. (2000b) Changes in the species composition of the ichthyofauna in the territory of the Czech Republic after 1990. *Biodiversity of Fishes in the Czech Republic* **III**, 21–28.

Lusk S., Lusková V., Halacka K., Šlechta V. & Šlechtová V. (1998) Trends and production of a fish communities of the barbel zone in a stream of the Czech Republic. *Folia Zoologica* **47**(Suppl. 1), 67–72.

Lusková V., Lusk S., Holcík J. & Halacka K. (2000) Wild Danube carp (*Cyprinus carpio carpio* m. *hungaricus* Heckel 1836) – its past, present and future in the Czech Republic. *Biodiversity of Fishes in the Czech Republic* **III**, 97–106.

Lusková V., Šlechtová V., Povz M., Šlechta V. & Lusk S. (1997) Genetic variability of *Chondrostoma nasus* populations in rivers of the Black Sea and Baltic Sea drainage systems. *Folia Zoologica* **46**(Suppl. 1), 27–36.

Lusková V., Šlechtová V., Šlechta V. & Lusk S. (1995) The population genetic diversity of *Chondrostoma nasus* from Moravian and Slovak rivers. *Folia Zoologica* **44**(Suppl. 1), 91–98.

Maitland P.S. & Lyle A.A. (1991) Conservation of freshwater fish in the British Isles: the current status and biology of threatened species. *Aquatic Conservation: Marine and Freshwater Ecosystems* **1**, 25–54.

Meffe G.K. (1986) Conservation genetics and management of endangered fishes. *Fisheries* **11**(2), 14–23.

Nei M. (1972) Genetic distance between populations. *American Naturalist* **106**, 283–292.

Nelson, K. & Soule M. (1987) Genetical conservation of exploited fishes. In R. Nyman & F. Utter (eds) *Population Genetics and Fisheries Management.* Seattle and London: University of Washington Press, pp. 345–368.

Ráb P., Rábová M., Bohlen J. & Lusk S. (2000) Genetic differentiation of the two hybrid diploid–polyploid complexes of loaches, genus *Cobitis* (Cobitidae) involving *C. taenia*, *C. elongatoides* and *C.* spp. in the Czech Republic. I. Karyotypes and cytogenetic diversity. *Folia Zoologica* **49**(Suppl. 1), 55–65.

Rogers J.S. (1972) Measures of genetic similarity and genetic distances. *Studies in Genetics. University of Texas Publication* **7213**, 145–153.

Šlechtová V., Lusková V., Šlechta V., Lusk S., Halacka K. & Bohlen J. (2000) Genetic differentiation of the two hybrid diploid–polyploid complexes of loaches, genus *Cobitis* (Cobitidae) involving *C. taenia*, *C. elongatoides* and *C.* spp. in the Czech Republic. II. Allozyme interpopulation and interspecific differences. *Folia Zoologica* **49**(Suppl. 1), 67–87.

Šlechová V., Šlechta V. & Berrebi P. (1993) Genetic diversity of *Barbus* in Czech and Slovak rivers, preliminary results. *Cahiers d'Ethologie* **13**, 207–210.

Šlechtová V., Šlechta V., Lusková V., Lusk S. & Berrebi P. (1998) Genetic variability of common barbel, *Barbus barbus* populations in the Czech Republic. *Folia Zoologica* **47**(Suppl. 1), 21–33.

Chapter 3
The status of threatened freshwater fishes in Estonia

T. SAAT*
Estonian Marine Institute, Tallinn and University of Tartu, Tartu, Estonia

R. JARVEKULG
Estonian Agricultural University, Tartu, Estonia

R. ESCHBAUM
Estonian Marine Institute, Tallinn and University of Tartu, Tartu, Estonia

J. TAMBETS
University of Tartu, Tartu, Estonia

Abstract

The Estonian freshwater fish fauna (together with migratory species and cyclostomes) comprises 44 species; most occur not only in inland waters but also in the brackish coastal waters of the Baltic Sea. Species composition, distribution and abundance of fish have been affected by anthropogenic activities (eutrophication, pollution, fishery, damming, ditching and dredging of rivers). Four species are protected by law – *Acipenser sturio*, *Thymallus thymallus*, *Aspius aspius* and *Silurus glanis*. The Estonian Red Data Book includes 13 freshwater and migratory species. One species, *Acipenser sturio*, is considered almost extinct (last record in 1996), four species are endangered (*Salmo salar*, *Thymallus thymallus*, *Silurus glanis* and anadromous *Coregonus maraena*); marine spawning *Coregonus widegreni* and sea trout, *Salmo trutta,* are considered vulnerable. However, several species that are rare and/or endangered in most of Europe are widespread in Estonia.

Keywords: Baltic Sea, conservation, distribution, inland waters.

3.1 Estonian fish fauna

The contemporary freshwater fish communities of Estonia are relatively young. They were formed after the last Würm-Weichselian-Valday glaciation by multiple colonisation of species from different refugia (Lõugas 1997).

The Estonian fauna includes three species of cyclostomes and 71 (68 native and 3 naturalised) species of fish (Saat 1992). There are 44 freshwater and migratory species (including 22 species of Cypriniformes; Table 3.1). *Abramis ballerus* (L.) is not included; this species inhabited the Emajõgi River basin in the 19th century (Dybowski 1862) but it has not been found during the 20th century.

*Correspondence: Estonian Marine Institute, 18b Viljandi Road, Tallinn, EE-11216, Estonia (email: tsaat@sea.ee).

Table 3.1 Freshwater and migratory fish and cyclostome species in Estonia

Scientific name	Lakes	Rivers	Baltic	Fishery	Remarks
Petromyzonidae					
Lampetra fluviatilis (L.)		C	C	++ r	
Lampetra planeri (Bloch)	R	C			
Petromyzon marinus L.		lit.	R		
Acipenseridae					
Acipenser sturio L.		lit.	R	Prohibited	
Clupeidae					
Alosa fallax (Lacépéde)		lit.	R		
Salmonidae					
Salmo salar L.		R	C	+ r	Stocking
Salmo trutta L. forma *fario*	R	C		+ r	Stocking
Salmo trutta L. 1758 forma *trutta*		C	C	+ r	Stocking
Oncorhynchus mykiss (Walbaum)	R	loc. C	C	+	Introduced
Coregonidae					
Coregonus albula (L.)	R	R		+ r	
Coregonus maraena (Bloch) + *Coregonus widegreni* Malmgren + endemic lake form	loc. C	R	C	++ r	Stocking
Osmeridae					
Osmerus eperlanus (L.)	loc. C	C	C	+++ r	
Thymallidae					
Thymallus thymallus (L.)	R	R	R	Prohibited	
Esocidae					
Esox lucius L.	C	C	C	+++ r	Stocking
Anguillidae					
Anguilla anguilla (L.)	loc. C	C	C	+++ r	Stocking
Cyprinidae					
Abramis brama (L.)	C	C	C	+++ r	
Alburnoides bipunctatus (Bloch)		C			
Alburnus alburnus (L.)	C	C	C	+	
Aspius aspius (L.)	R	R	lit.	Prohibited	
Abramis bjoerkna (L.)	C	C	C	++	
Carassius gibelio (Bloch)	C	R	C	++	Introduced
Carassius carassius (L.)	C	C	C	+	
Cyprinus carpio L.	R	R	loc. C	+	Introduced, stocking
Gobio gobio (L.)	C	C	C		
Leucaspius delineatus (Heckel)	C	loc. C			
Leuciscus cephalus (L.)	C	C	C	+	
Leuciscus idus (L.)	C	C	C	+++ r	
Leuciscus leuciscus (L.)	C	C	C	++	
Pelecus cultratus (L.)		lit.	R		
Phoxinus phoxinus (L.)	R	C	C		
Rutilus rutilus (L.)	C	C	C	+++	
Scardinius erythrophthalmus (L.)	C	C	C	+	
Tinca tinca (L.)	C	C	C	++ r	Stocking
Vimba vimba (L.)	loc. C	C	C	+++ r	
Cobitidae					
Cobitis taenia L.	R	C	loc. C		

Table 3.1 continued

Scientific name	Lakes	Rivers	Baltic	Fishery	Remarks
Misgurnus fossilis (L.)	loc. C	loc. C			
Balitoridae					
Barbatula barbatula (L.)	R	C			
Siluriformes					
Silurus glanis (L.)	R	R	lit.	Prohibited	
Gadidae					
Lota lota (L.)	C	C	C	++ r	
Gasterosteidae					
Gasterosteus aculeatus L.	R	loc. C	C		
Pungitius pungitius (L.)	C	C	C		
Percidae					
Gymnocephalus cernuus (L.)	C	C	C	+	
Perca fluviatilis L.	C	C	C	+++ r	
Stizostedion (Sander) lucioperca (L.)	C	C	C	+++ r	Stocking
Cottidae					
Cottus gobio L.	R	C	C		

R, rare; C, common; loc., locally; lit., only older literature data; +, ++ and +++, importance in fishery (marginal, important, very important); r, regulated fishery.

Some species are at the northern limit of their distribution area – mud loach *Misgurnus fossilis* (L.), sunbleak *Leucaspius delineatus* (Heckel), riffle minnow *Alburnoides bipunctatus* (Bloch) – or close to that limit (wels *Silurus glanis* L., spined loach *Cobitis taenia* L., asp *Aspius aspius* (L.), rudd *Scardinius erythrophthalmus* (L.), tench *Tinca tinca* (L.), vimba bream *Vimba vimba* (L.), and razorfish *Pelecus cultratus* (L.)) (see also Freyhof, Chapter 1; Lusk, Lusková, Halacka, Šlechtá & Šlechtová, Chapter 2).

3.2 Freshwater fish habitats

3.2.1 *Rivers*

There are over 1000 rivers and streams entering the sea and larger lakes (Peipsi-Pihkva, Võrtsjärv) in Estonia; most are small and have limited water discharge (Arukaevu 1986). The largest rivers are the Narva (length 77 km, watershed area 56 200 km^2, mean discharge 399 $m^3 s^{-1}$), Emajõgi (101 km, 9740 km^2, 70 $m^3 s^{-1}$), Pärnu (144 km, 6930 km^2, 64 $m^3 s^{-1}$) and Kasari (112 km, 3210 km^2, 30 $m^3 s^{-1}$). Rivers are usually ice-covered from late November to late March–early April. Spring floods peak in April, and low water periods usually last from mid-June to late September–early October, and from January to March (Pachel 1993). The flow rate also fluctuates annually; the last low water period (with limited spring floods affecting reproduction of spring spawning species) ended in the mid-1990s.

3.2.2 *Lakes*

There are over 1200 lakes and reservoirs in Estonia, 769 of which have an area of 2 ha or more (Kask 1964). Lakes are normally ice-covered from December to April. There are three large-standing water bodies (two of them are shared with Russia): Lake Peipsi-Pihkva (3555 km^2, maximum depth 15.3 m, mean depth 7.1 m), Lake Võrtsjärv (270 km^2, 6.0 m and 2.8 m) and the Narva Reservoir (191 km^2, 15.0 m and 1.8 m). The remaining lakes are comparatively small (only 11 of them have an area over 300 ha) and shallow (maximum depth up to 38 m, but usually <10 m). Most lakes are eutrophic or hypereutrophic. Between 1951 and 1991, the mean water transparency and the oxygen content near the bottom of lakes have decreased, and the total alkalinity as well as concentrations of SO_4^{2-} and Cl^- have increased, indicating eutrophication and pollution (Mäemets, Ott & Mäemets 1994).

3.2.3 *Coastal waters of the Baltic Sea*

Water salinity of the Estonian coastal waters is usually 5–7 (less in river estuaries). Most freshwater species inhabit these waters (Table 3.1), and many of them can reproduce there. Coastal waters, especially shallow bays, have become heavily eutrophicated. However, the situation has improved slightly (as in inland water bodies) after the decline of agriculture (and corresponding pollution) in the early 1990s. Coastal fish communities are dominated by perch and roach followed by several other cyprinids, ruffe, pike, pikeperch, *Stizostedion (Sander) lucioperca* (L.), and other species (T. Saat & R. Eschbaum, unpublished data).

3.3 Status of rare and endangered species

Several lamprey and fish species of Estonia are listed in annexes of the EU Council Directive 92/43/EEC (EU Habitats Directive), Bern Convention and in the 2000 IUCN Red List. Most of them are included in the latest version of the Estonian Red Data Book (ERDB) (Saat 1998) and four species are protected by the Law on Protected Natural Objects (adopted in 1994, amended in 1998) in Estonia (Table 3.2).

3.3.1 *Lampreys*

The river lamprey, *Lampetra fluviatilis* (L.), is common in Estonia. It spawns in 40–45 rivers, mostly falling in the Gulf of Finland and Pärnu Bay. This species is commercially exploited. Generally, catches have declined over the past 60–70 years, probably due to the loss of available spawning grounds. The annual catch was 41–102 t (average 67 t) in 1928–1938. In recent years, catches have been up to 10.4 t in 1996 (Saat, Tambets & Kangur, in press). Artificially inseminated lamprey eggs have been incubated in fish hatcheries since the 1950s and larvae have been distributed in rivers.

Table 3.2 Estonian cyclostome and fish species listed in the EU Habitat Directive (Annex No.), Bern Convention (Annex No.), 2000 IUCN Red List[a], ERDB[b] or protected by the law in Estonia[c]

Species	EU	Bern	IUCN	ERDB	Protection
River lamprey	II,V	III	LR/NT	–	–3
Brook lamprey	II	III	LR/NT	4	–3
Sea lamprey	II	III	–	–	–
Sturgeon	II, IV	II	CR	0	II
Twaite shad	II, IV	III	DD	–	–
Salmon (in freshwater)	II, V	III	–	1	–
Sea trout (in freshwater)	–	–	–	2	–
Vendace	V	III	DD	4	–
Whitefish	V	III	DD	2, 4	–
Grayling	V	III	–	1	III
Razorfish	–	III	–	–	–
Riffle minnow	–	III	–	4	–
Asp	II, IV	III	DD	5	III
Carp	–	–	DD	–	–
Crucian carp	–	–	LR/NT	–	–
Spined loach	II	III	–	5	–3
Mud loach	II	III	LR/NT	5	–
Wels	–	III	–	1	II
Bullhead	II	–	–	4	–3

[a]CR, critically endangered; LR, lower risk; NT, near threatened/close to vulnerable; DD, data deficient.
[b]0, Extinct or probably extinct; 1, endangered; 2, vulnerable; 3, rare; 4, care demanding; 5, indeterminate.
[c]Protection category; 3, Estonian position on adoption of the EU environmental acquis in membership negotiations was to apply for exemption of the species from Annex II.

The brook lamprey, *Lampetra planeri* (Bloch), is common in rivers but rare in lakes. It is frequently found sympatrically with the river lamprey.

The marine lamprey, *Petromyzon marinus* L., is rare in Estonian waters. There were some 20 confirmed catches during the 20th century, most of them from the western coast (Gulf of Riga, Väinameri); only a few specimens were recorded on the northern coast. There are some records from rivers (mostly the Pärnu River) in spring, including a female with ovulated eggs on 29 June 1943 (Mikelsaar 1984). However, there are no reliable data about the reproduction of sea lamprey in Estonia.

3.3.2 *Sturgeon* Acipenser sturio *L.*

Acipenser sturio is considered almost extinct. There have been about 30 sightings of sturgeon from Estonian waters since the end of the 19th century. At the beginning of the 20th century it was caught mainly from the Gulf of Finland and the Narva River. All latest records (eight fish since 1948) are from the western coast, mostly from the Gulf of Riga.

As there were small specimens (in 1954 and 1958; total length 54–55 cm, weight 480–500 g) among them, some reproduction had probably taken place in this region in the middle of the century. The latest specimen in Estonian waters was a female captured in the Väinameri on 24 May 1996. It was the biggest sturgeon ever recorded in Estonia (total length 290 cm, weight 136 kg, age > 45 years) (Paaver, in press).

3.3.3 *Twaite shad* Alosa fallax *(Lacépède)*

Twaite shad was abundant in Estonian waters (both in the gulfs of Finland and Riga) in the 1930s, especially in the River Narva (Mikelsaar 1984). Fishermen capture a few specimens of this species almost every year in the Gulf of Finland. This species was not recorded during extensive coastal fish surveys between 1991 and 1999. However, three specimens were captured in two areas of the Väinameri in the summer 2000.

3.3.4 *Salmon* Salmo salar *L.*

In the basin of the Gulf of Finland, salmon only reproduces naturally in some Estonian rivers. In the 1990s, it spawned in the rivers Vasalemma, Keila, Pirita, Jägala, Valgejõgi, Loobu, Selja, Kunda and Vääna. The populations of rivers Narva (formerly the most important salmon river) and Purtse were lost due to damming (Narva) and pollution (Purtse) in the 1950s and 1960s. In the Gulf of Riga basin, salmon ascends the River Pärnu and the River Vaidava, a tributary of the River Gauja in South Estonia (P. Komissarov & R. Järvekülg unpublished data).

Natural populations of salmon are rather small. Only 36–43 ha of spawning grounds are currently available in salmon rivers, and the potential annual smolt production is approximately 50 000 (Rappe 1999). Estimates of natural smolt production of Estonian rivers flowing into the Gulf of Finland and Main Basin (River Pärnu) in 1997 were 8000 and 3000 smolts respectively (Anonymous 1999).

Salmon were first stocked in Estonia in the 1920s. A total of 196 000 smolts and parr were stocked into the rivers Narva, Selja, Valgejõgi, Jägala, Pirita and Vääna in 1999.

Estonian annual salmon and sea trout catches have declined from 38–148 t in 1928–1939 to 4–12 t in 1956–1979. All these catches were taken in the coastal zone and rivers. An offshore drift-net fishery started in 1980 and peaked at 103 t between 1981 and 1991. During the last decade coastal catches of salmon have been 4–14 t annually, and offshore catches fell below 1 t in 1999 (Kangur, Drevs, Turovski & Paaver, in press).

3.3.5 *Sea trout* Salmo trutta *L.*

Wild sea trout populations are still found in >45 Estonian rivers and streams. The most productive rivers belong to the basin of the Gulf of Finland, smaller trout streams are situated in the Gulf of Riga basin and on western and northern coast of islands Saaremaa and Hiiumaa. Sea trout have been recorded in streams of south-east Estonia (the Gauja River tributary) >300 km from the sea. They can enter even very small streams or ditches, so long as the mouth of the stream is open and if there is a short period of increased water level in autumn.

3.3.6 *Vendace* **Coregonus albula** *(L.)*

This species occurs naturally in large lakes (Peipsi-Pihkva and Võrtsjärv). It has been successfully introduced in two larger of the medium-sized lakes: Saadjärv (708 ha) and Ülemiste (985 ha). In Lake Peipsi-Pihkva, vendace currently inhabits the larger and deeper northern part (Lake Peipsi), while in the southern part (Lake Pihkva), as well as in Lake Võrtsjärv, the stock has collapsed. Since 1990 the stock in Lake Peipsi has been at a very low level. Vendace occurs also in the eastern part (Russian waters) of the Gulf of Finland.

3.3.7 *Whitefish* **Coregonus maraena** *(Bloch),* **Coregonus widegreni** *Malmgren*, **Coregonus** *sp.*

Decline of whitefish stocks due to eutrophication and fishing in the coastal sea was evident in the 1970s. Several local stocks are close to extinction. The stocks of sea-spawning whitefish are, in general, in better condition than stocks of anadromous whitefish (the latter is regarded as endangered in the Red Data Book). Stocking has recently improved the state of anadromous whitefish spawning in the River Pärnu. An increase in whitefish abundance in the Gulf of Finland in recent years is probably due to stockings in Finnish waters.

An endemic species, *Coregonus* sp., inhabiting Lake Peipsi is rather abundant; commercial catches have been stable (about 60 t) during the last decades (T. Saat, V. Vaino, E. Afanasjev & N. Koncevaja, unpublished data).

3.3.8 *Grayling* **Thymallus thymallus** *(L.)*

Grayling is naturally distributed in the northern, eastern and south-eastern part of Estonia. During the 20th century its occurrence has continuously narrowed, especially after the 1950s when agricultural and industrial pollution increased rapidly. Pollution of rivers has declined during the last decade and since the mid-1990s, the state of some populations has improved and grayling is re-occupying its former distribution range. Nowadays grayling populations exist in 11–13 rivers of the former 24; four of them are rather weak and endangered.

Grayling, as a rather stenotopic fish, inhabits rivers and tributaries with clear, rather cold (maximum summer water temperature between 13.5°C and 21°C) well-oxygenated water, strong current, and a high flow rate (minimum water discharge of 0.7–1 $m^3 s^{-1}$ is needed during the summer low-water period). In smaller tributaries (minimum water discharge 0.2–0.5 $m^3 s^{-1}$) only spawning grounds and habitats for juveniles are found. It was never recorded in waters with signs of pollution (Järvekülg, in press).

3.3.9 *Razorfish* Pelecus cultratus *(L.)*

It is rare in the Estonian waters. It has occasionally been captured in the Gulf of Finland (including the Narva River), the Väinameri, in Pärnu Bay and off the southern coast of Saaremaa Island. In the Väinameri a few specimens were captured almost every May–June, usually with perch or herring trap nets. Periodically, e.g. in 1946 (Mikelsaar 1984) and in the warm summer of 1995, it was more abundant in Estonian waters (Saat, Lotman & Veldre, in press).

3.3.10 *Riffle minnow* Alburnoides bipunctatus *(Bloch)*

The riffle minnow is widely distributed in Estonian rivers (Järvekülg 1994 and unpublished data; Kangur 1999; Vetemaa, unpublished data). This species lives in larger river basins (Emajõgi, Kasari, Pärnu) and also in several smaller rivers with good water flow. The riffle minnow has not been found on any Estonian islands, or in lakes or coastal waters.

3.3.11 *Asp* Aspius aspius *(L.)*

Asp inhabits large lakes (Peipsi and Võrtsjärv) as well as some larger rivers of the basins of these lakes (Emajõgi; Väike-Emajõgi, Õhne, Narva). It is caught occasionally in about 10 smaller lakes of this region and occasionally in Estonian coastal waters of low salinity (Kirsipuu, Pihu & Saat, in press).

3.3.12 *Carp* Cyprinus carpio *L. and crucian carp* Carassius carassius *(Bloch)*

Crucian carp is common in many inland water bodies. In recent years it has spread widely in the Baltic Sea and now it occurs almost everywhere along the Estonian coast. Carp has been repeatedly stocked into lakes and the sea. Juveniles occur regularly in Saunja Bay (west Estonia; I. Taal unpublished data) which suggests some natural reproduction.

3.3.13 *Spined loach* Cobitis taenia *L.*

Spined loach is common in rivers and locally in the coastal sea (mostly in bays), sometimes in lakes.

3.3.14 *Mud loach* **Misgurnus fossilis** ***(L.)***

Mud loach is common in south-eastern Estonia (Lake Peipsi-Pihkva watershed area) but it also inhabits certain regions of the River Kasari in west Estonia.

3.3.15 *Wels* **Silurus glanis** ***L.***

Climatic change, fisheries and river pollution have led to the decline of the wels populations near the northern limit of its distribution. In Estonia a small population inhabits the Emajõgi river basin, including lakes Võrtsjärv and Peipsi. Formerly it inhabited the Kasari, Pärnu and Narva river basins; wels has been recorded in coastal waters of the Baltic Sea, in the Väinameri and in the Gulf of Riga both in Estonian and Latvian (Plikšs & Aleksejevs 1998) waters. Archaeological data (Lõugas 1997) indicate the presence of wels in Estonian waters since the Mesolithic; it was widely distributed in the Neolithic (6000–3500 years ago), and in the Bronze and Iron Ages (3500–1000 years ago, including records from Saaremaa Island).

3.3.16 *Bullhead* **Cottus gobio** ***(L.)***

Bullhead has been found in 77 rivers and streams. However, it is absent in streams of islands, as well as some river basins of the Estonian mainland. In warm-water streams it occurs mainly in rapids, below waterfalls or milldams where water is over-saturated with the oxygen. It is also absent in most cold-water brooks with a minimum water flow rate $<0.2\,m^3\,s^{-1}$ (Järvekülg 1994). In the Baltic Sea bullhead occurs over the whole coastal area. It is also encountered in some lakes but is only abundant in Lake Karujärv on Saaremaa Island.

3.4 Protection and monitoring

Several factors affect the abundance and distribution of freshwater fish in Estonia. Natural factors include climatic change, changes in river runoff, high abundance of fish-eating animals (cormorant *Phalacrocorax carbo* (L.), seals *Phoca hispida* (Schr.) and *Haliochoerus grypus* (Fabr.), otter *Lutra lutra* (L.) and mink *Mustela vison* (L.)). Among anthropogenic factors eutrophication and pollution of water bodies, overexploitation of the fisheries and damming of rivers are the most important.

Nevertheless, the Estonian freshwater fish fauna appears in good condition in the European context. Several species, which are rare and/or endangered in other parts of Europe (Lelek 1987) are widespread in Estonia (e.g. *Lampetra planeri*, *Alburnoides bipunctatus*, *Leucaspius delineatus*, *Cobitis taenia*, *Barbatula barbatula*, *Cottus gobio*) and in some cases commercially exploited (*Lampetra fluviatilis*, *Leuciscus idus*, *Vimba vimba*, *Lota lota*).

The four most endangered Estonian fish species are protected by law. According to the Law on Protected Natural Objects, the capture, keeping and killing of animal species of Protection Category II or III, disturbing the species to an extent that sets the animals in danger, and making transactions with the species, are prohibited. Destroying of their permanent habitats or causing damage to these to an extent that sets the preservation of the species at the concerned site in danger is prohibited. A protected area may be established for ensuring the preservation of the permanent habitat of a species.

Capture of most commercial species is regulated by the Fisheries Law, which sets closed areas, close seasons, minimum length, mesh size and gear restrictions. For example, the minimum legal total length of salmon in Estonian waters is 60 cm. An all-year-round closed area of 1000 m radius is established at the outlet of each salmon spawning river. Commercial salmon fishing in spawning rivers is not permitted. Downstream of dams and waterfalls fishing is prohibited within a distance of 100–500 m all-year-round. The close period for sport fishing in salmon spawning rivers lasts from 1 September to 30 November. Offshore salmon fishing is carried out in accordance with the fisheries rules of the International Baltic Sea Fisheries Commission. The main problem is enforcement of these regulations.

A nationally-funded monitoring programme includes a sub-programme for biodiversity monitoring in which all vertebrate groups except fish and cyclostomes are covered. However, there is ongoing monitoring of salmon and sea trout in spawning rivers funded by other sources. Populations of non-commercial species (including wels, asp and grayling protected by law) are not monitored at present.

Acknowledgements

This study was partly supported by the national funded research project TBGZH0515 (Biodiversity of the Estonian fauna; T. Saat). We thank anonymous referees for valuable comments.

References

Anonymous (1999) *ICES Cooperative Research Report No. 229, Part 2*, 446 pp.

Arukaevu K. (1986) *Eesti NSV Jõgede, Ojade ja Kraavide Nimestik (The List of the Estonian Rivers and Streams)*. Tallinn: Valgus, 72 pp. (in Estonian).

Dybowski B. (1862) Versuch einer Monographie der Cyprinoiden Livlands nebst einer synoptischen Aufzählung der europäischen Arten dieser Familie. *Archiv für die Naturkunde Liv-, Ehst- und Kurlands, II Serie* **3**, 133–362 (in German).

Järvekülg R. (1994) Eesti jõgede kalastik ja kalakooslused. In A. Järvekülg (ed.) *Eesti Jõgede ja Järvede Seisund ning Kaitse (The Status and Conservation of Estonian Rivers and Lakes)*. Tallinn: Teaduste Akadeemia Kirjastus, pp. 177–192 (in Estonian with English summary).

Järvekülg R. Grayling, *Thymallus thymallus* (L). In E. Ojaveer, E. Pihu & T. Saat (eds) *Fishes of Estonia*. Tallinn: Estonian Academy Publishers (in press).

Kangur M. (1999) Fishes of the North-Estonian rivers. In R.Volskis (ed.) *Hydrobiological Research in the Baltic Countries. Part 1. Rivers and Lakes* Vilnius: Institute of Ecology, pp. 363–368.

Kangur M., Drevs T., Turovski A. & Paaver T. Salmon, *Salmo salar* L. In E. Ojaveer, E. Pihu & T. Saat (eds) *Fishes of Estonia*. Tallinn: Estonian Academy Publishers (in press).

Kask I. (1964) *Eesti NSV Järvede Nimestik (The List of Lakes of Estonia)*. Tallinn: Eesti Riiklik Kirjastus, 223 pp. (in Estonian).

Kirsipuu A., Pihu E. & Saat T. Asp, *Aspius aspius aspius* (L.). In E. Ojaveer, E. Pihu E. & T. Saat (eds) *Fishes of Estonia*. Tallinn: Estonian Academy Publishers (in press).

Lelek A. (1987) *Threatened Fishes of Europe* (The Freshwater Fishes of Europe, Vol. 9). Wiesbaden: AULA-Verlag, 343 pp.

Lõugas L. (1997) *Post-glacial development of vertebrate fauna in Estonian water bodies*. Dissertationes Biologicae Universitatis Tartuensis 32, 184 pp.

Mäemets A., Ott I. & Mäemets A. (1994) Eesti väikejärvede seisundi muutused ja kaitse. In A. Järvekülg (ed.) *Eesti Jõgede ja Järvede Seisund ning Kaitse (The Status and Conservation of Estonian Rivers and Lakes)*. Tallinn: Teaduste Akadeemia Kirjastus, pp. 32–46 (in Estonian with English summary).

Mikelsaar N. (1984) *Eesti NSV Kalad (Fishes of the Estonian SSR)*. Tallinn: Valgus, 432 pp. (in Estonian).

Paaver T. Sturgeon, *Acipenser sturio* L. In E. Ojaveer, E. Pihu & T. Saat (eds) *Fishes of Estonia*. Tallinn: Estonian Academy Publishers (in press).

Pachel K. (1993) The state of rivers. Hydrological survey. In M. Pylvänäinen (ed.) *Water Pollution and Quality in Estonia*. Helsinki: Environmental Report 7, pp. 5–10.

Plikšs M. & Aleksejevs E. (1998) *Latvijas Daba. Zivis (Animals of Latvia. Fishes)*. Riga: Gandrs, 304 pp. (in Latvian).

Rappe C. (ed.) (1999) *Baltic Salmon Rivers*. Gothenburg: IBSFC and HELCOM, 69 pp.

Saat T. (1992) Estonian vertebrates. *Acta Musei Zoologici Universitatis Tartuensis* No.1, 24 pp.

Saat T. (1998) Kalad Cyclostomata + Pisces. In V. Lilleleht (ed.) *Eesti Punane Raamat (Estonian Red Data Book)*. Tartu: Eesti Teaduste Akadeemia Looduskaitse Komisjon, pp. 110–113 (in Estonian).

Saat T., Lotman K. & Veldre I. Razorfish, *Pelecus cultratus* (L.). In E. Ojaveer, E. Pihu & T. Saat (eds) *Fishes of Estonia*. Tallinn: Estonian Academy Publishers (in press).

Saat T., Tambets J. & Kangur M. Lampern, river lamprey, *Lampetra fluviatilis* L. In E. Ojaveer, E. Pihu & T. Saat (eds) *Fishes of Estonia*. Tallinn: Estonian Academy Publishers (in press).

Chapter 4
Status of the freshwater fishes and lampreys in Slovenia

M. POVZ*
Fisheries Research Institute, Zupanciceva, Ljubljana, Slovenia

Abstract

Slovenia's fresh waters comprise two main drainage basins, the smaller Adriatic and the larger Danube basins. They are inhabited by 85 fish species, including 68 indigenous fish and four lamprey species (Cyclostomata). Fifty-six species are listed in the Red Data List of Freshwater Fish Species and Lampreys of Slovenia. They fall into the following categories: two are classed as extinct, two as presumed extinct, 32 are endangered, nine are vulnerable and seven are rare. The major threats to the freshwater fishes include pollution, river regulation and species introductions and translocation.

Keywords: conservation, freshwater fish, Slovenia, status, threats.

4.1 Introduction

There is a paucity of information on the freshwater fishes of Slovenia. Most data concern systematic descriptions of the fish species and give only superficial information on their distribution. Data on status and abundance are mainly limited to the larger river basins. For example, Freyer (1842) listed the various fish species and the streams inhabited by them, whilst a more comprehensive report was provided by Heckel and Kner (1858). Beside precise descriptions of the fishes and their biology and ecology, they also gave the localities where each species had been recorded. Detailed lists of the fish in the River Drava was presented by Glowacki (1896) and the fishes of the main Slovenian river basins by Munda (1926), Svetina and Verce (1960), and Sket (1967). They occasionally added statements on the exact locality of the species, usually when a rare fish was being described. Seeley (1886) mentioned the fishes of the Adriatic and Danube river basins, but his data lacked geographical precision. By contrast, the literature on species popular with fishermen, e.g. marble trout, *Salmo marmoratus* Cuvier, Danube salmon, *Hucho hucho* (L.), pike, *Esox lucius* L., is reasonably abundant.

Thus it can be seen that the distribution and systematics of many of the freshwater fishes in Slovenia still require considerable attention, especially as there is a diverse range of habitats in the country. This chapter examines the present status of the fishes of Slovenia and some of the threats to the freshwater fish fauna.

*Correspondence: Fisheries Research Institute, Zupanciceva, SI-1000 Ljubljana, Slovenia (email: meta.povz@guest.arnes.si).

4.2 Study area

Although Slovenia is a small country (20 000 km^2) its freshwater systems comprise two main drainage basins, the smaller Adriatic and the larger Danube basin (Fig. 4.1). The Danube drainage is represented by the Drava, Sava and Mura catchments. The Adriatic drainage basin is represented by the River Soca and its tributaries as far as the Italian border, together with the Nadiza and smaller rivers and some brooks, flowing into the northern Adriatic Sea in Slovensko Primorje.

The River Drava rises in Italy, flows through Austria to reach northern Slovenia 285 km from its headwaters, and flows south-east towards the Croatian border. It is the largest river in Slovenia (catchment area of 3267 km^2), and is impounded by two dams and eight weirs for hydroelectric power (Zgonik & Zupan 1988). Water quality in the Drava and its tributaries is class II (beta-mesosaprobic) or III (alpha-mesosaprobic) (Zupan 1994).

The Mura originates in Austria in the Niedere Tauern and flows into the River Drava in Croatia. It reaches Slovenia after a distance of 39 km, and Croatia after 134 km. Ninety-five kilometres of its length is in Slovenia, and it has a watershed of 1386 km^2. No impoundments have been constructed either on the River Mura or on its tributaries in Slovenia. Water quality belongs to class II and III, several tributaries belong even to class III–IV (polysaprobic) (Zupan 1994).

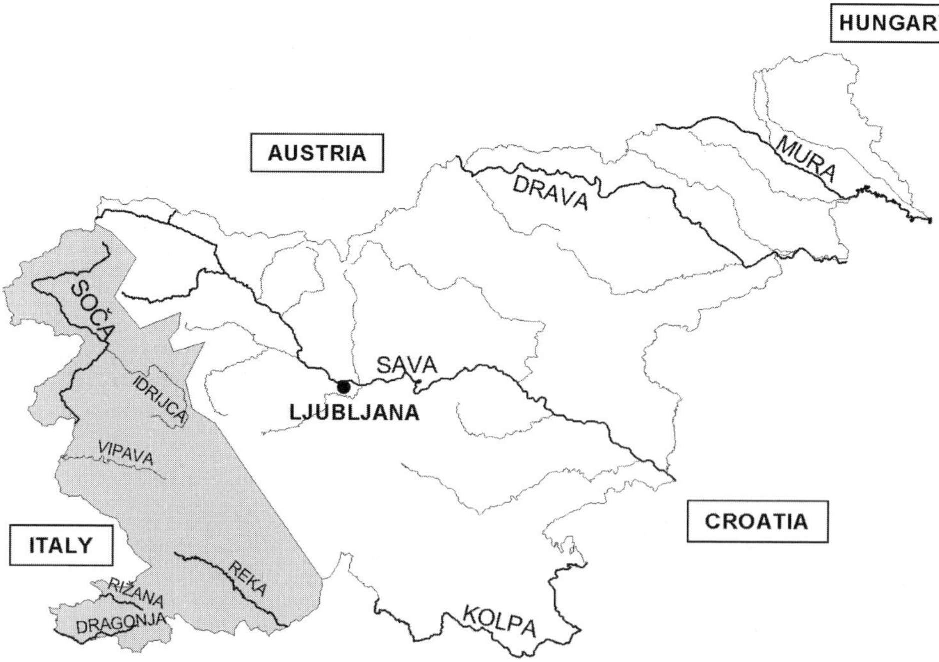

Figure 4.1 Slovene freshwater system with two main drainage basins, the smaller Adriatic (shaded area) and the larger Danube river basin

The River Sava originates in two arms, the so-called Sava Dolinka and the second arm, Sava Bohinjka, in north-west Slovenia, flows south-east to Croatia and leaves Slovenia after 219 km (Bat & Zupan 1996). The total watershed of the River Sava and its tributaries in Slovenia encompasses about 10 000 km^2. Four hydroelectric power plants, one nuclear power plant, and one weir have been constructed on the River Sava. Another hydroelectric power plant is being built and a further 14 plants are planned. Water quality in the River Sava deteriorates from the headwaters (class I – oligosaprobic) to class II–III and even class IV in the lower reaches. Water quality of the tributaries ranges from class II to IV (Zupan 1994).

The River Soca belongs to the Adriatic Sea basin. The river is 136 km long, with 94 km in the territory of Slovenia. The Soca catchment encompasses 2390 km^2, or 12% of the territory of Slovenia. Four hydroelectric power plants are situated in the upper reaches. With the exception of the headwaters, water quality of the River Soca and its tributaries ranges from class II to III (Zupan 1994).

4.3 National legislation for the protection of freshwater fish

Almost all indigenous freshwater fishes inhabiting Slovenia's watercourses are protected by the Decree on the Protection of Endangered Animal Species (Official Gazette (OG) RS, No. 57/93) (32 species) or the Decree on the Minimum Takeable Size of Fish and on the Close Season of Fishes, Crabs, Frogs and Mussels (OG RS, No. 14/93; pp. 632–633) (32 species). All endangered species (52 fishes and four lampreys) are included in the Red Data List of Freshwater Fishes and Lampreys of Slovenia (Povz, in press).

The freshwater fauna of most of the running and standing fishing waters of Slovenia is managed by 63 anglers' societies, although 10% was, by a special decree, allotted to the Fisheries Research Institute of Ljubljana. The fishing societies manage the freshwater fisheries using plans that are a constituent part of the Fishery Register, kept by the Fisheries Research Institute since 1954. The form in which it is known today was adopted in 1978. The Fishing Societies are united into nine Fishing Society Associations, and these are in the Fisheries Union of Slovenia.

The management of wildlife, and thus also of freshwater fishes, in Slovenia is regulated by the following acts: Freshwater Fishing Act (OG SRS, No. 25/1976); Decree on Determination of Fishery Districts and Sections (OG PRS, No. 17/1959); Decree on Separating out Individual Watercourses from Fishery Sections (OG PRS, No. 17/1959); and Declaration on Sport Fishing in Watercourses Bordering on Slovenia and Croatia.

4.4 Status of the freshwater fishes of Slovenia

Data on the distribution of freshwater fishes were collected between 1978 and 1999 from sport anglers, management plans of the angling societies and the Slovenian Fish Land Register (Ocvirk 1988), maintained by the Fisheries Research Institute in Ljubljana since 1978. Prior to 1978 data were obtained from annual reports on catches of individual freshwater fish species in the anglers' bulletin 'Ribic' and from archives of the angling societies of the Slovenian Anglers' Union. The studies were supplemented with data from reports and expert opinions on fish kills between 1976 and 1999, unpublished manuscripts on fishes in the Slovenian freshwaters between 1961 and 1962 (archives of the Fisheries Research Institute of Ljubljana) and from the literature.

4.4.1 *Species diversity*

There are 85 fish species inhabiting Slovenian fresh waters, including 68 indigenous fish and four lamprey species (Cyclostomata) (Table 4.1). Thirteen species have been introduced and 11 transplanted. Twelve endemic fish and lampreys are living in the Adriatic drainage system and 10 in the Danube drainage basin, but none of them live exclusively in Slovenia. Fifty-six species are listed in the Red Data List of Freshwater Fish Species and Lampreys of Slovenia (Povz, in press). They fall into the following categories (IUCN 1988): two are extinct (Ex), two are presumed extinct (Ex?), 32 are endangered (E), nine are vulnerable (V) and seven are rare (R).

4.4.2 *Adriatic drainage basin*

In the Adriatic basin, 40 species and subspecies of fish and two lamprey species are found, of which one lamprey lives exclusively in fresh water (Table 4.1). Among them, 24 species are indigenous, nine have been introduced and seven translocated from the Danube basin (Povz & Ocvirk 1990). Of the fish and lamprey species populating the waters of the Adriatic drainage basin, 15 are mentioned in the Red Data List, two are extinct (Ex), one is presumably extinct (Ex?), eight are endangered (E), three vulnerable (V) and one is classified as rare (R). Both lampreys are endangered.

Amongst the 22 indigenous fish species and two lampreys living in the Adriatic drainage basin in Slovenia, 11 fish and one of the lampreys are endemic (Table 4.1). None lives exclusively in Slovenian territory, their distribution extends far beyond the Slovenian borders towards Italy, to Dalmatia in Croatia, and further to Bosnia–Herzegovina, Montenegro and Albania. Of these, *S. marmoratus* (Povz, Jesensek, Berrebi & Crivelli 1996) and *Lethenteron zanandreai* Vladykov are threatened, while *Chondrostoma genei* Bonaparte and *C. soetta* (Bonaparte) became extinct after the introduction of *C. nasus* (L.) into the Adriatic drainage basin around 1960 (Povz 1983).

4.4.3 *Danube drainage basin*

The River Danube basin is inhabited by 71 species and subspecies of fishes and two lamprey species (Table 4.1). Fifty-seven fish species are indigenous and 12 introduced (Povz & Ocvirk 1990) and only four were translocated from the Adriatic drainage basin. None of the species migrates from the sea to fresh waters or vice versa.

Table 4.1 Freshwater fish and lampreys of Slovenia

	Danube river basin			Adriatic river basin		
Species	A	I	T	A	I	T
Petromyzontidae						
Petromyzon marinus L.				⊕		
Lampetra planeri (Bloch)	✚E					
Eudontomyzon mariae Berg	⊕					
Lethenteron zanandreai Vladykov				✚E		
Acipenseridae						
Acipenser ruthenus L.	⊕					
Clupeidae						
Alosa abone (Scopoli)	⊕					
Anguillidae						
Anguilla anguilla (L.)	⊕			⊕		
Salmonidae						
Thymallus thymallus (Linnaeus)	⊕			⊕		
Coregonus lavaretus (Linnaeus)	⊕					
Hucho hucho L.	✚E					
Salmo marmoratus Cuvier			⊕	✚E		
Salmo trutta fario L.	⊕					⊕
Salmo trutta lacustris L.	⊕					⊕
Salvelinus fontinalis (Mitchill)		⊕				
Salvelinus alpinus (L.)		⊕				
Oncorhynchus mykiss (Walbaum)		⊕			⊕	
Cyprinidae						
Hypophthalmichthys molitrix (Valenciennes)		⊕			⊕	
Aristichthys nobilis (Richardson)		⊕				
Carassius carassius L.	⊕			⊕		
Carassius auratus (L.)		⊕			⊕	
Barbus petenyi Risso	⊕					
Carassius gibelio (Bloch)		⊕			⊕	
Barbus caninus Valenciennes				✚		
Cyprinus carpio L.	⊕			⊕		
Barbus barbus (L.)	⊕					
Barbus plebejus Valenciennes			⊕	✚E		
Gobio gobio (L.)	⊕					
Gobio uranoscopus (Agassiz)	✚E					
Gobio albipinnatus Lukasch	⊕					
Gobio kessleri Dybowski	✚E					
Tinca tinca (L.)	⊕			⊕		
Pelecus cultratus L.	⊕					
Chondrostoma soetta (Bonaparte)				✚Ex		
Chondrostoma nasus (L.)	⊕					⊕
Chondrostoma genei Bonaparte				✚Ex		
Vimba vimba carinata L.	⊕					
Abramis brama L.	⊕					
Abramis ballerus L.	⊕					
Abramis sapa Pallas	⊕					
Blicca bjoerkna L.	⊕					

Table 4.1 continued

	Danube river basin			Adriatic river basin		
Species	A	I	T	A	I	T
Chalchalburnus chalcoides (Agassiz)	⊕					
Alburnus alburnus alburnus (L.)	⊕					
Alburnus a. arborella (Bonaparte)				✚E		
Alburnoides bipunctatus Bloch	⊕					
Rhodeus amarus Bloch	⊕					
Phoxinus phoxinus (L.)	⊕			⊕		
Leucaspius delineatus (Heckel)	⊕					
Scardinius erythrophthalmus (L.)	⊕					
Ctenopharyngodon idella (Valenciennes)		⊕			⊕	
Pseudorasbora parva (Schlegel)		⊕				
Rutilus pigus virgo (Heckel)	✚E					
Aspius aspius (L.)	⊕					
Rutilus rutilus (L.)	⊕					⊕
Rutilus aula (Bonaparte)				✚E		
Telestes souffia Risso	⊕			⊕		
Leuciscus leuciscus (L.)	⊕					
Leucicus cephalus cephalus (L.)	⊕					
Leucicus c. cabeda Risso			⊕	✚		
Leucicus c. albus Bonaparte				✚		
Leucicus idus (L.)	⊕					
Cobitidae						
Misgurnus fossilis (L.)	⊕					
Sabanejewia balcanica (Karaman)	⊕					
Cobitis elongatoides Bacescu & Maier	⊕					
Cobitis bilineata Canestrini				✚		
Cobitis elongata Heckel & Kner	✚E					
Balitoridae						
Barbatula barbatula (L.)	⊕					
Ictaluridae (Ameiuridae)						
Ictalurus nebulosus (Le Sueur)		⊕			⊕	
Siluridae						
Silurus glanis L.	⊕					⊕
Esocidae						
Esox lucius L.	⊕			⊕		
Umbridae						
Umbra krameri Walbaum	✚V					
Poeciliidae						
Gambusia affinis Baird & Girard					⊕	
Percidae						
Gymnocephalus schraetzer (L.)	✚E					
Gymnocephalus cernuus (L.)	⊕					
Gymnocephalus baloni Holcik & Hensel	✚E					
Zingel zingel (L.)	✚E					
Zingel streber (Siebold)	✚E					
Stizostedion lucioperca (L.)	⊕					⊕
Perca fluviatilis L.	⊕					⊕

Table 4.1 continued

Species	Danube river basin			Adriatic river basin		
	A	I	T	A	I	T
Centrarchidae						
Micropterus salmoides Lacépède		⊕			⊕	
Lepomis gibbosus L.		⊕			⊕	
Blenniidae						
Salaria fluviatilis (Asso)				⊕		
Gobiidae						
Padogobius martensi Günther				✚		
Cottidae						
Cottus gobio L.	⊕			⊕		
Gasterosteidae						
Gasterosteus aculeatus L.			⊕	⊕		
Gadidae						
Lota lota (L.)	⊕					

A, indigenous; I, introduced; T, translocated; ⊕, present; ✚, endemic; Ex, extinct; E, endangered; V, vulnerable.

The Red Data List of the fishes and lampreys from the River Danube basin comprises 42 fish, one of which is presumably extinct (Ex), 26 are endangered (E), nine vulnerable (V), and six are rare (R). One lamprey falls under the category of endangered (E) and the other is rare (R). It should be noted that single specimens of *Anguilla anguilla*, of unknown origin, are caught each year from the rivers Drava and Mura.

Of the 57 indigenous fish species and two lampreys living in the River Danube basin in Slovenia, 10 are endemic. None lives exclusively in Slovenia. Those living in the Slovenian waters of the River Danube basin are listed in Table 4.1 together with their conservation status.

4.5 Threats to freshwater fishes in Slovenia

The major threats to freshwater fishes are flow regulation, pollution, overfishing, introduction and translocation of fish species and competition. Conservation measures should thus include enforcement of suitable legal provisions, elimination of pollution, and proper flow regulation (Povz 1992, 1996).

Changes in water quality and the impact of stream regulation are reflected in the quality and quantity of fish populations in Slovenia. For example, the fish populations in the south-eastern streams of Slovenia, i.e. the Notranjsko area, are badly affected or even lost by the strong industrial pollution. Flow regulation through construction of dams in the River Soca and in the tributaries of the Pirano Bay in the north part of Adriatic Sea has resulted in the presumed extinction of *A. anguilla* in the Adriatic drainage basin in Slovenia. The freshwater lamprey *L. zanandreai* is also threatened in the Soca river basin (found at only one location in the last 20 years) through the loss of appropriate habitat, i.e. shoal areas and deposits of fine sands and silt, brought about by regulation. The only way to preserve this species is to designate the waters supporting *L. zanadreai* as nature reserves.

Overfishing was the main cause for the collapse of the *Salmo trutta m. lacustris* L. stocks in the Alpine Lake Bohinj in north-east Slovenia in the 1950s.

The introduction and translocation of fishes are prevalent in the waters of Slovenia, mainly by fishermen who introduced non-native fish species either intentionally, for the sake of angling, or accidentally, while restocking rivers. Such introductions and translocations have brought about marked changes in the indigenous fish fauna. A wide variety of consequences have become manifest, ranging from genetic pollution to extinction of species (Povz 1992). In the Slovenian part of the River Soca, fish introductions and translocations, respectively, have had a serious impact on the endemic species of the basin, particularly *C. genei* and *C. soetta*, and *S. marmoratus*. Extensive genetic pollution has developed in the case of *S. marmoratus* which hybridises with brown trout, *Salmo tutta*, with the F1 generation being fertile. According to Gridelli (1936), marble trout was the only trout of the Soca river basin until the transplantation of the brown trout in 1906. Its stocking has since been repeated despite impending extinction of *S. marmoratus*. The current distribution of genetically-pure marble trout extends over a few isolated tributaries that are not connected. The plight of this species has come to the fore in the past 5 years because of its importance as a sport fish. In the case of nase, the consequences are even more dramatic. The indigenous *C. genei* and *C. soetta* have been lost because of the translocation of the non-indigenous *C. nasus* from the Danube drainage basin. The measures that are being undertaken to preserve marble trout can unfortunately no longer be applied to *C. genei* and *C. soetta*, which use to populate the Soca river basin some 20 years ago.

Chondrostoma genei and *C. soetta* are undoubtedly the first fish species to become extinct in Slovenia. Measures to re-introduce them into their former distribution range are largely academic. Such an operation would require intensive removal of *C. nasus* and re-introduction of *C. genei* and *C. soetta* from northern Italian rivers. While re-introduction as such involves no particular problems, it would be difficult to eliminate *C. nasus* after its rapid expansion.

The populations of *Silurus glanis* L. have also increased in the Soca drainage basin since its accidental introduction with *C. nasus* around 1960. Its expansion has been slower than *C. nasus*, but each year, specimens weighing up to 60 kg are caught on the Vipava (tributary of the Soca). In 1992, a specimen reportedly weighing about 90 kg was caught.

Ctenopharyngodon idella (Val.) was also introduced into the Vipava at an unknown date. This is potentially damaging to the indigenous fauna because temperatures in the Vipava match those of its home range so it will probably reproduce naturally and it feeds on higher aquatic plants thus could destroy spawning places of the indigenous phytophilous fishes. A variety of other exotic fishes, e.g. *Lepomis gibbosus* L., *Micropterus salmoides* Lacépède, *Carassius gibelio* (Bloch), and *Ictalurus nebulosus* (Le Sueur), inhabit the Soca river basin. However, nothing is known about the influence of these species on the indigenous ichthyofauna, particularly on the smaller species.

These changes illustrate the extreme vulnerability of the fish communities in the region, especially those living in the short, swift-flowing rivers of the Adriatic basin.

4.6 Protection and conservation measures

Protection of fishes in Slovenia is provided under the Freshwater Fisheries Act (OG SRS 25/76) and in the Decree on the close season and minimal size of catchable fish, crayfish and mussels (OG SRS 7/78), the latter of which is currently being revised. Conservation measures include prevention of pollution, suitable flow regulation, reserves and artificial propagation and banning of introductions and translocations, and international agreements on the later two should be enforced immediately. Artificial propagation of indigenous salmonids species, i.e. genetically-pure marble trout, brown trout, Danube salmon and grayling, and some cyprinids, i.e. Danube roach and nase, is well developed and may become an important tool in the future. Finally, protection against pollution and flow regulation will only be possible if integrated management is applied to the development of the water resources. Special protection should also be afforded to the conservation of habitats where endangered species, such as *Umbra krameri* Walbaum (Povz 1995a, b), *Cobitis* species and all the lampreys, persist.

References

Bat M. & Zupan M. (1996) Sava – Enciklopedija Slovenije 10, Mladinska knjiga, pp. 407–408 (in Slovene).

Freyer H. (1842) Fauna der Krain bekante Saugethiere, Vögel, Reptilien und Fische. Gger'schen Gubernial – Buchdruckerei, pp. 46–50 (in German).

Glowacki J. (1896) Ribe savskega in dravskega porecja. I. Jahresberichte der K.K. Staats Untergymnasium in Cili. Verlag des k.k. Staats Untergymnasium in Cili, pp. 7–39 (in Slovene).

Gridelli E. (1936) I Pesci d'acqua dolce della Venezia Giulia. Tipografia Domenico del Bianco e figlio. Udine, 79 pp. (in Italian).

Heckel J. & Kner R. (1858) *Die Süsswasserfische des Östreichischen Monarchie*. Leipzig: Verlag von Wilhelm Engelman, 388 pp. (in German).

IUCN (1988) IUCN Red List of Threatened Animals. IUCN: Gland, Switzerland, and Cambridge, UK, pp. xvii–xviii.

Munda A. (1926) Ribe v slovenskih vodah. Ljubljana: Slovensko ribarsko društvo v Ljubljani, 63 pp. (in Slovene).

Ocvirk A. (1988) Ribiški kataster in ribiško gospodarjenje. *Ribic* **47**(7–8), 248–279 (in Slovene).

Povz M. (1983) Spremembe v arealu dveh vrst podusti /*Chondrostoma*, Pisces/ v Sloveniji. *Biološki Vestnik* **31**, 45–52 (in Slovene).

Povz M. (1992) The Red List of Endangered Pisces and Cyclostomata in Slovenia. *Varstvo narave* **17**, 51–59.

Povz M. (1995a) Threatened fishes of the world: *Umbra krameri* Walbaum, 1792 (Umbridae). *Environmental Biology of Fishes* **43**, 230–231.

Povz M. (1995b) The conservation status of freshwater fishes of the Mura River and its tributaries in the Republic Slovenia. *Polskie Archiwum Hydrobiologii* **41**, 311–321.

Povz M. (1996) Red data list of freshwater lampreys (Cyclostomata) and fish (Pisces) in Slovenia. In A. Kirchhofer & D. Hefti (eds) *Conservation of Endangered Freshwater Fish in Europe*. Basel: Birkhäuser Verlag, pp. 63–72.

Povz M. Sladkovodne riba, piškurfi (Pisces, Petromyzontidae). In *Konvencija o biološki Raznovrstnosti – Stange biotske Raznovrstnosti in Krajinske pestroski v Sloveniji*. RS Ministrstvo za okolje in prostor, ARSO (in Slovene) (in press).

Povz M., Jesenšek D., Berrebi P. & Crivelli A.J. (1996) The marble trout, *Salmo trutta marmoratus*, Cuvier 1817: in the Soca river basin, Slovenia. Arles: Tour du Valat, 65 pp.

Povz M. & Ocvirk A. (1990) Freshwater fish introductions and transplantations in Slovenia. *Ichthyos* **9**, 1–9.

Seeley H.G. (1886) *The Freshwater Fishes of Europe*. London, Paris, New York, Melbourne: Cassell & Co., 444 pp.

Sket B. (1967) Kljuci za dolocanje zivali: Sladkovodne ribe – Pisces. Inštitut za biologijo Univerze v Ljubljani, 63 pp. (in Slovene).

Svetina M. & Verce F. (1960) Sladkovodno ribištvo na Slovenskem. Ribiška zveza Slovenije, Ljubljana, 139 pp. (in Slovene).

Zgonik M. & Zupan M. (1988) Drava – Enciklopedija Slovenije **2**, 323–324 (in Slovene).

Zupan M. (1994) Kakovost voda v Sloveniji v letu 1993. R Slovenija. Ministrstvo za okolje in prostor, HMZ R Slovenije, pp. 13, 66, 69, 73, 104 (in Slovene).

Chapter 5
Freshwater fishes of Peloponnese (Greece): distribution, ecology and threats

R. BARBIERI,* A.N. ECONOMOU, M.Th. STOUMBOUDI
National Centre for Marine Research, Institute of Inland Waters, Athens, Greece

P.S. ECONOMIDIS
Aristotle University, Zoology Department, Thessaloniki, Greece

Abstract

The Peloponnese freshwater fish fauna comprises 15 native primary freshwater species, of which eight are endemic to Greece or to the Balkans. Brief descriptions of the distribution, life-history and biology of some native fishes are provided and their conservation status is discussed. On the basis of distribution data, abundance estimates, assessment of ecological requirements and evaluation of anthropogenic impacts, two species, *Leuciscus keadicus* and *Valencia letourneuxi*, were characterised as critically endangered, one, *Tropidophoxinellus spartiaticus*, as vulnerable and some others as locally threatened. The major threat to the species is water abstraction for irrigation that takes place mainly during the summer, and therefore coincides with the summer drought. For *V. letourneuxi* impacts also resulted from fluctuating discharge volumes. None of the species investigated were considered to be recruitment limited.

Keywords: freshwater fish conservation, life-history, Peloponnese, threats.

5.1 Introduction

The Peloponnese freshwater fish fauna comprises 15 native primary freshwater species, of which eight are endemic to Greece or to the Balkans, plus another five that were introduced. These figures do not include euryhaline and estuarine species that may occur in the lower parts of the rivers. There has been limited research on the taxonomy, distribution and conservation problems of the freshwater fishes of Peloponnese (Stephanidis 1971; Economidis 1991, 1995, 1996). So far, freshwater fish conservation has proved difficult, partly because fish have hardly featured in the management of freshwater bodies and partly because of deficiency of appropriate life history and ecological data. This chapter summarises the results of a project (Economou, Barbieri, Daoulas, Psarras, Stoumboudi, Bertahas, Giakoumi & Patsias 1999) to identify threatened species and vulnerable life-stages in western and southern Greece and to examine the nature of threats and the problems of fish conservation.

*Correspondence: National Centre for Marine Research, Institute of Inland Waters, Agios Kosmas, Hellinikon, 16604 Athens, Greece (email: robertab@ncmr.gr).

5.2 Materials and methods

Field data were collected during an extensive survey (Economou *et al.* 1999) from March 1997 to September 1998. Additional samples were collected in spring 1999 and in summer 2000. In total, 131 samples were collected from 80 localities. Samples were taken with a variety of techniques (beakers, scoop-nets, fry nets, gill nets, seines and electric fishing) targeting different life-history stages. Fishes were normally preserved in 4% formalin and subsequently identified following the nomenclature proposed by Kottelat (1997), measured (standard length, nearest mm) and ontogenetic development, growth, reproductive biology and feeding habits were examined in the laboratory. Visits were made to local environmental and reclamation services to collect information about agricultural practices and other human activities that may affect the fish habitats.

5.3 Results

Synoptic data on the habitat type, distribution, and conservation status of the fishes that occur in Peloponnese are shown in Table 5.1. Brief descriptions of the ecology and life-history of 10 species, the conservation status of which was ascertained with reasonable reliability, are provided below. Illustrations of the distribution maps of these species are presented in Figure 5.1.

5.3.1 Barbus albanicus

A fluvio-lacustrine species inhabiting lakes and deep portions of rivers in western Greece (not in Albania). It prefers sandy–muddy substrates and feeds chiefly on plants and epilithic algae or insect larvae buried in the sand. The species matures at age 2 or 3, may live up to 10 years reaching 300 mm. It breeds in late spring and early summer and digs on sand or gravel to deposit non-adhesive eggs about 3.5 mm in diameter from which larvae ranging from 7.9 to 9.3 mm hatch. Lake populations usually enter tributaries to spawn. Although this species requires large volumes of water, it can be considered as safe due to the stability of its current habitat and the absence of intense commercial exploitation.

5.3.2 Barbus peloponnesius

A strongly rheophilic barbel, living in stony–pebbly substrates and spending most of its time under stones. Its food consists mainly of epilithic organisms, which are taken by ‘liping’ on solid substrates. It matures at age 2 (possibly in some populations males mature at age 1+), at about 70 mm, and reproduces in May–July on gravels. In small streams it grows up to 160 mm but populations inhabiting large and stable rivers or cool waters may exceed 200 mm in length. Its larvae can be distinguished from the almost identical larvae of *Barbus albanicus* from pigment differences on the dorsal row of melanophores and from the presence in the latter species of serration on the last single dorsal ray. The species inhabits many small streams, which presently are almost

Table 5.1 Habitat type and conservation status of the fishes occurring in Peloponnese

Family/species	Habitat	Distribution/ endemism	Conservation status HD	BC	RDB
Cyprinidae					
Pseudophoxinus stymphalicus (Valenciennes)[a]	F	Bl		III	V/E
Leuciscus cephalus (L.)	F	W			T_{loc}
Leuciscus keadicus Stephanidis	F	P			
Leuciscus pleurobipunctatus (Stephanidis)	F	G	II		V/T_{loc}
Tropidophoxinellus hellenicus (Stephanidis)	F	G			
Tropidophoxinellus spartiaticus (Schmidt-Ries)	F	P			T/T_{loc}
Barbus albanicus Steindachner[b]	F	G	IIV		V/T_{loc}
Barbus peloponnesius Valenciennes[c]	F	W	IIV	III	V/T_{loc}
Cyprinus carpio L.	F	I			
Carassius auratus (L.)	F	I			
Ctenopharyngodon idella (Valenciennes)	F	I			
Salmonidae					
Salmo trutta L.[c]	F	W	II		E/V/R
Oncorhynchus mykiss (Walbaum)	F	I			
Gobiidae					
Knipowitschia sp.	B	?			
Valenciidae					
Valencia letourneuxi (Sauvage)[d]	F (B)	Bl	II*	II	E
Cyprinodontidae					
Aphanius fasciatus (Valenciennes)	F-E	W	II	II, III?	
Poeciliidae					
Gambusia affinis (Baird & Girard)	F-B	I			
Gasterosteidae					
Gasterosteus aculeatus L.	A	W			
Blenniidae					
Salaria fluviatilis (Asso)	F-B	W		III	
Anguillidae					
Anguilla anguilla (L.)	C	W			

Habitat: F, freshwater; B, brackish; E, euryhaline; A, anadromous; C, catadromous. Distribution/ endemism: P, Peloponnese; G, Greece; Bl, Balkans; W, widespread; I, introduced. Conservation status: HD, Habitats Directive 92/43/EEC (Appendices II, IV & V, *, priority species); BC, Bern Convention (Appendices II, III); RDB, Red Data Book of Greece (E, endangered; V, vulnerable; R, rare; T, threatened; T_{loc}, locally threatened).

[a]Referred to in the Bern Convention as *Leucaspius stymphalicus*. [b]Referred to in the Habitats Directive as *Barbus capito*. [c]Referred to in the Habitats Directive as *Barbus meridionalis*. [d]Referred to in the Habitats Directive as *Valencia hispanica*.

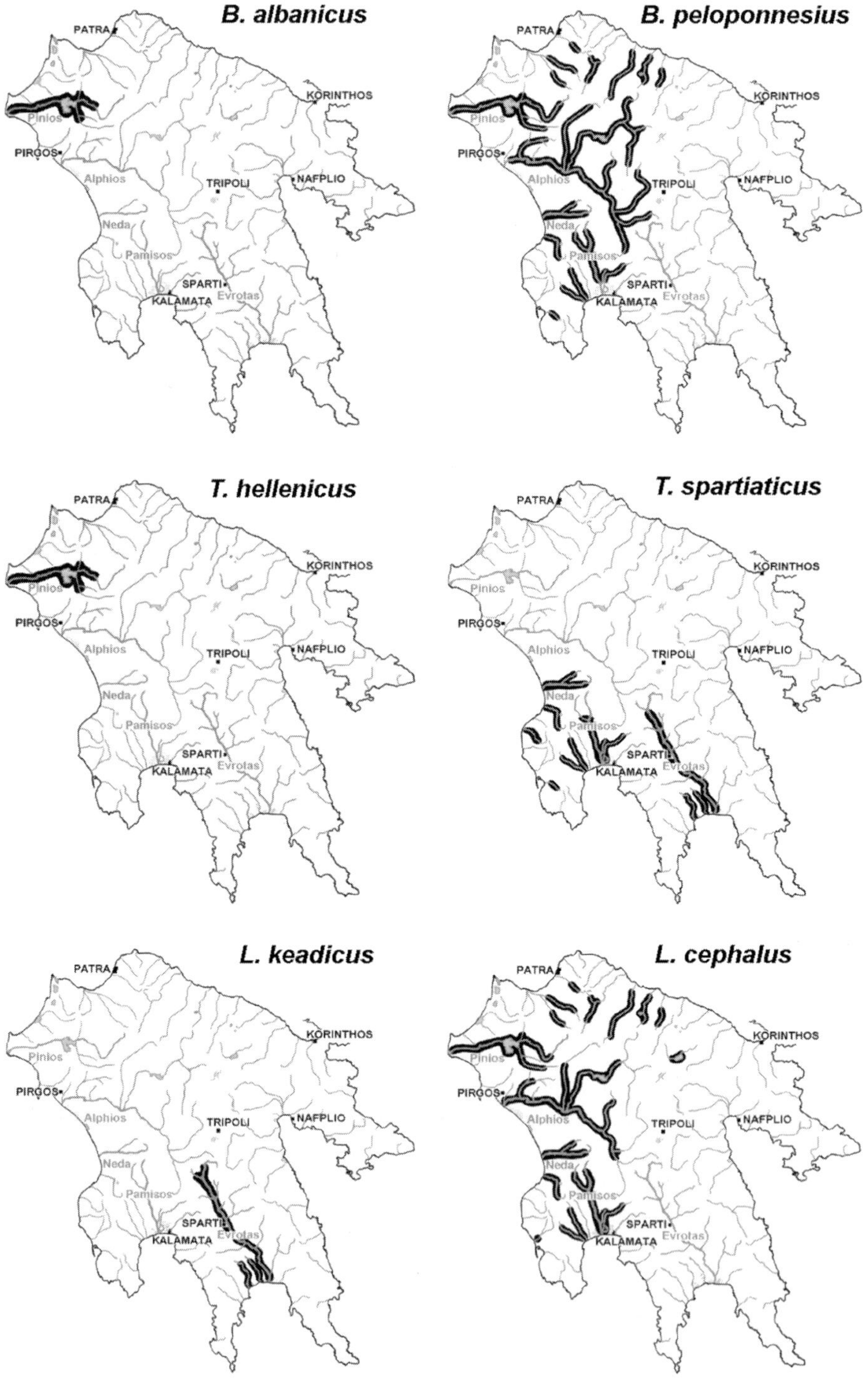

Figure 5.1 Geographical distribution of the main freshwater fish of Peloponnese

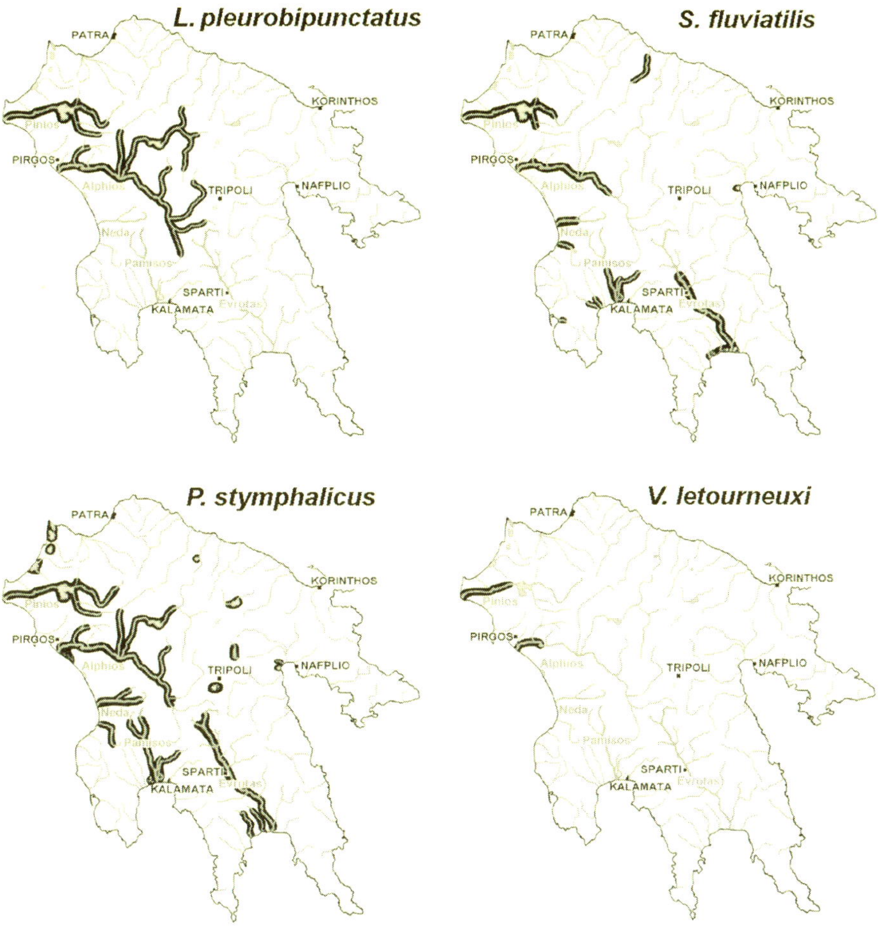

Figure 5.1 continued

completely dry during the summer drought periods. Nonetheless, the presence of *Barbus peloponnesius* in many systems and its broad tolerance to temperature and water flow fluctuations, reduce the danger of extinction of the species as a whole. This is not to say that conservation measures are not required. The genetic structure of the Peloponnese populations remains poorly explored, which justifies precautionary conservation actions to prevent loss of genetic information.

5.3.3 Tropidophoxinellus hellenicus

A fluvio-lacustrine, semipelagic, schooling species, found in lakes and deep portions of rivers. It lives 3 years, possibly four in the Pinios river system, rarely exceeding 110 mm in length. Matures at age 1+ at 45 mm, feeds both on plankton and epibenthic organisms, reproduces from April to mid-May and attaches its eggs (1.5 mm) on aquatic plants. Since this species inhabits large and relatively stable aquatic systems, it can be considered as safe.

5.3.4 Tropidophoxinellus spartiaticus

A rheophilic species, confined to the southern part of Peloponnese. It usually lives in slow flowing segments of rivers, grows up to 100 mm over 3 and rarely 4 years. It breeds in April–May, produces spherical adhesive eggs 1.6 mm in diameter from which weakly pigmented embryos, about 4.3 mm in length, hatch. Its larvae, as well as the larvae of *Tropidophoxinellus hellenicus*, are characterised by the heavy pigmentation of the lateral melanophores which discriminates them from similar *Leuciscus* larvae. Although this species does not require large volumes of water and specialised habitats, most of its populations live in unstable systems and can be regarded as vulnerable. Several populations, including that in Evrotas, are endangered due to water abstraction and pollution.

5.3.5 Leuciscus cephalus

A species with a wide European distribution, it is chiefly rheophilic, although it is also found in reservoirs and occasionally in lakes. The genetic structure of the Peloponnese populations gives support to the idea of two separate species (Doadrio & Carmona 1998; Zardoya, Economidis & Doadrio 1999). In large water bodies the species grows to 220 mm over 6 years, but in small streams no more than three age classes are usually found. It feeds on a wide variety of animal and plant foods, matures in the second year of life (in Lake Stymphalia most males mature in the first year) and breeds mainly in May. It spawns adhesive yellow–green eggs (1.8–2 mm in diameter) on gravel, from which unpigmented larvae about 6.3 mm emerge at hatching. As a whole this species can be considered as safe but locally threatened. The most endangered populations are in Lake Stymphalia and in the northern streams of Peloponnese, overexploited for their water resource.

5.3.6 Leuciscus keadicus

A species endemic to the River Evrotas (Economidis 1996). Its phylogenetic relationships with other *Leuciscus* species, especially those of the Iberian Peninsula, have attracted considerable research interest among European taxonomists, molecular biologists and genetists (Tsingenopoulos & Karakousis 1996; Doadrio & Carmona 1998; Zardoya *et al.* 1999). *Leuciscus keadicus* is a strongly rheophilic species, confined to open sites of the river with a fast flow and relatively cool waters. It grows up to 150 mm, feeds predominantly on aquatic insects, breeds in May and produces adhesive yellowish eggs around 2 mm in diameter from which unpigmented embryos about 5.1 mm in length hatch. In the past, *L. keadicus* was distributed along the entire river Evrotas and its tributaries. Evrotas is now progressively becoming a seasonal stream and the distribution of the species has declined concomitantly. In adverse hydrological years all tributaries and a large portion of the river dry completely during summer. In addition, pollution from the local agricultural industries continuously degrades water quality causing mass mortalities during dry periods. In view of the progressive loss of suitable habitats, *L. keadicus* can be regarded as critically endangered and a candidate for extinction in the next adverse hydrological year.

5.3.7 Leuciscus pleurobipunctatus

A rheophilic cyprinid, preferring protected cavities among the roots of riparian trees. It lives up to 3 and rarely 4 years, grows slowly to a maximum size of 110 mm. It starts reproducing in the second year of life at 60 mm and breeds between mid-April and mid-May, producing yellowish adhesive eggs about 1.9 mm in diameter. Up to the size of 11 mm, the larvae of *Leuciscus pleurobipunctatus* are almost indistinguishable from those of *Leuciscus cephalus*, except that the former has a darkly pigmented area in the dorsal lobe of the caudal fin. The feeding spectrum is wide and includes animal and plant food, chiefly insect larvae. Both its broad tolerance limits to temperature and to water supply variations and the stability of most of its current habitats, justify the view that this species is safe.

5.3.8 Pseudophoxinus stymphalicus

A marsh-living species commonly found in swamps, springs, drainage and irrigation channels, portions of the river with stagnant water, and even in irrigation wells. This species can reach the size of 60 mm and the age of 3 years, although in unproductive and warm habitats it may have an annual life cycle and a maximum size of 40 mm. It is a fractional spawner, matures in the first year of life and releases successive batches of eggs over a long spawning season that, in some populations, may extend from winter to late spring. The eggs, 1.1–1.6 mm in diameter, are deposited on aquatic plants. The larvae hatch at about 4.5–4.9 mm and are characterised by heavy pigmentation and a very weakly developed mediolateral row of melanophores. The diet includes a great variety of plants, insects and microcrustaceans. This species has high colonisation efficiency and is well adapted to unstable conditions. Its small body size, early reproduction, rapid alternation of generations, high reproductive effort, low oxygen demand and feeding plasticity, makes it able to resist prolonged periods of unfavourable conditions and to re-establish depleted populations quickly.

5.3.9 Valencia letourneuxi

A small-bodied, short-lived and early-maturing species of western Greece and south Albania. A synopsis of the available distribution, biological, life-history and ecological data is provided by Barbieri, Daoulas, Psarras, Stoumboudi & Economou (2000). The high dependence of *Valencia letourneuxi* on rich submerged vegetation for survival and reproduction, confines this species to small low density populations in a few springs and riverine sites. Some populations have already been lost permanently and the few remaining ones are recessing, due to the destruction of the specialised habitats required by the species. Major threats are imposed by land reclamation works, pollution, water abstraction and, predominantly, the irregular release of water through the dams that causes erosion of the banks and damage to the vegetation. Using the limited and punctuated geographical distribution and the low population densities as conservation status criteria, *V. letourneuxi* can be characterised as a critically endangered species.

5.3.10 Salaria fluviatilis

A small Mediterranean freshwater blenny with a wide distribution in the Peloponnese. It is a cryptobenthic carnivore species, growing to about 60 mm and exhibiting male parental care. Its larvae are extremely small at hatching by freshwater standards (3.5 mm) and share many of the characteristics of marine fish larvae (high transparency, light pigmentation, planktonic life). Successful reproduction occurs only in water bodies with a pelagic zone and availability of planktonic prey. In Greece *Salaria fluviatilis* is distributed in lakes and also rivers that either communicate with lakes or have extended areas with lentic water. In some countries (e.g. Elvira 1995; Keith & Allardi 1996) this species is considered as vulnerable or endangered, but in Greece it is safe.

5.4 Discussion

In comparison to the 81 autochthonous primary freshwater species of Greece, the fish fauna of Peloponnese comprises only 15 species, but includes several endemic forms. Typical lacustrine species are absent, which is to be expected since there is only one small and sporadically dry natural lake in the Peloponnese. Most species are entirely rheophilic, and from the life-history point of view, they share many of the traits that characterise r-strategists (Pianka 1970; Townsend & Hildrew 1994) (early maturation, short life-span, small adult size). The community structure (few species, high degree of endemism) and ecological and life-history characteristics reflect, at least partly, the variable nature of the hydrological regime. The River Alfios and the lower part of the River Pinios are the only stable aquatic systems. All other rivers and streams are highly variable environments, with conditions such as temperature, water velocity and dissolved oxygen fluctuating markedly on a spatial, seasonal and annual basis. Such ecological conditions generate survival uncertainty and favour communities dominated by r-strategists, either through genetic adaptation or through elimination of k-strategists.

In the last 30 years, the amplitude of environmental perturbations has increased due to anthropogenic actions. These perturbations induce an additive stress and make the ecological conditions even more adverse. The prospects for fish conservation are intimately linked with human impacts on the habitats, and also depend on their distributions and life strategies. On the basis of the available data, it was asserted that two geographically restricted, locally scarce and with narrow adult habitat requirements species, namely *L. keadicus* and *V. letourneuxi*, are heavily affected by habitat degradation and can be regarded as critically endangered. *Tropidophoxinellus spartiaticus* is a species with a relatively limited distribution, and although it is locally abundant, it can also be characterised as endangered, using as a conservation status criterion the increasing instability of its habitats. All other species have broader distributions and/or inhabit relatively stable habitats. Although they have also been affected by habitat degradation, from the conservation point of view they can be regarded as safe or locally threatened only.

Major impacts to fish populations have resulted from water abstraction that has dramatically altered many riverine habitats. Flow regulation, pollution, river management works and introduction of alien species also constitute potential threats, but their impacts are minor compared to those generated by water abstraction. In the future, the threats to the species are likely to increase in the face of the increasing water demands. In the longer term, Greece will be affected adversely by climatic warming that is expected to reduce water resources and surface runoff (Mimikou & Kouvopoulos 1991).

Although the evidence is not yet conclusive, it seems that the primary effect of water deficiency is on adult habitats. Since reproduction of almost all species occurs during the wet period, deterioration or destruction of spawning grounds and larval habitats does not seem to be a major reason of population decline. Indeed, data on the abundance and distribution of the early life-stages of fish do not generally support the concept of a recruitment bottleneck. Judging from the usually high number of larvae and fry in most localities studied, population replenishment is considered adequate in normal years. This view is consistent with the long-standing belief that the warmwater fishes have a high reproductive potential. Bennett (1970) argued convincingly that in such fishes a large reserve of fry is always available to support recruitment, and their populations expand and contract in relation to the capacity of the adult habitat. Severe reduction of water supply, which takes place during the summer intense water abstraction period, dictates that the ultimate regulation of population size takes place during this hydrologically adverse period of the year. One implication of this scenario is that introductions of fry are not necessarily required to stabilise populations.

A last issue to be considered is fish conservation policy. Public water policy encourages the creation of irrigation systems at the expense of free-flowing rivers and natural fish communities. Although national regulations and international treaties protect some fish species and their habitats, conservation has proved difficult, because of deficiency of appropriate biological data. At the same time, legal mechanisms that guide conservation efforts are either inadequate or poorly implemented. For instance, the critically endangered *L. keadicus* is neither listed in the Habitats Directive 92/43/EEC nor are its normal habitats included in the proposed network of NATURA 2000 sites, while *V. letourneuxi* is referred to in the forenamed Directive as *Valencia hispanica*. It is therefore necessary to adjust the perception and attitudes to aspects of fish conservation and to assign to fishes a similar conservation priority as those of other animal groups.

References

Barbieri R., Daoulas Ch., Psarras Th., Stoumboudi M.Th. & Economou A.N. (2000) The biology and ecology of *Valencia letourneuxi* Sauvage 1880 (Valenciidae) – prospects for conservation. *Mediterranean Marine Science* **1**, 75–90.

Bennett G.W. (1970) *Management of Lakes and Ponds*. London: Litton Educational Publishing, 375 pp.

Doadrio I. & Carmona J.A. (1998) Genetic divergence in Greek populations of the genus *Leuciscus* and its evolutionary and biogeographical implications. *Journal of Fish Biology* **53**, 591–613.

Economidis P.S. (1991) Check list of freshwater fishes of Greece (recent status of threat and protection). *Bulletin of the Hellenic Society for the Protection of Nature, Special Publication,* 48 pp.

Economidis P.S. (1995) Endangered freshwater fishes of Greece. *Biological Conservation* **72**, 201–211.

Economidis P.S. (1996) *Leuciscus keadicus* Stephanidis, 1971 (Pisces, Cyprinidae), a valid species from river Evrotas (Greece). *Cybium* **20**, 303–309.

Economou A.N., Barbieri R., Daoulas Ch., Psarras Th., Stoumboudi M.Th., Bertahas H., Giakoumi S. & Patsias A. (1999) Threatened endemic freshwater fishes of western Greece and Peloponnese – Distribution, abundance, threats and conservation measures. *Technical Report, National Centre for Marine Research, Institute of Inland Waters*, 341 pp. (in Greek).

Elvira B. (1995) Conservation status of endemic freshwater fish in Spain. *Biological Conservation* **72**, 129–136.

Keith P. & Allardi J. (1996) Endangered freshwater fish: The situation in France. In A. Kirkhhofer & D. Hefti (eds) *Conservation of Endangered Freshwater Fish in Europe.* Basel: Birkhauser Verlag, pp. 35–54.

Kottelat M. (1997) European freshwater fishes. An heuristic checklist of the freshwater fishes of Europe (exclusive of former USSR), with an introduction for non-systematists and comments on nomenclature and conservation. *Biologia, Bratislava,* section Zoology **52** (Suppl. 5), 1–271.

Mimikou A.N. & Kouvopoulos Y.S. (1991) Regional climate change impacts: 1. Impacts on water resources. *Hydrological Sciences Journal* **36**, 247–258.

Pianka E.R. (1970) On 'r' and 'K' selection. *American Naturalist* **104**, 592–597.

Stephanidis A. (1971) Poisson d'eau douce du Peloponnese. *Biologia Gallo-Hellenica* **3**, 163–212 (in French).

Townsend C.R. & Hildrew A.G. (1994) Species traits in relation to a habitat templet for river systems. *Freshwater Biology* **31**, 265–275.

Tsingenopoulos C. & Karakousis Y. (1996) Phylogenetic relationships of *Leuciscus keadicus*, an endemic cyprinid species from Greece, with other Greek species in the genus *Leuciscus. Folia Zoologica* **45**, 87–93.

Zardoya R., Economidis P.S. & Doadrio I. (1999) Phylogenetic relationships of Greek Cyprinidae: Molecular evidence for at least two origins of Greek cyprinid fauna. *Molecular Phylogenetics and Evolution* **13**, 122–131.

Chapter 6
The ichthyofauna of the small Mediterranean-type drainages of Portugal: its importance for conservation

N. MESQUITA and M.M. COELHO*
Centro de Biologia Ambiental/Departamento de Zoologia e Antropologia, Faculdade de Ciências de Lisboa, Lisboa, Portugal

Abstract

The ichthyofauna of the small Mediterranean-type drainages of Portugal (SMTDP) was almost unknown. The present work studied their species composition and distribution, to access their particularities and needs for conservation. These drainages are of considerable ichthyofaunistic importance, because of the presence of SMTDP endemic species with very restricted distribution, i.e. two species of *Leuciscus* and one species of *Chondrostoma*, and the presence of other endangered Iberian endemic cyprinid species. Historical events (concerning the origin, formation and evolution of the drainage systems) and ecological ones (frequent bottlenecks due to the completely dry up of long extensions of the river's course during the summer) may explain the observed specificities of the SMTDP cyprinid ichthyofauna. Preservation of the habitats that support the populations during the droughts and measures to prevent the dispersion of exotic species are particularly important to protect the fish biodiversity of these rivers.

Keywords: conservation, endemic cyprinids, intermittent streams, Portugal.

6.1 Introduction

Information available on the ichthyofauna of the small Mediterranean-type drainages of Portugal (SMTDP) is scarce, and concern essentially only the two larger ones, the rivers Mira and Arade (e.g. Almaça 1978a, b; Collares-Pereira 1985; Coelho, Bogutskaya, Rodrigues & Collares-Pereira 1998; Magalhães & Collares-Pereira 1998). In addition to these two drainages, the SMTDP are formed by other smaller drainages between the Mira and Arade and between the Arade and Guadiana (Fig. 6.1), for which there was almost no information available.

The endemic cyprinid species present in the SMTDP belong to the genera *Barbus*, *Chondrostoma* and *Leuciscus*. The taxonomy of the Iberian Peninsula cyprinids, initially based on morphological characters (e.g. Lozano-Rey 1947; Almaça 1967;

*Correspondence: Centro de Biologia Ambiental/Departamento de Zoologia e Antropologia, Faculdade de Ciências de Lisboa, Campo Grande, Bloco C2-3° Piso, 1749-016 Lisboa, Portugal (email: mmcoelho@fc.ul.pt).

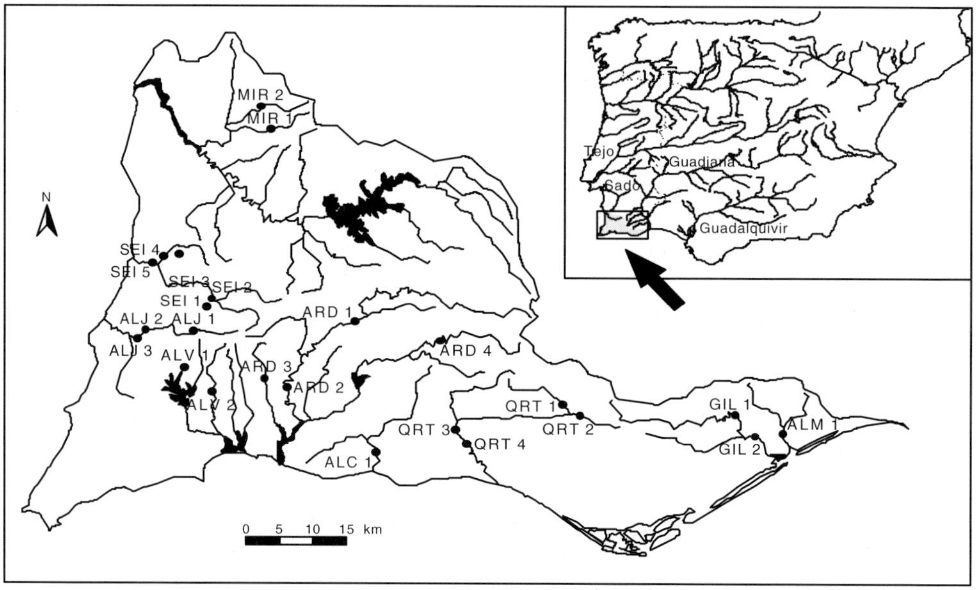

Figure 6.1 Location of the sampling sites in the small Mediterranean-type drainages of Portugal: Mira (MIR), Seixe (SEI), Aljezur (ALJ), Alvor (ALV), Arade (ARD), Alcantarilha (ALC), Quarteira (QRT), Gilão (GIL) and Almargem (ALM)

Doadrio 1988; Nelva, Collares-Pereira & Coelho 1988), considered the species present in the SMTDP to be *Leuciscus* (*Leuciscus pyrenaicus* Günther), *Chondrostoma* (*Chondrostoma lusitanicum* Collares-Pereira and *Chondrostoma lemmingii*, Steindachner) and *Barbus* (*Barbus sclateri* Günther) which exist in the larger neighbouring drainages, Sado and Guadiana (Fig. 6.1). However, allozyme studies for *L. pyrenaicus* (Coelho, Brito, Pacheco, Figueiredo & Pires 1995) and *C. lusitanicum* (Coelho, Alves & Rodrigues 1997) found high levels of genetic differentiation for the populations from the Mira and Arade drainages, suggesting the existence of different species for these two genera in the SMTDP. Recently, mitochondrial DNA studies on *Leuciscus* corroborated the high levels of genetic differentiation (Brito, Briolay, Galtier, Bouvet & Coelho 1997). Later, Coelho *et al.* (1998) formally described two new species of *Leuciscus*: *Leuciscus torgalensis* for the Mira, and *Leuciscus aradensis* for the Arade. There is also evidence of high genetic differentiation in the *Chondrostoma* genus between the Mira and Arade populations of *C. lusitanicum* and the northern ones (Sado and Tejo; Fig. 6.1) based on mitochondrial DNA studies, supporting the presence of a different, still undescribed, *Chondrostoma* species in these SMTDP (Mesquita, Carvalho, Shaw, Crespo & Coelho 2001). No population genetic studies have been performed on the species *C. lemmingii* and *B. sclateri* in this area, but similar scenarios are likely.

The purpose of the present study was to describe the species composition and the species distribution boundaries in the SMTDP, in particular for the endemic cyprinids, to establish the ichthyofaunal specificities and needs for conservation.

6.2 Materials and methods

To study the ichthyofauna of the SMTDP, 24 sites were sampled in April of 1999 (Fig. 6.1). At each site fish were captured by electric fishing (300 V, 3–4 A, DC) along a 45-m stretch of river during approximately 20 min. All fish specimens collected were identified and weighted (±0.1 g), fin-clips and scales were collected for genetic analysis and the fish were returned to the water.

Due to the very high morphological similarities, but considerable genetic divergence between the endemic cyprinid taxa within the same genus (Coelho *et al.* 1995, 1997, 1998; Brito *et al.* 1997; Mesquita *et al.*, in press), the species assignment was examined through sequencing analysis of the mitochondrial DNA cytochrome *b* gene (N. Mesquita & M.M. Coelho, unpublished data).

To characterise the fish community, abundance (fish. 100 m^{-2}), biomass (g 100 m^{-2}), the Shannon diversity index (H', Ludwig & Reynolds 1988), the evenness index based on the Shannon diversity index (J', Ludwig & Reynolds 1988) and the species richness index (S, number of species recorded in each sampling site) were estimated per sampling site.

6.3 Results

Of the 14 species sampled in the SMTDP, 50% were Iberian endemic cyprinid species, *B. sclateri*, *C. lemmingii*, the undescribed new species of *Chondrostoma*, *Leuciscus alburnoides* Steindachner, *L. aradensis*, *L. pyrenaicus* and *L. torgalensis*, 36% were other native species, *Anguilla anguilla* (L.), *Atherina* sp., *Cobitis paludica* De Buen (Iberian endemic), *Gasterosteus aculeatus* L. and *Chelon labrosus* (Risso), and 14% were introduced species, *Lepomis gibbosus* L. and *Gambusia holbrooki* Girard.

At the majority of the sampling sites, fish communities were dominated by the Iberian endemic cyprinid species, not only in terms of number of species but also in abundance (7.5–98.7%, average: 62.7%; Fig. 6.2) and biomass (1.8–99.3%, average: 58.9%). When this was not observed, the families Anguiliidae and Cobitididae represented the majority of both abundance and biomass. Cyprinids were not caught only in the Alcantarilha drainage and in one locality in the Aljezur drainage, (Fig. 6.2). Exotic species were caught in five of the 24 sites, which were located in four drainages (Seixe, Arade, Alcantarilha and Quarteira), and represented a relatively small part in the fish community abundance (1.5–18.0%, average: 7.3%) and biomass (0.0–7.4%, average: 2.2%) (Fig. 6.2). The larger SMTDP (Mira and Arade), and Quarteira and Gilão were the drainages with the highest Shannon diversity index values, and Seixe, Aljezur and Almargem were the drainages with the lowest Shannon diversity index values (Fig. 6.2). In general, high and low diversity index values were associated, respectively, with high and low evenness and richness indices values (Fig. 6.2).

With respect to the Iberian endemic cyprinid species, the *Barbus* species assigned to the SMTDP was *B. sclateri* (abundance: 3.7–75.0%, average: 26.3%; biomass: 11.6–89.2%, average: 39.9%), which was present in the Mira, Seixe, Arade, Quarteira and Gilão drainages, and also in the larger southern Iberian Guadiana and Guadalquivir

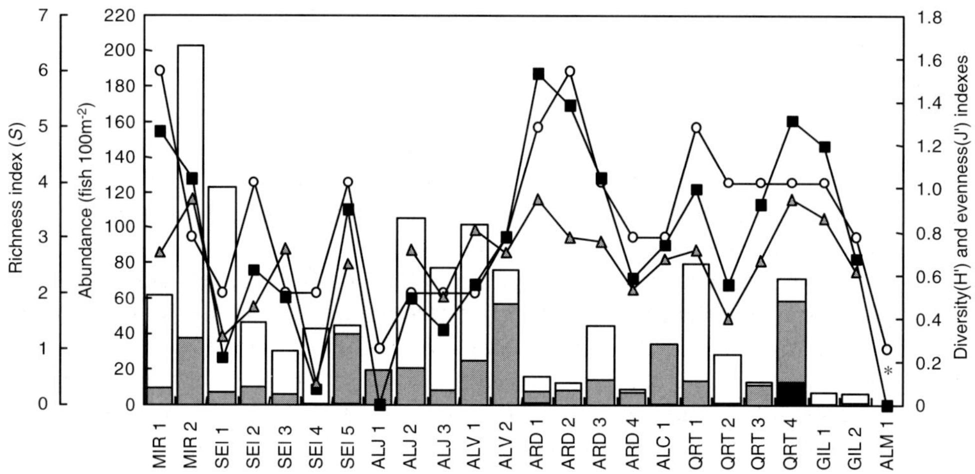

Figure 6.2 Endemic cyprinid species (□), other native species (▩) and exotic species (■) abundance (fish. 100 m^{-2}), and diversity (H'; -■-), evenness (J'; -▲-) and richness (S; -○-) indexes for each sampling site. Mira (MIR), Seixe (SEI), Aljezur (ALJ), Alvor (ALV), Arade (ARD), Alcantarilha (ALC), Quarteira (QRT), Gilão (GIL) and Almargem (ALM) drainages. *Only cyprinid alevins of *C. lemmingii* (identified by sequencing analysis) were caught, outside the sampling sector

drainages (Fig. 6.1). The undescribed species of *Chondrostoma* was present only in Mira and Arade drainages (western SMTDP) (abundance: 2.0–45.7%, average: 25.4%; biomass: 3.5–53.3%, average: 20.6%), while *C. lemmingii* (abundance: 2.5–18.5%, average: 10.5%; biomass: 0.8–8.4%, average: 5.7%) was present only in the eastern SMTDP (Quarteira, Gilão and Almargem drainages) and in the eastern neighbouring Guadiana drainage. The distribution of *L. torgalensis* (abundance: 33.0–40.1%, average: 36.6%; biomass: not available) was restricted to the Mira, the drainage from which it was described (Coelho *et al.* 1998), while *L. aradensis* (abundance: 5.0–98.7%, average: 53.0%; biomass: 1.0–99.3%, average: 48.2%) was present, in addition to Arade, in the other small western drainages (Seixe, Aljezur and Alvor) and in Quarteira. *Leuciscus pyrenaicus* (abundance: 14.3% and biomass: 16.3%) was restricted to the most eastern SMTDP (Gilão), in the vicinity of the Guadiana drainage, where it is also present. *Leuciscus alburnoides* (a unisexual complex with an hybrid origin; reviewed in Alves, Coelho & Collares-Pereira, in press) was recorded for the first time in the Quarteira drainage, although with very reduced values of abundance (0.5%) and biomass (0.2%).

6.4 Discussion

In general, the SMTDP ichthyofauna was dominated by the Iberian endemic cyprinid species. When compared with its neighbour Mediterranean-type drainage, the Guadiana (Pires, Cowx & Coelho 1999), one of the larger and richest Iberian drainages, the

SMTDP fish community presented smaller levels of abundance (fish. 100 m^{-2}), biomass (g 100 m^{-2}) and species diversity and richness. However, the levels were similar to those obtained for the larger northern Sado drainage (Magalhães, Collares-Pereira & Coelho 1999; Collares-Pereira, Moreira da Costa & Mesquita 2000; Bernardo, Ilhéu & Costa in press) (see Fig. 6.1 for location of drainages). Each SMTDP does not present more than six native species, nevertheless, the total SMTDP includes 12 native species, and in particular, regarding the *Leuciscus* genus, contribute four of the five *Leuciscus* species existing in Portuguese fresh waters.

The SMTDP area can be divided in two very distinct regions with respect to cyprinids, comprising the eastern and the western drainages, and in a transition zone, including the Quarteira drainage. The eastern SMTDP ichthyofauna (Gilão and Almargem drainages) includes *C. lemmingii* and *L. pyrenaicus*, both inhabiting the eastern neighbour Guadiana drainage. The western SMTDP ichthyofauna (Mira, Seixe, Aljezur, Alvor and Arade drainages) comprise distinct *Leuciscus* and *Chondrostoma* species: the undescribed *Chondrostoma*, *L. torgalensis* and *L. aradensis*. Finally, the Quarteira drainage comprises a different fish assemblage, between both the western and the eastern SMTDP, comprising both *L. alburnoides* and *L. aradensis*, and *C. lemmingii*. In the present study *L. alburnoides* was recorded for the first time in the SMTDP, leading to an increase of the known distribution area of this species.

Historical events, concerning the origin and evolution of the drainage systems, and ecological ones, related to their particular intermittent hydrological conditions, may explain the differences between the SMTDP cyprinid ichthyofauna. The speciation of the Iberian cyprinids is considered to have occurred through an allopatric process via subdivision (Almaça 1978b), which started only during the Early Quaternary with the establishment of the present-day drainages (Coelho 1992). Recently, phylogeographic mitochondrial DNA studies (Brito *et al.* 1997; Mesquita *et al.* 2001; Coelho, Mesquita & Cunha, unpublished data) supported this vicariance speciation process, in a way that the phylogenetic relationships almost exactly agree with the geographic distribution of drainages. In relation to the SMTDP, little is known about the evolution and isolation of these drainages. In terms of ecological conditions, the SMTDP are typical Mediterranean-type drainages, presenting a heterogeneous hydrological system throughout the year. The higher flows occur between November and February and decrease until some sections of the river dry up, often becoming a series of isolated pools during the summer. This intermittency results in periodically high fish mortality and subsequent bottlenecks, which might explain the high levels of genetic divergence observed (Coelho *et al.* 1997; Mesquita *et al.* 2001).

The fish fauna in the SMTDP has a very important conservation value due to the presence of distinct species of *Leuciscus* and *Chondrostoma*. These endemics are more vulnerable to extinction due to their extremely restricted distribution area and to the specific intermittent hydrological conditions. *B. sclateri*, *C. lemmingii* and *C. lusitanicum* (including the undescribed *Chondrostoma* distribution area) are listed as Rare in the Portuguese Red Data Book (SNPRCN 1991); for *L. aradensis* and *L. torgalensis* a similar status has already been suggested (Coelho *et al.* 1995) and the same status is now proposed for the undescribed species of *Chondrostoma*. As highlighted by Coelho *et al.* (1997), it is very important to protect the particular genetic diversity of these rivers, in particular the refuge habitats that support the populations during the droughts.

It is commonly known that human pressure on rivers and the presence of exotic species represent important threats for the endemic species survival (Rahel 2000; Moritz, McGuigan & Bernatchez, Chapter 27), which become even more relevant in the intermittent river systems. Recent studies in the Guadiana drainage (e.g. Pires *et al.* 1999; Collares-Pereira, Cowx, Ribeiro, Rodrigues & Rogado 2000; Ribeiro, Cowx & Collares-Pereira 2000) showed that, although adapted to the intermittent regime, habitat degradation and the introduction of alien species could represent important threats to the survival of the endemic cyprinids, especially during the drought periods. Although exotic species exhibited a distribution restricted to some SMTDP and represented a relatively small portion of fish community abundance and biomass, it is important to prevent their further increase and dispersion to protect the endemic species.

Acknowledgements

The authors thank L.M. da Costa, P. Salgueiro, A.F. Filipe, T. Marques and I. Antunes for their collaboration in the field work and M.F. Magalhães for her help in the fish community analysis. This work was funded by Direcção Regional de Ambiente e Ordenamento do Território do Algarve, by Centro de Biologia Ambiental and by a grant from Fundação para a Ciência e a Tecnologia to N. Mesquita (PRAXIS XXI BD/19751/99).

References

Almaça C. (1967) Estudo das populações portuguesas do Gén. *Barbus* Cuvier, 1817 (Pisces, Cyprinidae). *Revista da Faculdade de Ciências de Lisboa* **14**, 151–400 (in Portuguese).

Almaça C. (1978a) Répartition géographique des Cyprinidae ibériques et secteurs ichthyogéographiques de la Péninsule Ibérique. *Vestník Ceskoslovenské Spolecnosti Zoologické* **62**, 241–248 (in French).

Almaça C. (1978b) Spéciation et subspéciation chez les Cyprinidae ibériques et nord-africains. *Bulletin Office Nationale des Pêches Tunisie* **2**, 23–30 (in French).

Alves M.J., Coelho M.M. & Collares-Pereira M.J. Evolution in action through hybridization and polyploidy in an Iberian freshwater fish: a genetic review. *Genetica* (in press).

Bernardo J.M., Ilhéu M. & Costa A.M. Ictiofauna do Rio Sado: características dos povoamentos e avaliação da integridade. In I. Moreira, G. Saraiva & F.N. Correia (eds) *Conservação, Valorização e Gestão Ambiental de Sistemas Fluviais (Aplicação à Bacia Hidrográfica do Rio Sado)*. Lisboa: INAG (in Portuguese) (in press).

Brito R.M., Briolay J., Galtier N., Bouvet Y. & Coelho M.M. (1997) Phylogenetics relationships within genus *Leuciscus* (Pisces, Cyprinidae) in Portuguese freshwaters, based on mitochondrial DNA cytochrome *b* sequences. *Molecular Phylogenetics and Evolution* **8**, 435–442.

Coelho M.M. (1992) Genetic differentiation of the Iberian cyprinids *Chondrostoma polylepis* Steind., 1865 and *Ch. willkommii* Steind., 1866. *Archiv für Hydrobiologie* **125**, 487–498.

Coelho M.M., Alves J. & Rodrigues E. (1997) Patterns of genetic divergence in *Chondrostoma lusitanicum* Collares-Pereira, in intermittent Portuguese rivers. *Fisheries Management and Ecology* **4**, 223–232.

Coelho M.M., Bogutskaya N.G., Rodrigues J.A. & Collares-Pereira M.J. (1998) *Leuciscus torgalensis* and *L. aradensis*, two new cyprinids for Portuguese freshwaters. *Journal of Fish Biology* **52**, 937–950.

Coelho M.M., Brito R.M., Pacheco T.R., Figueiredo D. & Pires A.M. (1995) Genetic variation and divergence of *Leuciscus pyrenaicus* and *L. carolitertii* (Pisces, Cyprinidae). *Journal of Fish Biology* **47**(Suppl. A), 243–258.

Collares-Pereira M.J. (1985) Ciprinídeos do Alentejo. *1º Congresso sobre o Alentejo*. Évora, Vol. II, pp. 537–545 (in Portuguese).

Collares-Pereira M.J., Cowx I.G., Ribeiro F., Rodrigues J.A. & Rogado L. (2000) Threats imposed by water resource development schemes on the conservation of endangered fish species in the Guadiana River Basin in Portugal. *Fisheries Management and Ecology* **7**, 167–178.

Collares-Pereira M.J., Moreira da Costa L. & Mesquita N. (2000) Estudo do significado ambiental e ecológico dos transvases de água Guadiana-Sado previstos no Empreendimento de Fins Múltiplos de Alqueva – Componente Ictiofaunística. Lisboa: CBA – FCUL, 22 pp. + Annexes (in Portuguese).

Doadrio I. (1988) *Leuciscus carolitertii n. sp.* from the Iberian Peninsula (Pisces: Cyprinidae). *Senckenbergiana Biol.* **68**, 301–309.

Lozano-Rey L. (1947) Peces ganoides y fisostomos. *Memorias Real Academia de Ciencias Exaceas Fisicas y Naturales de Madrid* **11**, 839 pp. (in Spanish).

Ludwig J.A. & Reynolds J.F. (1988) *Statistical Ecology – A Primer on Methods and Computing*. New York: John Wiley & Sons, 337 pp.

Magalhães M.F. & Collares-Pereira M.J. (1998) Avaliação do estado de conservação da ictiofauna dulciaquícola do Parque Natural do Sudoeste Alentejano e Costa Vicentina: populações de boga-portuguesa *Chondrostoma lusitanicum*, barbo-do-sul *Barbus sclateri* e escalo-do-Mira *Leuciscus torgalensis*. Relatório Final, Instituto da Conservação da Natureza, Lisboa: CBA – FCUL, 179 pp. + Annex (in Portuguese).

Magalhães M.F., Collares-Pereira M.J. & Coelho M.M. (1999) Peixes. In M. Santos-Reis & A.I. Correia (eds) *Caracterização da flora e da fauna do montado da Herdade da Ribeira Abaixo (Grândola – Baixo Alentejo)*. Lisboa: Centro de Biologia Ambiental, pp. 169–176 (in Portuguese).

Mesquita N., Carvalho G., Shaw P., Crespo E. & Coelho M.M. (2001) River basin-related genetic structuring in an endangered fish species, *Chondrostoma lusitanicum*, based on mtDNA sequencing and RFLP analysis. *Heredity* **86**, 253–264.

Nelva A., Collares-Pereira M.J. & Coelho M.M. (1988) Systematique et repartition du genre *Chondrostoma* Agassiz, 1835 (Pisces, Cyprinidae). *Archiv für Hydrobiologie* **113**, 93–112 (in French).

Pires A.M., Cowx I.G. & Coelho M.M. (1999) Seasonal changes in fish community structure of intermittent streams in the middle reaches of the Guadiana basin, Portugal. *Journal of Fish Biology* **54**, 235–249.

Ribeiro F., Cowx I.G. & Collares-Pereira M.J. (2000) Life history traits of the endangered Iberian cyprinid *Anaecypris hispanica* and their implications for conservation. *Archiv für Hydrobiologie* **149**, 569–586.

SNPRCN (Serviço Nacional de Parques, Reservas e Conservação da Natureza) (1991) *Livro Vermelho dos Vertebrados de Portugal, Vol. II – Peixes Dulciaquícolas e Migradores*. Lisboa: Secretaria de Estado do Ambiente e Defesa do Consumidor, 55 pp. (in Portuguese).

Chapter 7
Conservation of salmonid fishes in Austria: current status and future strategies

S. WEISS*
Unidade de Genética Animal e Conservação, Vila do Conde, Portugal

F. UIBLEIN
Institut für Zoologie, Universität Salzburg, Salzburg, Austria

A. JAGSCH
Institut für Gewässerökologie, Fischereibiologie und Seenkunde Mondsee, Austria

Abstract

Austrian river systems and their biota have been subject to manifold changes brought on by hydropower development, river channel regulation, exotic introductions and intensive fishery management. Consequently, the distribution and abundance of nearly all fish species, including salmonids, has been drastically reduced. Recently, overwintering cormorants, *Phalacrocorax carbo* have added a level of predation pressure thought to be uncharacteristic for this region. Of particular concern is the cormorant's potential effect on European grayling, *Thymallus thymallus*. An interdisciplinary status report on four salmonid species in Austria is provided to highlight common and unique features of their vulnerability as well as potential for improved management. For two species, European grayling and brown trout *Salmo trutta*, genetic and ecological data are integrated to assess the species' status and to provide scientifically-based guidelines for an economically-efficient and conservation-oriented management regime. The potential role of genetic data is evaluated in consideration of present social and legal constraints. These considerations are synthesised into a list of proposed conservation and management actions.

Keywords: Austria, Danube, *Hucho hucho*, *Salmo trutta*, *Salvelinus umbla*, *Thymallus thymallus*.

7.1 Introduction

Austrian river systems and their biota have been subject to manifold changes brought on by hydropower development, river channel regulation, exotic introductions and intensive fishery management (Jungwirth, Schmutz & Weiss 1998). Consequently, the distribution and abundance of nearly all fish species, including salmonids, has been drastically reduced. Recently, overwintering cormorants, *Phalacrocorax carbo* (L.) have added a level of predation pressure thought to be uncharacteristic for this region.

*Correspondence: Centro de Estudos de Ciência Animal-ICETA, Unidade de Genética Animal e Conservação, Campus Agrário de Vairão, R.Monte-Crasto-Vairão, 4485-661 Vila do Conde, Portugal (email: sjweiss@mail.icav.up.pt).

Of particular concern is the cormorant's potential effect on European grayling, *Thymallus thymallus* L. (Uiblein, Jagsch, Honsig-Erlenburg & Weiss 2001). The following contribution provides an interdisciplinary synopsis on the conservation status of four salmonid species in Austria, including recommendations for improved management.

7.2 Case studies

7.2.1 *Brown trout* **Salmo trutta** *L.*

In Austria, brown trout dominate headwater fish assemblages and occur in the middle reaches of most rivers as well as sub-alpine lakes. Populations exist in alpine lakes but stem from translocations. Historically, brown trout in sub-alpine lakes routinely reached large sizes (>20 kg) and were recognised as *Salmo lacustris* Linnaeus a perspective that survives today in the common usage of '*morph*' nomenclature rightfully criticised by Kottelat (1997). More importantly, this nomenclature is still used, although based on polyphyletic and/or environmentally plastic characters. This has promoted the transfer of phenotypically similar stocks between divergent evolutionary lineages. The disappearance of such populations, through overfishing, habitat alteration, and stocking is the most significant change to this species' status in the region.

Austrian rivers belong almost exclusively to the Danube drainage, where river populations of brown trout have been tentatively renamed, *Salmo labrax* Pallas (Kottelat 1997). Regardless of resistance to adopt new nomenclature, available genetic data define a distinct and divergent lineage of brown trout in the Danube basin. However, both mtDNA (Bernatchez, Guyomard & Bonhomme 1992) and allozyme data (Riffle, Storch & Schreiber 1995) revealed the presence of northern lineages in the upper Danube. Such introgression in Austria is extensive, and primarily the result of stocking (Weiss, Schlötterer, Waidbacher & Jungwirth 2001). Nonetheless, significant phylogeographic structure by the more diverse Danubian lineage remains and several populations with no signs of introgression were found in Kalkalpen National Park (Weiss, Linhares & Haunschmidt, in press).

Stocking of hatchery-reared salmonids is intensive and obligatory wherever fishing licenses are sold. Experimental studies demonstrated the poor performance of hatchery strains and the potential for negative demographic effects on naturally reproducing populations (Weiss & Schmutz 1999a, b). Such studies are fuelling debate on how to improve management of brown trout in Austria.

7.2.2 *European grayling* **Thymallus thymallus** *L.*

Grayling inhabit about two-thirds of Austrian rivers, and several sub-alpine lakes. Local reports of declining stocks have initiated research to assess these claims and recommend management action (Uiblein, Jagsch, Kössner, Weiss, Gollman & Kainz 2000). Conservation issues concerning grayling differ markedly from brown trout due to their differing ecological demands and evolutionary histories. Uiblein *et al.* (2000) reported deeply divergent lineages of grayling north and south of the Alps, a pattern not seen in brown trout. Overlapping habitat preferences for grayling and introduced rainbow trout *Oncorhynchus mykiss* and agonistic interactions between the two species suggest competition (Uiblein *et al.* 2000), and there are concerns that cormorants may be exhibiting significant predation pressure on some populations.

Abundance of overwintering cormorants has stabilised at about 4000 individuals (Aubrecht & Winkler 1997). Diet analyses demonstrated that percid and cyprinid fishes are their dominant prey (Schratter & Trautmannsdorf 1993) and only *ad hoc* studies suggest that cormorants are reducing grayling abundance. Uiblein *et al.* (2000) investigated four river reaches, but only one was characterised by a senescent population as well as significant bird-related injuries (Fig. 7.1(a) and (b)). Pellet examination at nearby roost sites ($n = 93$) found European perch, *Perca fluviatilis* L., was the dominant prey (40%) while grayling were identified in only a single pellet. Such data demonstrate the ambiguous results of short-term, local snapshots of cormorant/fish interactions. Long-term studies in Austria are lacking but Suter (1995) reported no effects of cormorants on a commercial grayling fishery in a large, relatively natural system in Switzerland but the interpretation has met controversy (Staub, Egloff, Kramer & Walter 1998).

The lack of objective information offers few conclusions. Presently, both cormorants and grayling are managed to the needs of the community. Limited numbers of cormorants are being legally culled in six of the eight rural provinces; whereby one province (Burgenland) reports no problem and the Tyrol expects new legislation permitting culling.

7.2.3 *Arctic charr* Salvelinus umbla *L.*

Arctic charr is indigenous to Austria's sub-alpine lakes in the north and occurs in alpine lakes throughout the country via translocations that began in medieval times (Pechlaner 1984). These populations form a disjunct and southern-most glacial refuge. While most large sub-alpine lakes in the eastern Alps contain charr, some populations have been extirpated and the species is listed as endangered. Eutrophication is the primary concern but stocking threatens the genetic integrity of various sympatric and allopatric forms (Jagsch 1987).

Recent reports provided data on genetic variation in Central Europe (Brunner, Douglas & Bernatchez 1998; Englbrecht 2000), but most populations have been the subject of stocking thus complicating efforts to unravel their evolutionary history. Brunner *et al.* (1998) reported little mtDNA variation in Swiss populations but high levels of variation and drainage-specific alleles across six microsatellite loci. These data suggest that Central European charr stem from a recent common ancestral population but have differentiated significantly in the last 12 000 years.

Most charr fisheries in Austria are commercial, making attempts to apply conservation measures difficult. The long-term prospects for populations in shallow lakes are bleak, where eutrophication processes are difficult to reverse. Populations founded in

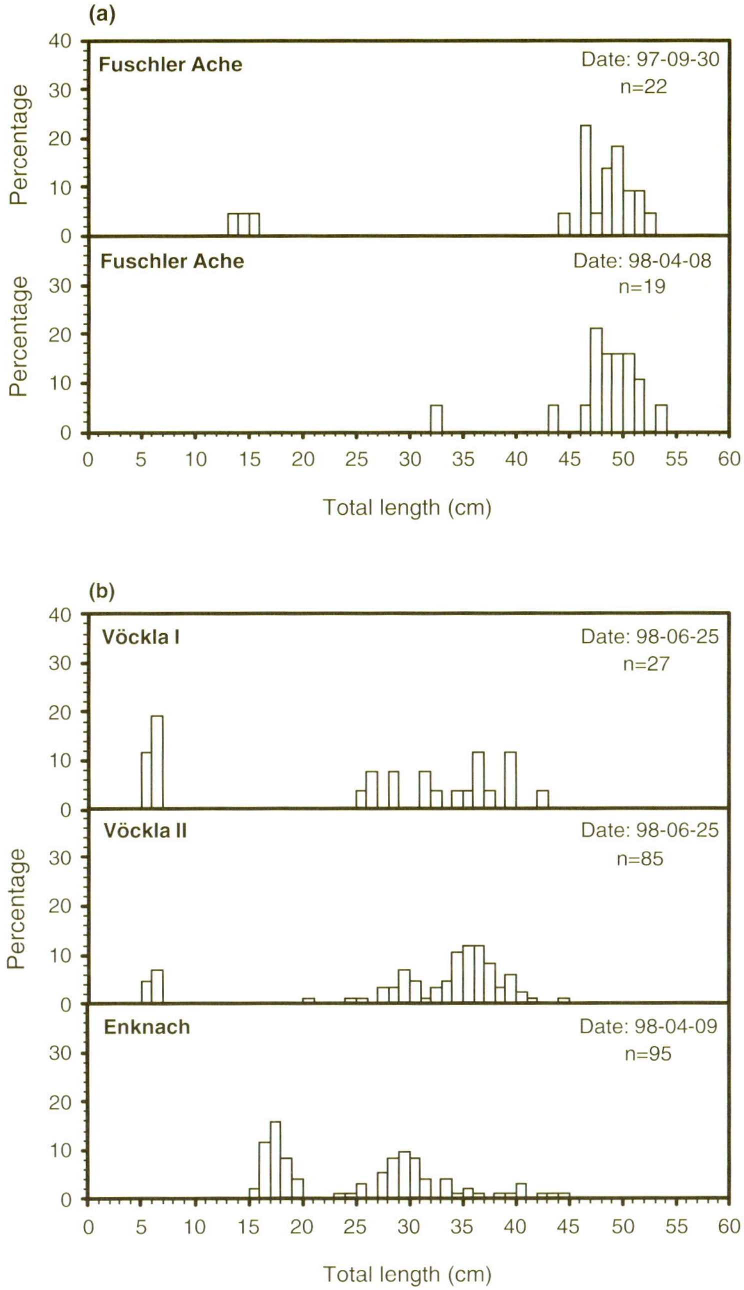

Figure 7.1 (a) and (b) Length frequency histograms of grayling sampled from five sites near cormorant roosts. Only the Fuschler Ache (a) displayed significant senescence reported to be the result of bird predation (adapted from Uiblein *et al.* 2000)

alpine lakes may represent a significant pool of genetic resources for rehabilitation efforts.

7.2.4 Huchen '*Danube salmon*' Hucho hucho *L.*

Huchen, the world's largest salmonid, is among the most threatened freshwater fish (Holcík 1990). Historically, this endemic species was abundant in the mid to lower reaches of Danube tributaries. Naturally sustained populations in Austria are difficult to document as most stable populations (Pielach, Mur, Drau, and lower Gail rivers) are supplemented by stocking. No population genetic data exist, and ecological data are scarce. Rearing and release of huchen have been conducted without regard to potential genetic differences between river basins. Present conservation efforts are limited to restricting harvest, stock enhancement and attempts to provide passage around river barriers to aid spawning and foraging movements.

7.3 Conservation strategies

Danubian populations of brown trout occupy the largest and most diverse glacial refuge for the species. On a large scale, these populations are not threatened. But promoting conservation-oriented stewardship of a widespread sport fish may be more successful in establishing some balance between short-term human interests and aquatic ecosystem integrity than conservation plans targeting more obscure species, however worthy these actions may be.

Grayling may be a more sensitive indicator of disturbance than brown trout. Thus, uncovering the causes of their apparently recent decline may hold widespread importance. For charr in Central Europe, there is a high danger of losing a significant number of populations within this highly disjunct refuge. More attention must be brought to the uniqueness of these populations.

Huchen should be a conservation priority for the upper Danube as their preservation requires the rehabilitation of the most endangered habitats and associated faunal elements. However, despite high economic value, interest and knowledge of the species has not extended far beyond a handful of resource users who carry little of the political clout necessary to protect habitats from encroachment.

7.3.1 *Genetic data and fish conservation in Austria*

For genetic conservation of brown trout, population-level or metapopulation management is the soundest policy to pursue within a country (Laikre 1999). Within the highly fragmented and private framework of Austrian fisheries, conservation must be promoted at the smallest scale. Genetic data provide fishery associations with a means of identifying with conservation and promote discussion on overhauling fisheries management throughout the country.

7.3.2 *Proposals for promoting salmonid conservation in Austria*

(1) Remove obligatory stocking laws: prioritise habitat enhancement and other management actions geared towards self-sustaining populations.
(2) Restrict cross-border trade of live fish: EU-policy prohibits restrictions on the trade of food fishes but this contradicts international agreements to conserve local biodiversity.
(3) Limit translocations to within the smallest watershed area practical: management should be geared toward a conservative 'do-no-harm' principle.
(4) Construct new, or improve old, fish passes: habitat fragmentation is the most significant environmental problem for fish in Austria.
(5) Limit stocking of rainbow trout, *O. mykiss*, and investigate potential competition: in Kalkalpen National Park, plans are underway to remove rainbow trout. This will only be possible for small systems.
(6) Investigate the potential threat of cormorants: as long as the causes of recent grayling declines remain obscure, all threats need to be monitored closely.
(7) Expand the role of genetics in conservation: basic data are needed for Arctic charr and huchen. Multi-locus analyses are needed to assess the role of hatchery operations in maintaining genetic diversity. Once such data are properly disseminated, market forces will improve breeding practices and promote use of local strains of fish.

References

Aubrecht G. & Winkler H. (1997) Analyse der Internationalen Wasservogelzählungen (IWC) in Österreich (1970–1995). Trends und Bestände. *Biosystematics and Ecology Series* **13**, 175 pp. (in German).

Bernatchez L., Guyomard R. & Bonhomme F. (1992) DNA sequence variation of the mitochondrial control region among geographically and morphologically remote European brown trout *Salmo trutta* population. *Molecular Ecology* **1**, 161–173.

Brunner P.C., Douglas M.R. & Bernatchez L. (1998) Microsatellite and mitochondrial DNA assessment of population structure and stocking effects in Arctic charr *Salvelinus alpinus* (Teleostei: Salmonidae) from central Alpine lakes. *Molecular Ecology* **7**, 209–223.

Englbrecht C. (2000) Untersuchungen zur Phylogeographie und Populationsstruktur zweier paläarktischer Fischarten, der Mühlkoppe (*Cottus gobio* L.) und des alpinen Seesaiblings (*Salvelinus alpinus* L.). Dissertation, Ludwig–Maximilians-Universität München, 123 pp. (in German).

Holcík J. (1990) Conservation of the huchen, *Hucho hucho* (L.), (Salmonidae) with special reference to Slovakian rivers. *Journal of Fish Biology* **37**(Suppl. A), 113–121.

Jagsch A. (1987) Arctic charr in some of the lakes of the Eastern Alps (Austria). *ISACF Information series* **4**, 64–72.

Jungwirth M., Schmutz S & Weiss S (eds) (1998) *Fish Migration and Fish Bypass Channels*. Oxford: Fishing News Books, Blackwell Science, 438 pp.

Kottelat M. (1997) European freshwater fishes: an heuristic checklist of the freshwater fishes of Europe (exclusive of former USSR), with an introduction for non-systematists and comments on nomenclature and conservation. *Biologia* **52**(Suppl. 5), 1–271.

Laikre L. (ed.) (1999) Conservation genetic management of brown trout (*Salmo trutta*) in Europe. Report by the concerted action on identification, management and exploitation of genetic resources in the brown trout (*Salmo trutta*) 'TROUTCONCERT'; EU FAIR CT97-3882, 91 pp.

Pechlaner R. (1984) Historical evidence for the introduction of Arctic charr into high-mountain lakes of the Alps by man. In: L.L. Johnson & B.L. Burns (eds) *Biology of the Arctic Charr*. Winnipeg: University of Manitoba Press, pp. 547–557.

Riffle M., Storch V. & Schreiber A. (1995) Allozyme variability of brown trout (*Salmo trutta* L.) populations across the Rhenian-Danubian watershed in southwest Germany. *Heredity* **74**, 241–249.

Schratter D. & Trautmansdorff J. (1993) Kormorane *Phalacrocorax carbo sinensis* an Donau und Enns in Österreich: analyse der Speiballen. *Ornithologische Verhandlungen* **25**, 129–150 (in German).

Staub E., Egloff K., Kramer K. & Walter J. (1998) The effect of predation by wintering cormorants (*Phalocrocorax carbo*) on grayling (*Thymallus thymallus*) and trout (Salmonidae) populations: two case studies from Swiss rivers. Comment. *Journal of Applied Ecology* **35**, 607–610.

Suter W. (1995) The effect of predation by wintering cormorants *Phalacrocorax carbo* on grayling *Thymallus thymallus* and trout (Salmonidae) populations: two case studies from Swiss rivers. *Journal of Applied Ecology* **32**, 29–46.

Uiblein F., Jagsch A., Honsig-Erlenburg W. & Weiss S. (2001) Status, habitat use and vulnerability of European grayling (*Thymallus thymallus*) in Austrian waters. *Journal of Fish Biology* **59**(Suppl. A).

Uiblein F., Jagsch A., Kössner G., Weiss S., Gollman P. & Kainz E. (2000) Untersuchungen zu lokaler Anpassung, Gefährdung und Schutz der Äsche (*Thymallus thymallus*) in drei Gewässern in Oberösterreich. *Österreichs Fischerei* **53**, 88–165 (in German).

Weiss S., Schlötterer C., Waidbacher H. & Jungwirth M. (2001) Haplotype (mtDNA) diversity of brown trout *Salmo trutta* in tributaries of the Austrian Danube: massive introgression of Atlantic basin fish – by man or nature? *Molecular Ecology* **10**, 1241–1246.

Weiss S., Linhares D. & Haunschmidt R. Vorläufige Untersuchungen der genetischen Diversität der Bachforelle (*Salmo trutta* L.) im National Park Kalkalpen. *Österreichs Fischerei* (in German) (in press).

Weiss S. & Schmutz S. (1999a) Performance of hatchery-reared brown trout and their effects on wild fish in two small Austrian streams. *Transactions of the American Fisheries Society* **128**, 302–316.

Weiss S. & Schmutz S. (1999b) Response of wild brown and rainbow trout to the stocking of hatchery-reared brown trout. *Fisheries Management and Ecology* **6**, 365–375.

Section II
Biology of rare and endangered species

Chapter 8
Biology of rare and endangered non-migratory fish species: problems and constraints

P.S. ECONOMIDIS*
Department of Zoology, Laboratory of Ichthyology, Aristotle University, Thessaloniki, Greece

Abstract

Rare, non-migratory freshwater fish often frequent small and isolated catchments in arid and semi-arid areas, usually in habitats that form a mosaic of patches. Factors responsible for the fragmentation include: fluctuations in water level; environmental extremes (floods and droughts); demands for water resources; natural variation in productivity related to food availability and population size. Species are characterised by small body size, drab colour patterns, opportunistic feeding behaviour and a high tolerance of extreme environmental conditions. They invest energy in reproduction rather than body size so they have fast growth, precocious maturity, and a short life span. Populations typically are small and genetic processes influence diversity. They are also sensitive to loss of spawning and nursery areas and by the restriction of movements linking local populations which support gene flow. To conserve these fragile populations there is a need to implement appropriate policies, which must target the specific populations.

Keywords: conservation, life history strategy, non-migratory fish, threats.

8.1 Introduction

Non-migratory fish species have drawn the attention of conservation agencies because they usually exhibit a high rate of decline linked to environmental perturbation. This decline can rapidly lead the populations to becoming firstly vulnerable, and then to fall into the category of rare and even extinct. Relevant questions include: What are non-migratory freshwater fishes? Why do they have problems? When are they 'rare' and 'endangered'? What is the meaning of 'rare' and 'endangered'? What are their survival strategies? Why and when are they stressed? Are their problems recent or persistent?

Following the terminology adopted by international agencies (IUCN 1994) rare is a small population, not endangered but at risk, while endangered is a population at a critical level and in immediate danger of extinction in the wild. Furthermore, problems with their survival should be evident from environmental pressures acting on the population, e.g. critical environmental changes, pollution or development works. These environmental problems impose constraints on the well being and survival of the species. This chapter reviews the fundamental biology of rare and endangered freshwater fish species and examines the constraints on their continued survival.

*Correspondence: Department of Zoology, Laboratory of Ichthyology, Aristotle University, Box 134, GR-54006, Thessaloniki, Greece (email: PSEcon@bio.auth.gr).

8.2 The fish species and the habitat history

8.2.1 *The fish species*

Non-migratory, rare and endangered freshwater fishes occur throughout the world, but the majority are distributed in temperate zones, mainly in arid and/or semi-arid regions. In general, they are found in Australia, the Middle and Near East, south and north-west Africa, North America (south-western USA and Mexico) and Europe (mainly Mediterranean drainages). However, rare non-migratory freshwater fish species can be also found in large rivers, but normally occupying habitats isolated and/or distant from each other. In European inland waters, such species include *Romanichthys valsanicola* Dumitrescu, Banarescu & Stoica, in the River Arges (tributary of the Romanian sector of the Danube catchment) (Perrin, Bless & Nalbant 1993; see Banarescu, Chapter 17), *Zingel asper* (L.) in the River Rhône in France (Keith & Allardi 1996; see Mari, Labonne & Gaudin, Chapter 15), *Cottus petiti* Bacesu & Bacescu-Mester, in the River Lez in southern France (Persat, Beaudou & Freyhof 1996). The majority of rare non-migratory species frequent small and usually isolated catchments with still or slow-flowing waters or waters that exhibit huge fluctuations in water level (floods and droughts) between years. These fishes normally have weak swimming ability and present some phenotypic convergent features related to specific habitat requirements that have evolved over many generations. Such features are small body size, almost homogenous grey or olive green body colour, the frequent presence of a lateral dark strip and/or some spots, the usually black peritoneum, and sometimes an incomplete lateral line.

8.2.2 *Habitat*

The drying up of isolated watersheds in arid or semi-arid zones is often the final step before a species becomes extinct. The lack of any recent connection with the 'mother' system is typical. The large Mediterranean Islands (Corsica, Sardinia, Crete, Cyprus, etc.), which lack primary freshwater fish species, are typical examples. Rhodes is a unique exception, because it has a primary freshwater fish species, *Ladigesocypris ghigii* (Gianferrari). The speciation of small non-migratory fish species, and hence the starting point of the composition of any contemporary ichthyofaunal complex, is related to several factors. For southern Europe these factors include: the land re-arrangement during recent geological time essentially influenced by the glaciations of the Quaternary and the consequent sea level regression and/or transgression; climate changes towards an unstable precipitation regime with the related water budget; and habitat change that has caused rapid environmental disturbance and alterations of aquatic habitats. Consequently, the recent picture is considered a result of the combined action of the evolution of the natural landscape and human influence during historical and recent times. Human influence during historical times is often neglected in the estimation of impact. For instance, almost nothing is known about the freshwater fish fauna in the Ancient Greek inland waters (mainly in Attiki, Peloponnesus, Beotia). The same is also true for Roman Italy, Iberia and the Middle East during their reciprocal historical processes. It is likely that several freshwater fish species were present in the aquatic systems, which supplied water for large settlements and agriculture, and they had a distribution at that time different from today. For example, the tiny cyprinid, *Pseudophoxinus stymphalicus* (Valenciennes), lives in the River Helidonou, which belongs to the main drainage of the territory the modern city of Athens (Stephanidis 1971), but its current distribution bears no relation to the situation in ancient times because the drainage has completely changed. Furthermore, it is not easy to know how much present day environments, in which we find fish, resemble the environments in which they evolved (Matthews 1998).

8.3 Biology, ecology and genetics

Life history characteristics of any species play an essential role in their survival. However, some characteristics are more important, especially feeding, age and growth, reproduction, population structure, and genetics.

8.3.1 *Feeding*

Non-migratory, rare and endangered freshwater fishes are normally generalists and feed at almost all trophic levels. Feeding is typically opportunistic and fish search for food in any locality. They prefer animal items (zooplankton, worms, insects larvae), because these have a higher energy content. But, when they become scarce, many species, especially cyprinids, can shift to feeding on plant material and thereby buffer variation in availability of animal food (Persson 1991). Furthermore, as food resources vary continuously in space and time fish have to cope by shifting from pelagic to benthic feeding, from particulate to filter feeding or by migrating to other habitats (Lammens & Hoogenboezem 1991). Such movements are more free when there is not strong predation in the water column from piscivorous fishes or birds. The species living in still water feed upon plant material and/or plankton, while the running water species feed upon invertebrates mainly insects, and periphyton algae. For small fish in general, Miller (1996) suggested they be defined as nektonic, epibenthic, cryptobentic. These definitions although useful are not wholly applicable to rare non-migratory species. In several examples, the opportunistic feeding goes up to the anthropophily as some of them take food provided by man.

8.3.2 *Age and growth*

Rare and endangered freshwater fishes exhibit some characteristics of growth which are related to their special mode of life. Within a population, there is wide phenotypic plasticity, which confers the potential for a rapid response to changes in environmental conditions in an individual fish's growth rate and development to sexual maturity (Mann 1991). In temperate zones, growth seasonality is a function of water temperature, day length and food availability (Mann 1991). Consequently, the species have to achieve in time an optimal body and gonad weight to ensure appropriate reproductive potential. Environmental pressures force the fish to reach a suitable body size in a shorter time, to avoid mortality during a protracted somatic growth period. Therefore, these species have fast early growth and a short span of life.

8.3.3 *Reproduction*

The main features are early maturation normally attained in the first year of life; some species are annual. Often the period of reproduction is prolonged and batch spawning during a breeding season is common. Repeated reproduction dominates when juvenile mortality is high or variable (Matthews 1998). Additionally, fish of this category have a high gonadosomatic index (GSI) because of huge investment in reproduction. Mills (1991) noted that earlier maturing individuals were significantly larger than immature fish of the same age. For example, in European minnow, *Phoxinus phoxinus* (L.), in some few individuals, reproductive effort was extremely high and accompanied by a loss in somatic condition. Furthermore, at spawning the gonads accounted for 15–25% of total body weight, and the mass of eggs spawned exceeded 15% of total body weight. These results suggest that small fishes invest much more energy in reproductive effort than somatic growth. Finally, stillwater-living species of this category lay their eggs on vegetation, while rheophilous species spawn upon stones or gravel or sand.

8.3.4 *Population structure*

Individual growth rate is dependent on resource supply, consequently resource competition can have a direct influence upon population structure. Inter-specific competition can lead to a juvenile competitive bottleneck with profound effects on population dynamics or population size structure (Persson 1991). It seems that this kind of competition is developed among rare, non-migratory fish populations.

Populations of such species often contain a small number of individuals. In a drainage there are often several independent populations, which may be completely isolated from each other and may be evolving separately. Nevertheless, separate populations are, in many cases, of greater importance than the total number of individuals in a single drainage. This is because they contain the majority of the initial variability, which is important for the viability of the species, as they can give proper responses to the variable environment. The lack of movement and mixing of populations prevents gene flow and reduces the variability (bottleneck effect).

8.3.5 *Genetics*

The scale and the rate of man-made changes is often beyond the genetic capacity of species to respond. Conservation is essential to maintain the genetic continuity and adaptability of species in the face of continual environmental change (Thorpe, Gall,

Lannan & Nash 1995). Genetic processes also influence diversity. However, little is known about future environmental conditions and about the adaptive value of particular genetic attributes (alleles and allelic combinations) under these conditions. Ryman (1991) underlined that there are no realistic alternatives to this goal. However, biological resource conservation is genetically driven, and thus the application of genetic principles is essential.

8.4 Problems and constraints

Any environmental disturbance, with associated physicochemical effects, forms a set of problems and this process leads to a set of constraints for living organisms, including fish. The usual environment for these species can be summarised as habitats that take the form of a mosaic of patches of different characteristics that are frequently disturbed, extreme seasonal variations in temperature and oxygen concentrations, and sometimes poor water quality. Water levels are related to seasonal rainfalls and their year to year oscillations including extended droughts, humans' needs for water, natural water availability and the availability of food, foraging mainly by land vertebrates, and the critical population size of the species in view, are the main problems for a rare fish population.

Many aquatic systems supporting rare, non-migratory fish species are fed by water coming from permanent springs and/or intermittent sources in the headwaters. There is a strong negative correlation between fish diversity and variability in stream discharge (Matthews 1998). A permanent spring could be used as a good refuge for stressed fish, intermittent water sources are not. Furthermore, low annual temperature variation of cool or cold water can support, even in hot climates, a local assemblage of cool or clean-water species (Matthews 1998), some of which could be endemic. So, any disturbance of these well equilibrate systems would be fatal for fish. Many springs have dried up because the water is used for domestic supply, especially in arid areas.

Problems may appear when the physical structure of a natural habitat (submerged rocks and/or woody debris, macrophytes, etc.) is destroyed. This structure is used by fish as shelter from predators or swift currents, as foraging and as spawning sites (Matthews 1998). The heterogeneity and the complexity of physical structure and microhabitats, their connectivity, and typical flow regime, form a natural environment. When this frame is disturbed it could be detrimental to fish.

Serious impacts on rare fish assemblages because of drought have been registered in most arid areas (Matthews 1998). The reduction in flow decreases available water and increases thermal oscillations. In small intermittent streams, water temperature approximates to the air temperature after stream flow ceases in late summer. At this time, the hyporeic layer becomes a nitrate sink rather than a source. Consequently, drought has potentially severe effects on stream fish by killing them directly, and additionally by destruction of invertebrates, algae and other normal components (e.g. Cowx, Young & Hellawell 1984).

Constraints appear when fish have to face environmental disturbances expressed by unusual characteristics of the medium and/or the habitat morphology in relation to the life history of the species. These comprise any loss of reproductive habitats, including spawning and nursery areas, any confinement of movements in the system, because they disrupt local gene flow. Additionally, any lack of knowledge about fish species, their basic biology, ecology, genetics and evolutionary biology limits the effectiveness of current policies (Thorpe *et al.* 1995), and inhibits appropriate mechanisms of conservation. There is an inadequacy of national and international institutional and legal systems to implement appropriate action programmes (Thorpe *et al.* 1995).

Fish species living in stressed environments may have the capacity to persist through extended periods of drought (Matthews 1998). Native fishes of arid lands have probably already passed through an 'evolutionary filter' with adaptation of their life history or physiological tolerances for drought conditions (Matthews 1998). However, for continued existence through drought periods downstream or upstream refugia from which colonists return in wetter periods may be required. Fish populations of a species distributed across different watersheds, might function as metapopulations, with survivors in one watershed 'rescuing' those in others by dispersal after a drought ends (Matthews 1998). The great swimming ability of these species permits rapid recolonisation from permanent waters and rapid reproduction to repopulate inundated areas. Irrespectively, some species are tolerant to drought, so long as standing water remains.

8.5 Survival strategies

Highly variable environments, where temperature, water velocity and dissolved oxygen fluctuate markedly on a spatial and a temporal basis, favour communities driven by r-selection strategies (Barbieri, Economou, Stoumboudi & Economidis, Chapter 5). This variable environment also favours early maturation and batch spawning, and investment in reproduction rather than growth, which promotes the small body size. The majority of such species rarely exceed 10 cm in total length. However, the small size could also be regarded as a derived condition, with functional constraints in a small patch size (Miller 1996). The wide phenotypic plasticity also favours a short span of life. All these adaptations permit a fast reaction to favourable environmental factors when they reappear, a recovery of the depleted population and colonisation of new habitats. Such fish are not prevented from thriving in a large and constant system. Mann (1991) noted in cyprinid species, that individuals within a population show wide phenotypic plasticity, which confers the potential for a rapid response to changes in environmental conditions, especially in growth rate and sexual maturity. In cyprinid and non-cyprinid species, fast-growing fish of a particular cohort often attain sexual maturity at earlier age than slower growing conspecifics (Mann 1991).

Special reference must be made to floodplain habitats, which in spite of the unstable conditions they form annually, also maintain abundant populations and high species richness. These habitats are typically found in South Europe during the rainy period in late winter–early spring, and there is synchrony with their presence and the main annual reproductive events of many fish species. The phenomenon favours a rather high reproductive potential which could support adequate recruitment of the populations (Barbieri *et al.*, Chapter 5). In such cases stocking with fry, a common management action, is not necessary to maintain populations because these suffer during the summer droughts. Other measures which sustain habitats are more appropriate.

8.6 Conservation

Conservation is a problem on a global scale, which needs financial support, proper institutional support at the national level, careful studies, correct methods and continuing activity. Rare freshwater fish must be classified as an important national and global heritage. Also, the attention of national and international bodies must be drawn to the precarious nature of rare freshwater fishes to give them a conservation status comparable to that seen for birds, mammals and other vertebrates. The real power to manage and conserve lies mainly within national authorities, guided by responsible and proper education. Conservation management programmes must be considered on a species-by-species, case-by-case basis, but within a broad ecosystem context (Thorpe *et al.* 1995). The loss of potentially useful and ecologically-important species must be addressed aggressively if further irreparable damage is to be prevented. The conservation of aquatic genetic resources must become a national and international priority, to avert the loss of many important aquatic species. Conservation of species is closely related to the conservation of ecosystems and their constituent communities. Furthermore, habitat protection and rehabilitation are the principal long-term means through which successful fish conservation will be achieved. Environmental legislation must be tightened up and a code of practice for stock management needs to be drawn up (Collares-Pereira *et al.* 2000).

In the case of serious alterations, habitat rehabilitation and management should be major goals in the conservation of most species and communities. Special consideration should be given to the genetic integrity of translocations and maximum genetic diversity should be sought by selecting material widely in space and time (Maitland & Lyle 1996).

8.7 Slogans

Recently, some aspects of nature and/or species conservation have taken the form of slogans. These slogans are addressed either passively or actively as short propositions or orders to the decision makers and/or to the open public. However, new ways of moving beyond this superficial approach and providing descriptions of the species and the threats they might face because of management prescriptions at an ecosystem level should be presented (Brackett 1999). An evaluation following the IUCN and SSC norms is absolutely necessary. A rare fish species, living in an isolated aquatic system and having several distinguishing features would be more easily placed under a protection frame if it bears a name, since, as noted by Brackett (1999), knowledge about species will remain the foundation on which conservation action will be based. The classical slogan 'think globally, act locally' must also have a wide distribution for promoting international collaboration. Furthermore, it is necessary to improve technical capacity, to share proper information and to act more responsibly. Under these conditions, more meetings, where representatives can put protection points across to governments, and better legal instruments for conservation actions are needed to save species at risk (Brackett 1999). Special features of fauna and ecosystems could be more efficiently managed if they were under the umbrella of the Natural Heritage. Therefore, any agency or individual scientist has to insist on this concept for all rare fish species. Actions such as releasing stamps and special books for elementary school, and increasing public awareness, are necessary for fish. The enormous success of bird protection is a good model on which to promote the 'fish case'. Public concern for some charismatic aquatic species, such as whales, dolphins and turtles, is a good reason for conserving many less visible species, which are almost unknown to the public. In the case of small non-migratory fish, the interest should be the conservation rather than the exploitation of a population, because sound scientific principles give priority to conservation overexploitation. Additionally, knowledge is required, as in the key cases of threatened species in Europe, e.g. saramugo, *Anaecypris hispanica* (Collares-Pereira, Cowx, Rodriguez & Rogado, Chapter 19) of Portugal and Spain, samaruc, *Valencia hispanica* (Caiola & de Sostoa, Chapter 13) of Spain, and gizani, *Ladigesocypris ghigii* (Stoumboudi, Corsini-Foka, Barbieri, Economu & Economidis, Chapter 18) of Greece, because this is a useful tool to apply to similar fish. Therefore, it is essential to know everything about the species, especially its biology and genetics. But, before this knowledge is available, priority must be given to saving the habitat, and by this the species. In the absence of any concrete scientific information, empirical knowledge must promote conservation and prevent management decisions that inadvertently place aquatic resources at risk.

References

Brackett D. (1999) From the chairman. *Species: Newsletter of the Species Survival Commission IUCN – The World Conservation Union*, No 31–32, 1–2.

Collares-Pereira M.J., Cowx I.G., Ribeiro F., Rodrigues J.A. & Rogado L. (2000) Threats imposed by water resource development schemes on the conservation of endangered fish species in the Guadiana River in Portugal. *Fisheries Management and Ecology* **7**, 167–178.

Cowx I.G., Young W.O. & Hellawell J.M. (1984) The influence of drought on the fish and invertebrate populations on an upland stream in Wales. *Freshwater Biology* **14**, 165–177.

IUCN (1994) Draft IUCN Red List categories, Gland: IUCN, 14 pp.

Keith P. & Allardi J. (1996) Endangered freshwater fish: the situation in France. In A. Kirchhofer & D. Hefti (eds) *Conservation of Endangered Freshwater Fish in Europe*. Basel: Birkhäuser Verlag, pp. 35–54.

Lammens E.H.R.R. & Hoogenboezem W. (1991) Diets and feeding behaviour. In I.J. Winfield & J.S. Nelson (eds) *Cyprinids Fishes: Systematics, Biology and Exploitation*. London: Chapman & Hall, pp. 353–376.

Maitland P.S. & Lyle A.A. (1996) Threatened freshwater fishes of Great Britain. In A. Kirchhofer & D. Hefti (eds) *Conservation of Endangered Freshwater Fish in Europe*. Basel: Birkhäuser Verlag, pp. 9–21.

Mann R.H.K. (1991) Growth and reproduction. In I.J. Winfield & J.S. Nelson (eds) *Cyprinids Fishes: Systematics, Biology and Exploitation*. London: Chapman & Hall, pp. 456–482.

Matthews W.J. (1998) *Patterns in Freshwater Fish Ecology*. London: Chapman & Hall, 756 pp.

Miller P.J. (1996) The functional ecology of small fish: some opportunities and consequences. *Symposium of the Zoological Society of London* **69**, 175–199.

Mills C.A. (1991) Reproduction and life history. In I.J. Winfield & J.S. Nelson (eds) *Cyprinids Fishes: Systematics, Biology and Exploitation.* London: Chapman & Hall, pp. 483–508.

Perrin J.-F., Bless R. & Nalbant T. (1993) *Romanichthys valsanicola*, l' expédition de la dernière chance (Octobre 1992, rivière Vilsan, Roumanie). *Revue française de Aquariologie* **20**, 37–42.

Persat H., Beaudou D. & Freyhof J. (1996) The sculpin of the Lez spring (South France) *Cottus petiti* (Bacescu and Macescu-Mester 1964), one of the most threatened fish in Europe. In A. Kirchhofer & D. Hefti (eds) *Conservation of Endangered Freshwater Fish in Europe*. Basel: Birkhäuser Verlag, pp. 321–328.

Persson L. (1991) Interspecific interactions. In I.J. Winfield & J.S. Nelson (eds) *Cyprinids Fishes: Systematics, Biology and Exploitation.* London: Chapman & Hall, pp. 530–551.

Ryman N. (1991) Conservation genetics considerations in fisheries management. *Journal of Fish Biology* **39**(Suppl. A), 211–224.

Stephanidis A. (1971) On some freshwater fish of Greece. *Biologia Gallo-Hellenica* **3**, 213–241 (in Greek).

Thorpe J.E., Gall G.A.E., Lannan J.E. & Nash C.E. (1995) *Conservation of Fish and Shellfish Resources: Managing Diversity*. London: Academic Press, 206 pp.

Chapter 9
Conservation needs of *Sandelia bainsii*, an endangered African anabantid

J.A. CAMBRAY*
Albany Museum, Grahamstown, South Africa

Abstract

The Eastern Cape rocky, *Sandelia bainsii*, is an endangered species endemic to several rivers in the Eastern Cape of South Africa. This species is unlike its northern anabantid relatives and has acquired several distinct temperate specialisations. These include several apomorphic traits such as substratum nests, adhesive demersal eggs, reduced size of egg oil globules and male inspection of eggs. The plesiomorphic anabantid reproductive style is free spawning with the release of buoyant eggs. *Sandelia bainsii* is a male guarder and very aggressive during the spawning period. The males are thus very easily caught by anglers during this period. If the males are removed this leaves the adhesive eggs, which are in a nest, exposed to predators. The species is highly threatened by angling, excessive water abstraction during droughts, poor water quality and alien fish and plants, which have resulted in serious reductions of the species over the past 30 years. There is an urgent need to manage the remaining riverine populations and establish sanctuaries to keep this large African anabantid from extinction in the near future.

Keywords: conservation, reproductive strategies, *Sandelia bainsii*, sanctuaries.

9.1 Introduction

Anabantoidei, or labyrinth fishes, are a monophyletic group of freshwater inhabiting percomorphs that comprise more than 80 species (Nelson 1994). Within the Anabantoidei the Afro-Asian family Anabantidae consists of four genera, *Anabas*, *Ctenopoma*, *Microctenopoma* and *Sandelia*. The three African anabantoid genera are *Ctenopoma*, *Microctenopoma* and *Sandelia* with four recognised monophyletic lineages (clades) (Elsen 1976; Norris & Teugels 1990). In the one clade there are two *Sandelia* species, *Sandelia bainsii* Castelnau and *Sandelia capensis* (Cuvier), and the three other clades are grouped within the *Ctenopoma* and *Microctenopoma*. The *Ctenopoma* and *Microctenopoma* species occur at lower latitudes in Africa and have their greatest diversity in the rain forest areas of central Africa, (the Zaïre River Basin), whereas the two *Sandelia* species occur in South Africa. There is thus a distinct gap in African anabantid distribution with *S. bainsii* occurring in the small, sub-temperate, Eastern Cape rivers and *S. capensis* occurring from the Eastern to Western Cape. *Sandelia bainsii* has a very restricted distribution in the Eastern Cape of South Africa, occurring

*Correspondence: Albany Museum, Somerset St., Grahamstown, 6139 South Africa (email: J.Cambray@ru.ac.za).

only in short sections of the Kowie, Great Fish, Keiskamma, Gulu, Buffalo and Nahoon Rivers most of which undergo erratic flows and are subject to the prolonged droughts in the area. *Sandelia bainsii* is rare in the upper, lower middle and lower reaches but more common in the upper middle and middle reaches of the Tyume (Keiskamma River system) and Kowie Rivers (Bok 1983; Mayekiso & Hecht 1988a). Recent land use patterns, water transfer schemes, water abstraction, damming, introduction of alien fish and plants, as well as poor water quality, are threats to the continuing survival of *S. bainsii* in the wild (Cambray 1995; 1996; 1997a; Lorenz, Ribbink & Kirby 1995). *Sandelia bainsii* is endangered and several populations are on the verge of extinction (Cambray 1996, 1997a, b).

This chapter reviews and analyses previous research on the reproductive and early life history behaviour of *S. bainsii*, which is relevant to designing both in-stream and off-stream sanctuary areas.

9.2 Reproductive and early life history behaviour

The reproductive traits and early life history characters of *S. bainsii* highlight a number of temperate specialisations which separates it from it's northern African relatives in the genera *Ctenopoma* and *Microctenopoma* (Table 9.1). The males of *S. bainsii* have nuptial colouration, anterior contact organs, make substratum nests, eggs have reduced oil globules relative to other African anabantids, eggs are adhesive and demersal, males 'inspect' the eggs, males aggressively guard the developing embryos in their nests and free embryos (larvae) have cement cells on their heads (Cambray 1980; 1997a; Britz & Cambray 2001).

Sandelia bainsii is a multiple spawner, which was implied from analysis of recruitment patterns and the presence of mature eggs in the ovaries (Mayekiso 1986). This author suggested this reproductive trait reduces the risk associated with spawning in the erratic flow regimes in Eastern Cape rivers. This multiple spawning behaviour is also a common trait in other Eastern Cape freshwater fish species (Cambray 1994). However, *Anabas testudineus* Bloch (Breder & Rosen 1966), several *Ctenopoma* species (Berns & Peters 1969) and the sister species of *S. bainsii*, *S. capensis* (Siegfried 1963) also exhibit this multiple spawning trait. Since *S. bainsii* is able to spawn over an extended period, October to February, this may allow for several batches of eggs per season under ideal conditions. Harvesting from one sanctuary after the first spawned fish have developed into juveniles and stocking of another sanctuary may prove to be an important management technique. This would also reduce predation pressure on the larval fish from the second spawning of *S. bainsii*.

Sandelia bainsii does not reach sexual maturity until its second year (Mayekiso & Hecht 1988b). This suggests a delay of several years before fish were able to reproduce if young-of-the-year are stocked into the sanctuaries.

Males are very aggressive when they have chosen a nesting site. They chase both males and females from their nest site until they are ready to 'invite' a female into the area to spawn. After spawning the female quickly leaves the area due to the aggressive behaviour of the male (Cambray 1997a). Suitable hiding structures, such as vegetation

Table 9.1 Summary of knowledge on the biology of the African anabantid, *S. bainsii*

	Life history traits and other relevant information on *S. bainsii*	Reference
Distribution	Kowie, Fish, Keiskamma, Gulu, Buffalo, Nahoon Rivers	Cambray 1997a, 2000
Food	Insects, crabs or small fish	Mayekiso & Hecht 1990
Mode of feeding	Lurking predator	Cambray, unpublished
Growth and longevity	Males grow faster, live longer and reach a larger size than females – attain 260 mm SL	Mayekiso & Hecht 1988b, Cambray 1997a
Reproduction	Mature in their second year, at 80–90 and at 60–69 mm SL	Mayekiso & Hecht 1990
	Breed in summer, October to February	Mayekiso & Hecht 1990
	Multiple spawner which was implied from analysis of recruitment patterns and presence of mature eggs in the ovaries	Mayekiso 1986
	Contact organs (develop visible ctenii on cheek scales), anterior only	Cambray 1980
	Sexual dichromatism, male nuptial colouration	Cambray 1997a
	Males very aggressive when nesting	Cambray 1997a
	Male guarder	Cambray 1997a
	Cleared substratum nest site	Cambray 1997a
	Eggs 1.3 mm in diameter and have a small oil globule of 0.6 mm in diameter	Cambray 1997a
	Adhesive demersal eggs	Cambray 1997a
	Reduced oil globules, relative to other African anabantids	Cambray 1997a
	Male inspects eggs	Cambray 1997a
	Cement glands on free embryo heads	Britz & Cambray 2001

or logs, are necessary near the spawning structures. Females can retreat into these areas after a spawning episode and wait for the male to invite her for the next spawning.

Under laboratory conditions the males are larger and live longer than females (unpublished data) confirming Mayekiso's (1986) field observations. The shorter life span of females may, in part, be attributed to the aggressive nature of the males. Male contact organs damage the females during the spawning embrace (Fig. 9.1). Cambray (1997a) also noted other aggressive, possibly damaging, behaviour during spawning. In sanctuaries shelter structures for females would be important when the males chase them from the nesting sites.

Typical habitat features required for successful spawning include rocks or woody debris placed at suitable intervals (Cambray 1980) and should be matched in any sanctuary area (Fig. 9.2).

Analysis of the apomorphic reproductive characteristics of *S. bainsii* are considerably different from the plesiomorphic characters of the Anabantidae (Table 9.2). After

Figure 9.1 Spawning embrace of male *S. bainsii*. It is at this stage of the embrace, when eggs are squeezed from the female, that the contact organs are gripped onto the female and cause damage to scales. Drawing from actual video-footage of a pair of spawning fish

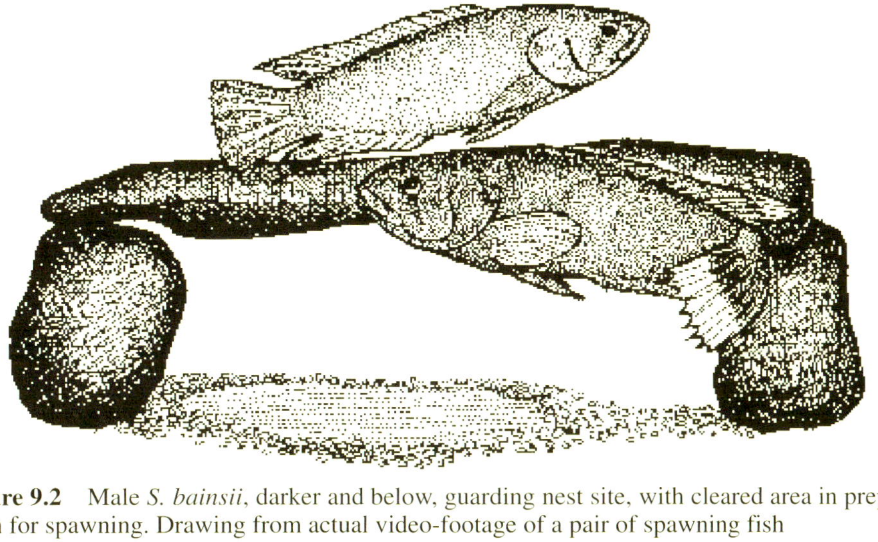

Figure 9.2 Male *S. bainsii*, darker and below, guarding nest site, with cleared area in preparation for spawning. Drawing from actual video-footage of a pair of spawning fish

Table 9.2 Analysis of the apomorphic reproductive characters of *Sandelia bainsii*

Characters	Anabantidae plesiomorphic characters	*S. bainsii* apomorphic characters
Male behaviour	Free spawner	Male guarder
Nest position	No nest	Substratum nest
Egg behaviour	Floating eggs	Adhesive demersal eggs
Oil globule	Large oil globule	Reduced oil globule
Male/egg behaviour	No male egg inspection	Male inspects substratum eggs

hatching, free embryos/larvae of other African anabantid genera, *Ctenopoma* and *Microctenopoma*, have a large oil globule in their yolk sac. The oil globule causes buoyancy and thus keeps larvae close to the water surface, either in a bubble nest or free floating which is essential in areas with low dissolved oxygen levels. The relatively smaller oil globules in *S. bainsii* allow for the sinking of the eggs to the substratum nest but offer some positive buoyancy to the free embryo once it hatches (Cambray 1997a).

All anabantoid larvae, with the exception of the mouth-brooding species, possess cement cells which aid the free embryos/larvae in attaching to different substrates. Cement cells were first described for African anabantids by Morike (1977) for *Ctenopoma damasi* (Poll & Damas) and *Ctenopoma muriei* (Boulenger). Staining showed that there are numerous large cement cells on the dorsal and frontal parts of the head and the dorsal half of the eyes for the two *Sandelia* species, however, there are no cement cells on the yolk sac as in some anabantoids (Britz & Cambray 2001). When embryos of *S. bainsii* hatch, the free embryos adhere to the substrate at the nest site where they are guarded by the male. When the free embryos disperse from the nest site, aided by the positive buoyancy of the oil globule, the cement cells on the head of *S. bainsii* would be effective in keeping the developing larval fish off the substrate and attached to rocks or vegetation. In still waters when the fish are attached they are motionless, which would be an effective predator avoidance trait and would also conserve energy for development. If there is a strong water flow the larvae can be seen floating in the water current and the cement glands effectively keep the free embryo attached to rocks or vegetation. This may be important in the wild when the nests are located in the main stream flow as the rivers in the Eastern Cape are subject to flash floods. When the larvae begin to disperse they attach to plants or other substrates above the substratum.

Thus, in a sanctuary it would be important to have good aquatic vegetation substrate or other submerged structures such as logs close to the spawning structures for the developing larvae to attach to and hide from predators after the male finishes guarding the young. These features would increase the survival rates of the larval fish.

9.3 Characteristics of a *Sandelia bainsii* sanctuary

In setting up sanctuaries for *S. bainsii*, consideration has to be given to the habitat in which the species has evolved its apomorphic reproductive traits. Mayekiso and Hecht

(1988a) suggested that *S. bainsii* mainly occur in the middle and upper middle reaches of rivers. Most of the rivers in which this species occurs have rocky reaches interspersed with large riverine pools. The rivers are subject to erratic flows and high turbidity during these periods. However, most of these rivers have undergone changes in the past 100 years. For example, the Kowie River has had many anthropogenic changes. A major problem was experienced in the early 1990s with the invasion of red water fern, *Azolla filiculoides* Lam. (Cambray 1993), which spread along the system. The impact of this weed on the aquatic biota was compounded by nutrient loading, which fuelled its rampant growth, and water abstraction by farmers, which reduced the flushing flow in the river. These factors combined to make the river suitable for the rapid colonisation of the riverine pools by the weed. Fortunately a biological control, a weevil, *Stenopelmus rufinasus* Glllenhal. was introduced in 1999 and this has effectively controlled the weed (Hill & Cilliers 1999). During this period of invasion populations of *Sandelia* and one of its prey fish species, *Barbus pallidus* Smith, were drastically reduced along the system.

Thus sanctuaries out of the main stream may be a solution, but these would not be riverine environments. However, the sanctuaries would in some ways mimic the riverine pools as they would be mainly small ponds. In addition, main instream areas will need to be monitored for recruitment.

No angling should be permitted, especially during the spawning season October to February, when the males may be aggressively guarding their nests. It has been found that in the past when *S. bainsii* was more abundant that they were a bycatch for people night angling for eels (*Anguilla* spp.).

Males require suitable spawning substratum which needs to be a hard substrate such as rocks with surrounding and overhead structure (Fig. 9.3). This structure provides a territory and if they are placed around the sanctuary at a depth of 50 cm, they would provide the males with good spawning areas. Cambray (personal observation) has

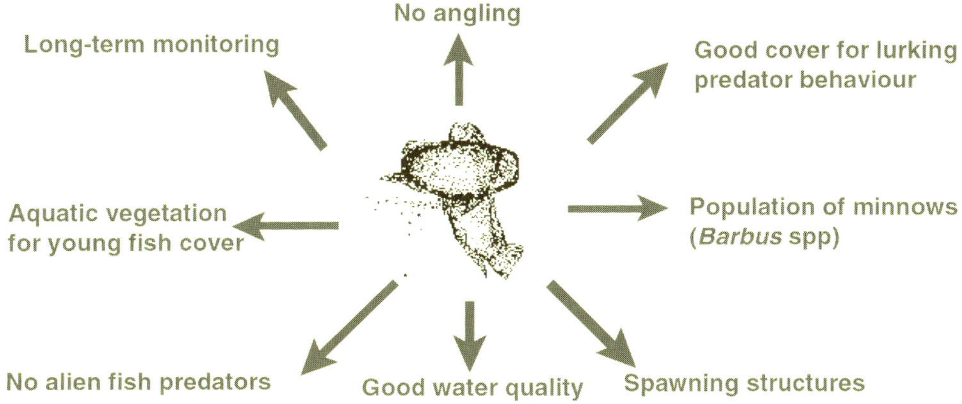

Figure 9.3 Some of the essential characteristics of a *S. bainsii* sanctuary

noted that individual females will mate with more than one male which would be advantageous considering the probably low numbers of fish available to be stocked into the sanctuaries.

Long-term monitoring of the sanctuary areas would be essential, including spawning success, shelter usage during spawning, forage fish populations, juvenile fish cover and food as well as the genetics of the population. It has been suggested that some populations have already passed through a genetic bottleneck (Lorenz *et al.* 1995). At present there is more ongoing genetic research on the species from all the remaining populations which will help when sanctuaries are established.

In addition to sanctuaries, further research is required on all river systems which still have viable populations of this species. There is an urgent need for the government of the Eastern Cape Province to allocate funds and manpower to prevent the extinction of this large African anabantid.

Acknowledgements

The National Research Foundation of South Africa is thanked for financial support and a special travel grant which made it possible for me to participate at the Freshwater Fish Conservation Conference in Portugal, October 2000. WWF-SA has supported the project through the Ubuntu Welfare and Development Trust. Companies such as Total SA, Mazda, Eheim and Tetra are also thanked for their contributions to the project. I would also like to thank various aquarist groups, such as the Anabantoid Association of Great Britain and the International Labyrinth Society for their positive support of the *S. bainsii* conservation campaign. Ms Eve Cambray is thanked for her editorial skills, fieldwork and encouragement.

References

Berns S. & Peters H. (1969) On the reproductive behaviour of *Ctenopoma muriei* and *Ctenopoma damasi* (Anabantidae). In Report East African Freshwater Fisheries Organization, Jinja, pp. 44–49.

Bok A. (1983) *The demography, breeding biology and management of two mullet species (Pisces: Mugilidae) in the eastern Cape, South Africa.* PhD Thesis, Rhodes University, Grahamstown, 258 pp.

Breder S.P. & Rosen P.W. (1966) *Reproductive Behaviour in Fishes.* New York: TFH Publications, 458 pp.

Britz R. & Cambray J.A. (2001) Structure of egg surfaces and adhesive organs in anabantoids. *Ichthyological Exploration of Freshwaters* **417**.

Cambray J.A. (1978) *A contribution to the character phylogeny of the Anabantidae: with particular reference to the respiratory organs of* Sandelia bainsii. Hon. Project, Rhodes University, Grahamstown, 115 pp.

Cambray J.A. (1980) Contact organs in fishes of the genus *Sandelia* (Anabantidae). *South African Journal of Science* **76**, 277–279.

Cambray J.A. (1990) The early ontogeny of the Cape kurper, *Sandelia capensis* (Pisces: Anabantidae). *Annals of the Cape Provincial Museums (Natural History)* **18**, 159–182.

Cambray J. (1993) Blaauwkrantz nature reserve, community involvement and the Eastern Province rocky (*Sandelia bainsii*). *The Naturalist* **37**, 8–10.

Cambray J.A. (1994) The comparative reproductive styles of two closely related African minnows (*Pseudobarbus afer* and *P. asper*) inhabiting two different sections of the Gamtoos River system. *Environmental Biology of Fishes* **41**, 247–268.

Cambray J.A. (1995) *Sandelia bainsii* will it be next? *Ichthos* **45**, 2–3.

Cambray J.A. (1996) Threatened fishes of the world: *Sandelia bainsii* Castelnau, 1861 (Anabantidae). *Environmental Biology of Fishes* **45**, 150.

Cambray J.A. (1997a) The spawning behaviour of the endangered Eastern Cape rocky, *Sandelia bainsii* (Anabantidae), in South Africa. *Environmental Biology of Fishes* **49**, 293–306.

Cambray J.A. (1997b) Captive breeding and sanctuaries for the endangered African anabantid *Sandelia bainsii*, the Eastern Cape rocky. *Aquarium Sciences and Conservation* **1**, 159–168.

Cambray J.A. (2000) Community involvement and the Eastern Cape rocky (*Sandelia bainsii*, Pisces, Anabantidae) from the perspective of a researcher. *South African Journal of Wildlife Research* **30**(3), 114–117.

Elsen M. (1976) La vessie gazeuse et l'organe labyrinthique des anabantidae. *Bulletin de Académie r. de Belgique* **62**, 49–79 (in French).

Hill M.P. & Cilliers C.J. (1999) *Azolla filiculoides* Lamarck (Pteridophyta: Azollaceae), its status in South Africa and control. *Hydrobiologia* **415**, 203–206.

Lorenz M., Ribbink A.J. & Kirby R. (1995) A measure of the genetic variability in *Sandelia bainsii* and *Sandelia capensis* using crude proteins and RAPDs. *Zoological and Aquatic Sciences in South Africa*, ZSSA & SASAQS Symposium, Rhodes University, Grahamstown, p. 93 (Abstract).

Mayekiso M. (1986) Some aspects of the ecology of the Eastern Cape rocky, *Sandelia bainsii* (Pisces: Anabantidae) in the Tyume River, Eastern Cape, South Africa. Unpublished MSc Thesis, Rhodes University, Grahamstown, 115 pp.

Mayekiso M. & Hecht T. (1988a) Conservation status of the anabantid fish, *Sandelia bainsii* in the Tyume River, South Africa. *South African Journal of Wildlife Research* **18**, 101–108.

Mayekiso M. & Hecht T. (1988b) Age and growth of *Sandelia bainsii* Castelnau (Pisces: Anabantidae) in the Tyume River, Eastern Cape (South Africa). *South African Journal of Zoology* **23**, 295–300.

Mayekiso M. & Hecht T. (1990) The feeding and reproductive biology of a South African (sic) Anabantid fish, *Sandelia bainsii*. *Revue Hydrobiologie tropicale* **23**, 219–230.

Morike D. (1977) Vergleichende Untersuchungen zur Ethologie zweier Labyrinthfischarten, *Ctenopoma muriei* (Boulenger 1906) und *Ctenopoma damasi* (Poll 1939) (Anabantoidei: Pisces). Doctoral Dissertation, Tubingen University, 178 pp.

Nelson J.S. (1994) *Fishes of the World*, third edition. New York: John Wiley and Sons, 600 pp.

Norris S.M. & Teugels G.G. (1990) A new species of *Ctenopoma* (Teleostei: Anabantidae) from South Eastern Nigeria. *Copeia* **1990**, 472–479.

Siegfried W.R. (1963) Observations on the reproduction and feeding of the Cape Kurper *Sandelia capensis* in the De Hoop Lake, Bredasdorp. Department of Nature Conservation Inv. Report, Cape Town, No. 3, 5–12.

Skelton P.H. (1993) *A Complete Guide to the Freshwater Fishes of Southern Africa*. Halfway House, South Africa: Southern Book Publishers (Pty) Ltd., 388 pp.

Chapter 10
Aspects of the ecology, parasites and future conservation of twaite shad, *Alosa fallax*, and allis shad, *Alosa alosa*, in south-eastern Ireland

D. DOHERTY* and T.K. McCARTHY
Zoology Department, National University of Ireland, Galway, Ireland

Abstract

Twaite shad, *Alosa fallax*, and allis shad, *Alosa alosa*, both occur in Irish coastal waters, although only the former species has been confirmed to reproduce in Ireland. Relatively little is known about these species in Ireland, so samples of twaite shad ($n = 66$) and allis shad ($n = 12$) from the Waterford area in southern Ireland were used to investigate the morphometric and meristic characteristics, as well as the parasitology and general biology. The two species could be distinguished by meristic characters or by a combination of shape-related morphometric characters. Similarities in the diets and parasite assemblages of the two species were found. The potential threats to both the twaite and allis shads in Ireland, where they are categorised as 'vulnerable' and 'endangered', respectively, were noted.

Keywords: allis shad, conservation, Ireland, parasites, twaite shad.

10.1 Introduction

The genus *Alosa* is represented in the British Isles by twaite shad, *Alosa fallax* Lacépède, and allis shad, *Alosa alosa* L. These species have not been extensively studied in Irish waters. O'Maoileidigh, Cawdrey, Bracken & Ferguson (1988) examined the landlocked dwarf Killarney twaite shad, referred to as *Alosa fallax killarnensis* by Tate Regan (1911) from south-western Ireland, and made a comparison with the anadromous twaite shad from the south-eastern River Barrow. Fahy (1982) analysed variation in bycatch of *Alosa fallax* from net fisheries in the Wexford estuary. There have also been occasional records from locations around the Irish coast (Fig. 10.1) and these have been referred to in general accounts of Irish rare and endangered fishes (Bracken & Kennedy 1967; Maitland 1993; Whilde 1993). However, many of these records are historical and there is poor understanding of their current distribution. In Ireland, the only known spawning population of anadromous twaite shad is found at the upper tidal limit of the River Barrow.

*Correspondence: Zoology Department, National University of Ireland, Galway, Ireland (email: denis.doherty@mail.esb.ie).

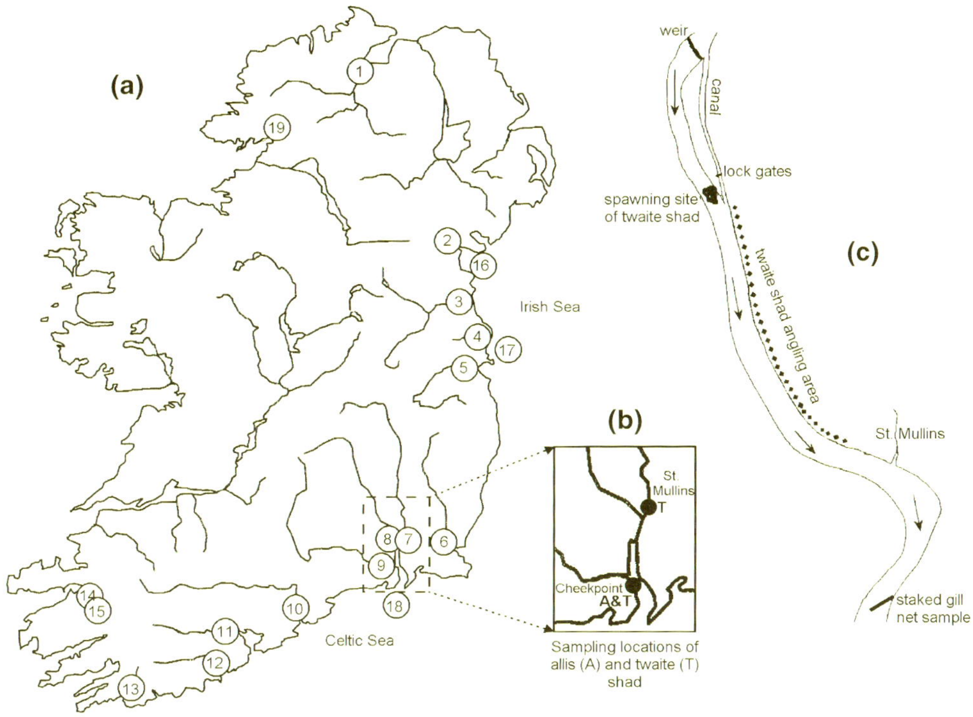

Figure 10.1 (a) Distribution of twaite shad recorded in Ireland: 1, River (R.) Foyle; 2, Castletown R.; 3, R. Boyne; 4, Nanny R.; 5, R. Liffey; 6, R. Slaney; 7, R. Barrow; 8, R. Nore; 9, R. Suir; 10, R. Blackwater (Munster); 11, R. Lee; 12, R. Bandon; 13, R. Ilen; 14, R. Laune; 15, Lough Leane; 16, Dundalk Bay; 17, East of Lambay Island; 18, Off Waterford Harbour; 19, R. Eske. Allis shad have been captured occasionally from the Irish and Celtic Seas (Bracken & Kennedy 1967). (b) The location of the sampling sites for allis and twaite shad near Cheekpoint on the Waterford estuary and from St Mullins on the River Barrow. (c) A detailed map of the St Mullins area on the R. Barrow, which shows the area fished during the annual shad fishing competition, the spawning site below the weir and the location from which a staked gill netted sample was obtained

According to Whilde (1993), allis shad are considered 'endangered', while twaite shad, are 'vulnerable'. Both species are much less common than formerly over their European range, including Ireland (Maitland 1993). European populations of allis and twaite shad are in decline due to overfishing in estuaries, obstruction of river migration by pollution, dams and weirs, and habitat destruction by pollution and engineering works.

The present study aimed to:

(1) compare morphometric and meristic characteristics of anadromous *Alosa alosa* and *Alosa fallax* from south-eastern Ireland;
(2) to present new data on their diets and parasite fauna; and
(3) to review their conservation status in Irish waters.

10.2 Materials and methods

10.2.1 *Study area*

The combined Barrow, Suir and Nore river system is the second largest in Ireland, with a catchment area of 9028 km^2 (Fig. 10.1) and a long-term mean discharge of 123 $m^3 s^{-1}$, draining the greater part of south-eastern Ireland (Lucey 1998). The River Barrow is the second longest Irish river, after the River Shannon, and has a catchment area of 2350 km^2, and a mean discharge of 40 $m^3 s^{-1}$. There are no lakes of any significance in the catchment. The length of the freshwater section of the River Barrow is 142 km, with an additional 48 km of estuary. The River Barrow is canalised with 23 lock and weir gates. The spawning area of twaite shad is immediately downstream of the lowest weir at the upper tidal limit near St Mullins (Fig. 10.1). The area of river bank fished by anglers is also indicated and from where some samples of twaite shad were obtained.

Despite the varied geology of the south-eastern region of Ireland the River Barrow is for the most part, floored with limestone and consequently has relatively calcium-rich waters. Waterford Harbour is Ireland's third most important commercial cargo port, which makes it an important area for human activity.

10.2.2 *Sampling*

A total of 66 (19 males and 47 females) twaite shad and 12 (4 males and 8 females) allis shad were sampled. Twaite shad were caught ($n = 33$) by angling at St Mullins on the River Barrow (Fig. 10.1) during the 1996 and 1997 spawning seasons, or by multimesh gill nets ($n = 27$) in 1997. An additional six twaite shad were obtained from salmon drift net fishermen (using a stretched mesh size of 119–125 mm) in Waterford estuary during 1996, together with 12 allis shad taken during 1996, 1997 and 1998. All fish were labelled and deep-frozen within 4 h of landing for later examination.

10.2.3 *Data collection and analysis*

A total of 25 morphometric measurements (Table 10.1) were recorded to the nearest 0.01 mm for each fish, using the same methodology as Boulva (1972) and O'Maoileidigh *et al.* (1988). The meristic characters recorded were gill raker number, branchiostegal ray number and fin ray counts (dorsal, pectoral, pelvic and anal). Scales were removed and stored for ageing according to Aprahamian (1988) and O'Maoileidigh *et al.* (1988).

To allow for the overall effect of fish size, all the morphometric measurements were transformed for fish size using the equation of Senar, Leonart & Metcalfe (1994): $Y_i' = \log_{10} Y_i - b(\log_{10} L_i - \log_{10} X)$, where $L_i = LF$ for fish i, LF, the fork length, X, mean LF (all fish), Y_i', size corrected morphometric variable value for fish i, Y_i, the uncorrected variable for fish i, and b, the pooled regression coefficient of $\log_{10} Y$ on $\log_{10} LF$ for all fish combined.

Table 10.1 Summary of the morphometric and meristic characters (n = 25) (measured nearest 0.01 mm) for twaite and allis shad sampled from the River Barrow and Waterford estuary

		Twaite shad, n = 66				Allis shad, n = 12			
		Mean	Standard error	Minimum	Maximum	Mean	Standard error	Minimum	Maximum
Morphometric characters									
Total length	TL	401.5	11.4	231.2	480.9	501.5	11.4	476.8	522.2
Standard length	SL	321.0	6.3	177.3	396.6	411.1	8.7	363.3	462.3
Head length	HL	79.8	1.1	55.5	96.6	95.7	1.8	83.6	103.6
Head width at nostrils	HW (nos)	17.7	0.3	11.5	22.8	20.6	0.4	18.5	23.6
Head width at eyes	HW (eyes)	21.8	0.5	12.9	31.4	27.4	0.6	23.0	30.2
Eye diameter	ED	12.8	0.2	9.1	15.9	14.6	0.3	13.5	16.6
Gape	Gape	31.2	0.5	22.1	44.0	37.4	0.5	35.2	40.9
Mouth height	MH	37.8	0.6	25.5	46.7	50.3	1.2	43.6	57.5
Mouth width	MW	29.6	0.5	19.4	36.9	31.8	0.9	27.6	37.3
Body height at insertation of pectoral fin	BHPF	79.4	1.3	51.5	100.4	98.2	5.0	49.2	118.8
Caudal peduncle height	CPH	26.8	0.4	12.9	32.9	33.9	0.5	31.5	36.2
Caudal peduncle width	CPW	13.4	0.3	8.1	25.7	17.0	0.4	14.6	19.7
Caudal peduncle length	CPL	33.3	0.9	16.3	65.9	45.3	5.1	33.4	99.4
Insertation of pelvic fin to insertation of anal fin length	PFAFL	83.6	1.9	23.3	102.7	97.5	6.1	39.2	126.0
Insertation of dorsal fin to insertation of ventral fin length	DFVFL	108.7	15.4	41.9	1101.6	129.1	9.9	110.4	236.5

Table 10.1 continued

		Twaite shad, $n = 66$				Allis shad, $n = 12$			
		Mean	Standard error	Minimum	Maximum	Mean	Standard error	Minimum	Maximum
Insertation of dorsal fin to insertation of ventral fin width	DFVFW	42.4	1.1	26.7	86.5	56.5	1.1	50.3	62.1
Base of dorsal fin length	BDFL	48.9	1.2	29.2	64.2	62.7	3.0	56.8	69.2
Post-orbital	PO	48.3	1.1	29.3	59.2	59.1	2.8	53.0	66.2
Snout length	SnL	13.3	0.5	8.4	19.2	19.7	3.0	11.2	24.9
Pectoral fin length	PFL	54.1	1.2	35.4	67.2	65.6	2.5	60.2	71.5
Pelvic fin length	PelFL	35.9	0.6	23.5	41.2	41.0	1.3	37.5	43.0
Base of anal fin length	BAFL	53.4	1.3	34.0	72.6	77.7	2.9	70.8	85.1
Predorsal fin length	PreDFL	157.6	3.8	98.8	197.4	183.1	8.6	166.7	198.7
Prepelvic fin length	PrePelFL	158.9	3.3	108.2	191.8	192.1	10.1	167.9	216.7
Preanal fin length	PreAFL	244.3	5.2	165.0	293.0	291.3	14.2	265.7	322.6
Meristic characters									
Branchiostegal ray number	BrStR	8.9	0.1	7.0	9.0	8.6	0.2	7.0	9.0
Gill raker number	GR	42.1	0.2	39.0	45.0	127.4	1.2	120.0	133.0
Dorsal fin ray count	DFR	17.8	0.1	17.0	20.0	18.5	0.3	18.0	19.0
Pectoral fin ray count	PecFR	16.0	0.2	14.0	18.0	15.3	0.3	15.0	16.0
Pelvic fin ray count	PelFR	9.0	0.0	8.0	9.0	9.0	0.0	9.0	9.0
Anal fin ray count	AFR	20.6	0.1	19.0	22.0	22.8	0.6	21.0	24.0

A list of the abbreviations used is given in Fig. 10.3 and some of their locations are shown in Fig. 10.2.

The effectiveness of transformation was checked by ensuring that the size corrected variables were not themselves correlated with fork length. Principal component analysis (PCA) was used to analyse the transformed morphometric data. This highlights the possible heterogeneity between and among populations.

For the meristic characters, the significance of inter- and intra-population differences were initially tested using a one-way analysis of variance. Subsequently discriminant function analysis was also applied to the transformed morphometric data to:

(1) classify the fish into one of severally mutually exclusive groups;
(2) establish the most important characters for distinguishing between the two species; and
(3) calculate the percentage overlap between each group based upon the generalised distance (Mahalanobis, D^2).

Fish samples, stored at −20°C, were dissected and examined for parasites following Conneely and McCarthy (1984). Dietary analysis followed standard protocols (Hyslop 1980) and the results are presented as percentage numerical and gravimetric taxonomic composition.

10.3 Results

In the present study, the twaite shad ranged from 202 to 429 mm in length, 182 to 1178 g in weight and 2+ to 7+ years in age. Allis shad were larger and older, ranging from 386 to 504 mm in length, 988 to 1775 g in weight and 5+ to 9+ in age. A Mann Whitney U test ($U = 55.5$, $P \leq 0.05$), showed that male twaite shad were significantly younger than females with a mean age of 2.95 years, compared to 5.6 years for females. The growth for female and male twaite shad (Fig. 10.2) was similar to that recorded for some French anadromous populations (Quignard & Douchement 1991b). Due to the low number of allis shad sampled it was not possible to determine a growth curve. The high proportion of female twaite shad ($n = 47$ out of 60 fish sampled) may indicate that spawning had commenced at the time of sampling. Eight out of 12 migrant, non-mature allis shad sampled from Waterford estuary, were also female fish.

Pre-spawned River Barrow twaite shad samples had a higher gonadosomatic index (GSI) of 16.5 ± 7.3 (18.8 ± 7.5 for females and 11.4 ± 2.9 for males) than the post-spawned estuarine sample 3.1 ± 4.2 (3.6 ± 4.5 for females and 0.4 ± 0 for males). Mature females on the spawning site at St Mullins on the River Barrow had a mean age of 4.1 years compared to younger mature males which were 2.95 years old. Post-spawned estuarine-captured female twaite shad were older, at 4.8 years, than 3.0 years for male fish.

Considerable differences were found between each morphometric characters recorded for each species (Table 10.1). With the exception of pelvic fin ray count, all meristic measurements take for the two species were significantly different (ANOVA, $P < 0.05$) Sexual dimorphism occurred in the two shad species, the females being significantly larger in size.

PCA was initially used for the 25 untransformed morphometric characters. Due to the length composition of both species samples and since all the variables (untransformed)

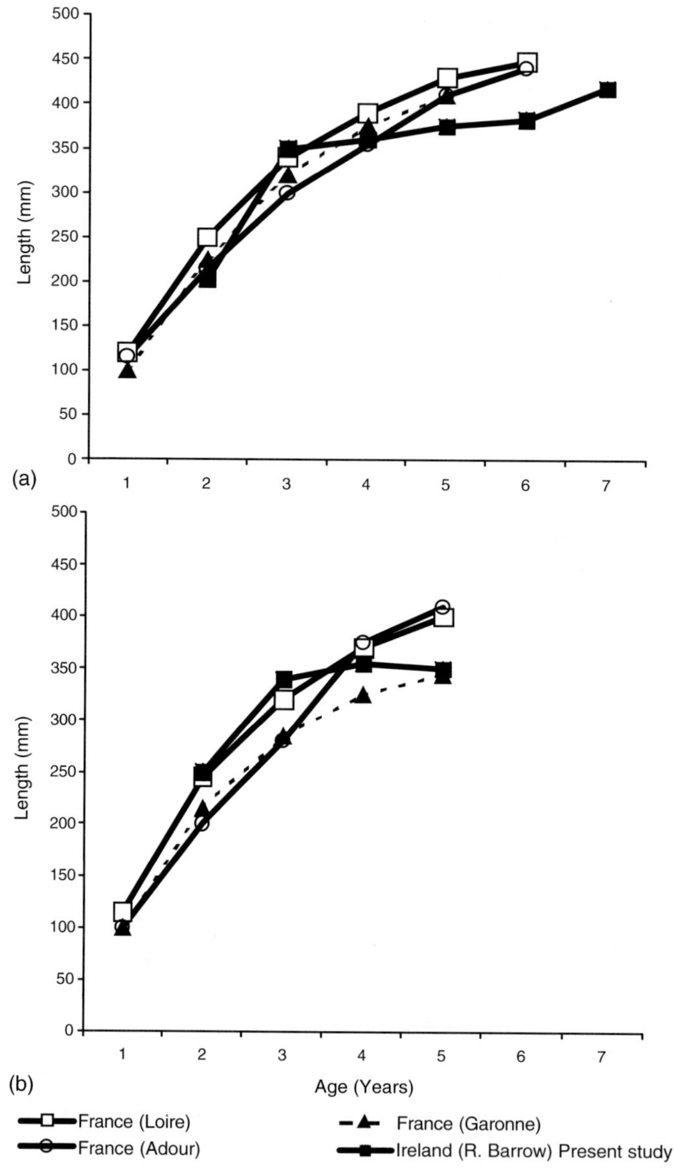

Figure 10.2 Growth of some European populations (Quignard & Douchement 1991b) of female (a) and male (b) twaite shad compared to those from the Waterford estuary and the River Barrow

were correlated with fish length, the component coefficients for the first principal component were all of the same sign and similar magnitude. Components having all coefficients of the same sign can be regarded as being indicative of size variation, whereas any component having both positive and negative coefficients is indicative of shape variation (Jolicoeur & Mosimann 1960). Thus, the first principal component accounted for size variation between twaite and allis shad. Due to the small sample size it was not possible to compare morphometric characters of both species within the same length range.

PCA was then used for the transformed morphometric characters and indicated that principal component 1 coefficients were both positive and negative highlighting shape variation between species (Fig. 10.3).

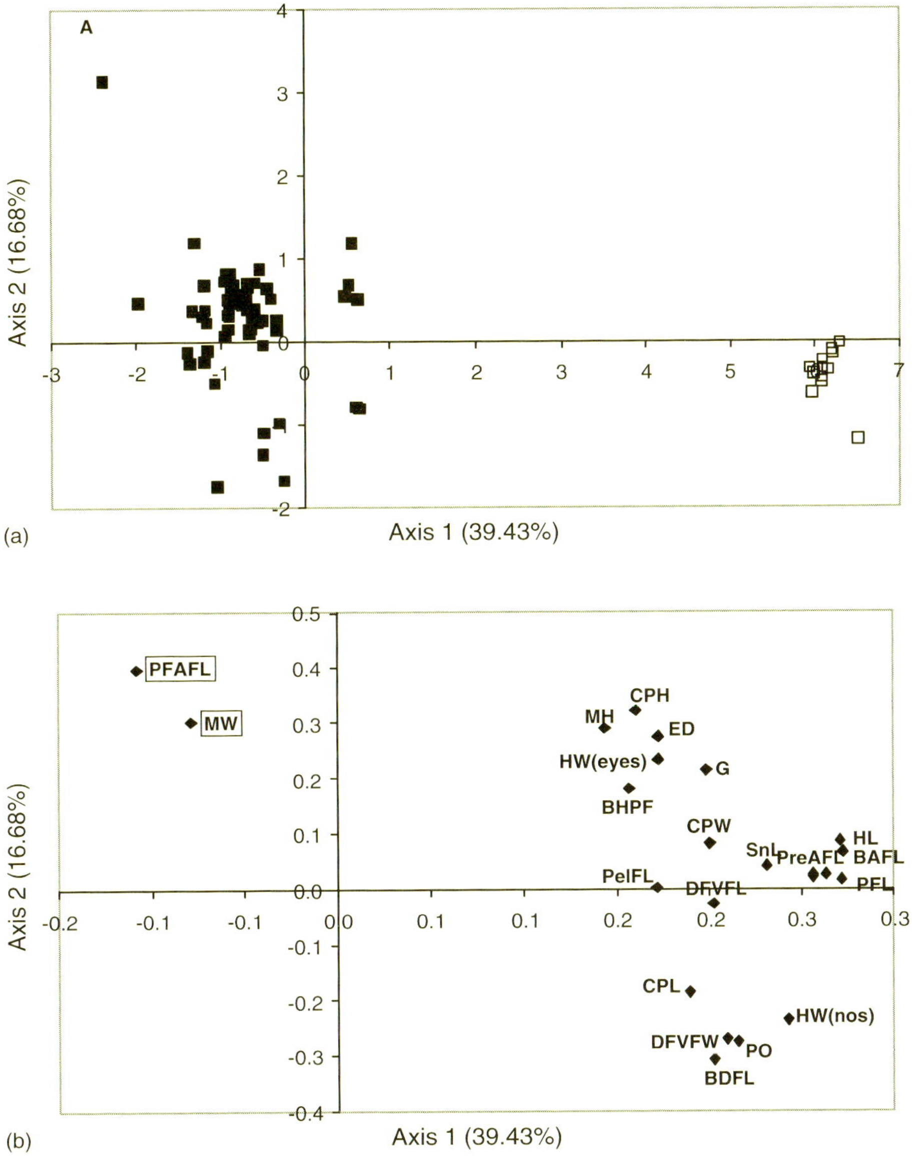

Figure 10.3 (a) PCA of transformed morphometric data from samples of twaite (n = 66) and allis (n = 12) shad. (b) Representation of each morphometric character relative to axis 1 and 2 (for abbreviations see Table 10.1)

Linear discriminant function analysis was carried out on the transformed morphometric data with fish being classified *a priori* to their respective species samples (Table 10.2). The overall percentage of fish classified correctly according to the discriminating function was 100% with a Mahalanobis squared distance (D^2) between the two samples of 1813.25. The function coefficients (Table 10.2), for each transformed morphometric character showed that caudal peduncle length (CPL), the insertation of the pelvic fin to the insertation of the anal fin length (PFAFL) and caudal peduncle fin height (CPH), were the most important characters in discerning twaite from allis shad. The smaller size of the function coefficients of other characters indicates that the addition of many of these characters was trivial.

The dietary analyses indicated a variation in feeding intensity as none of the spawning twaite shad ($n = 60$) were feeding, whereas all of the post-spawned twaite shad ($n = 6$) sampled from Waterford estuary were actively feeding upon marine mysids *Praunus neglectus* (G.O. Sars), which formed 99.9% by number and 96.8% by weight of the stomach contents. The remainder of the diet was made up by small sprats

Table 10.2 Discriminant function analysis showing the function coefficients of each morphometric character (excluding total and standard length) and the group constant value

Morphometric character	Twaite shad	Allis shad	Function coefficients
Head length	−8.9	−16.6	7.7
Head width at nostrils	−181.4	−310.9	129.5
Head width at eyes	−559.8	−521.6	−38.2
Eye diameter	5.9	−9.8	15.7
Gape	568.4	839.4	−271.0
Mouth height	572.2	656.5	−84.3
Mouth width	−451.2	−730.8	279.6
Body height at insertation of pectoral fin	227.5	208.9	18.6
Caudal peduncle height	1755.2	2562.6	−807.4
Caudal peduncle width	−932.0	−1434.0	502.0
Caudal peduncle length	1865.2	2828.5	−963.3
Insertation of pelvic fin to insertation of anal fin length	1819.2	2697.2	−878.0
Insertation of dorsal fin to insertation of ventral fin length	211.0	335.1	−124.1
Insertation of dorsal fin to insertation of ventral fin width	626.9	957.9	−331.0
Base of dorsal fin length	2.5	3.8	−1.3
Post-orbital	57.0	83.7	−26.7
Snout length	−216.9	−308.2	91.3
Pectoral fin length	79.9	84.7	−4.8
Pelvic fin length	26.7	64.9	−38.2
Base of anal fin length	78.5	126.0	−47.5
Predorsal fin length	15.4	42.2	−26.8
Prepelvic fin length	5.4	16.9	−11.5
Preanal fin length	−7.7	−11.8	4.1
Constant	−3708.2	−7705.2	3997.0

The Mahalanobis D^2 value between the groups was 1813.25.

Sprattus sprattus (L.). The relatively larger allis shad, although fewer in number ($n = 12$), also fed extensively upon the marine mysid *Praunus neglectus*. Piscivory was important when stomach weights were considered, as three marine fish species (whiting *Merlangus merlangus* (L.), sprats and smelt, *Osmerus eperlanus* L.), were recorded. A small number of shrimps *Crangon allmani* Kinahan were also recorded.

The parasite fauna was similar for both *Alosa* species (Table 10.3). Eight species were recorded for *Alosa alosa*, and of these, all co-occurred along with *Eubothrium fragile* Rudolphii in *Alosa fallax*. Differences in levels of infection are summarised in Table 10.3. No acanthocephalan parasite species were recorded from either twaite or allis shad. Differences were noted at the infra-community level in the parasite assemblages of the two shad species. Twaite shad harboured between 3 and 8 parasite species, $\bar{x} = 6.1 \pm 1.1$, compared to allis shad with 2 and 5, $\bar{x} = 3.3 \pm 1.2$. Likewise parasite burdens differed between the two species of shad with twaite shad harbouring between 43 and 1156 parasites, $\bar{x} = 435.4 \pm 261.6$, compared to allis shad with 42 and 436, $\bar{x} = 149.3 \pm 112.7$. None of the parasite abundances were significantly correlated with host length, weight or age for either shad species. Some differences were noted between the anadromous spawning twaite shad sampled from freshwater compared with those from the estuary (Table 10.3).

10.4 Discussion

Growth of Irish twaite shad is broadly similar to other European populations, although Quignard and Douchement (1991b) reported that growth rates and longevity of both twaite and allis shad varied with locality. The ages and weights of the small sample of Waterford allis shad indicated that these shad were the slowest growing and smallest in length when compared with other European allis shad populations (Quignard & Douchement 1991a). However, this may be due to the small sample size or because these fish were occurring at the edge of their European range. The age profiles of twaite and allis shad sampled from the River Barrow and Waterford estuary showed that the allis shad were larger and older (5–9 years) fish compared to the twaite shad (2–7 years).

It is doubtful that a spawning allis shad population occurs within either the Nore, Suir or Barrow rivers, that discharge into Waterford Harbour as no mature allis shad were captured and no documented or anecdotal evidence is known to exist. The allis shad samples in the present study most likely represent transitory non-spawning migrant fish, with the nearest known reproducing population being recorded in northern and western France (Quignard & Douchement 1991a).

According to Quignard & Douchement (1991b), the GSI of French and Portuguese twaite shad populations increased rapidly 1 or 2 months prior to spawning with high values reported at spawning time due to the loss of somatic weight. In the present study significant differences occurred in the GSI between twaite shad sampled at the spawning site compared to those sampled later in the season within Waterford estuary. Quignard and Douchement (1991b) stated that twaite shad GSI may differ from year to year and from site to site (e.g. males (and females) of the Loire River presented 7.8 (12.2); Ardour River 9.1 (16.4); Rhône River 7.7 (18.7)).

Table 10.3 Infection parameters for parasites recorded from twaite and allis shad sampled from the River Barrow and Waterford estuary

	Spawning freshwater twaite shad ($n = 60$)	Post-spawned estuarine twaite shad ($n = 6$)	Twaite shad ($n = 66$)			Allis shad ($n = 12$)		
	Abundance	Abundance	Abundance	Prevalence	Mean intensity	Abundance	Prevalence	Mean intensity
Monogenea								
Mazocraes alosae Hermann	43.4	58.2	48.6	44.7	44.7	83.7	100.0	83.7
Digenea								
Pronoprymna ventricosa Rudolphii	353.1	76.2	327.9	100.0	327.9	2.7	16.7	16.0
Diplostomum gasterostei Williams	1.3	7.2	2.0	1.8	7.0	0.5	16.7	3.0
Diplostomum spathaceum Rudolphii	10.5	8.0	11.0	10.3	12.1	0.8	33.3	2.3
Hemiurus appendiculateus Rudolphii	5.6	5.0	5.6	5.5	7.6	27.9	41.7	67.0
Nematoda								
Hysterthylacium osculatum	7.6	2.2	7.6	7.1	11.5	2.8	41.7	4.9
Hysterthylacium aduncum Rudolphii	36.9	9.2	34.7	34.4	42.8	30.9	83.3	37.1
Cestoda								
Eubothrium fragile Rudolphii	1.1	0.5	1.2	1.0	1.9	Absent	Absent	Absent
Crustacea								
Clavellisa emarginata Kroyer	2.6	2.5	3.9	2.6	10.2	0.1	8.3	1.0

In the twaite and allis shad sampled significant differences were noted in the meristic character sets. Size differences did occur, the allis shad being larger, but a combination of shape differences in the morphometric character sets allowed for differentiation between the two shad species on the basis of their morphometric and meristic character sets. O'Maoileidigh *et al.* (1988), in a comparative study of the two forms of Irish twaite shad reported that apart from the dwarfed appearance of the lacustrine Killarney shad and the increased number of gill rakers carried on the first branchial arch, a broad overlap was noted in the meristic character sets. Furthermore, no consistent shape differences were observed between these two forms of twaite shad.

The variation in feeding intensity of spawning French populations of twaite and allis shad recorded by Quignard and Douchement (1991a, b), and previously noted in River Barrow twaite shad by O'Maoileidigh (1989), was also evident in the present study. Despite the large number of invertebrates ingested, piscivory appeared to be particularly important in larger specimens of both species. In the River Severn, twaite shad were found to feed upon a range of prey items with a preference for fish and secondary items including mysids, shrimps, isopods, detritus and other aquatic insects (Aprahamian 1989). According to Quignard and Douchement (1991a), large allis shad are predatory planktonic feeders with the main food being copepods (*Calanus* sp.), decapods (*Pandalus* sp.), euphasids (*Meganyctiphanes* sp.) and young fish. Bracken & Kennedy (1967) reported the diet of adult twaite and allis shad captured in the Irish Sea consisted of plankton and small euphasids.

The lack of work in marine Irish parasitology is illustrated by the limited published data and the parasite species recorded in this investigation. All are new Irish host records although they all have been previously reported from European twaite and allis shad populations. Differences noted in parasite species infection levels of the two shad species may be due to the specificity of the parasite or aspects of the host fishes such as dietary preferences or movements at sea. Parasite species recorded in the present study which were specific to the shad are: *Pronoprymna ventricosa*, *Hemiurus appendiculateus*, *Eubothrium fragile*, *Mazocraes alosae* and *Clavellisa emarginata*. The intestinal cestode, *Eubothrium fragile*, which was absent from the Waterford estuary allis shad may either be specific to twaite shad or may have a limited geographical distribution in the Irish Sea (Kennedy 1981). Variation in levels of infection of *Hemiurus appendiculateus* may arise from differential predation upon copepods, their intermediate hosts. Likewise variation in feeding on amphipods acting as intermediate hosts for *Pronoprymna ventricosa* may account for the high levels of this species in twaite shad (Dogiel, Petrushveski & Polyanski 1961; Bray & Gibson 1980). The adult nematode *Hemiurus aduncum* and the encysted nematode *Hemiurus osculateum* (final hosts are marine mammals such as seals), recorded in both shad species, may be acquired through feeding upon various crustaceans (such as *Praunus neglectus*), or through ingestion of paratenic fish hosts (Moravec 1989). The two gill parasites *Mazocraes alosae* and *Clavellisa emarginata* recorded in the present study for both shad species have direct life cyles and thus are acquired through the activities of the parasite and the host fish. Aprahamian (1985) noted the absence of skin parasites except for an occassional *Caligus elongatus*, which were recovered from a small proportion of the River Severn twaite shad spawning population. No external skin parasites were recorded on either species in the present study.

In Ireland, twaite and allis shad are not, like many other rare and endangered fish species, protected under any specific national legislation with the most relevant aimed at the conservation of fish as an exploitable resource rather than as an element of the native fauna. Twaite and allis shad are occasionally captured by commercial salmon drift net fishermen (March–July), within Waterford estuary and for the most part are returned unharmed to the estuary. A relatively large annual catch (0.4 t) of twaite shad was reported by Fahy (1982), on the River Slaney in County Wexford. Although adequate information is not available it appears that the bycatch is not to this level in Waterford estuary. In the River Slaney, twaite shad are captured by ring and snap nets which use smaller mesh sizes (44–125 mm) compared to the drift nets (119–125 mm) used within Waterford estuary. In Ireland generally, the capture of either shad species as bycatch by netsmen is not usually reported. Thus, the extent of the migrant estuarine twaite and allis shad elsewhere is unknown.

Evidence concerning the annual fluctuations in abundance of the twaite shad population can be obtained by analysis of records of the annual twaite shad competition on the River Barrow, which involves a catch and return policy, and those of the Irish Specimen Fish Committee (ISFC). The ISFC, managed by the Central Fisheries Board, maintains records of large rod caught fish, referred to as specimen (>0.964 kg), twaite shad. Angling between March and June (99.7% of the catch), influenced by tidal cycles, indicated considerable annual variation in the number ($n = 283$) of specimen twaite shad reported captured ($\bar{x} = 13.5 \pm 13.96$, range 0–41) during 1980 to 2000. No specimen allis shad (>1.814 kg) were recorded in Irish waters by the ISFC in the past 45 years (Anonymous 2000).

During the first half of the 20th century, twaite and allis shad were the main commercial riverine clupeids in Europe with high catches, >100 t, in the North and Baltic Sea basins. Nowadays they have declined or disappeared from many areas of Europe (Assis 1990; Bervoets, Coeck & Verheyen 1990; Maitland & Lyle 1990; Alexandrio 1996; Nicola, Elvira & Almodóvar 1996; Elvira, Nicola & Almodóvar 1998). The general decline of the shads throughout their European range has been attributed to overfishing in estuaries, obstruction of river migration, habitat destruction by pollution and engineering works. Threats to the Irish fish fauna include deteriorating water quality, species introductions and inadequate, or difficult to enforce, legislation. These threats, especially water quality problems, apply to the localised River Barrow twaite shad spawning population. The significance of bycatch levels for both species of shad within Waterford estuary and in adjacent coastal waters are difficult to quantify as no reliable records are kept.

Acknowledgements

The assistance of the Zoology staff of the National University of Ireland, Galway, the Central Fisheries Board, and the Cheekpoint drift net fishermen is gratefully acknowledged. Parasite identifications were verified by staff of the Natural History Museum, London.

References

Alexandrio P. (1996) Genetic and morphological differentiation among some Portuguese populations of allis and twaite shad *Alosa alosa* (L. 1758) and twaite shad *Alosa fallax* (Lacépède, 1803). *Publicaciones Especiales, Instituto Espanol de Oceanografia* **21**, 15–24.

Anonymous (2000) Irish Specimen Fish 2000. *Annual Report of the Irish Specimen Fish Committee*. Dublin: Central Fisheries Board, Ireland, pp. 1–56.

Aprahamian M.W. (1985) The effect of the migration of *Alosa fallax fallax* (Lacépède) into freshwater, on branchial and gut parasites. *Journal of Fish Biology* **27**, 521–532.

Aprahamian M.W. (1988) The biology of the twaite shad *Alosa fallax fallax* (Lacépède), in the Severn estuary. *Journal of Fish Biology* **33**(Suppl. A), 141–152.

Aprahamian M.W. (1989) The diet of juvenile and adult twaite shad *Alosa fallax fallax* (Lacépède) from the Rivers Severn and Wye (Britain). *Hydrobiologia* **179**, 173–182.

Assis C.A. (1990) Threats to the survival of anadromous fishes in the River Tagus, Portugal. *Journal of Fish Biology* **37**(Suppl. A), 225–226.

Bervoets L., Coeck J. & Verheyen R.F. (1990) The value of lowland rivers for the conservation of rare fish in Flanders. *Journal of Fish Biology* **37**(Suppl. A), 223–224.

Boulva J. (1972) Morphometrics of three sympatric Arctic codfishes of the genera *Arctogadus* and *Gadus*. *Journal of the Fisheries Research Board Canada* **29**, 243–249.

Bracken J. & Kennedy M. (1967) Notes on some Irish estuarine and inshore fishes. *Irish Fisheries Investigations* Series B **3**, 1–28.

Bray R.A. & Gibson D.I. (1980) The Fellistomidae (Digenea) of fishes from the NE Atlantic. *Bulletin of the British Museum of Natural History (Zoology)* **37**, 199–293.

Conneely J.J. & McCarthy T.K. (1984) The metazoan parasites of freshwater fishes in the Corrib catchment area, Ireland. *Journal of Fish Biology* **24**, 363–375.

Dogiel V.A., Petrushveski G.K. & Polyanski Y.J. (1961) *Parasitology of Fishes*. Edinburgh: Oliver & Boyd Ltd, 384 pp.

Elvira B., Nicola G.G. & Almodóvar A. (1998) A catalogue of fish passes at dams in Spain. In M. Jungwith, S. Schmutz & S. Weiss (eds) *Fish Migration and Fish bypass Channels*. Oxford: Fishing News Books, Blackwell Science, pp. 203–207.

Fahy E. (1982) A commercial net fishery taking twaite shad *Alosa fallax* (Lacépède) in the estuary of the River Slaney. *Irish Naturalists Journal* **20**, 498–500.

Hyslop E.J. (1980) Stomach content analysis – a review of methods and their application. *Journal of Fish Biology* **17**, 411–429.

Jolicoeur P. & Mosimann J.E. (1960) Size and shape variation in the painted turtle: a principal component analysis. *Growth* **24**, 691–699.

Kennedy C.R. (1981) The occurrence of *Eubothrium fragile* (Cestoida: Pseudophyllidae) in twaite shad, *Alosa fallax* (Lacépède), in the River Severn. *Journal of Fish Biology* **19**, 171–178.

Lucey J. (1998) The Barrow, the Nore and the Suir. In C. Moriarty (ed.) *Studies of Irish Rivers and Lakes*. Dublin: Marine Institute, Ireland, pp. 99–117.

Maitland P.S. (1993) Threatened freshwater fishes of the British Isles, with special reference to Ireland. In J.D. Reynolds (ed.) *The Conservation of Aquatic Systems*. Dublin: Royal Irish Academy, pp. 84–100.

Maitland P.S. & Lyle A.A. (1990) Practical conservation of British fishes: current action on six declining species. *Journal of Fish Biology* **37**(Suppl. A), 255–256.

Moravec F. (1989) *Parasitic Nematodes of Freshwater Fishes in Europe*. Amsterdam: Kluwer Academic Publishers, 473 pp.

Nicola G.G., Elvira B. & Almodóvar A. (1996) Dams and fish passage facilities in the large rivers of Spain: effects on migratory species. *Archive für Hydrobiologia* **113**(Suppl.), 375–379.

O'Maoileidigh N. (1989) *A study of fish populations in the Killarney lakes*. Unpublished PhD Thesis, University College, Dublin, Ireland, 307 pp. (plus illustrations).

O'Maoileidigh N., Cawdrey S., Bracken J.J. & Ferguson A. (1988) Morphometric, meristic character and electrophoretic analyses of two Irish populations of twaite shad, *Alosa fallax* Lacépède. *Journal of Fish Biology* **32**, 355–366.

Quignard J.P. & Douchement C.L. (1991a) *Alosa alosa* (Linnaeus 1758) In H. Hoestlandt (ed.) *The Freshwater Fishes of Europe*, Vol. 2. Germany: GmbH Wiesbaden, AULA-Verlag, pp. 89–126.

Quignard J.P. & Douchement C.L. (1991b) *Alosa fallax fallax* (Lacépède 1803) In H. Hoestlandt (ed.) *The Freshwater Fishes of Europe*, Vol. 2, Germany: GmbH Wiesbaden, AULA-Verlag, pp. 225–253.

Senar J.C., Leonart J. & Metcalfe N.B. (1994) Wing variation between resident and transient wintering siskins *Carduelis spinus*. *Journal of Avian Biology* **25**, 50–54.

Tate Regan C. (1911) *The Freshwater Fishes of the British Isles*. London: Metheun & Co. Ltd, 207 pp.

Whilde T. (1993) *Threatened Mammals, Birds, Amphibians and Fish in Ireland. Irish Red Data Book 2: Vertebrates*. Belfast: UK, HMSO, 224 pp.

Chapter 11
Growth characteristics of an anadromous and a landlocked population of allis shad, *Alosa alosa*, from the River Mondego, Portugal

L.S. GORDO*
Departamento de Zoologia e Antropologia & Instituto de Oceanografia, Faculdade de Ciências, Lisboa, Portugal

Abstract

Specimens of *Alosa alosa* from a landlocked population (Aguieira reservoir) and an anadromous population from the River Mondego, Portugal, were sampled between February 1998 and November 1999. No difference in length was detected between sexes in the landlocked population, but females from the anadromous population attained a larger size than males. All fish were aged based on otolith readings and von Bertalanffy growth equations were estimated separately for males and females. Growth rates were statistically different between populations and between the sexes from the anadromous population. No significant differences were detected between sexes in the landlocked population.

Keywords: allis shad, *Alosa alosa*, anadromous, landlocked populations, Portugal.

11.1 Introduction

Allis shad, *Alosa alosa* (L.), is an anadromous species with a wide distribution area in the Northeast Atlantic, occurring from Norway and the British Isles southward to Morocco and including the Mediterranean coasts of Spain, France and Algeria (Lelek 1987). This widespread distribution has been reduced in most countries because of anthropogenic activities (damming, overfishing, channelisation and poor water quality) that affect the upstream riverine migration, preventing this species from reaching its spawning grounds (Maitland 1995). Allis shad has now disappeared from Germany (Meyer & Brunken 1997), the Netherlands (de Groot 1989) and Norway (Quignard & Douchement 1991). A self-sustaining population no longer exists in the UK (Aprahamian & Aprahamian 1990), although a number of projects have been initiated to conserve the species (Maitland 1995).

In France, it still occurs in the Loire (Boisneau, Mennesson-Boisneau & Baglinière 1990), Garonne (Le Corre, Linhares, Castro, Alexandrino, Sabatié & Baglinière 1998)

*Correspondence: Departamento de Zoologia e Antropologia & Instituto de Oceanografia, Faculdade de Ciências de Lisboa, Bloco C 2, Campo Grande, 1749-016 Lisboa, Portugal (email: lsgordo@fc.ul.pt).

and Adour (Prouzet, Martinet & Badia 1994) basins located in the west and southwest of the country. In Portugal, allis shad populations still occur in the Minho, Lima, Mondego and Tagus rivers (Alexandrino 1996), but have been steadily declining due to construction of dams without fish passages facilities, river pollution and the destruction of spawning grounds by increased sediment deposition.

Until recently, the existence of exclusively freshwater populations of allis shad were limited to two landlocked populations: one in Castelo do Bode reservoir (Tagus basin, Portugal) (Eiras 1981, 1983) and another in Lake El Kansera, Morocco (Furnestin 1952) which according to Sabatié (1993) is now extinct. However, another landlocked population has been identified in the Aguieira reservoir (Mondego catchment, Portugal) (Collares-Pereira, Cowx, Luis, Pedrosa & Santos-Reis 1999). The Aguieira dam, located about 70 km from the mouth river, was built in 1981 and enclosed the landlocked population.

The biology of landlocked populations of lampreys and several fish families, e.g. Salmonidae and Osmeridae, is reasonably well known (McDowall 1988). However, little comparative information exists on the biological features of landlocked and anadromous populations in the same river, and this is the main goal of this chapter with respect to the allis shad populations of the River Mondego.

11.2 Materials and methods

Alosa alosa from the landlocked population (Aguieira reservoir) and the anadromous population, which exists downstream of Aguieira dam (called the Mondego population), were sampled monthly between February 1998 and November 1999. Seine nets (20-mm knot to knot mesh size) were used to catch the juveniles, whilst trap and trammel nets (40 and 100 mm) were used to capture sub-adults and adults. Only 77 specimens were available for biological purposes from the anadromous population because of problems with the fishermen, but 242 individuals were sampled from Aguieira reservoir.

In the laboratory, all individuals were measured to the nearest millimetre below (later grouped in 1-cm length classes) and weighted to the nearest 0.1 g. Sagittal otoliths were removed, cleaned and stored in plastic tubes. Finally, sex was recorded. Since this species is known to exhibit sexual dimorphism, age data was separated by sex.

Otoliths were read under a stereoscope microscope with reflected light. Length at age was determined by back-calculation based on otolith measurements taken between the focus and the edge, as well as the distances between the focus and the successive annuli were recorded. The relationship between the total length of the fish (L_t) and the otolith radius (OR) was established by linear regression. To model fish growth, von Bertalanffy growth equations (VBGE) were applied by sex and area using the iterated least square method developed in the software FiSAT (Gayanilo, Sparre & Pauly 1995). Hotelling's T^2 test (Bernard 1981) was used to compare statistically the growth parameters estimated for each sex and area. The length–weight relationship was also estimated by sex and area.

11.3 Results

11.3.1 *Length composition*

In the anadromous population, the smallest fish caught was 8 cm total length and the largest 83 cm, with females attaining a larger length than the males (Fig. 11.1). However, due to the small number of fish caught, many length classes were poorly represented or even absent. In Aguieira reservoir, the length distribution was basically bimodal, with a first group of individuals smaller than 24 cm and another group between 44 and 57 cm total length.

11.3.2 *Age and growth*

The relationships between otolith radius (*OR*, mm) fish length (L_t, mm) for each population and sex separately were:

(a) Mondego females $L_t = 517.08OR - 474.07$, $r = 0.96$, $n = 29$,
males $L_t = 399.91OR - 282.05$, $r = 0.96$, $n = 30$.

(b) Aguieira females $L_t = 309.03OR - 167.31$, $r = 0.97$, $n = 116$,
males $L_t = 313.59OR - 176.75$, $r = 0.98$, $n = 126$.

These relationships were used to determine the back-calculated length for age (Table 11.1) and the von Bertalanffy growth equations for each population and sex as follows:

(a) Mondego females $L_t = 112.06\,(1 - \exp(-0.177\,(t - 0.139)))$, $r = 0.99$, $n = 29$,
males $L_t = 73.98\,(1 - \exp(-0.264\,(t + 0.349)))$, $r = 0.99$, $n = 30$.

(b) Aguieira females $L_t = 66.17\,(1 - \exp(-0.304\,(t + 0.472)))$, $r = 0.99$, $n = 116$,
males $L_t = 67.03\,(1 - \exp(-0.289\,(t + 0.507)))$, $r = 0.99$, $n = 126$.

Comparing the growth curves, no significant differences were found between sexes for Aguieira reservoir ($T^2 = 7.507$, $T_0^2 = 9.94$, $P > 0.05$) but growth of females from the anadromous population was statistically different from that of males ($T^2 = 431.52$, $T_0^2 = 10.12$, $P < 0.05$). Statistical difference in growth rate were also found when females and males from Mondego were compared with the females and males from Aguieira reservoir ($T^2 = 1323.8$, $T_0^2 = 10.04$, $P < 0.05$ for the females; $T^2 = -393.8$, $T_0^2 = 9.997$, $P < 0.05$ for the males).

The total length (L_t) and the total weight (W_t) relationship for each species from each population were:

(a) Mondego females $W_t = 0.0042\,L_t^{3.2009}$, $r = 0.99$, $n = 40$,
males $W_t = 0.0036\,L_t^{3.2485}$, $r = 0.99$, $n = 37$.

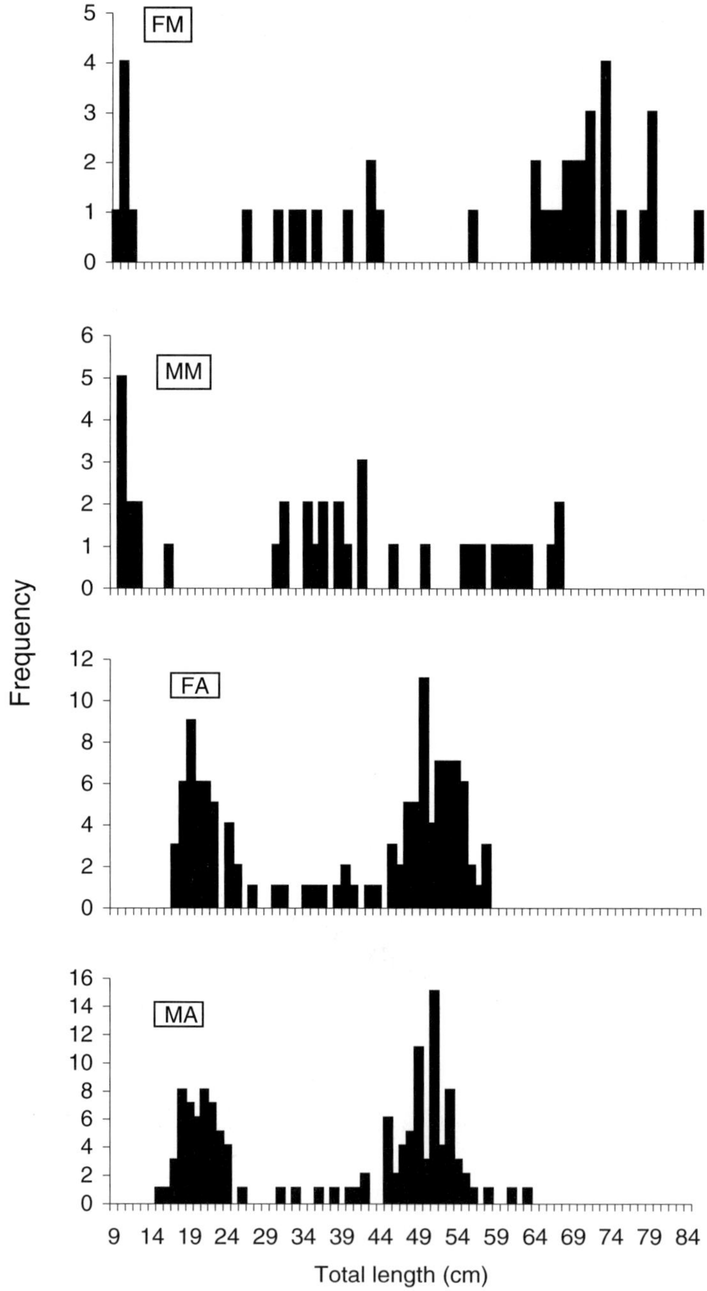

Figure 11.1 Length–frequency distribution of females and males of the anadromous population (FM and MM, respectively) and of females and males from Aguieira reservoir (FA and MA, respectively)

Table 11.1 Length for age (L_n (mm) ± standard deviation) for male and females allis shad in the River Mondego anadromous population and the landlocked Aguieira reservoir

	Mondego		Aguieira	
Age	Males	Females	Males	Females
1	223.0 ± 31.55	155.9 ± 51.96	236.7 ± 36.26	237.5 ± 36.57
2	341.5 ± 26.00	319.8 ± 35.74	345.2 ± 32.69	353.6 ± 37.71
3	436.0 ± 19.14	443.9 ± 37.87	430.0 ± 28.02	431.2 ± 26.41
4	504.0 ± 11.83	548.6 ± 29.62	484.8 ± 24.82	486.1 ± 28.62
5	560.2 ± 0.0	642.8 ± 27.92	536.2 ± 31.38	539.8 ± 26.30
6		733.6 ± 13.96	568.3 ± 0.0	
7		782.4 ± 0.0		

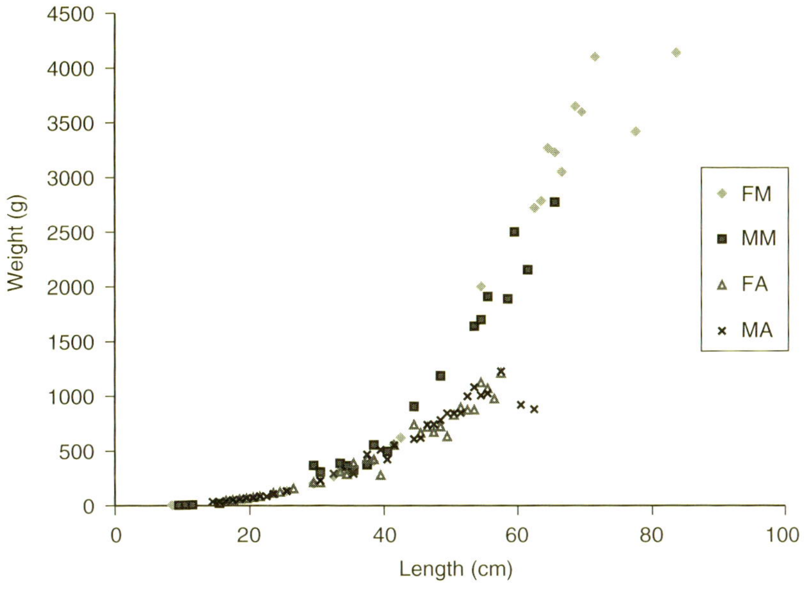

Figure 11.2 Length–weight relationships of females and males of the anadromous population (FM and MM, respectively) and of females and males from Aguieira reservoir (FA and MA, respectively)

(b) Aguieira females $W_t = 0.0289\,L_t^{2.6142}$, $r = 0.99, n = 116$,
males $W_t = 0.0254\,L_t^{2.6553}$, $r = 0.99, n = 126$.

Both females and males from the anadromous population were heavier for the same length than the landlocked females and males (Fig. 11.2), suggesting better condition.

11.4 Discussion

Allis shad have exhibited an accentuated sexual dimorphism in all anadromous populations studied to date (e.g. Eiras 1983; Sabatié, Alexandrino & Baglinière 1996). The Mondego anadromous population also showed sexual dimorphism, with the females attaining larger lengths than the males. However, in the landlocked population no sexual dimorphism was detected. Moreover, when the length distributions of the two populations were compared, few individuals were represented in the length classes between 25 and 43 cm (corresponding to a mean theoretical age of 2 and 3 years). In the anadromous population this lack of fish could be associated with there being few migratory fish of these sizes (Sabatié *et al.* 1996), since they would be immature and still at sea. However, no such problems should have been found with the landlocked population. The low representation of these age groups was probably due to poor recruitment in 1995, 1996 and 1997, possibly linked to adverse environmental conditions.

The maximum age of fish in Aguieira reservoir was 6 years. This age was similar to that found by Collares-Pereira *et al.* (1999) (5+ based on 35 specimens) but greater than the Castelo do Bode reservoir population (3 years and 53 specimens analysed (Eiras 1983); 4 years based on 29 individuals (Alexandrino 1996)). In the anadromous population of the Mondego River, the maximum age was 7 years, and complies with the findings of Alexandrino (1996) for the same river.

Data on age and growth corroborate the analysis of length frequency data, i.e. males and females from the landlocked population exhibited no differences in growth rate but females from the anadromous populations in the River Mondego grow faster than males.

Studies on landlocked populations of alewife *Alosa pseudoharengus* (Wilson) (Nigro & Ney 1982), dolly varden *Salvelinus malme* (Walbaum) (Blackett 1973), and lampreys *Lampetra planeri* (Bloch) and *Mordacia mordax* (Richardson) (Potter 1980) (to mention a few species), showed that landlocked individuals have a shorter life span, are faster growing and mature earlier than anadromous populations. These aspects were characteristic of the allis shad population of Aguieira reservoir, but the usual pattern of accentuated sexual dimorphism that occurs in anadromous populations is absent in this landlocked population, a point that has not been referred to in any studies dealing with landlocked populations.

The absence of sexual dimorphism of the landlocked population may be due to an adaptation to the new freshwater habitat. Moreover, these individuals exhibit poor condition when compared with anadromous fishes. This seems to be a characteristic of landlocked populations, *cf.* allis shad from Castelo do Bode reservoir (Eiras 1983). The poor condition of landlocked populations may be an adaptation to the new environment since they do not need to attain a larger weight (and consequently a better condition) that would allow them to perform the upstream migration to the spawning grounds. The limited food availability in the freshwater environment, or poor quality of the food items ingested, may also explain the poor condition of the individuals of landlocked populations.

It is commonly known that there are many factors that affect migratory fishes and are responsible for their decline in abundance. In the River Mondego construction of dams and habitat degradation are the two main negative factors. In Portugal, dams are essential for water supply, irrigation and hydropower generation, but fish passages are rarely found; among the 150 large dams that exist in the country, less than 12 have fish passages and these do not always function. Thus, to conserve the migratory fishes of the region, priority should be to construction of fish passes. This should be coupled with habitat improvement, especially around the spawning areas.

Finally, the existence of landlocked populations is so rare, despite being the result of human intervention, that their conservation should be a priority both for the state and local authorities. Special attention should be paid to protecting the Aguieira population as it is more vulnerable to anthropogenic factors, including low water levels in the reservoir, intense eutrophication or introduction of alien species such as large mouth bass, *Micropterus salmoides* (Lacépède). Meanwhile, environmental education actions should be implemented among the local fishermen communities to conserve the stocks and avoid overfishing, especially of the smaller individuals.

Acknowledgements

I would like to thank Célia Teixeira and M. João Correia for their help in the field work. This work was funded by Fundação para a Ciência e Tecnologia, project PRAXIS/3/3.2/CA/1981/95.

References

Alexandrino P.J.B. (1996) Estudo de populações de sável (*Alosa alosa* L.) e savelha (*Alosa fallax* Lacépède). Análise da diferenciação interespecífica, subestruturação e hibridação. PhD Thesis. University of Porto, 185 pp. (in Portuguese).

Aprahamian M.W. & Aprahamian C.D. (1990) Status of the genus *Alosa* in the British Isles; past and present. *Journal of Fish Biology* **37**(Suppl. A), 257–258.

Bernard D.R. (1981) Multivariate analysis as a mean of comparing growth in fish. *Canadian Journal of Fisheries and Aquatic Sciences* **38**, 233–236.

Blackett R.F. (1973) Fecundity of resident and anadromous dolly varden (*Salvelinus malma*) in southeastern Alaska. *Journal of the Fisheries Research Board of Canada* **30**, 543–548.

Boisneau P., Mennesson-Boisneau C. & Baglinière J.L. (1990) Description d'une frayère et compartement de reproduction de la grande alose (*Alosa alosa* L.) dans le cours supérieur de la Loire. *Bulletin Français de Pêche et Pisciculture* **316**, 15–23 (in French).

Collares-Pereira M.J., Cowx I.G., Luis T.S., Pedrosa N. & Santos-Reis M. (1999) Observations on the ecology of a landlocked population of allis shad in Aguieira Reservoir, Portugal. *Journal of Fish Biology* **55**, 658–664.

de Groot S.J. (1989) The former allis shad and twaite shad fisheries of the lower Rhine, The Netherlands. ICES CM1989/M **19**, 1–5.

Eiras J.G.C. (1981) Contribuição para o conhecimento da biologia de *Alosa alosa* L. Estudo de algumas modificações somáticas, fisiológicas e bioquímicas durante a migração anádroma no rio Douro. PhD Thesis, University of Porto, 228 pp. (in Portuguese).

Eiras J.G.C. (1983) Some aspects of the biology of a landlocked population of anadromous shad *Alosa alosa* L. *Publicações do Instituto de Zoologia Dr. Augusto Nobre* **180**, 1–16.

Furnestin J. (1952) L'alose d'El Kansera, cas d'une adaptation totale d'une espèce anadrome aux eaux douces. *Comptes rendues de l'Academie de Sciences de Paris* **234**, 2018–2020 (in French).

Gayanilo F.C.Jr., Sparre P. & Pauly D. (1995) The FAO-ICLARM Stock Assessment Tools (FiSAT) User's Guide. *FAO Computerized Information Series (Fisheries)* **8**, 1–126.

Le Corre M., Linhares D., Castro F., Alexandrino P., Sabatié R. & Baglinière J.L. (1998) Premiers éléments de caractérisation génétique de l'alose du Rhône (*Alosa fallax rhodanensis* Roule, 1924). *Bulletin Français de Pêche et Pisciculture* **350–351**, 635–645 (in French).

Lelek A. (1987) Threatened Fishes of Europe. In: A. Lelek (ed) *The Freshwater Fishes of Europe*, Volume 9, Wiesbaden: Aula-Verlag, pp. 1–343.

Maitland P.S. (1995) Freshwater fish of annexes II and IV of the EC Habitats directive. EEC contract B4-3040/94/000439/MAR/B2, 181 pp.

McDowall R.M. (1988) *Diadromy in Fishes: Migrations between Freshwater and Marine Environments*. London: Croom Helm, 308 pp.

Meyer L. & Brunken H. (1997) Historisches Vorkommen, aktuele Verbreitung und Einschätzung der zukünftigen Bestandsentwicklung wandernder Fischarten und Rundmäuler (Osteichthyes et Cyclostomata) im Allersystem (Niedersachsen). *Braunschweiger Naturkundliche Schriften* **5**, 281–303 (in German).

Nigro A.A. & Ney J.J. (1982) Reproduction and early-life accomodations of landlocked alewives to a southern range extension. *Transactions of the American Fisheries Society* **111**, 559–569.

Potter I.C. (1980) The Petromyzoniformes with particular reference to paired species. *Canadian Journal of Fisheries and Aquatic Sciences* **37**, 1595–1615.

Prouzet P., Martinet J.-P. & Badia J. (1994) Caractérisation biologique et variation des captures de la grande alose (*Alosa alosa*) par unité d'effort sur le fleuve Adour (Pyrénées Atlantiques, France). *Aquatic and Living Resources* **7**, 1–10 (in French).

Quignard J.P. & Douchement C. (1991) *Alosa alosa* (Linnaeus, 1758). In H. Hoestland (ed.) *The Freshwater Fishes of Europe*, Volume 2, Wiesbaden: Aula-Verlag, pp. 89–126.

Sabatié M.R. (1993) Recherches sur l'écologie et biologie des aloses du Maroc (*Alosa alosa* Linné, 1758 et *Alosa fallax* Lacépède, 1803). Exploitation et taxonomie des populations atlantiques; bioécologie des aloses de l'oued Sebou. PhD Thesis, University of Brest, 326 pp. (in French).

Sabatié M.-R., Alexandrino P. & Baglinière J.-L. (1996) Comparaison des caractéristiques biologiques de la grande alose (*Alosa alosa*) dans l'oued Sebou (façade nordatlantique du Maroc) et dans le fleuve Lima (Portugal). *Cybium* **20**, 59–73 (in French).

Chapter 12
Larval habitat of the sea lamprey (*Petromyzon marinus*) in the River Mondego (Portugal)

P.R. ALMEIDA*
Departamento de Biologia, Universidade de Évora, Évora, Portugal

B.R. QUINTELLA
Instituto de Oceanografia, Faculdade de Ciências, Universidade de Lisboa, Lisboa, Portugal

Abstract

The relationship between spatial distribution of sea lamprey, *Petromyzon marinus*, ammocoetes and environmental parameters was investigated between October 1998 and February 2000 in the River Mondego, Portugal, downstream from the Açude-Ponte dam. Larvae distribution was strongly dependent upon sediment characteristics, especially particle size composition. Smaller individuals (20 mm < total length (TL) ≤ 60 mm) were commonly found on silty, sand bottoms. Ammocoetes with a TL between 60 and 140 mm prefer a more heterogeneous substrate, with equal contributions of gravel and silt (gravel–silty-sand). Larger ammocoetes (140 mm < TL ≤ 200 mm) prefer coarse-grained sediments, substratum composed of sand or gravelly-sand. Finally, ammocoete abundance was higher in areas where sea lamprey nests were observed. The use of this kind of information in the formulation of conservation measures concerning the rehabilitation of river basins for sea lampreys is discussed.

Keywords: ammocoete, conservation, larval habitat, *Petromyzon marinus*, Portugal.

12.1 Introduction

Although the location of larval lamprey (i.e. ammocoetes) populations within a river system can usually be predicted with some accuracy by an experienced observer, it is more difficult to specify in precise physico-chemical terms the essential characteristics of an ammocoete habitat (Hardisty & Potter 1971).

Typical ammocoete habitat is an area protected from major fluctuations in water levels or stream flow, and where current velocity is usually slow. Such conditions are commonly found in eddies or backwaters or at bends in a river, where the accumulation of soft silt and sand, which provide a suitable substrate for the burrowing ammocoetes, is likely to occur. In such places, often partially shaded by trees, diatoms may form an incrustation on the interface between the silt and the water, probably contributing to the stability of these larval microenvironments, also called 'ammocoete beds' (Hardisty 1979). Above all, the existence of suitable conditions for ammocoete

*Correspondence: Departamento de Biologia, Universidade de Évora, 7002-554 Évora, Portugal (email: pmraposo@mail.telepac.pt).

colonisation is dependent on stream gradients, which will, in turn, determine the overall velocity of the current, the type of substrate particles that are deposited and also the accumulation of organic debris (Hardisty & Potter 1971).

Although the influence of the sediment particle size in the distribution of ammocoetes is generally accepted (Kainua & Valtonen 1980; Morman, Cuddy & Rugen 1980; Young, Kelso & Weise 1990a; Ojutkangas, Aronen & Laukkanen 1995), the question concerning the substrate preferences of the different larval length-classes is still a subject of debate. The main objective of this work was to determine the habitat preferences of different length-classes of sea lamprey, *Petromyzon marinus* L., ammocoetes in the River Mondego, Portugal.

This river supports a viable population of sea lamprey but is subject to a considerable stress from human activities, especially habitat modification. The sea lamprey's inability to negotiate the Açude-Ponte dam during the spawning migration has confined it to the lower 35 km of the river (Almeida, Silva & Quintella 2000). This makes the River Mondego sea lamprey population one of few populations of diadromous species to survive a severe reduction in its natural habitat. Knowledge of the environmental variables that influence ammocoete distribution in river basins will help improve the success of conservation actions concerning the rehabilitation of anadromous sea lamprey populations.

12.2 Materials and methods

12.2.1 *Study site*

The River Mondego is an important Portuguese river (234 km long), located in the central region of the country and with a river basin of approximately 6644 km^2 (Fig. 12.1) (Loureiro, Almeida, Machado & Teixeira 1986). It is a highly modified river, especially the 35 km downstream from the Açude-Ponte dam, with an 11 km, man-made stretch with 12 riffle areas, constructed some 40 years ago (Fig. 12.1). These riffles result from the presence of submerged stone weirs installed across the river to avoid downstream sediment carriage. This freshwater zone is also characterised by the presence of shallow water sectors with several small islands (Almeida *et al.* 2000).

The sampling area was selected along the last 35 km of the River Mondego in the freshwater zone, following a previous survey (October and November 1998), which identified the main ammocoete beds downstream from the Açude-Ponte dam (Fig. 12.1). A set of different habitats, representative of the observed variability present in a 830 $\times$ 125-m^2 river stretch, was selected to provide information on ammocoete habitat preferences. Physical limitations (i.e. depth and visibility) forced site selection to the river margins.

12.2.2 *Field sampling*

Samples were collected monthly, at seven sampling sites (total area of 175 m^2), from February 1999 to February 2000. In October 1999 sampling was not possible because

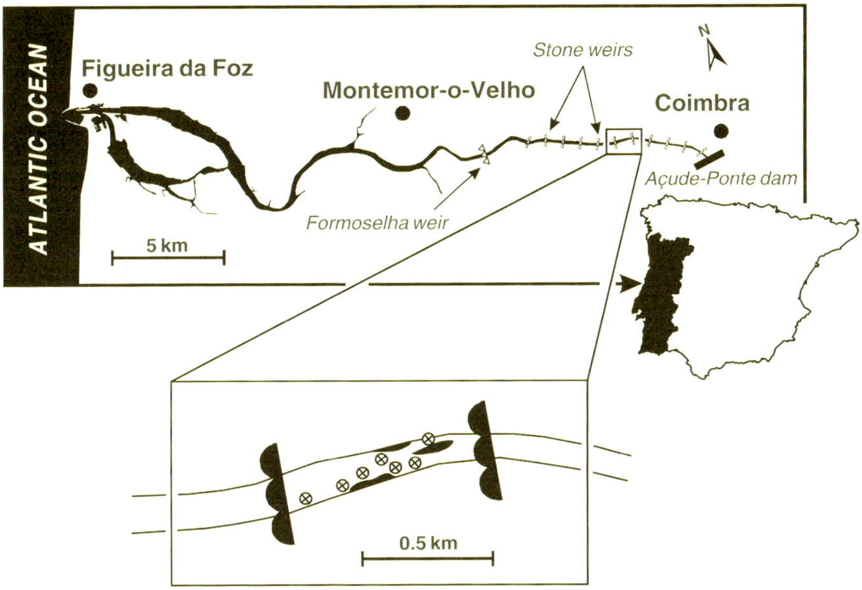

Figure 12.1 Location of the study area. Circles (⊗) indicate the position of the sampling sites in the study area

of floods. The ammocoetes were collected by electric fishing (Hans Grassl EL62, 600 V DC, 10 A). Fishing was possible only in areas where depth did not exceed 1.0 m. All animals caught in the first sweep of the sampling area were counted and the total length (TL) of each animal was measured to the nearest millimetre. At the end of each sampling session all ammocoetes were returned live to the river.

Catch per unit of effort (CPUE) was estimated as the number of animals caught per 10 min of fishing time. For each sampling site, temperature (°C), dissolved oxygen ($mg\,L^{-1}$), conductivity (μS) and total dissolved solids ($g\,L^{-1}$) were recorded, using a calibrated Datasonde Surveyor 4 (Hydrolab). Current velocity ($m\,s^{-1}$) was measured using a Valeport current metre (model 105). To determine the granulometric composition and the organic matter content (OMC) of the bottom substratum, the first 15 cm of sediment were collected using a corer (inner diameter: 12 cm) in each of the sampling sites during the 13 field campaigns.

12.2.3 *Laboratory analysis*

The sediment samples were dried at 60°C for 24 h, and a 100-g sub-sample (dry weight) was washed in a 63 μm sieve to determine the finer fraction (silt). The sediment retained in the sieve (sand and gravel) was dried again at 60°C for 24 h and passed through a column of two sieves (63 and 2000 μm) of the AFNOR type. The sediment fractions retained in each sieve were then weighed individually.

The sampling sites were classified according to the Roux scale as gravel (>2000 μm), sand (2000–63 μm) and silt (<63 μm) (Roux 1964), and represented in a Shepard diagram (Buchanon & Kain 1971).

The OMC (%) was estimated from the reduction in weight of a 70 g portion of sediment, after ignition at 480°C over 24 h (Almeida, Moreira, Costa, Assis & Costa 1993).

12.2.4 *Data analysis*

Detrended Canonical Correspondence Analysis (DCCA, Ter Braak 1987) was used to characterise the relation between the spatial variability of the ammocoetes length-classes (expressed in CPUE) in the River Mondego and the selected environmental gradients, i.e. percentage composition of gravel, sand, silt and OMC, and current velocity. The abiotic variables directly related to the water (i.e. temperature, dissolved oxygen, conductivity and total dissolved solids) were not used in the DCCA.

The result of this analysis is an ordination diagram where symbols represent the ammocoetes length-classes, and vectors correspond to the environment variables. The vectors indicate the direction of maximum variation of the correspondent environmental variable. Environmental variables, whose projection in a particular axis is extended, are strongly correlated with the referred ordination axis (Ter Braak 1987).

A specific length-class is related with a certain environmental variable by projection of its point on the correspondent vector, which represents approximately the optimum point of response of the length-class on the considered environmental variable (Ter Braak 1986).

Nine, 20-mm length-classes were considered within the interval 20–200 mm. The individuals falling into the 0–20 mm length-class were not used in this analysis, as electric fishing is inefficient at capturing of ammocoetes <20 mm.

The statistical significance of the relation between the ammocoete length-classes and the set of environmental variables was accessed through a Monte Carlo global permutation test (9999 permutations) (Ter Braak 1987).

12.3 Results

During the study the abundance of the ammocoetes changed within the sampling sites. As a result, in some of the monthly campaigns ammocoetes were absent from some sampling stations. The characteristics of the sampling site sediments also changed throughout the year, highlighting the dynamic nature of riverine ecosystems, and the response of sea lamprey larvae to their changing environment.

A high proportion (81%) of all ammocoetes ($n = 426$) caught were from sampling sites where sand was the most abundant component of the substratum. These samples corresponded to 79% of all sampling sites and could be broken down into four predominant particle size categories: sand (33%), gravel–silty-sand (18%), silty-sand (16%) and gravelly-sand (12%) (Fig. 12.2).

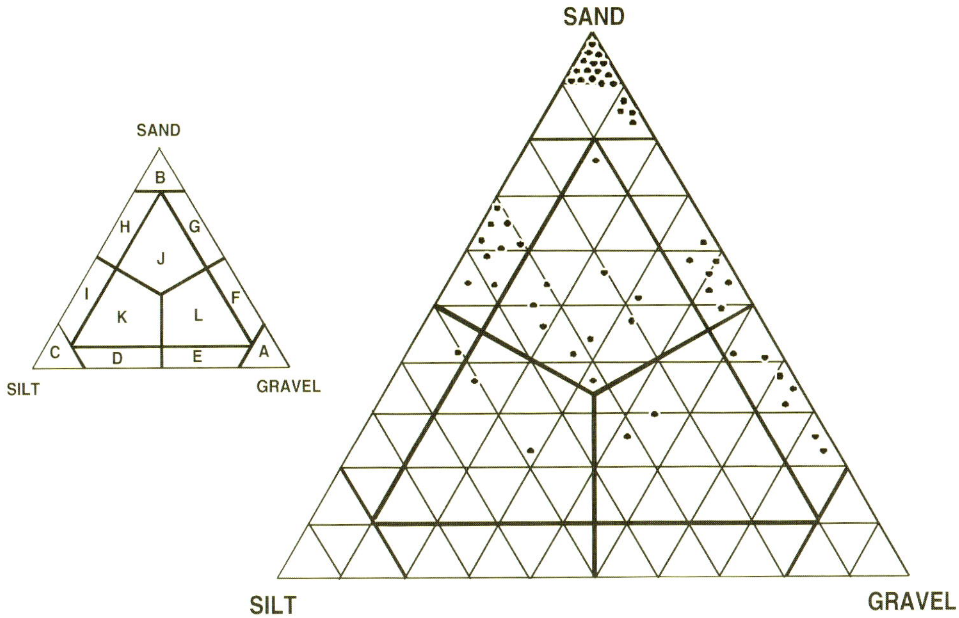

Figure 12.2 Shepard diagram of the sampling sites (CPUE > 0) based on the granulometric composition of their sediment. The reduced diagram indicates the position of the different types of sediments. A, gravel; B, sand; C, silt; D, gravelly-silt; E, silty-gravel; F, sandy-gravel; G, gravelly-sand; H, silty-sand; I, sandy-silt; J, gravel–silty-sand; K, gravel–sandy-silt; L, sand–silty-gravel

Table 12.1 Abiotic characteristics of the water–sediment interface and OMC of the sediment in the study area

	Mean	SD	Maximum	Minimum
Water temperature (°C)	17.1	5.0	24.9	8.4
Dissolved oxygen (mg L^{-1})	8.4	2.0	11.3	3.8
Total dissolved solids (g L^{-1})	0.08	0.01	0.10	0.07
Specific conductivity (μS)	120	10	150	100
Current velocity (cm s^{-1})	1	3	17	0
OMC (% dry weight)	1.2	3.5	8.5	0.3

Throughout the sampling period, water temperature and dissolved oxygen exhibited seasonal variation. During the summer, water temperature rose to a maximum of 24.9°C and fell to a minimum of 8.4°C in the winter. Dissolved oxygen reached a maximum of 11.3 mg L^{-1} during the winter and a minimum of 3.8 mg L^{-1} in the summer (Table 12.1). Total dissolved solids and conductivity exhibited little variation between sampling sites or with season, maintaining mean values of 0.08 g L^{-1} and 120 μS cm^{-1}, respectively. Current velocity was slow with a mean value of 10 cm s^{-1}

and varied between sites from 0 to 17 cm s^{-1}. OMC in the sediment reached a maximum value of 8.5%, but the substrata at the sampling sites were generally low in organic matter (1.2%).

The DCCA showed that different ammocoete length-classes preferred sediments with different particle size composition (Fig. 12.3). The first two DCCA ordination axes explained 12.3% of the population spatial variability, and 66.1% of the relation between length-class abundance and the selected environmental variables (i.e. gravel, sand, silt, OMC and current velocity) (Table 12.2).

The relatively high correlation coefficients between ammocoete and environmental variables, obtained for the first two axes (0.74 and 0.42 respectively), suggest that environmental variables explain the variability associated with ammocoete length-class distribution. The high correlation between silt with the first canonical axis, and gravel and sand with the second canonical axis, identified the main environmental gradients for each axis (Table 12.2). On the other hand, current velocity and OMC showed a low correlation with both axes, and therefore add only a small contribution to the ordination. The global permutation test (F-ratio = 6.096) showed that the first canonical axis was

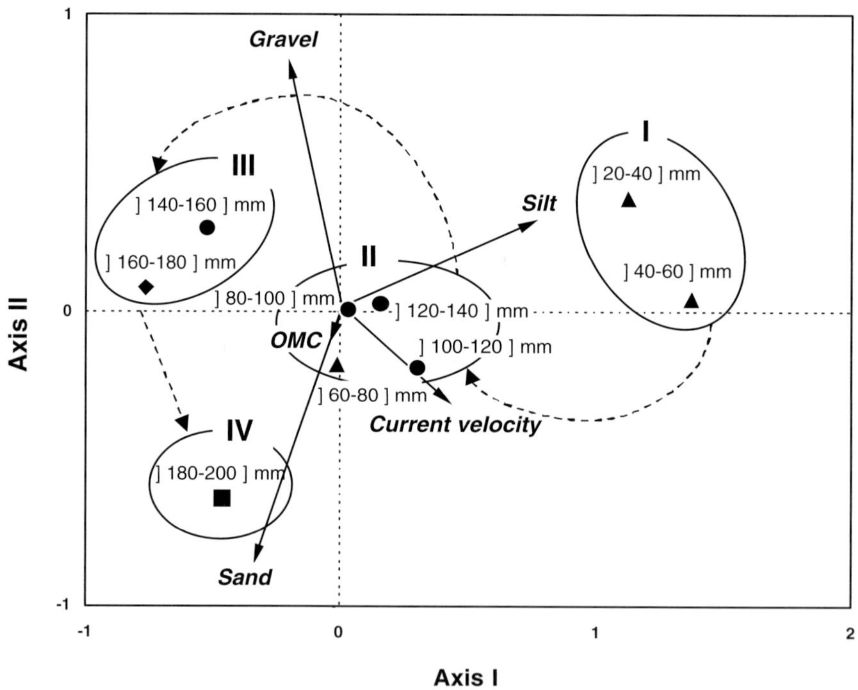

Figure 12.3 Sea lamprey ammocoete length-classes distribution according to the sediment particle size, OMC and current velocity. DCCA ordination diagram, with symbols corresponding to the nine length-classes, and arrows representing the environmental variables. The roman numerals identify the groups with similar preferences. The dash arrows represent the colonisation sequence of the different sediment type. Age-classes are also represented (▲, 1st year class; ●, 2nd year class; ◆, 3rd year class; ■, 3rd and 4th year class)

Table 12.2 Results of the ordination by DCCA of ammocoete length-classes data: eigenvalues, ammocoete environment correlation coefficients, and intraset correlation of environmental variables with the first two canonical axes

	Axis I ($\lambda_1 = 0.24$)	Axis II ($\lambda_2 = 0.05$)
Ammocoete environment	0.74	0.42
Gravel	−0.19	0.36
Sand	−0.30	−0.38
Silt	0.64	0.13
OMC	−0.03	−0.04
Current velocity	0.35	−0.14

statistical significant ($P < 0.01$). The test based on the sum of all canonical eigenvalues led to an F-ratio of 3.025, demonstrating that the relation between ammocoete length-classes and environmental variables was highly significant ($P < 0.001$).

The first and second canonical axes separated four groups (Fig. 12.3). The first group, comprising smaller individuals (20 mm < TL ≤ 60 mm), showed a preference for sediment with a higher percentage of silt. The selection of the second group (60 mm < TL ≤ 140 mm), corresponded to a more heterogeneous substrate, where gravel and silt seem to have an equal contribution to the sediment composition. According to the Roux scale, the former sediment type is included in the granulometric category silty-sand substrate, and the second may be classified as a gravel–silty-sand substrate (Fig. 12.3).

The larger ammocoetes (140 mm < TL ≤ 180 mm and 180 mm < TL ≤ 200 mm) preferred coarse-grained sediments. The former tended to occur in substrata where, besides sand, the gravel was a well representated component, while the ammocoetes belonging to the latter size group were most commonly found on sandy bottoms. In the Roux classification, the 140–180-mm long ammocoetes were associated with gravelly-sand substratum, while the 180–200-mm long larvae were strongly correlated with sediments where sand was dominant (Fig. 12.3).

Although some of the length-classes (e.g. 100 mm < TL ≤ 120 mm) seemed to prefer areas with higher current velocity (in the range 0.00–0.17 m s^{-1}), most of the animals showed a preference for locations with a low water velocity (Fig. 12.3).

Considering the age groups corresponding to each of the length-classes (unpublished data), the larger individuals of the first two age groups (60 mm < TL ≤ 80 mm and 140 mm < TL ≤ 160 mm) were associated with the length-classes of the following age groups (Fig. 12.3).

12.4 Discussion

An appropriate river substrate is an essential environmental characteristic for the development of larval lampreys, not only because it allows burrow construction, but also because it helps to maintain a vital water flux. The detritivorous larvae depend on a unidirectional flow of water through their branchial chamber, for the provision of food and exchange of respiratory gases and metabolic wastes (Hardisty & Potter 1971).

It is widely recognised that the availability of optimal river substrate particle size is one of the most important factors limiting the distribution of ammocoetes (Hardisty 1979; Kainua & Valtonen 1980; Morman *et al.* 1980; Potter 1980; Young *et al.* 1990a; Young, Houston, Weise & Kelso 1990b; Ojutkangas *et al.* 1995; Beamish & Lowartz 1996). This work confirmed this observation and showed that distinct ammocoete length-classes prefer sediments with different particle size composition. On the other hand, Beamish and Lowartz (1996) suggested that optimal particle size and abundance, for American brook lamprey, *Lampetra appendix* Dekay, did not change with larval length. Kainua and Valtonen (1980) also stated that the differences in the preferred particle size of the burrowing areas of different sized *Lampetra fluviatilis* (L.) ammocoetes, were small.

According to the present study, ammocoetes with a TL between 20 and 60 mm prefer silty-sand substrates, i.e. sediments with a comparatively higher percentage of sand, but also with an relatively large portion of silt. Considering the larval burrowing habit, it is understandable that smaller ammocoetes are usually associated with fine-grained sediments. Soft sediments allow younger larvae, with a reduced swimming capacity, to propel the head and branchial region below the surface. Potter (1980), stated that small ammocoetes are mainly present in habitats where the substrate is composed of fine particles, confirming this preference.

A gravel–silty-sand substrate was the selection of the 60–140-mm ammocoete length-class. In this substrate, gravel and silt seem to have an identical contribution to the composition of this more heterogeneous sediment. The larger body of the ammocoetes in this class gives them the opportunity to colonise a wider range of sediment types.

Kainua and Valtonen (1980) found the largest larvae preferred smaller particles (soft banks), whereas the larger ammocoetes (140 mm $<$ TL $\leqslant$ 200 mm) sampled in the River Mondego preferred coarse-grained sediments (gravelly-sand and sand), suggesting an inverse particle size selection with size.

The data suggest that the ammocoete distribution was associated with slow current areas, but since granulometric composition in the lotic environment is closely related with current velocity, it is probable that sediment particle size was the main environmental variable that determined the abundance of sea lamprey larvae. Young *et al.* (1990b) supported this observation by suggesting that ammocoete distribution is largely a function of sediment particle size or variables correlated with particle size, namely current velocity.

Since the selection of the burrowing sediment is a size-dependent characteristic, the differences observed in the preferences for distinct sediment types within the same age group, probably resulted from a reorganisation of the ammocoetes distribution pattern at the end of each annual growing season. This behaviour could be a strategy developed by this species to avoid high densities in the areas colonised by the younger individuals, and therefore reduce intraspecific competition for space and food.

Morman *et al.* (1980) indicated that ammocoete distribution was strongly associated with the spawning areas, the larval density being inversely related to the distance downstream from the spawning areas. This was also found in the River Mondego, and during a preliminary survey (October and November 1998) with the purpose of selecting the most convenient sampling area, it was observed that the abundance of ammocoetes was higher in areas around sea lamprey nests (unpublished data).

This study showed that particle size composition of the sediment is one of the most important factors that contributes to the success of an ammocoete population in a river basin. It is thus essential that river stretches classified as potential ammocoete beds, should comprise areas with sediments were sand is the most representative component, but silty-sand zones are also present. It is important to understand that during the colonisation process, the selection of the habitat will be made by the adults in their spawning migration (Almeida *et al.* 2000), which means that all rehabilitation actions should be engaged close to the areas identified as potential spawning grounds for the species.

Acknowledgements

Thanks are due to our colleagues C.A. Assis and J.L. Costa for the useful comments on early drafts of the typescript. This research was supported by the Water Institute (INAG) of the Ministry of Environment.

References

Almeida P.R., Moreira F., Costa J.L., Assis C.A. & Costa M.J. (1993) The feeding strategies of *Liza ramada* (Risso, 1826) in fresh and brackish water in the River Tagus, Portugal. *Journal of Fish Biology* **42**, 95–107.

Almeida P.R., Silva H.T. & Quintella B. (2000) The migratory behaviour of the sea lamprey *Petromyzon marinus* L., observed by acoustic telemetry in River Mondego (Portugal). In A. Moore & I. Russell (eds) *Advances in Fish Telemetry*. Suffolk: CEFAS, pp. 99–108.

Beamish F.W.H. & Lowartz S. (1996) Larval habitat of American brook lamprey. *Canadian Journal of Fisheries and Aquatic Sciences* **53**, 693–700.

Buchanon J.B. & Kain J.M. (1971) Measurement of the physical and chemical environment. In A.D. Holme & N.A. Mcintyre (eds) *Methods for the Study of Marine Benthos*. Oxford: Blackwell Scientific Publications, pp. 30–58.

Hardisty M.W. (1979) *Biology of the Cyclostomes*. London: Chapman and Hall Ltd, 428 pp.

Hardisty M.W. & Potter I.C. (1971) The behaviour, ecology and growth of larval lampreys. In M.W. Hardisty & I.C. Potter (eds) *The Biology of Lampreys*, Vol. I. London: Academic Press, pp. 85–125.

Kainua K. & Valtonen T. (1980) Distribution and abundance of European river lamprey (*Lampetra fluviatilis*) larvae in the tree rivers running into Bothnian Bay. *Canadian Journal of Fisheries and Aquatic Sciences* **37**, 1960–1966.

Loureiro J., Almeida M., Machado M. & Teixeira E. (1986) Bacia Hidrográfica do Rio Mondego. Monografias hidrológicas dos principais cursos de água de Portugal continental. Lisboa: Direcção Geral dos Recursos e Aproveitamentos Hidráulicos, pp. 239–278 (in Portuguese).

Morman R.H., Cuddy D.W. & Rugen P.C. (1980) Factors influencing the distribution of sea lamprey (*Petromyzon marinus*) in the great lakes. *Canadian Journal of Fisheries and Aquatic Sciences* **37**, 1811–1826.

Ojutkangas E., Aronen K. & Laukkanen E. (1995) Distribution and abundance of river lamprey (*Lampetra fluviatilis*) ammocoetes in the regulated river Perhonjoki. *Regulated Rivers: Research and Management* **10**, 239–245.

Potter I.C. (1980) The Petromyzontiformes with particular reference to paired species. *Canadian Journal of Fisheries and Aquatic Sciences* **37**, 1595–1615.

Roux R.M. (1964) Les sédiments de l'étange de Reine. *Recueils des Travaux de la Station Marine d'Endoume* **35**, 257–285 (in French).

Ter Braak C. (1986) Canonical correspondence analysis. A new eigenvector technique for multivariate direct gradient analysis. *Ecology* **67**, 1167–1179.

Ter Braak C. (1987) CANOCO – a Fortran program for canonical community ordination by correspondence analysis, principal correspondence analysis and redundancy analysis. TNO Institute of Applied Computer Science, Wageningen, 82 pp.

Young R.J., Kelso J.R.M. & Weise J.G. (1990a) Occurrence, relative abundance, and size of landlocked sea lamprey (*Petromyzon marinus*) ammocoetes in relation to stream characteristics in the great lakes. *Canadian Journal of Fisheries and Aquatic Sciences* **47**, 1773–1778.

Young R.J., Houston K.A., Weise J.G. & Kelso J.R.M. (1990b) The effect of environmental variables on the population dynamics of sea lamprey, *Petromyzon marinus*. *Canadian Technical Report of Fisheries and Aquatic Science* **1736**, 33 pp.

Chapter 13
Life history and feeding ecology of a captive population of the endangered Valencia toothcarp, *Valencia hispanica*

N. CAIOLA* and A. DE SOSTOA
Animal Biology Department (Vertebrates), University of Barcelona, Barcelona, Spain

Abstract

The life history and feeding ecology of *Valencia hispanica* from a captive population from the Ebro Delta was studied. Maximum age was 3+ in both males and females. The largest sizes found were 64 mm *v.* 56 mm for females and males respectively. Seasonal growth started in May and continued until August. *Valencia hispanica* spawns once a year, in May, releasing small batches of oocytes (10–30). The life history pattern of this population is characterised by slow growth, low reproductive effort, relatively late maturity and a single annual spawning. Valencia toothcarp feeds exclusively on invertebrates and feeding intensity declines in autumn and winter. The diet varies both with size and season, the latter dependent on food availability. Some recommendations for the maintenance of captive populations, to be used in future reintroductions for the conservation of the species, are discussed.

Keywords: diet, growth, Iberian Peninsula, reproduction, *Valencia hispanica*.

13.1 Introduction

Valencia toothcarp, *Valencia hispanica* (Val.), is an endangered species listed in the Spanish Red Data Book (Blanco & Gonzalez 1992) and is one of the most threatened fish in Europe (Baillie & Groombridge 1996). The species is endemic to the Iberian Peninsula and is restricted to small, isolated marshes and coastal lagoons of the Mediterranean coast of Spain, along the coast line of Catalonia and Valencia (de Sostoa 1983; Perdices, Machordom & Doadrio 1996). Habitat loss and the introduction of exotic fish species are major contributors to its decline (de Sostoa 1983; Perdices *et al.* 1996). Intensive human pressure in its natural habitats has fragmented the species into small isolated populations (Planelles & Risueño 1999).

In an attempt to recover the Valencia toothcarp, an EU LIFE-NATURE project was formulated. The main objectives are captive breeding (for future reintroduction) and understanding the basic biology of the species. This study examines the life history (age, growth and reproduction) and feeding ecology of a captive breeding population

*Correspondence: Animal Biology Department (Vertebrates), Faculty of Biology, University of Barcelona, Avda. Diagonal 645, 08028 Barcelona, Spain (email: caiola@bio.ub.es).

of the Valencia toothcarp and compares these data with a wild population from the same geographical area (Caiola, Vargas & de Sostoa 2001).

13.2 Materials and methods

13.2.1 *Study site*

The work was carried out in an artificial pool of the captive breeding station in the Ebro Delta Natural Park (Catalonia, Spain), where the species was maintained in monoculture. The sampled pool is $20 \times 10\,m^2$, has a maximum depth of 1 m and a sandy bottom. Salinity was constant throughout the study period (7‰). During the study water temperature varied between 12.5°C (December) and 31.5°C (August). Macrophyte coverage was non-existent in winter and dense in summer, when the dominant species was *Ceratophyllum* sp. The banks were covered by *Juncus maritimus* Lam. and *Phragmites australis* (Cav.). Water was supplied from a channel network used for inundating rice fields.

13.2.2 *Sampling*

From November 1996 to October 1997, 212 specimens were captured with a semi-circular drag net with 3.0-mm stretched mesh size with a diameter of 100 cm. Specimens were preserved in ice and transported to the laboratory where they were immediately processed. The total length (L_t) (mm), weight (W_t), gonad weight (W_g) and eviscerated wet weight (W_{ev}) were measured (weights to the nearest 0.001 g), and sex was determined from external characteristics and confirmed by gonad examination. Guts were preserved in 70% ethanol. Prey present in the digestive tracts of the specimens was counted and identified by family, whenever possible, to determine the number of prey categories per gut. Prey were generally little digested, which made identification relatively easy.

At the time of sampling, three water and sediment samples were collected monthly using a PVC core 100-cm high with a diameter of 16 cm to assess the invertebrate community. The core was pushed 30 cm into the sediment in a place where the water depth reached 50 cm. Consequently the sediment and water samples always had a constant volume. The water was then separated from the sediment and filtered throughout a 250-μm nylon sieve. A 500-mL aliquot of each sediment sample was filtered throughout a 500-μm steel sieve. The invertebrates retained in the net were collected and preserved in 4% formalin.

Due to water management problem at the installation, no sampling was performed in April and September.

13.2.3 *Age and growth*

To determine the age and growth, a small number of scales was removed from the left side of the body below the dorsal fin and dry mounted between two slides for stereomicroscopic study. Total scale and annuli radii were measured (to the nearest 0.1 mm) from the focus to the anterior margin of the scale and annuli, respectively. Monthly mean L_t were calculated, to determine the seasonal growth pattern. As no differences were detected between males and females (ANOVA, $P > 0.05$), data for both sexes were combined.

The $L_t - W_t$ and $L_t - W_{ev}$ relationships were calculated for each sex according to $W_t = \mathrm{a}L_t^{\mathrm{b}}$ and $W_{ev} = \mathrm{a}L_t^{\mathrm{b}}$, where a and b are the regression constants. The Fulton condition factor (Bagenal & Tesch 1978) was calculated monthly for each sex as $K = 10^5 (W_t/L_t^{\mathrm{b}})$ and $K_{ev} = 10^5 (W_{ev}/L_t^{\mathrm{b}})$, where b is the slope of the regression line between weight and total length.

13.2.4 *Reproduction*

The monthly mean gonadosomatic index (GSI) for each sex was determined as $\mathrm{GSI} = (W_g/W_t) \times 100$. The total number of eggs in each ovaries (preserved in Gilson's fluid) was counted and the size distribution of oocytes in each was determined by direct measurement of eggs.

The sex ratio was determined for the whole sample only as insufficient data were available for monthly estimates.

13.2.5 *Diet*

Seasonal feeding activity was determined from the proportion of empty guts in each month. Frequency of occurrence (F_i) and prey abundance (P_i) were calculated following Hyslop (1980), and Amundsen, Gabler & Staldvik (1996), respectively, according to: $F_i = (N_i/N) \times 100$, N_i is the number of guts that contain prey i and N is the total number of observed guts that were not empty; and $P_i = (\Sigma S_i/\Sigma S_{ti}) \times 100$, where S_i is the number of prey i in each gut and S_{ti} is the total gut content in only those predators with prey i in their guts.

To analyse seasonal and intraspecific differences in diet, months were categorised into seasons (winter = November–February; spring = March–May; summer = June–August; autumn = September–October) and total length categorised in length classes (A, $L_t \leq 32$ mm; B, $L_t > 32$ mm), where 32 mm was the maximum L_t of immature individuals.

13.3 Results

13.3.1 *Age and growth*

An annulus is laid down on the scales of *V. hispanica* once a year between the end of March and May, thus 1 April was taken as the day of birth. Four age groups (0+ to 3+) were found, both in males and females. Only 2.4% (4) of the individuals were 3+,

10.4% (22) 2+, 25% (53) 1+ and 62.3% (132) 0+. The highest proportion of 0+ individuals (above 50%) was found in June and July.

The maximum length of females (64 mm) and males (56 mm) were similar (Kruskal–Wallis, $P > 0.05$). Growth was distinctly seasonal (Fig. 13.1), with rapid growth between May and August (warm periods), and little evidence of growth between October and March (cold period) (Kruskal–Wallis, $P < 0.05$).

There were strong correlations ($P < 0.05$) between the $L_t - W_t$ and $L_t - W_{ev}$ relationships (Table 13.1), although the $L_t - W_t$ relationship was weaker because it reflects the influence of other parameters, such as the reproductive period or stomach fullness. There were no significant differences between the slopes of the length–weight relationships for sexes (ANCOVA, $P > 0.05$) or months (ANCOVA, $P > 0.05$).

Condition (K and K_{ev}) followed a discrete seasonal pattern (ANOVA, $P < 0.05$) in both males and females (Fig. 13.2). This was more evident in K than in K_{ev}, suggesting that the variation was related to stomach fullness rather somatic level.

Table 13.1 Coefficients of regression and correlation between total and eviscerated weight (W_t and W_{ev}) and total length (L_t) for males, females and the whole sample of *V. hispanica*

	n	a	b	*r*
$W_t = a L_t^b$				
Males	133	2×10^{-6}	3.29	0.99
Females	46	4×10^{-6}	3.17	0.99
Total	212	4×10^{-6}	3.23	0.99
$W_{ev} = a L_t^b$				
Males	133	1×10^{-6}	3.34	0.99
Females	46	3×10^{-6}	3.21	0.98
Total	212	2×10^{-6}	3.32	0.99

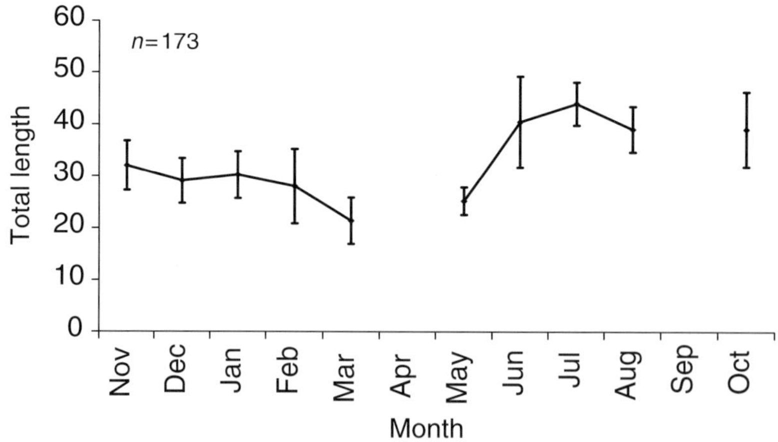

Figure 13.1 Seasonal variation in total body length (L_t) of *V. hispanica* along the study period (mean values ± standard deviation)

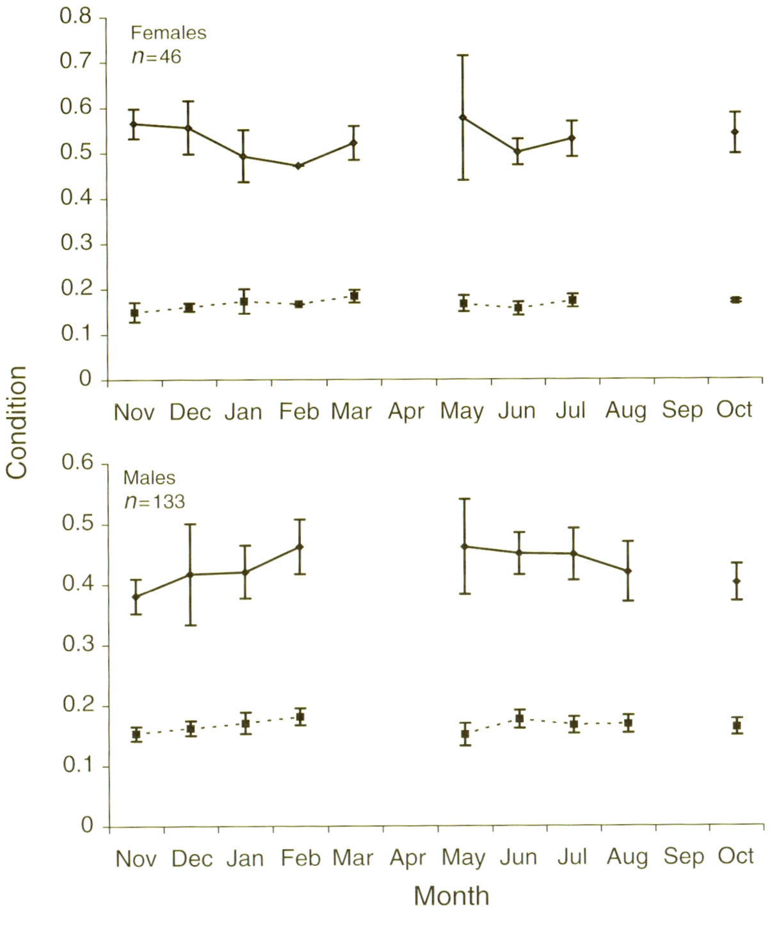

Figure 13.2 Seasonal changes in the condition factor, using total weight (K – solid line) and eviscerated weight (K_{ev} – broken line) for males and females of *V. hispanica*, throughout the study period (mean values ± standard deviation)

13.3.2 *Reproduction*

The overall sex ratio deviated significantly from 1 : 1 (χ^2, $P < 0.05$), with males dominant (133 males *v.* 46 females).

Although the gonad development of males and females differed in absolute gonad weight (Kruskal–Wallis, $P < 0.05$), a similar seasonal pattern was found for both sexes (Fig. 13.3). In females two periods of gonad development occurred: quiescence between July and February, and active gonad development between March and May. Two GSI periods were also identified in males (Fig. 13.3): the quiescent period began later (October), consequently gonad activity began and ended later (May–August).

No mature females (containing oocytes with a diameter greater than 0.9 mm) belonging to the 0+ age group were found as it is unlikely that females can reproduce in their first year. The first maturation occurred when females reached the 1+ age group.

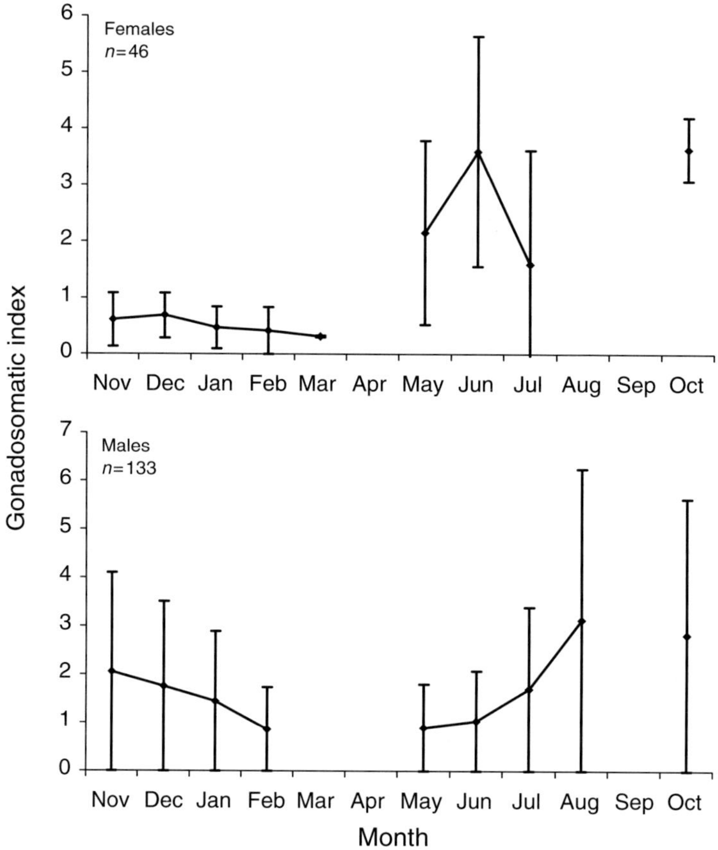

Figure 13.3 Seasonal changes in the GSI of males and females of *V. hispanica*, during the study period (mean values ± standard deviation)

Recruitment oocytes were detected throughout the year. These were characterised by a small diameter (0.1–0.5 mm) and translucent appearance. At the onset of gonad activity (in March), white opaque oocytes with a diameter >0.5 mm appeared (Fig. 13.4). In May, the distribution became slightly bimodal when the number of amber coloured, mature oocytes of a large diameter (0.9–1.4 mm) increased. The reproductive cycle ended in June (Fig. 13.4), when the number of oocytes with a diameter over 0.5 mm disappeared from female gonads.

13.3.3 *Food and feeding*

Feeding intensity of *V. hispanica* varied significantly between months (Kruskal–Wallis, $P < 0.05$). The percentage of empty guts was higher in autumn (50%) and winter (32%) than in spring (23%) and summer (25%).

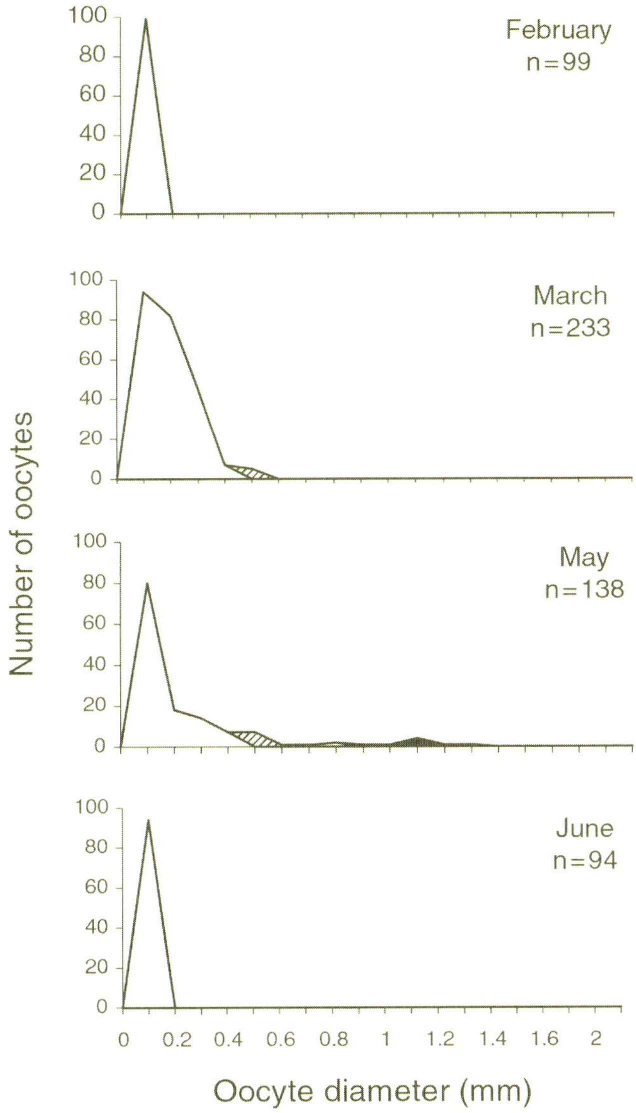

Figure 13.4 Size frequency distribution of oocytes from four *V. hispanica* females of similar length (40–45 mm L_t) before, during and after the spawning period. White areas represent immature oocytes, hatched areas represent maturating oocytes and black areas represent mature oocytes

A total of 12 prey categories, all invertebrates, were found in the gut contents of *V. hispanica* (Table 13.2). Mollusc of the genus Physidae (23%), chironomid larvae (21%) and adults (21%), *Daphnia* (21%) and copepods (20%) were the main items in the diet. Consumption of these items varied seasonally (Table 13.3), related to differences in availability. *Daphnia* was the prey consumed by more individuals in

Table 13.2 Overall, seasonal and intraspecific diet composition of *V. hispanica*

			Season				Sex		Length-class	
Prey category	AF	Total (n = 212)	Winter (n = 60)	Spring (n = 40)	Summer (n = 72)	Autumn (n = 40)	Male (n = 133)	Female (n = 46)	$L_t \leqslant 32$ mm (n = 37)	$L_t > 32$ mm (n = 175)
Physidae	322	23.1 (29.0)	37.0 (69.1)	10.0 (21.4)	6.1 (28.5)	60.0 (76.1)	37.5 (39.6)	40.0 (27.9)	3.5 (14.1)	27.2 (72.1)
Chironomid larvae	185	20.9 (48.5)	1.8 (7.7)	23.3 (20.9)	44.8 (32.2)	30.0 (42.8)	26.4 (73.0)	28.0 (64.7)	5.0 (21.4)	27.2 (59.5)
Chironomid adults	152	20.9 (28.1)	1.8 (5.2)	10.0 (27.2)	32.6 (75.6)	30.0 (27.2)	22.9 (72.8)	20.0 (52.1)	0.0 (0.0)	22.8 (43.8)
Daphnidae	110	20.8 (68.3)	0.0 (0.0)	60.0 (87.9)	18.3 (40.7)	0.0 (0.0)	20.6 (26.6)	20.0 (57.2)	29.2 (79.5)	9.3 (59.4)
Copepoda	98	20.3 (43.6)	0.0 (0.0)	23.3 (66.6)	18.3 (22.2)	0.0 (0.0)	16.0 (29.4)	6.0 (50.0)	26.1 (27.8)	4.6 (25.0)
Gammaridae	76	16.6 (71.3)	29.1 (26.9)	3.3 (100.0)	2.0 (100.0)	10.0 (50.0)	11.4 (36.5)	12.0 (91.6)	11.5 (22.9)	14.4 (79.3)
Sphaeromatidae	70	11.1 (75.1)	3.6 (18.2)	3.3 (100.0)	24.4 (26.5)	10.0 (50.0)	1.1 (88.2)	16.0 (31.8)	0.0 (0.0)	11.2 (25.3)
Cypridae	45	2.0 (26.7)	5.5 (36.4)	13.3 (60.0)	22.4 (62.5)	10.0 (12.5)	1.1 (50.0)	8.0 (33.3)	23.9 (55.1)	6.5 (41.7)
Hydrobidae	7	7.6 (33.3)	3.6 (21.4)	0.0 (0.0)	0.0 (0.0)	10.0 (60.0)	3.4 (26.6)	12.0 (52.1)	4.4 (59.9)	1.3 (20.0)
Oligochaetae	6	3.4 (19.5)	1.8 (3.7)	10.2 (40.8)	0.0 (0.0)	0.0 (0.0)	2.2 (7.4)	12.0 (20.0)	2.9 (15.5)	5.2 (21.3)
Other insects	2	1.3 (7.4)	1.8 (33.3)	12.2 (42.3)	2.0 (5.5)	0.0 (0.0)	3.4 (27.2)	4.0 (33.0)	0.0 (0.0)	2.6 (7.4)
Cardidae	1	1.3 (6.8)	0.0 (0.0)	6.1 (4.2)	0.0 (0.0)	0.0 (0.0)	2.2 (6.8)	4.0 (33.3)	0.0 (0.0)	2.6 (6.7)

First number, frequency of occurrence (%); number in parentheses, prey specific abundance (%); n, number of digestive tracts; AF, absolute frequency.

Table 13.3 $R \times C$ independence G-test to compare the absolute frequencies of the major food items in the diet of *V. hispanica*

	Season			Sex			Length-class		
Prey category	G	df	Res	G	df	Res	G	df	Res
Physidae	111.52	3	Aut Win Spr Sum	2.06	1	F M	641.23	1	A B
Chironomidae (L)	70.31	3	Win Spr Sum Aut	1.71	1	F M	63.54	1	A B
Chironomidae (A)	93.17	3	Win Spr Sum Aut	2.18	1	F M	399.12	1	A B
Daphnidae	45.93	3	Aut Win Spr Sum	1.85	1	F M	45.02	1	A B
Copepoda	54.21	3	Aut Win Spr Sum	46.75	1	F M	125.16	1	A B

Rules join non-significant sets ($P < 0.05$). Win, winter; Spr, spring; Sum, summer; Aut, autumn; M, male; F, female; A, $L_t \leqslant 32$ mm; B, $L_t > 32$ mm.

spring ($F = 60\%$) and larvae and adult chironomids in summer ($F_i = 45\%$ and $F_i = 33\%$, respectively), when they were most abundant in the pools (unpublished data). Gammarid amphipods and sphaeromatid isopods were, however, the most abundant prey items during spring and summer ($P_i = 100\%$ in spring for both prey and $P = 100\%$ in summer for Gammarid amphipods). Physidae molluscs were the predominant prey consumed in winter ($F_i = 37\%$; $P_i = 69\%$) and autumn ($F_i = 60\%$; $P_i = 76\%$).

No differences in diet between sexes were found (Table 13.3). There was a shift in diet between immature individuals (small specimens) and adults. Immature individuals consumed mainly small size prey: *Daphnia* ($F_i = 29\%$, $P_i = 79\%$) and copepods ($F_i = 26\%$), whilst adults ate predominantly adult chironomids and Physidae molluscs ($F_i = 27\%$; $P_i = 72\%$).

13.4 Discussion

The captive breeding population of *V. hispanica*, were shorter lived and grew slower than a wild population from the same geographical area (Caiola, Vargas & de Sostoa 2001b). Moreover, only a few individuals of the captive population reached the adult 2+ and 3+ age classes, indicating high adult mortality.

In the wild females were bigger than males, but no such difference was found in the captive population, suggesting differential mortality between sexes or loss of genetic variability. Controlling adult mortality and genetic variability are key issues in captive breeding for recovery programmes (Planelles & Risueño 1999).

The growth of fish is influenced by water temperature, food and nutrient availability, light regime, oxygen, salinity, pollutants, predator density and intraspecific social interactions (Helfman, Collette & Facey 1997). The maximum growth rate of *V. hispanica* from the captive population occurred during the summer, at the end of the reproductive period, when temperature was mild and food availability was higher, which allows most energy to be directed to somatic growth.

Reproduction occurs when temperatures are high, there is little precipitation and there are many hours of sunshine (Panareda 1977). Although these conditions were similar in the captive breeding pools and in the study site of the wild population, the reproduction patterns were quite different (Caiola *et al.* 2001b). The wild population exhibited multiple spawning, while the captive population spawned once, in May. Multiple spawning is a common strategy among small fishes that ensures the survival of at least part of the offspring. Fish size is a factor that limits fecundity because small fish can only have small gonads (Byrne 1978), and this constrains the number of oocytes that they can hold (Vargas & de Sostoa 1997). This can be compensated if oocyte release takes place over a protracted period, thus allowing greater fecundity than in a single spawning period (Byrne 1978). Multiple spawning increases offspring survival by reducing the risk of predation on eggs and larvae. If unfavourable conditions occur at any time during reproduction, only part of the total spawning episode is affected. The restriction of spawning to only one batch implies a reduced fecundity of Valencia toothcarp in the captive breeding pools.

The influence of both environmental and reproductive factors is reflected in condition and reproductive effort. At the beginning of the reproductive period, part of the energy is used for growth and part for gonad development, and as the reproductive cycle progresses more energy is invested in gonad development (Vargas & de Sostoa 1997). In contrast to the wild population, where reproductive effort was especially higher for females, males and females from the captive breeding population showed similar patterns in condition presumably due to a lower investment in oocyte formation. The small monthly variations in condition were probably caused by trophic influences.

The deviation in the sex ratio from 1 : 1 is not unusual in fish populations with marked sexual dimorphism, and may be related to different mortality rates or to the role played by each sex during reproduction (Vargas & de Sostoa 1997). In contrast to the wild population, males dominated the captive population. These differences in sex ratio may be due to slight changes in environmental conditions. In fishes, sex chromosomes are relatively rare, establishing the possibility for environmental sex determination to occur, with environment being understood to include climate, food availability and social interactions (Helfman *et al.* 1997).

The greatest proportion of empty guts from the captive population of *V. hispanica* occurred in autumn, when prey abundance was lowest. This coincided with a decrease in the water level in the pools due to the closing of the channel network that supplies water. The higher prey consumption in spring and summer was probably a consequence of greater food availability and linked to higher energy requirements for growth and reproduction.

The Valencia toothcarp feeds exclusively on invertebrates and its feeding pattern depends on prey availability. The wild population has a more specialised diet, with little differences between seasons (Caiola, Vargas & de Sostoa 2001a) than the captive population. There was a marked variation in the diet of adult and immature individuals. Bigger fish consume more and larger prey and have a wider diet range because they can consume prey that is not available to small fish (Grossman, Coffin & Moyle 1980). Thus, it is possible that larger fish maximise energy intake by increasing the consumption of larger prey. Also ontogenic differences in diet cause a decrease in intraspecific competition (Grossman *et al.* 1980).

The native freshwater fish fauna of the Iberian Peninsula has declined markedly over the last two decades (Aparicio, Vargas, Olmo & de Sostoa 2000). Reintroduction programmes are generally recommended to recover endangered species. Reintroduction can be done from the remaining wild populations or from captive ones. However, the latter is more appropriate because the success of an action of this type depends, among other things, on the number of specimens reintroduced, which should be high (Doadrio, Perdices & Machordom 1996; Perdices *et al.* 1996). Taking such numbers from stable populations of a threatened species is clearly inappropriate. To obtain specimens for these actions it is necessary to establish captive breeding stocks.

This study illustrates some problems in the maintenance of a captive breeding livestock. The high adult mortality, low reproductive effort, low fecundity and diet constraints, which probably lead to low condition, are a result of the inadequate design and management of this captive breeding programme. The main problems detected at the installations during this study were deficient water intake during the months when water supply is low or non-existent due to the closing of the paddy fields irrigation channel network, and a lack of scientific-based criteria in breeding and stock management (substrate, water quality, diet, rearing, etc.). Therefore, it is recommended that these installations follow a suitable plan for the establishment of a captive breeding programme. According to Reid (1990), such a plan should take into account several issues: organisation (funding, staffing, installations and licensing), fish stock (space, substrate, heat and light, water quality and diet) and breeding (broodstock, reproduction, rearing and disorders). Each species has distinct requirements, therefore previous studies (biology, behaviour, genetics, etc.) should be used to design both fish stock and breeding management programmes.

Acknowledgements

To Josep Maria Queral and Iolanda Bel i Díaz for their assistance in the field work. To Ebro Delta Natural Park (Departament de Medi Ambient) for installations and laboratory facilities. Financial support was provided by EU LIFE-NATURE Project No. B4-3200/96/502.

References

Amundsen P.A., Gabler H.M. & Staldvik F.J. (1996) A new approach to graphical analysis of feeding strategy from stomach contents data – modification of the Costello (1990) method. *Journal of Fish Biology* **48**, 607–614.

Aparicio E., Vargas M.J., Olmo J.M. & de Sostoa A. (2000) Decline of native freshwater fishes in a Mediterranean watershed on the Iberian Peninsula: a quantitative assessment. *Environmental Biology of Fishes* **59**, 11–19.

Baillie J. & Groombridge B. (1996) *IUCN Red List of Threatened Animals*. Gland: IUCN, 368 pp. + annexes.

Bagenal T.B. & Tesch F.W. (1978) Eggs and early life history. In T.B. Bagenal (ed.) *Methods for Assessment of Fish Production in Freshwaters*. IBP Handbook No. 3. Oxford: Blackwell Scientific Publications, pp. 101–136.

Blanco J.C. & Gonzalez J.L. (1992) *Libro rojo de los vertebrados de España*. Madrid: Ministerio de Agricultura, Pesca y Alimentación. Colección técnica ICONA, 714 pp. (in Spanish).

Byrne D.M. (1978) Life history of the spotfin killifish, *Fundulus luciae* (Pisces: Cyprinodontidae), in Fox Creek Marsh, Virginia. *Estuaries* **1**, 211–217.

Caiola N., Vargas M.J. & de Sostoa A. (2001a) Feeding ecology of the endangered Valencia toothcarp, *Valencia hispanica* (Actinopterygii: Valenciidae). *Hydrobiologia* **448**, 97–105.

Caiola N.A., Vargas M.J. & de Sostoa A. (2001b) Life history pattern of the endangered Valencia toothcarp, *Valencia hispanica* (Actinopterygii: Valenciidae) and its implications for conservation. *Archive für Hydrobiologie* **150**, 473–489.

de Sostoa A. (1983) *Las comunidades de peces del delta del Ebro*. PhD Thesis, University of Barcelona, 622 pp. (in Spanish).

Doadrio I., Perdices A. & Machordom A. (1996) Allozymic variation of the endangered *Aphanius iberus* and its applications to conservation. *Environmental Biology of Fishes* **45**, 259–271.

Grossman G.D., Coffin R. & Moyle P.B. (1980) Feeding ecology of the bay goby (Pisces: Gobidae). Effects of behavioural, ontogenic and temporal variation on diet. *Journal of Experimental Biology and Ecology* **44**, 47–59.

Helfman G.S., Collette B.B. & Facey D.E. (1997) *The Diversity of Fishes*. Oxford: Blackwell Science, 528 pp.

Hyslop E.J. (1980) Stomach content analysis – a critical review of methods and their application. *Journal of Fish Biology* **17**, 411–429.

Panareda J.M. (1977) Climatologia. *Dryas: Estudio de Protección del Medio Físico del Delta del Ebro* **1**, 1–45 (in Spanish).

Perdices A., Machordom A. & Doadrio I. (1996) Allozymic variation and relationships of the endangered cyprinodontid genus *Valencia* and its implications to conservation. *Journal of Fish Biology* **49**, 1112–1127.

Planelles M. & Risueño P. (1999) El samaruc (*Valencia hispanica* Valenciennes, 1846) en la comunidad valenciana: declive, esfuerzos para su recuperación y situación actual. In Generalitat Valenciana (ed.): *Peces ciprinodóntidos ibéricos fartet y samaruc*. Valencia: Conselleria de Medio Ambiente, pp. 253–277 (in Spanish).

Reid G. (1990) Captive breeding for the conservation of cichlid fishes. *Journal of Fish Biology* **37**, 157–166.

Vargas M.J. & de Sostoa A. (1997) Life history pattern of the Iberian toothcarp *Aphanius iberus* (Pisces: Cyprinodontidae) from a Mediterranean estuary, the Ebro Delta (Spain). *Netherlands Journal of Zoology* **47**, 143–160.

Chapter 14
From the protection of sturgeon, *Acipenser sturio*, to river management in general

G. PUSTELNIK,* O. GUERRI
Etablissement Public Interdépartemental Dordogne, Castelnaud-La-Chapelle, France

M. LEPAGE
Association Girondine pour l'Expérimentation et le Développement des Ressources Aquatiques, Saint-André de Cubzac, France

Abstract

The European sturgeon, *Acipenser sturio*, is an endangered migratory fish which has disappeared from the majority of the rivers in its natural distribution range. Today the species is known to reproduce only in the Gironde catchment, in southwest France, where an EU LIFE-NATURE project has been carried out since 1994. This programme highlights the direct link between disappearance of the species, overfishing and habitat disturbance, leading the programme managers to worry about the sociological aspects of the species' vulnerability. Despite legal protection of the species, the fish conservation practices suffer from a weak administrative framework. New legislation is needed to approach the management of freshwater and migratory fishes within a much more global and integrated scheme for the management of the rivers.

Keywords: Atlantic sturgeon, conservation, integrated management, legislation, migratory fish.

14.1 Introduction

The Atlantic sturgeon, *Acipenser sturio* L., is a migratory fish which was formerly prolific along the entire seaboard of Western Europe, from the Baltic to the Mediterranean (Rochard, Castelnaud & Lepage 1990) (Fig. 14.1). It has now disappeared from most of the rivers it once frequented. Today, the Gironde catchment is the only area in which the species is known to breed. *Acipenser sturio* is threatened with extinction, and protected by international (Bern & Washington Conventions, IUCN Red Book, European Union Habitats Directive – 92/43/CEE) and French legislation. Efforts to save the sturgeon are now under way in the Gironde, Garonne and Dordogne River basins, within the framework of the European Union LIFE-NATURE programme. They are briefly described in the present chapter.

*Correspondence: EPIDOR, BP 13, F-24250 Castelnaud-La-Chapelle, France (email: epidor@cg24.fr).

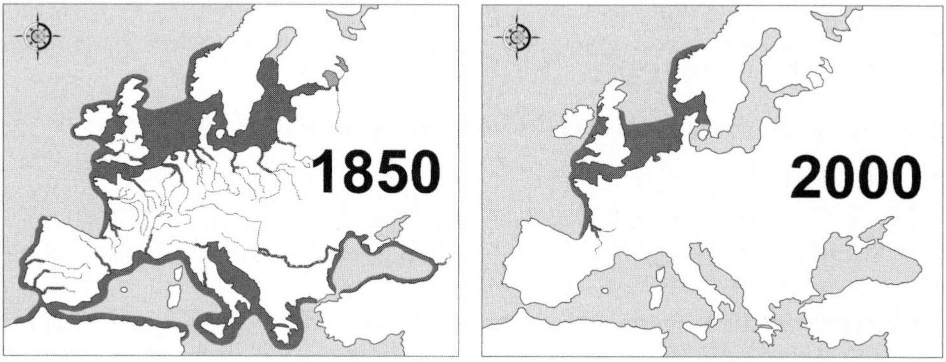

Figure 14.1 Marine and continental past and present distribution of *Acipenser sturio*

14.2 An overview of the LIFE project

During this programme considerable knowledge was gained regarding the river and estuarine habitats, marine distribution and feeding behaviour of the sturgeon, enabling specific recommendations to be presented to the authorities (Elie 1997; EPIDOR & CEMAGREF 1997; Rochard, Lepage & Meauze 1997; Brosse, Rochard, Dumont & Lepage 2000; Rochard, Lepage, Dumont, Trembley & Gazeau 2001). Tracking arrangements were put in place to study the evolution of the remaining sturgeon population, effected through regular samplings in the Gironde estuary, with tagging and ageing operations. Demographic models were drawn up, helping to form an overall picture of the current status of *Acipenser sturio* in the basin. Techniques were developed for the different phases of breeding sturgeon in captivity (Williot, Rochard, Castelnaud, Rouault, Brun, Lepage & Elie 1997). Facilities were put in place for the acclimatisation of captive sturgeon, for artificial reproduction, for rearing the larvae, and for the holding of juvenile and young adult fish (thermostable building, closed circuits, large dimension breeding tanks). A stock of nearly 200 individual sturgeon has been established, some caught in the wild, others derived from artificial reproduction carried out in 1995. This stock is for both conservation purposes and to serve as the basis for future operations to restore sturgeon populations, not only in the Gironde River basin but also possibly in other European river basins. As *Acipenser sturio* has totally disappeared from other European basins, stocking with animals from the Gironde would not present any genetic risk.

The research completed to date shows that, for the remaining population in the Gironde River basin, the adjoining Dordogne and Garonne River basins offer spawning and feeding habitats which are satisfactory in both quantity and quality. There are, however, several uncertainties, notably:

(1) the vulnerability of the habitats, particularly in the Gironde estuary, resulting from the high demand for gravel extraction;
(2) the impact of sediment contamination by cadmium in the Gironde estuary, dating back several decades; and

(3) the survival of the adults and young adults at sea, because these fish are apparently easy to catch and they generally frequent the shallow offshore areas where coastal fishing activities are most prolific.

Whether incidental or otherwise, the number of sturgeons caught each year is thought to range from several dozen to several hundreds yet the total population estimated in the summer 1996 ranged between 2267 and 8862 individuals (Rochard, Lepage, Gazeau & Lambert 1997), and no natural reproduction is known to occur since 1994. Incidental or intentional captures are thought to be the principal factor in the mortality of Atlantic sturgeon.

It is apparent that the survival of the sturgeon lies in society's capacity to solve problems which have been prevalent for almost half a century. The social aspect of its recovery, already recognised as a key element in the LIFE project, has presently become a prime concern. Efforts to heighten awareness of the problem, directed at the general public during the first years of the LIFE programme, have now been strengthened and re-focused towards the elected representative, the Administration and, above all, the coastal and estuarine fishermen, who have the power of decision as regards the future of captured sturgeon.

Despite the knowledge gained, the status of *Acipenser sturio* in Europe continues to worsen. Nowadays, the main elements that characterise the European sturgeon are:

- only one population remains functional, in the Gironde River basin, even if occasional sighting of individual strays concern the last survivors of local populations that are now extinct – for instance, off Spain's Guadalquivir estuary in 1992 (Elvira & Almodovar 1993) and in the Baltic Sea in 1996 (Paver 1999);
- only two natural reproductive events have been recorded since 1985, i.e. in 1988 and 1994;
- the return of spawners became increasingly rare: 150 individuals caught between 1971 and 1980, 36 from 1981 to 1990, and 10 from 1991 to 2000;
- in the year 2000, the stock probably did not exceed 10 000 individuals (Fig. 14.2).

If there is no improvement over the next 6 years (the anticipated period for the return of the last migrant exodus in 1994), the species will probably become extinct.

14.3 Fish conservation plans – a social challenge

While the difficulties which the sturgeon faces are acute and urgent, they are part of a broader picture stemming from a problem common to all migratory and freshwater fish which need help not only in technical and scientific terms but also from the social, political and financial dimensions. When the necessary means are allocated, it is generally easy to pinpoint the issues and define the technical solutions needed to protect or manage a population of migratory fish (decrease fishing pressure, restocking, establishing reserve area, improve water quality, etc.). But often, implementing these solutions runs into buffers in the social context.

Pustelnik and Dulude (1992) described the restoration of the Atlantic salmon, *Salmo salmo* L., as a 'technical problem'. They supported the idea that the failure of restora-

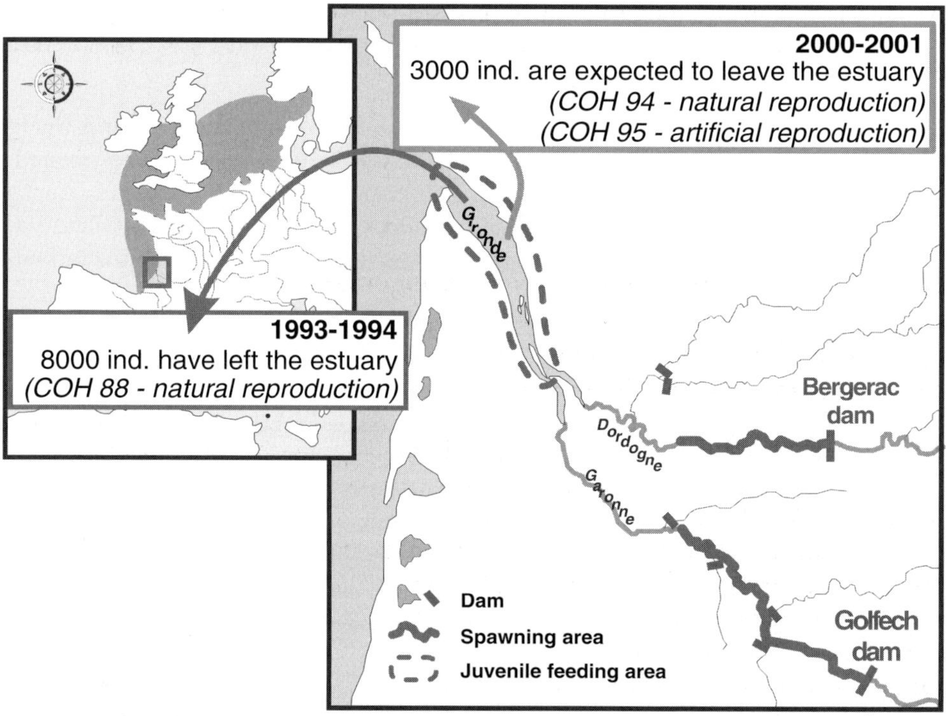

Figure 14.2 Scheme of the last exodus of *Acipenser sturio* from the Gironde estuary. COH indicates year of cohort

tion plans up to then, was linked to decisions on the nature of the investments made, to the choice of strategies adopted which sometimes favoured basic work and sometimes the type of results, to short- rather than long-term criteria of evaluation, and above all to attitudes which still paid court to the old working habits of the Administration, the fish farmer, the commercial fisherman or any other partner involved in the restoration plan, whereas these habits were incompatible with the initial objectives.

Pustelnik and Guerri (1999) also stated that the sturgeon problem was more social than technical. This led to the concept of a social charter for migratory fish. The charter requires an undertaking by governments, political decision-makers, administrators and financiers to recognise the care of migratory fish as one of the elements in effecting integrated procedures for river management – in turn leading them to draw up coherent strategies on the local, national and international scales. The whole thrust of such a project is to go beyond the limits currently imposed by the everyday realities of managing watercourses. These limits are of three kinds, and every country can contribute. They are:

- 'habitual practices' perpetuated by administrative cuts, sector-based approaches, the clumsiness and complexity of the bureaucracy; it is generally complicated to bring people to exchange their points of view to elevate the discourse to the main point of species interest;

- the 'corporatism' which diminishes all notion of collective interest and lends weight to individual interests (e.g. agricultural use of land and water, industrial use of land and water, recreational use of land and water);
- the 'financial rules', with a disproportion between funds allocated to management of the environment and those devoted to development, and which also make it difficult to finance long-term measures.

14.4 A new law to support river management

In France, the legislation is inadequate to manage the aquatic environment and species. The laws that are focused on the concept of economic resource, define sector-based management objectives, as seen through the eyes of users who are looking much more for access to exploitable reserves than for the balance of sustainable development. It is evident, as a logical consequence, that the mechanisms for implementing these laws are disconnected. They leave too much freedom for sector-based procedures and they are inadequate as regards to an integrated management. It does not seem possible to make any significant progress on the issue of saving or restoring fish species without developing transversal approaches, adapted from the start to the social demand, which would lead to integrated strategies and the provision of implements enabling them to be applied. The conservation of fish, whether freshwater or migratory, would benefit greatly from a new apparatus, complementary to the present mechanisms. This is the aim of a proposed new Law on Rivers currently being put forward by the Etablissements Publics Territoriaux de Bassin (broadly, the official working groups for the catchment areas). Its objectives are:

(1) the recognition of the intrinsic value of the aquatic environment, taking account of their sensitivity and vulnerability, beyond the concept of economic resource and short-term considerations;
(2) the establishment of management procedures on the catchment scale which would be independent of the present administrative barriers;
(3) to reconcile the management of the environment and the river's economic development; and
(4) the formulation of integrated strategies, tailored to each basin, covering long-term objectives.

14.5 Conclusion

During this LIFE project, significant effort was made to acquire knowledge of the European sturgeon's biology and ecology. However, the situation for the species has worsened. Although the formation of a stock of captive sturgeon now holds out hope of an eventual restoration, the chances of success lie more in the social and political aspects of river management. New legislation is needed to implement the recovery measures formulated from the scientific studies. The formulation, acceptance and implementation of a new framework will take several years and it will not settle the urgent problem of the sturgeon's survival nor, indeed, that of other species threatened with extinction. In the short term, support for such a project from the political representatives would nevertheless be a valuable recognition of past setbacks, thus giving a necessary kick-start to mobilise the forces for the urgent actions now required.

References

Brosse L., Rochard E., Dumont P. & Lepage M. (2000) Premiers résultats sur l'alimentation de l'esturgeon européen, *Acipenser sturio* Linneaus, 1758, dans l'estuaire de la Gironde. Comparaison avec la faune benthique. *Cybium* **24**(Suppl.), 46–61 (in French).

Elie P. (ed.) (1997) Restauration de l'esturgeon européen *Acipenser sturio*. Contract Life n° B4-3200/94/754, rapport final du programme d'exécution. Opération I–II–III. Programme LIFE 94, Etude CEMAGREF – Bordeaux n°24, 381 pp. (in French).

Elvira B. & Almodovar A. (1993) Notice about the survival of sturgeon (*Acipenser sturio* L., 1758) in the Guadalquivir estuary (S.W. Spain). *Archives für Hydrobiologie* **129**, 253–255.

EPIDOR & CEMAGREF (1997) Restauration de l'esturgeon européen *Acipenser sturio*. Contract Life no. B4-3200/94/754, rapport final. Programme LIFE 94, EPIDOR, 26 pp. (in French).

Paver T. (1999) Historic and recent records of native and exotic sturgeon species in Estonia. *Journal of Applied Ichthyology* **15**, 129–132.

Pustelnik G. & Dulude P. (1992) La relation entre les intervenants dans une procédure de restauration du saumon atlantique: le cas de la rivière Dordogne (France). Colloque international Fédération Québécoise du Saumon Atlantique, Québec, Dec. 1992, 47 pp. (in French).

Pustelnik G. & Guerri O. (1999) Analysis of partnership and conservation requirements for a threatened species, *Acipenser sturio* L.: towards the implementation of a recovery plan. *Boletín. Instituto Español de Oceanografía* **16**(1–4), 209–216.

Rochard E., Castelnaud G. & Lepage M. (1990) Sturgeons (Pisces: Acipenseridae) threats and prospects. *Journal of Fish Biology* **37**(Suppl. A), 123–132.

Rochard E., Lepage M., Dumont P., Trembley S. & Gazeau C. (2001) Downstream migration of juvenile European sturgeon *Acipenser sturio* L. in the Gironde estuary. *Estuaries* **24**, 108–115.

Rochard E., Lepage M., Gazeau C. & Lambert P. (1997) Tableau de bord de la population, estimation de l'abondance des différentes classes d'âge. In P. Elie (ed.) *Restauration de l'esturgeon européen* Acipenser sturio. Contract LIFE rapport final du programme d'exécution. Etude CEMAGREF – Bordeaux n°24, pp. 351–376 (in French).

Rochard E., Lepage M. & Meauze L. (1997) Identification et caractérisation de l'aire de répartition marine de l'esturgeon européen *Acipenser sturio* à partir de déclarations de captures. *Aquatic Living Resources* **10**, 101–109 (in French).

Williot P., Rochard E., Castelnaud G., Rouault T., Brun R., Lepage M. & Elie P. (1997) Biological characteristics of European Atlantic sturgeon, *Acipenser sturio*, as the basis for a restoration program in France. *Environmental Biology of Fishes* **48**, 359–370.

Chapter 15
A conservation strategy for *Zingel asper*, a threatened endemic percid of the Rhône Basin

S. MARI*
Reserves Naturelles de France, Quetigny, France

J. LABONNE and P. GAUDIN
River Hydrosystems Laboratory, Lyon I University, Villeurbanne, France

Abstract

Zingel asper is an endemic fish of the Rhône basin, the distribution of which is now restricted to very few stretches of river. An EU LIFE-NATURE programme was initiated in 1998 to establish a conservation strategy for this endangered species. As little was known about its biology, studies were carried out on its population genetics, population dynamics, and micro-habitat use. The species, although considered sedentary, was found to migrate between habitat patches and utilise whole river sections. These movements were related to diel shifts in habitat use and life history traits. Artificial rearing was shown to be feasible and studies in an experimental river should provide more information about reproductive biology and the ability of the fish to negotiate obstacles. This information should support guidelines for rational management for *Z. asper* being developed in close co-operation with the water authorities and angling community.

Keywords: conservation, endangered fish, habitat fragmentation, *Zingel asper*.

15.1 Introduction

The 'apron du Rhône', *Zingel asper* L., is an endemic percid of the Rhône basin. This fish species lives in bedrock and shallow rivers approximating to the barbel and grayling zones. The species is endangered (IUCN Red List of Threatened Animals, Baillie & Groombridge 1996), and is quoted in the Habitats Directive of the European Commission (annexes II and IV) and in the Bern Convention (annex II).

During the last century, the Rhône River basin was altered by hydropower plant construction, pollution, and gravel extraction, and *Z. asper* lost 83% of its former distribution (Boutitie 1984). Despite studies led by the Environment Ministry from 1993 to 1997, important gaps in the biological knowledge of this species still remained (present distribution, population ecology, habitats requirements, and life history traits). This lack of information has prevented identification of the precise causes of the species' demise.

In this context, an EU LIFE-NATURE conservation programme was initiated in 1998 by the French Nature Reserves network (RNF), involving scientists from Lyon I

*Correspondence: Reserves Naturelles de France, 3 rue de la Forge – B.P. 100, 2803 Quetigny Cedex, France (email: Lifeapron@aol.com).

University. The main objectives of the programme were to protect the remaining populations and their habitats, essentially by having *Z. asper* considered in the local water management policy, and to develop guidelines for management of the species. The programme included scientific studies mainly on physical habitat selection, population dynamics and genetics, and additional experiments on captive breeding of fish were planned. The expected results of these studies will provide tools to manage the populations and increase public awareness of the precarious state of the species.

15.2 Distribution and regression of the species

The French River Rhône basin covers 97 000 km^2, equivalent to about 15% of France. At the beginning of the century, *Z. asper* was found throughout the basin, inhabiting the barbel zone and the lower part of the grayling zone of all tributaries. The river length colonised was 1700 km, but the species was considered as naturally rare. Its present distribution is limited to 380 km of river (Fig. 15.1) but, in 1997, only one large residual population was known, in the River Beaume, a tributary of the River Ardeche. This population is spread over 13 km at the downstream reach of the river and its size was estimated at between 2000 and 4000 individuals (Perrin 1988). Its upstream dispersal is prevented by an old mill weir. Another large population was discovered in 1999 on the upper part of the River Durance (a tributary of the River Rhône; Moullec, Chappaz & Cavali 2000), where it occupies a 25 km length of river. Elsewhere, some fish have been found locally, either by the French Fishery High Council (CSP) or by scientists or fishermen, but only as isolated specimens. A recent distribution study (OFEFP 1999) in the Swiss sector of the River Doubs (upper part of the River Rhône basin) illustrated this fragmentation as some stretches seemed to be inhabited by a very small number of individuals: in about 40 km of the River Doubs (from Goumois to Bremoncourt) only 107 fish were observed, with a maximum of 10 individuals per surveyed station.

Before 1998, with the exception of an old publication dealing mainly with captive breeding (Leger & Stankovitch 1921), very little was known about the distribution, life cycle, and behaviour of this species. It was described as a sedentary and lonely fish, with a maximum size of around 23 cm ($\approx$100 g) (Fig. 15.2), living preferentially in deep waters (Dottrens 1951; Spillmann 1961), and usually found in the upper part of the barbel zone (Huet 1949). A peculiarity of this species is its natural low density (in the case of the population of River Beaume: 30–120 fish ha^{-1} (0.6–2.45 kg ha^{-1}) with a mean weight of 20 g). When several individuals are found together, they evenly disperse in the river, suggesting territoriality (Boutitie 1984).

As for many fish species, threats to populations of this species were commonly identified as: habitat degradation, pollution, impoundments, and obstacles to migration. The main problem is to determine the impact of each perturbation, and to detect the vulnerable phases of the life cycle. This is essential to understand fully the causes of species decline. In the case of *Z. asper*, a sedentary species, no common factor could be found when comparing the different rivers where it became extinct. For example, in low disturbed rivers like the rivers Gardon or Ceze in the south of the Rhône basin, physical habitat, water quality, and the flow regime were not strongly altered.

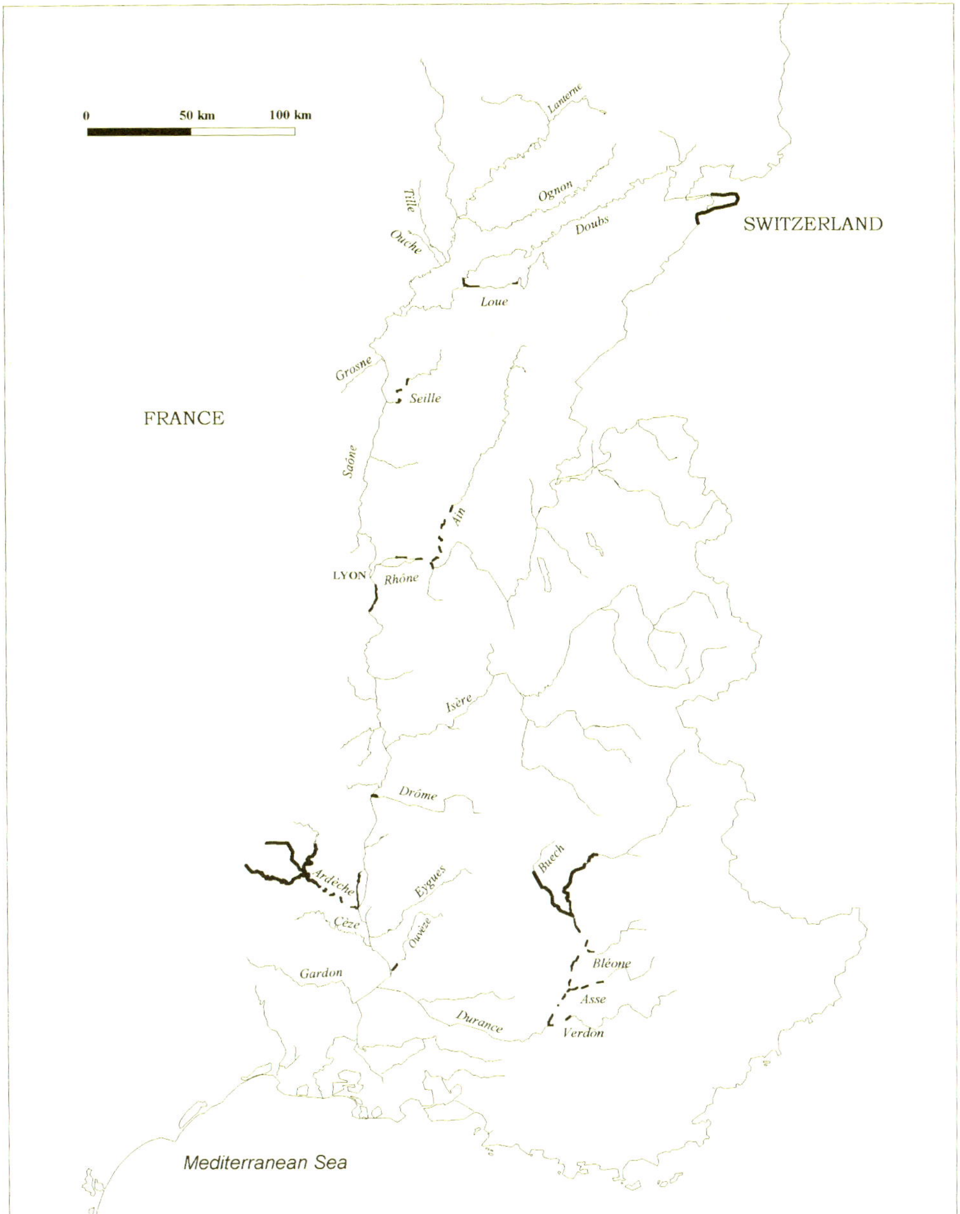

Figure 15.1 Distribution of *Z. asper* in the Rhône basin. Thin lines show non-colonised rivers, medium lines represent the species distribution in 1900, and large lines show recorded distribution in 1984

Figure 15.2 Adult female of *Z. asper* in the wild. This gravid fish was 17-cm long and weighed 55 g (photograph: Xavier Desmier)

This suggests that correlations between habitat degradation and species distribution may be imprecise. As a consequence, a global population study was initiated to elucidate the population ecology, and to understand which demographic traits could be affected by environmental modifications.

15.3 Preliminary scientific studies

15.3.1 *Field studies*

The main sampling site was on the River Beaume, where three sampling stations corresponding to three suitable habitats along a 1.8-km stretch of river were chosen. Upstream dispersal was limited by a weir. Fish were located by night using headlights as the eyes of *Z. asper* reflect light beams. It was then possible to capture individuals with landing nets. To standardise effort, four people prospected the station for 2 h, but capture probability varied according to habitat type, discharge, or water transparency. Fish were tagged using passive integrated transponders, and a fin clip provided samples for genetic microsatellite analysis. Three habitat variables (depth, water velocity, and particle size) were sampled at the point of each captured fish to describe the habitat preferences of the species in the River Beaume. This protocol was carried out four times a year, once in each season.

Strong habitat preferences for the species were found for the three physical variables studied (water velocity, depth, and roughness) (Allouche, Gaudin & Labonne 1999), which resulted in an aggregated distribution along the river. Fish concentrated in run and riffle heads, and pools were inhabited only when connected to a run. Lotic and lentic channels appeared to be avoided. In the spawning period, a shift in habitat occurred at the meso-scale, when a proportion of the fish left the run to relocate on riffles for spawning (Fig. 15.3). Thus direct connection between runs and riffles is necessary for successful spawning. Home-range data in the spawning period and low numbers of spawners in each station suggested sexual competition between males (Labonne & Gaudin 1999, 2000). Up to 7 males m^{-2} were sometimes observed on spawning site in the riffles, a situation not observed at other times. Gravid females preferred deeper habitats with lower velocity (Fig. 15.3). There was no clear evidence of synchronisation of female movements towards riffles for spawning.

The population demography showed strong heterogeneity in spatial dynamics. The fish aggregated in each station were considered as a functional dynamic unit. Some

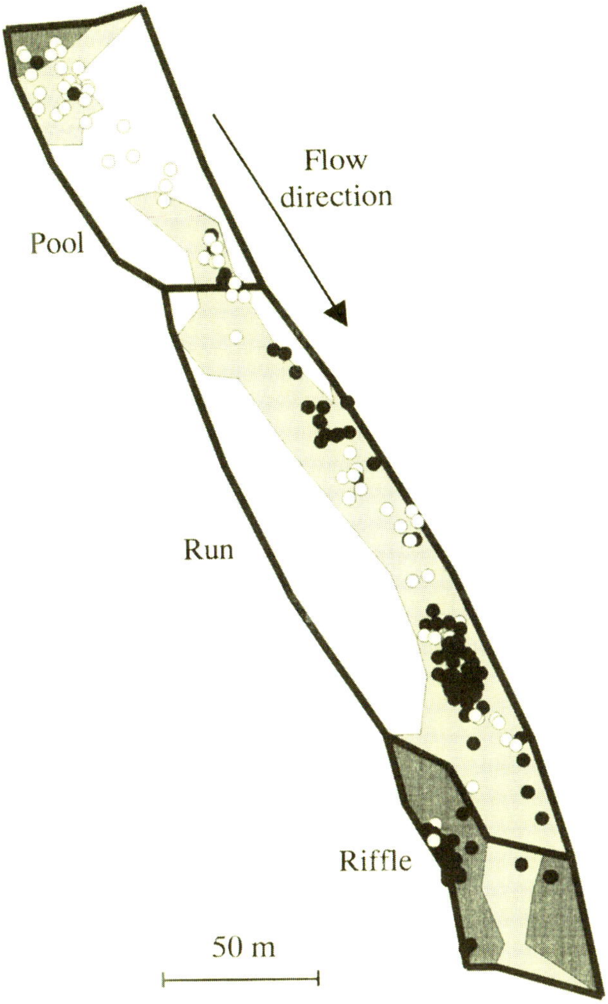

Figure 15.3 Distribution of *Z. asper* in one station during the spawning period. (Empty circles for female, full circle for male; white area is 0–0.05 m s^{-1} water velocity, light grey area for 0.05–0.2 m s^{-1} water velocity, and dark grey area for up to 0.2 m s^{-1} water velocity)

stations were considered as sources, whilst others were sinks (Pulliam 1988; Schlosser 1998). Significant adult movements were observed between stations. A relatively high proportion of fish (0–40% depending on the station) showed non-sedentary behaviour. Some of these fish were in constant displacement. High immigration rates were recorded in some stations. Estimation of survival rate (around 0.5) remained difficult because of high temporal variations due to environmental conditions.

The populations in the rivers Durance and Beaume were genetically distinct (Durand & Laroche 2000). This could be due to geographical factors as well as recent reproductive isolation. Genetic homogeneity between fish of the same river confirmed that the term 'population' is acceptable. Occasional specimens found in other tributaries of the rivers Ardeche and Durance were genetically similar to one of these two main populations.

A major objective of the preliminary studies was to identify the potential role of fish movements in population viability. This is deemed essential because the only threat affecting the whole Rhône basin is habitat fragmentation by dams, weirs, channels, and habitat loss. Dams were first built on the River Rhône in 1952, and earlier on some tributaries (1931 on the River Ain). On the 522 km of the French River Rhône, the presence of 19 hydropower plants, 14 large locks, and 24 dams, partitions the river into reaches shorter than 35 km and sections with natural slope and discharge now represent less than 5% of the 522 km. Moreover, the main tributaries (e.g. rivers Ardeche and Durance) are disconnected from the main river, small hydropower plants have been developed on most of them, and thresholds were built to reduce erosion processes, increasing fragmentation of the head waters.

Consequently the major fragmentation of the Rhône basin, whatever the spatial scale, which is a major problem for migratory fish like shad, *Alosa alosa* L., eel, *Anguilla anguilla* L., sea lamprey, *Petromyzon marinus* L., has probably also a negative effect for the so-called sedentary species because it induces reproductive isolation. The study of migration patterns of a species is thus an essential element in any conservation strategy.

15.3.2 *Experimental studies*

The first attempt to breed *Z. asper* artificially was in March 2000. The main aim was to assess the feasibility of captive breeding; any restocking perspective was excluded. Four females and six males were caught in the River Beaume during the spawning season. Ova of each female were stripped and mixed dry with the sperm from two males. One of the four fertilisations hatched correctly and a second had a low hatching success (10%). One female died after stripping. Altogether 1500 eggs were incubated in the laboratory at 13.5°C for 14 days. The larvae hatched at about 7-mm total length. The transition to exogenous feeding began 7 days after hatching. Young fish refused artificial food but accepted zooplankton and *Artemia* nauplii. When 2-cm long, fish could be fed with small frozen Chironomidae larvae. After 1 year, about 600 young fish were still alive, so captive rearing of small numbers of *Z. asper* is potentially feasible.

Further understanding of the habitat preferences and behaviour of *Z. asper*, especially the younger life stages up to 3 months of age, will be determined using a purpose-built, 35-m long and 2–4 m wide, artificial channel, which has a range of important habitats, viz. run, riffle, and pool. The discharge is 50–70 L s^{-1}, depth from 5 to 75 cm, substratum roughness varies from sand to stones, and water velocity, which is low in the pool, reaches 50 cm s^{-1} in the riffle. Basic behaviour such as rheophily, attraction by light and use of the water column will be studied during ontogenesis and more complex behaviour will also be observed.

Because the studies on population dynamics showed the importance of movements and connectivity between habitat units, a miniature fish pass was constructed in the artificial stream to determine the natural water velocity threshold limiting the upstream movement of *Z. asper*. This will be used to determine the maximum water velocity that a fish can tolerate over short distances and to test the attraction of different model fish passes to the ascent of fish (i.e. impact of slope, vertical barrier, type of material following Larinier, Porcher, Travade & Gosset 1996).

15.4 Conservation actions

These preliminary results already provide the basis to reverse the decline of this species and manage the natural populations. The decline seems to result mainly from habitat fragmentation that isolates populations, and makes them more sensitive to local disturbances (floods, droughts or pollution). This information was used to draw the attention of the local authorities, institutions responsible for water management and fishery organisations to collaborate in a conservation action plan for *Z. asper*. This task is difficult because of the poor economic value of the fish, but it is nevertheless essential to ensure the conservation of the remaining populations and restore good conditions for their survival. Since the species was largely unknown to the public, anglers and river managers, a workshop was organised in April 1999 to educate a wide diversity of interests (managers, naturalists, fish guards, and anglers) across the Rhône basin. Information on the basic ecology, habitat preferences, and sampling methods was presented, and good feedback on both the past and present distribution of the species was obtained from participants. Moreover, it helped to make river managers, anglers and scientists exchange information. Illustrative materials were produced, in co-operation with the Regional Environment Direction (DIREN), to provide suitable materials for distribution to water and river managers and anglers. A video for freshwater managers of the Rhône basin was also produced to describe the current understanding of the ecology of the species and the objectives of the LIFE project.

It is hoped the first effects of this awareness campaign will be felt before the end of the LIFE programme (end of 2001). In the River Beaume, it is expected that the species will be included in the river management plan (led by the local councillors), effort will be invested in surveying the river thoroughly, and increased attention will be given by the water authorities when authorising work (administration and fishery council) on the river. In the future, the procedures adopted in the River Beaume will be implemented on the Durance basin, but special attention will be paid to the problems caused by flow regulation for hydro-electric power production. In wider Rhône basin, specific measures for the conservation of this species should be implemented through local and regional action plans, including the re-establishment of connectivity for non-migratory fish or the global rehabilitation of the river. Only after this agreement in local and national management river policy, should a restocking programme be considered, but in accordance with maintaining the genetic integrity of the populations.

Finally it should be recognised that the conservation of a species such as *Z. asper* involves many stakeholders, often with high economic stakes, e.g. production of energy, industry or tourism, so the task will take time and considerable effort to maintain awareness of the precarious state of the species. Step by step, people and institutions have to be convinced, because no legal measure will totally guarantee protection.

References

Allouche S., Gaudin P. & Labonne J. (1999) Preferences d'habitat de l'Apron du Rhône, *Zingel asper* (L., 1758): Etude de la population de la Beaume, affluent de l'Ardeche. Rapport RNF, University of Lyon 1, Quetigny, 14 pp. (in French).

Baillie J. & Groombridge B. (1996) *IUCN Red List of Threatened Animals*. Gland: IUCN, Switzerland, Gland: 368 pp. +annexes.

Boutitie F. (1984) L'Apron – *Zingel asper* (LINNE), Percidae – Poisson rare menacé de disparition (Biologie – Répartition – Habitat). DEA, University of Lyon l, 22 pp. (in French).

Dottrens E. (1951–1952) *Les Poissons d'eau Douce*. Vol. 2. Neuchatel: Delachaux & Niestle, 43 pp. (in French).

Durand J.D. & Laroche J. (2000) Structure génétique et conservation des populations d'Apron. Rapport de fin de travaux. Rapport RNF, University of Lyon l, Quetigny, 22 pp. (in French).

Huet M. (1949) Aperçu des relations entre la pente et les populations piscicoles des eaux courantes. *Revue suisse d'Hydrobiologie* **11**, 332–351 (in French).

Labonne J. & Gaudin P. (1999) Rapport intermédiaire du volet scientifique du programme LIFE Apron: Etude de la dynamique de populations d'aprons de la Beaume. Rapport RNF, University of Lyon l, Quetigny, 19 pp. (in French).

Labonne J. & Gaudin P. (2000) Eléments de dynamique des populations d'aprons sur la rivière Beaume. Rapport de fin de travaux Rapport RNF, University of Lyon l, Quetigny, l9 pp. (in French).

Larinier M., Porcher J.P., Travade F. & Gosset C. (1996) Passes a poissons: expertise et conception des ouvrages de franchissement. Paris: Conseil Supérieur de la Pêche, 336 pp. (in French).

Leger L. & Stankovitch S. (1921) Fécondation artificielle et développement de l'Apron. *Travaux du Laboratoire d'Hydrobiologie et de Pisciculture de Grenoble*, pp. 335–338 (in French).

Moullec P., Chappaz R. & Cavali L. (2000). L'apron, *Zingel asper*, dans le bassin de la Durance sur l'axe Serre-Ponçon – Sisteron (Département des Hautes-Alpes). Biologie et répartition de l'espèce. Rapport d'avancement. CSP Paris, 41 pp.+annexes (in French).

OFEFP (Office fédéral de l'Environnement, des Forets et du Paysage) (1999) Concept de protection de l'apron (*Zingel asper*): recensement des effectifs dans le Doubs, 44 pp. (in French).

Perrin J.F. (1988) Maintien en aquarium de l'Apron du Rhône, *Zingel asper* (L.), espèce menacée d'extinction. *Revue française d'Aquariologie* **15**, 17–20 (in French).

Pulliam H.R. (1988) Sources, sinks and population regulation. *The American Naturalist* **132**, 652–661.

Schlosser I.J. (1998) Fish recruitment, dispersal, and their trophic interactions in a heterogeneous lotic environment. *Oecologia* **113**, 260–268.

Spillmann Ch.J. (1961) *Faune de France – Poissons d'Eau Douce*. Vol. 5, Paris: Le Chevallier Ed, 304 pp. (in French).

Chapter 16
Integrated approach to the conservation and restoration of Danube salmon, *Hucho hucho*, populations in Austria

S. SCHMUTZ,* A. ZITEK, S. ZOBL, M. JUNGWIRTH
Department of Hydrobiology, Fisheries and Aquaculture, University of Agricultural Sciences, Vienna, Austria

N. KNOPF, E. KRAUS
Regional Government of Lower Austria, St Pölten, Austria

T. BAUER and T. KAUFMANN
Freiwasser, Environmental Engineering Consultant, Vienna, Austria

Abstract

The Danube salmon, *Hucho hucho*, is among the most endangered fish species in Europe. In Austria, as a consequence of river channelisation and hydropower development, only 10% of its former distribution (about 4500 km river length) harbours self-sustaining populations. This chapter focuses on the development and implementation of an integrated approach to protect and restore Danube salmon populations in NATURA 2000 areas within the EU LIFE-NATURE project. The project is located on two tributaries of the Danube in Lower Austria (rivers Pielach and Melk) that are disconnected from the latter by weirs and artificial drops blocking spawning runs. The objectives of the project are to develop, implement and monitor rehabilitation measures, in particular the re-establishment of the continuum (fish passes) and the improvement of habitat conditions (channel rehabilitation). Pre-implementation monitoring showed that the largest population of Danube salmon currently inhabits the Pielach River (21 km), with an estimated population size of 532 individuals and 162 adults. The population is divided by several migratory obstacles into six sub-populations with sizes ranging from 0 to 50 adults. In the Melk and Mank River (24 km), only nine and 17 adults were present. Remaining populations are small and fragmented and lie close to or below requirements of minimum population sizes. Rehabilitation measures aim to provide free passage and population extensions within a Danube tributary river network of 78 km to enhance the viability of the populations and the long-term survival of the species.

Keywords: current status, Danube salmon, distribution, habitat restoration, minimum population size.

16.1 Introduction

The Danube salmon, *Hucho hucho* L., is one of the most endangered fish species in Europe. Austria represents the centre of its natural distribution, but hydroelectric

*Correspondence: Department of Hydrobiology, Fisheries and Aquaculture, University of Agricultural Sciences, Vienna, Max Emanuelstr. 17, 1180 Vienna, Austria (email: schmutz@mail.boku.ac.at).

power developments and river channelisation have left only a few populations. The Danube salmon is listed in Annex II of the Habitats Directive, forcing Austria to nominate protected areas and set actions for the rehabilitation of the species under the European Union (EU) NATURA 2000 network of protected areas (Habitat and Bird Directive). Within these areas, LIFE-NATURE projects, have been promoted to conserve natural habitats and wild fauna and flora of EU interest.

This chapter focuses on:

(1) the description of the former and current distribution of the Danube salmon in Austria;
(2) the development and implementation of an integrated approach protecting and restoring Danube salmon populations in NATURA 2000 areas within the LIFE project 'Living space of Danube salmon' (LIFE99 NAT/A/006054; from 1999 to 2003); and
(3) the characterisation of the status of Danube salmon in the project area prior to implementation of a comprehensive monitoring programme.

16.2 Status of Danube salmon

The Danube salmon is the largest salmonid and one of the biggest freshwater fish species in the world (length >140 cm, weight >60 kg, age >15 years). It is a rheophilic species, spawning on gravel (lithophilic) in early spring after migrating upstream in the main river or into tributaries. Its piscivorous behaviour is well pronounced even in the juvenile stage. *Chondrostoma nasus* (L.) and *Thymallus thymallus* (L.) are its main prey fish.

Austria lies in the heart of the natural distribution of this species, endemic to the Danube catchment. Beyond Austria, larger populations within the natural zoogeographical distribution were once found in Bavaria (Germany), former Yugoslavia, Slovakia and western Ukraine, although only a few of them still remain (Harsanyi 1982; Holcik, Hensel, Nieslanik & Skacel 1988; see also Freyhof, Chapter 1).

Before human alterations, the species inhabited the Austrian Danube and its larger tributaries belonging to the grayling and barbel zones over a total of >2500 km (Table 16.1, Fig. 16.1). The largest stock occurred in the River Danube and strongly depended on intact connectivity to the tributaries for spawning; it had already declined dramatically by the late 19th century (Jungwirth 1984a). The long-term viability of populations in smaller tributaries, due to their small sizes, probably relied on exchanges with other populations to maintain genetic diversity and to balance population fluctuations. According to meta-population theory, the type and intensity of connectivity between sub-populations are the decisive factors in regulating population exchange. Schmutz and Jungwirth (1999) discussed several potential meta-population structures of the former situation in Austria:

(1) a single, large, genetically homogeneous population within the Danube and separate populations within the tributaries with very limited exchange;
(2) several sub-populations in the Danube defined by their areas of reproduction within the tributaries; or

Table 16.1 Historic distribution of the Danube salmon in Austria

Main river	Tributary	River segment	References
Donau		Total length	Jungwirth 1980, 1984a; Harsanyi 1982; Holcik *et al.* 1988;
Inn		Mouth to Landeck	Jungwirth 1980; Harsanyi 1982; Holcik *et al.* 1988; Jungwirth, Schmutz & Waidbacher 1989; Haidvogl & Waidbacher 1997
	Salzach (r): Saalach, Lammer	Mouth to Mittersill	Kollmann 1898; Jungwirth 1980; Harsanyi 1982; Holcik *et al.* 1988; Wiesbauer, Bauer, Jagsch, Jungwirth & Uiblein 1991; Haidvogl & Waidbacher 1997
Großache		Lower course	Kähsbauer 1961
Große Mühl		Lower course	Koller 1907; Harsanyi 1982; Holcik *et al.* 1988
Traun		Mouth to Traunfall	Jungwirth 1980; Harsanyi 1982; Holcik *et al.* 1988; Haidvogl & Waidbacher 1997
	Alm (r)	Lower course	Harsanyi 1982; Holcik *et al.* 1988
	Krems (UA)(r)	Lower course	Jungwirth 1980; Holcik *et al.* 1988
	Ager (l): Aurach, Vöckla	Lower course	Hawlitschek 1895; Koller 1907; Holcik *et al.* 1988
Enns		Mouth to Haus	Jungwirth 1980; Harsanyi 1982; Holcik *et al.* 1988; Woschitz 1991; Haidvogl & Waidbacher 1997
	Palten (r)	Lower course	Krafft 1874; Haidvogl & Waidbacher 1997
	Salza (r)	Mouth to Fachwerk	Hampel 1882; Haidvogl & Waidbacher 1997
	Steyr (l)	Lower course	Hawlitschek 1895; Kähsbauer 1961; Jungwirth 1980; Harsanyi 1982; Holcik *et al.* 1988
Ybbs		Mouth to Waidhofen	Harsanyi 1982; Jungwirth 1980; Holcik *et al.* 1988
	Url (l)	Lower course	
	Ferschnitz (r)	Lower course	Jungwirth 1984b
Erlauf		Mouth to Scheibbs	Jungwirth 1980
Melk		Mouth to Oberndorf	Jungwirth 1980; Harsanyi 1982; Holcik *et al.* 1988
	Mank (r)	Lower course	Jungwirth 1984b
Pielach		Mouth to Kirchberg	Jungwirth 1980; Harsanyi 1982; Holcik *et al.* 1988
Krems (LA)		Lower course	Wedland 1965
Traisen		Mouth to Lilienfeld	Hawlitschek 1895; Jungwirth 1980; Holcik *et al.* 1988; Haidvogl & Waidbacher 1997

Table 16.1 continued

Main river	Tributary	River segment	References
Kamp		Lower course	Anonymous 1907; Harsanyi 1982; Holcik *et al.* 1988
Schwechat		Lower course	Starmühlner 1969
Mur		Jugoslavian border to Murau	Jungwirth 1980; Harsanyi 1982; Holcik *et al.* 1988; Haidvogl & Waidbacher 1997
	Pöls (l)	Mouth to Pöls	Woschitz 1991
	Mürz (l)	Mouth to Kindberg	Woschitz 1991
	Kainach (r)	Lower course	Kreissl 1991; Woschitz 1991
	Sulm (r): Schwarze and Weiße Sulm	Mouth to Brunnpolfing	Wallner 1917; Kreissl 1991; Woschitz 1991
Drau		Jugoslavian border to Lienz	Hartmann 1898; Jungwirth 1980; Harsanyi 1982; Holcik *et al.* 1988; Haidvogl & Waidbacher 1997
	Isel (l)	Lower course	Hartlieb 1948; Michor 1990
	Möll (l)	Lower course	Hartmann 1898; Jungwirth 1980; Holcik *et al.* 1988; Jungwirth, Moog, Schmutz & Wiesbauer 1990
	Lieser (l)	Lower course	Jungwirth *et al.* 1990
	Gail (r)	Mouth to Kötschach-Mauten	Hartmann 1898; Honsig-Erlenburg & Schulz 1989; Holcik *et al.* 1988; Jungwirth *et al.* 1990
	Gurk (l)	Mouth to Straßburg	Hartmann 1898; Jungwirth 1980; Holcik *et al.* 1988; Jungwirth *et al.* 1990
	Glan	Lower course	Jungwirth *et al.* 1990
	Vellach (r)	Lower course	Hartmann 1898; Holcik *et al.* 1988; Jungwirth *et al.* 1990
	Lavant (l)	Mouth to St Andrä	Hartmann 1898; Holcik *et al.* 1988; Jungwirth *et al.* 1990; Haidvogl & Waidbacher 1997

l, Left-sided tributary; r, right-sided tributary, UA, Upper Austria, LA, Lower Austria (compare with Fig. 16.1).

(3) the Danube functions exclusively as a habitat for sub-adults and adults (sink), while reproduction and development of larvae and juveniles take place in the tributaries (source).

As a consequence of river channelisation and hydropower development, most of the populations collapsed. Today only 10% are self-sustaining populations (based on natural reproduction and recruitment, Fig. 16.1). The largest population inhabits a 110-km reach of the River Mur between the cities of Murau and Leoben (Styria), consisting of

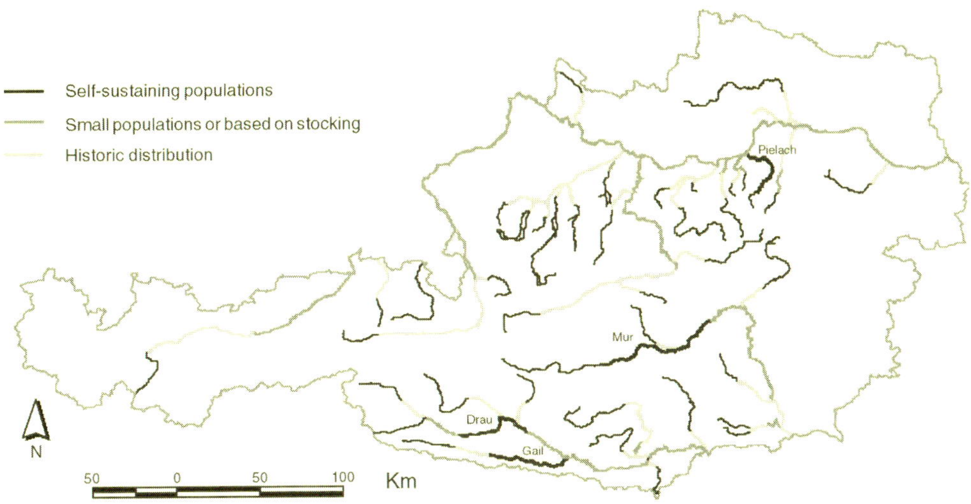

Figure 16.1 Historic and current distribution of the Danube salmon in Austria (*cf.* Table 16.1)

about 2000 adult individuals (Kaufmann, Muhar, Raderbauer, Rathschüler, Schmutz, Waidbacher & Zauner 1991). The most reproductively viable population is in the River Pielach, which is covered by the LIFE project (see below). Smaller populations still exist in the upper River Drau and one of its tributaries (lower River Gail). All other occurrences (shown in Fig. 16.1) are based on extremely low densities and/or stocking. In the Danube, a comparably high density has been re-established by continuous stocking interventions over the last two decades, but access to former spawning areas in the tributaries is nearly completely blocked. Due to the extinction of most populations, the Danube salmon is classified as 'threatened with extinction' in Austria (Spindler, Zauner, Mikschi, Kummer, Wais & Spolwind 1997) and listed as an endangered species according to the IUCN criteria and the Bern Convention (Appendix III).

The size and abundance of Danube salmon made it a species of high commercial interest for many centuries. Nowadays it is a high-value species in recreational fisheries and is protected by strict fishery regulations. Comprehensive conservation activities started in the late 1970s with the development of artificial propagation methodologies to produce fish for stocking (Jungwirth 1979) and were followed in the 1980s by local habitat enhancements (Jungwirth, Muhar & Schmutz 1995) and building of fish passes (Jungwirth & Pelikan 1989).

16.3 Development of an integrated approach for the conservation and recovery of Danube salmon in a NATURA 2000 area

In Austria, the running water rehabilitation philosophy is based on the concept of Ecological Integrity (EI), the 'maintenance of all internal and external processes and attributes interacting with the environment in such a way that the biotic community corresponds to the natural state of the type-specific aquatic habitat' (ÖNORM/M6232

1995; Chovanec, Jäger, Jungwirth, Koller-Kreimel, Moog, Muhar & Schmutz 2000). An interdisciplinary approach is therefore inherent in river rehabilitation methodology and is widely used in developing integrated river management plans (Muhar 1994).

The integrated approach of the project is based on an inter- and transdisciplinary network (Fig. 16.2). The regional authorities for nature conservation and river engineering co-operate as both project leaders and responsible institutions for the implementation of rehabilitation measures. Civil engineering/environmental consultants are in charge of the detailed planning of restoration measures and an NGO (WWF) is involved in wetland protection and rehabilitation activities. Together, the teams negotiate the necessary measures with landowners and users. Local authorities implement the river engineering measures through private construction companies that are supervised by the planning consultant. A scientific institute (HFA-BOKU) acts as an advisor throughout the course of the project. In addition the HFA runs the accompanying monitoring programme. Other authorities (agriculture, waste water, fishery, tourism), NGOs, and the landowners and other users are assembled in an advisory board interacting with the project team on a regular basis. As small-scale hydropower schemes are one of the major concerns, one of the hydropower operators is integrated in the project team as a consultant. Public information campaigns take place regularly in the form of

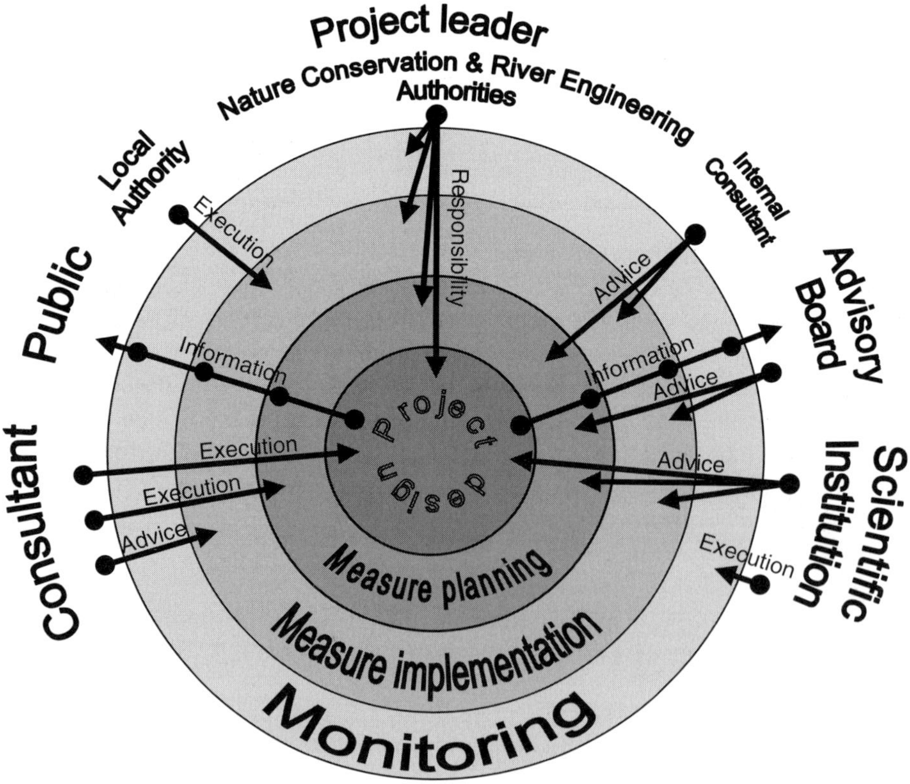

Figure 16.2 Project design, involved parties and their functional interactions

press releases, folders, local presentations, excursions, etc. Hence, the integrated approach of the project is based on a balanced team of various disciplines and a functional network between authorities, executive and advisory bodies, and the public.

16.3.1 *Study site*

The LIFE project is located in Lower Austria and consists of a river network of a free-flowing section of the Danube (Wachau, 33 km length), the near-natural tributary, River Pielach (21 km), the channelised tributary, River Melk (16 km), and the semi-natural River Mank (8 km), a tributary of the latter (Fig. 16.3). The main characteristics of these rivers are given in Table 16.2. All three rivers are at the transition from hyporhithral to epipotamal, whereas lowland river characteristics prevail in the Melk. For example, mean summer water temperatures at the Melk are about 2°C higher (19.5°C in July–August 2000) than at the Pielach (17.5°C in July–August 2000).

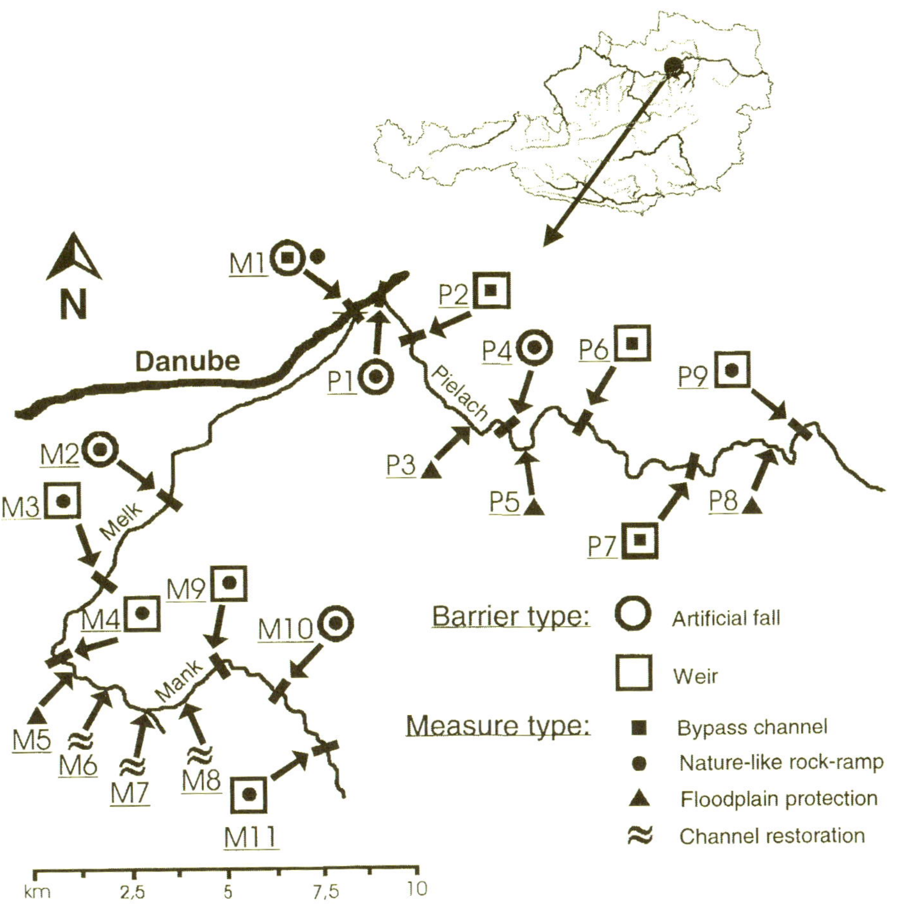

Figure 16.3 Study area: location of migratory barriers and restoration measures (*cf.* Table 16.3)

Table 16.2 Main characteristics of the rivers Danube, Melk, Pielach and Mank

Characteristic	Danube	Melk	Pielach	Mank
Stream order	9	5	4	4
Altitude (maA)	209–196	233–206	252–205	270.5–233
Gradient (‰)	0.4	1.6	2.23	4.6
Mean flow ($m\,s^{-1}$)	~1900	~3	~6.5	0.5
Flow regime	Moderate nival	Winter pluvial	Pluvial-nivo	Winter pluvial
Fish region	Epipotamal	Epipotamal	Hyporhithral/ epipotamal	Hyporhithral/ epipotamal
Typical fish species	*Chondrostoma nasus* *Barbus barbus* *Hucho hucho* *Leuciscus cephalus*	*Chondrostoma nasus* *Barbus barbus* *Hucho hucho* *Leuciscus cephalus*	*Chondrostoma nasus* *Hucho hucho* *Barbus barbus* *Thymallus thymallus* *Leuciscus cephalus*	*Chondrostoma nasus* *Barbus barbus* *Hucho hucho* *Thymallus thymallus* *Leuciscus cephalus*

The river continuum is interrupted by artificial falls as a result of river engineering measures (1–4 m height) and by weirs built for hydroelectric power production (1.2–3.5 m height). In the study area a total of 13 obstacles were identified (Table 16.3). As a consequence, former spawning runs of Danube salmon and other species are blocked at the mouth of the Melk. A few Danube salmon and hundreds to thousands of *Chondrostoma nasus*, *Barbus barbus* (L.) and *Vimba vimba* (L.) still immigrate from the Danube into the lowest section of the Pielach every year to spawn below the first larger barrier (1.6 km upstream of the mouth).

Hydropower use has a long tradition in this area and goes back to Roman times (Berger 1999). Today, large stretches in the Pielach, Melk and Mank rivers within the study area are affected by water abstraction (37%, 5%, 16%) and impoundment (15%, 26%, 11%). While the Melk and most of the Mank are channelised, the Pielach has retained some of its natural morphological characteristics, such as meandering sections, side arms, dynamic gravel bars, large woody debris and small oxbows. In addition, inundation areas and floodplain forests accompany the river along most of its course. Based on a nation-wide survey (Muhar, Kainz & Schwarz 1998), the Pielach was classified as the most nature-like, lowland river (altitude <250 m) in Austria. About one-third of its length within the project area is still in 'high' (23.6%) and 'good' (10.3%) hydromorphological condition.

16.3.2 *Restoration and protection objectives*

The objectives of the LIFE project are to develop, implement and monitor restoration measures between 1999 and 2003. The main focus is on re-establishing the continuum between the Danube and the tributaries and within the tributaries; the result should be

Table 16.3 Actual situation, location (distance from mouth), height of barriers, length of stretches and type of restoration measure at the Pielach, Melk and Mank (compare to Fig. 16.3)

River	Code	Actual situation	Distance from mouth (m)	Height, h and length, l (m)	Measure
Pielach	P1	Artificial fall[a]	0	$h = 1$	Nature-like rock-ramp
Pielach	P2	Weir	1600	$h = 2$	Bypass channel
Pielach	P3	Nature-like stretch	3200–5500	$l = 2300$	Purchase of 5 ha land
Pielach	P4	Artificial fall	5500	$h = 1.5$	Nature-like rock-ramp
Pielach	P5	Nature-like stretch	6300–6550	$l = 220$	Purchase of 1.6 ha land
Pielach	P6	Weir	8300	$h = 3.5$	Bypass channel
Pielach	P7	Weir	13 300	$h = 2.5$	Bypass channel
Pielach	P8	Nature-like stretch	15 600–17 500	$l = 1900$	Purchase of 5 ha land
Pielach	P9	Weir	17 500	$h = 1.2$	Nature-like rock-ramp
Melk	M1	Artificial fall	250	$h = 4$	Bypass channel + nature-like rock-ramp
Melk	M2	Artificial fall	7000	$h = 1$	Nature-like rock-ramp
Melk	M3	Weir	9500	$h = 3$	Nature-like rock-ramp
Melk	M4	Weir	12 200	$h = 1.5$	Nature-like rock-ramp
Melk	M5	Oxbow	13 300	$l = 950$	Purchase of 0.5 ha land
Melk	M6	Regulated channel	14 000–15 200	$l = 1200$	Channel restoration
Mank	M7	Regulated channel	15 200–15 400	$l = 200$	Restoration of river mouth + purchase of 2 ha land
Mank	M8	Regulated channel	15 400–16 900	$l = 1500$	Channel restoration
Mank	M9	Weir	17 800	$h = 1.2$	Nature-like rock-ramp
Mank	M10	Artificial fall	20 000	$h = 2$	Nature-like rock-ramp
Mank	M11	Weir	22 300	$h = 2.5$	Bypass channel or nature-like rock-ramp

[a]Migration barrier at low flow condition in the Danube.

a re-connected river network 78 km long. Further objectives are to improve habitat conditions in the Melk and Mank and to protect the nature-like meandering section at the Pielach.

16.3.3 *Rehabilitation and protection measures*

The re-establishment of the river continuum with respect to fish migration will be realised by converting obstacles to rock ramps or circumventing them with nature-like bypass channels (measures code P1, P2, P4, P6, P7, P9 in Table 16.3, Fig. 16.3). Relevant technologies have been developed in Austria over the last two decades (Parasiewicz, Eberstaller, Weiss & Schmutz 1998) and have been successfully applied to various case studies (e.g. Eberstaller, Hinterhofer & Parasiewicz 1988; Jungwirth 1996; Mader, Unfer & Schmutz 1998; Schmutz, Giefing & Wiesner 1998).

To regain natural or at least residual flows in abstracted stretches and to supply bypass channels with adequate flows, the water abstraction rights are partly or completely taken over by providing compensation payments to the hydropower owners.

Habitat improvements will be initialised by rehabilitating the regulated Melk and Mank channel at three sites (in total about 2 km, measure code M6, M7, M8 in Table 16.3, Fig. 16.3) and further efforts will take place within the framework of continuous maintenance activities of the local water engineering authority. The rivers will be rebuilt and adapted to the former meandering shape according to the natural type-specific reference conditions; they will also be widened to regain freedom for dynamic channel developments.

Protection of the nature-like meandering section at the Pielach and its adjacent floodplains will be supported by buying about 12 ha of riparian land (Mühlau, Neubacher Au). This will enable further dynamic development of eroding banks and meanders specific to this morphological river type.

16.4 Monitoring programme

The objectives of the monitoring programme are to observe and evaluate the efficiency of rehabilitation measures. Accordingly, the monitoring programme focuses on the efficiency of implemented migration facilities. In addition, the populations of the Danube salmon, its main prey species, *Chondrostoma nasus* and *Barbus barbus*, and the general development of the fish community will be observed before and after implementation.

16.4.1 *Monitoring methods*

The development of fish populations was monitored before, and will be monitored after, the implementation of the various measures at both representative and rehabilitated sites. Prior to sampling the fish, a habitat mapping survey is carried out to identify major mesohabitat types and spatial distributions. In 1999, 22 and 15 mesohabitats (pool, run, riffle) were sampled at the Pielach and Melk rivers, respectively, by electric fishing using the removal method. Due to electric fishing selectivity, small-sized species were excluded from estimations. Fishes >200 mm were tagged with VI tags and the fin clipping technique was used for Danube salmon <200 mm. Populations were estimated by weighting mesohabitat values proportionally to their occurrence in the specific river sections. In the River Mank, due to its small channel size and clear water conditions, a combination of visual observation and snorkelling techniques was used to assess population densities and structures of the entire study area. Further details are given in Zitek, Zobel, Schmutz and Jungwirth (2000).

Table 16.4 Species composition of the rivers Pielach, Melk and Mank

		River		
Family	Species	Pielach	Melk	Mank
Balitoridae	*Barbatula barbatula*	f	f	f
	Cobitis aurata	–	r	–
	Cobitis taenia	r	–	–
Cottidae	*Cottus gobio*	d	f	0.2
Cyprinidae	*Alburnoides bipunctatus*	f	f	–
	Aspius aspius	–	0.1	–
	Barbus barbus	18.5	6.9	7.8
	Carassius auratus	–	r	–
	Carassius auratus gibelio	–	0.3	–
	Chondrostoma nasus	18.2	28.3	13.5
	Cyprinus carpio	0.9	0.1	–
	Gobio gobio	f	d	3.9
	Leuciscus cephalus	42.0	51.6	59.8
	Leuciscus leuciscus	r	r	–
	Phoxinus phoxinus	d	d	d
	Rutilus rutilus	–	2.9	–
	Scardinius erythrophthalmus	–	1.4	–
	Vimba vimba	r	r	–
Esocidae	*Esox lucius*	–	0.2	–
Gadidae	*Lota lota*	0.5	0.5	–
Gobiidae	*Proterorhinus marmoratus*	–	r	–
Percidae	*Perca fluviatilis*	0.1	0.5	–
	Zingel zingel	0.6	0.02	–
Salmonidae	*Hucho hucho*	3.6	1.5	2.8
	Oncorhynchus mykiss	2.0	0.5	1.8
	Salmo trutta	1.9	4.3	3.5
	Salvelinus fontinalis	0.2	0.2	–
	Thymallus thymallus	11.5	0.8	6.7
Number of species		19	26	11

Categories for small-sized species: d, dominant; f, frequent; r, rare.

16.4.2 *Monitoring results of the situation prior to measure implementation*

In total, the fish fauna of the rivers Pielach, Melk and Mank comprises 19, 26 and 11 species, respectively (Table 16.4). *Leuciscus cephalus* (L.) dominates in all three rivers, followed by *Chondrostoma nasus* and *Barbus barbus*. The lower number of species and the higher proportion of *Thymallus thymallus* in the Pielach (11.5%) and Mank (6.7%) underline their more rhithralic character compared with the Melk (0.8%).

The largest population of Danube salmon occurs in the Pielach (total number of individuals in the study area is 532) accounting for 3.6% of the total fish density (17% of total biomass). Of those, about 162 individuals were adults (>70 cm total length). The length frequency distribution of the population showed a high proportion of

0+ along with representatives of all successive size classes (Fig. 16.4). Danube salmon rarely exceeds a total length of 1 m in the Pielach. The population is divided by several migratory obstacles into six sub-populations with sizes ranging from 0 to 50 adults (median: 30, Table 16.5).

In the Melk and Mank, a total of 250 and 50 Danube salmons were present, with adults accounting for only nine and 17 individuals, respectively (Table 16.6).

Although the estimated total population sizes of the three dominant species per river were high, the values between the barriers were below 1000 individuals except for *Leuciscus cephalus* (Tables 16.5 and 16.6). Figure 16.5 shows the length frequency distribution of the three dominant species in the Pielach. All three were dominated by large adults and showed significant deficiencies of juveniles or even a complete lack of sub-adults (*Chondrostoma nasus*).

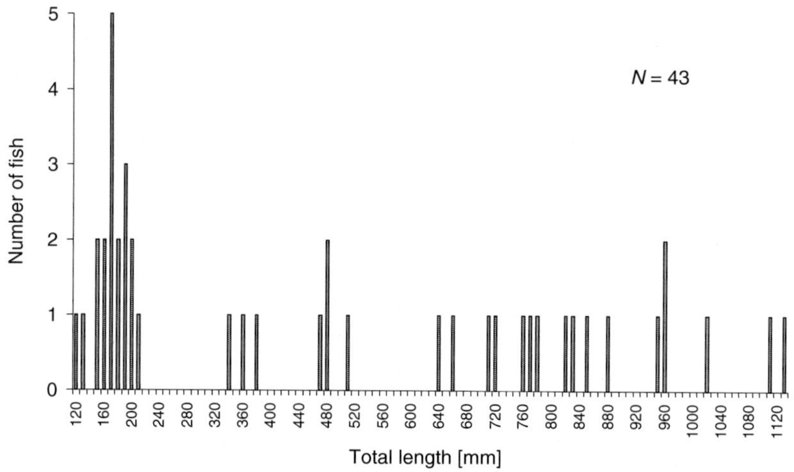

Figure 16.4 Total length distribution of the Danube salmon captured (N = 43) in the Pielach River

Table 16.5 Number of adult fish between migratory barriers in the Pielach (impoundments are included) (compare Fig. 16.3, Table 16.3)

Section	1	2	3	4	5	6	Median	Total
Barrier code	P1–P2	P2–P4	P4–P6	P6–P7	P7–P9	P9		
Length (m)	1760	3703	2884	4918	4420	864	3294	18549
Species								
Leuciscus cephalus	402	1235	710	1391	1397	200	973	5336
Barbus barbus	129	530	315	707	620	116	423	2418
Hucho hucho	0	47	24	35	50	6	29	162
Chondrostoma nasus	190	362	241	478	428	165	301	1864

Table 16.6 Number of adult fish between migratory barriers in the Melk River (impoundments are excluded) (compare Fig. 16.3, Table 16.3)

Section	1	2	3	4	Median	Total
Barrier code	M1–M2	M2–M3	M3–M4	M4–M6		
Length (m)	4855	2325	1728	2391	2358	11 299
Species						
Leuciscus cephalus	1081	531	107	819	675	2539
Barbus barbus	34	0	144	829	89	1006
Hucho hucho	0	0	0	9	0	9
Chondrostoma nasus	501	239	621	74	370	1435

16.5 Discussion

The decline of fish populations and the extinction of sensitive species in cultural landscapes are rarely caused by a single human impact; they mostly involve the consequences of many activities and their complex interactions. Rehabilitation and protection strategies must target every potentially responsible party and operate within an integrated strategy. From the onset, the LIFE project strove to integrate administrative, executive, scientific and public bodies into a common approach. This represents the first comprehensive and integrated effort to restore and protect Danube salmon populations in Austria.

The main threats to Danube salmon are from hydropower development and river engineering. Rehabilitation activities should therefore focus on these issues. The project design accomplished this by giving the administrative bodies responsible for river engineering activities and for nature conservation a co-operative leadership role. Scientific knowledge is provided by a research institution that advises the project team through the planning and implementation phases and that is also responsible for monitoring the ecological effects of rehabilitation measures.

The general situation of the Danube salmon in Austria is well known, as this species receives wide publicity. A reduction to 10% of its natural distribution justifies its classification as a species 'threatened with extinction' (Spindler *et al.* 1997). This is underlined by there being only two large populations, one in the River Pielach, remaining. A single fish kill could easily reduce the genetic variability of the species. The risk of further population declines is high since the existing populations are fragmented by migratory barriers: this yields small, isolated sub-populations lacking the ability to balance natural and artificial population fluctuations. Access to spawning grounds in upstream areas and tributaries is prohibited by weirs and numerous artificial falls. Reduced flows caused by water abstraction for hydropower use additionally affect large-sized species such as Danube salmon that have high requirements with regard to habitat dimension. Further threats include hydropeaking and flushing of impoundments. The situation has reached the stage where even cormorant predation on the prey fish of Danube salmon is considered to be a serious threat in the few remaining populations.

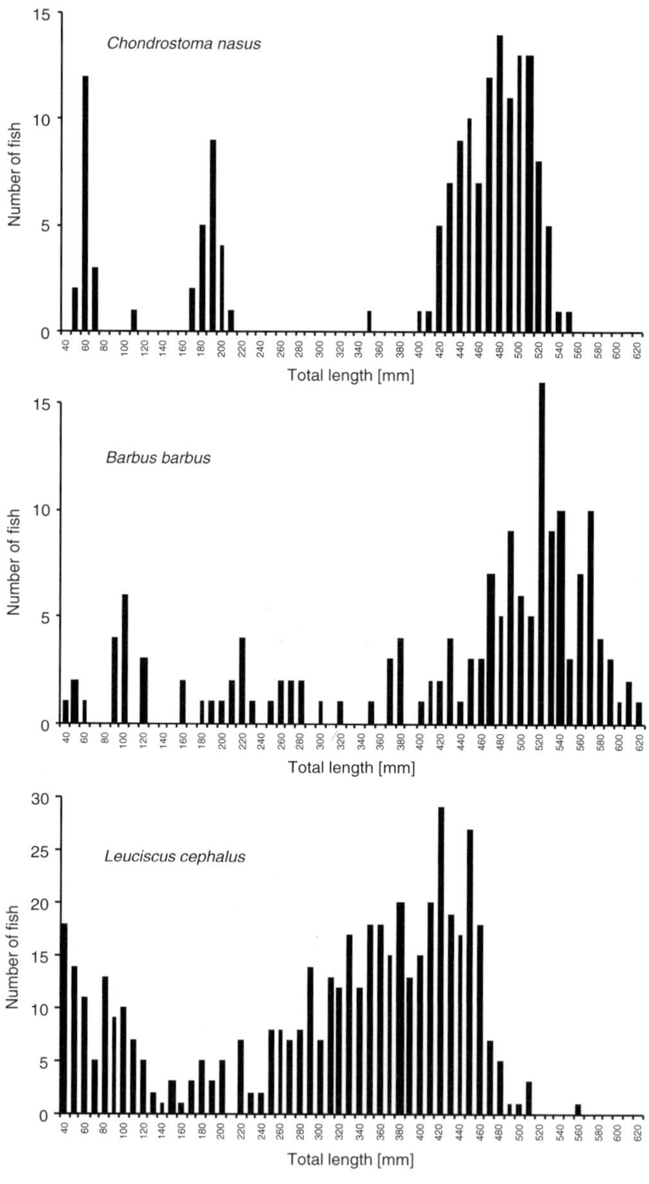

Figure 16.5 Total length distribution of the three dominant species – *Chondrostoma nasus*, *Barbus barbus* and *Leuciscus cephalus* – captured in the Pielach River

Although the general situation of the Danube salmon is well known, detailed information about population sizes, specific threats and restoration opportunities are mostly lacking. In the original project proposal it was roughly estimated that a population size of several hundred adults existed in the River Pielach and about 50 in the Melk and Mank. This represents a considerable overestimation even if the population of the

Pielach upstream of the project area of about 60 adults is included (G. Unfer, personal communication). According to Franklin's 50/500 rule (Franklin 1980), 162 adults in the studied stretch of the Pielach lies at the lower level for minimum population size requirements. In addition the sub-division into six sub-populations of 0–50 adults is far below the above-mentioned thresholds. As a consequence, genetic variability may drop and spawning migrations and the finding of adequate mating partners may be hindered. The populations of Danube salmon in the Melk (nine adults) and Mank (17 adults) are fighting for survival, but have persisted over many years and do represent an important resource for recolonisation of rehabilitated habitats. Although piscivorous species with comparably low natural population densities, such as the Danube salmon, are more likely run the risk of approaching or falling below minimum population sizes, even abundant species, such as *Chondrostoma nasus* and *Barbus barbus*, can be threatened in channelised and/or multiply fragmented rivers.

The dominance of the eurytopic species *Leuciscus cephalus* in all three rivers is an indication of degradation. This species takes advantage of its more resident behaviour compared to the migratory *Chondrostoma nasus* and *Barbus barbus*, which would prevail under natural conditions.

In the Pielach the Danube salmon, accounting for about 17% of the total biomass, is probably close to the carrying capacity of the river. This assumption is supported by the target size of prey fish of adult salmon, at between 15 and 35 cm, which is significantly lacking in the population structure of the dominant prey species. Hence, the protection and rehabilitation of Danube salmon relies not only on measures directly focusing on the target species, but also on those that focus on prey species. The reaction of prey species to opening of the continuum and habitat improvements therefore also has to be of special concern.

Due to disconnection, the tributaries have lost their function as spawning and nursery areas for the Danube population. Re-opening the river continuum and observing the subsequent migratory exchange processes during the monitoring programme will demonstrate the importance of the tributaries as crucial elements of the Danube river system.

References

Anonymous (1907) Zum Artikel: 'Über die geographische Verbreitung des Huchens' von H. Koller, 1907. *Österreichische Fischerei-Zeitung* **IV/15**, 205 pp. (in German).

Berger J. (1999) 88 Jahre altes Eisen mit verblüffendem Wirkungsgrad. *Wasserkraft & Energie* **5**, 24–37 (in German).

Chovanec A., Jäger P., Jungwirth M., Koller-Kreimel V., Moog O., Muhar S. & Schmutz S. (2000) The Austrian way of assessing the ecological integrity of running waters – a contribution to the EU Water Framework Directive. *Hydrobiologia* **422/423**, 445–452.

Eberstaller J., Hinterhofer M. & Parasiewicz P. (1998) The Effectiveness of two nature-like bypass channels in an upland Austrian river. In M. Jungwirth, S. Schmutz & S. Weiss (eds) *Fish Migration and Fish Bypasses*. Oxford: Fishing News Books, Blackwell Science, pp. 363–383.

Franklin I.R. (1980) Evolutionary change in small populations. In M.E. Soulé & B.A. Wilcox (eds) *An Evolutionary-Ecological Perspective*. Sunderland, USA: Sinauer, pp. 135–149.

Haidvogl G. & Waidbacher H. (1997) Ehemalige Fischfauna an ausgewählten österreichischen Fließgewässern. Wien, Univ. für Bodenkultur/Abteilung für Hydrobiologie, Fischereiwirtschaft und Aquakultur, 85 pp. (in German).

Hampel L. (1882) Die Fischerei im Flußgebiet der Salza in Steiermark. *Mitteilungen des österreichischen Fischereivereins* **2**, 146–149 (in German).

Harsanyi A. (1982) *Der Huchen*. Hamburg und Berlin: Paul Parey Verlag, 175 pp. (in German).

Hartlieb R. (1948) *Der Huchenfischer*. Wien: Hubertusverlag, 143 pp. (in German).

Hartmann V. (1898) Die Fische Kärntens. XXV. Jahrbuche des naturhistorischen Landes-Museums von Kärnten, 48 pp. (in German).

Hawlitschek A. (1895) Die Salmen der österreichischen Gewässer. *Mitteilungen des österreichischen Fischereivereines* **15/57**, 27–34 (in German).

Holcik J., Hensel K., Nieslanik J. & Skacel L. (1988) The Eurasian huchen, *Hucho hucho*: largest salmon of the world. *Perspectives in Vertebrate Science* **5**, 239 pp.

Honsig-Erlenburg W. & Schulz N. (1989) Die Fische Kärntens. *Carinthia II* **179/99**, 141–224.

Jungwirth M. (1979) Ovulation inducement in prespawning adult Danube salmon (*Hucho hucho*, L.) by injection of acetone-dried carp pituitary (CP). *Aquaculture* **17**, 129–135.

Jungwirth M. (1980) Der Huchen – derzeitiger Stand einer gefährdeten Fischart. Österreichische Fischereigesellschaft, Jubiläumsschrift, Wien, pp. 105–114 (in German).

Jungwirth M. (1984a) Die fischereilichen Verhältnisse in Laufstauen alpiner Flüsse, aufgezeigt am Beispiel der österreichischen Donau. *Österreichische Wasserwirtschaft* **36**, 104–111 (in German).

Jungwirth M. (1984b) Auswirkungen von Fließgewässerregulierungen auf Fischbestände. Wasserwirtschaft-Wasservorsorge, BMLF, Wien, 188 pp. (in German).

Jungwirth M. (1996) Bypass channels at weirs as appropriate aids for fish migration in rhithral rivers. *Regulated Rivers: Research and Management* **12**, 483–492.

Jungwirth M. & Pelikan B. (1989) Zur Problematik von Fischaufstiegshilfen. *Österreichische Wasser und Abfallwirtschaft* **41**, 80–89 (in German).

Jungwirth M., Schmutz S. & Waidbacher H. (1989) Fischökologische Fallstudie Inn. Sudie im Auftrag des Fischereirevierausschuß Innsbruck Stadt and Land, 93 pp. (in German).

Jungwirth M., Moog O., Schmutz S. & Wiesbauer H. (1990) Endbericht zur RUVP Obere Drau – Aquatischer Lebensraum. HFA-BOKU, Wien, 343 pp. (in German).

Jungwirth M., Muhar S. & Schmutz S. (1995) The effects of recreated instream and ecotone structures on the fish fauna of an epipotamal river. *Hydrobiologia* **303**, 195–206.

Kähsbauer P. (1961) *Catalogus Faunae Austriae*. Teil XXI aa: Cyclostomata, Teleostomi (Pisces). Wien: Springer Verlag, pp. 1–11 (in German).

Kaufmann T., Muhar S., Raderbauer J., Rathschüler O., Schmutz S., Waidbacher H. & Zauner G. (1991) *Fischökologische Studie Mur*. Wien: Universität für Bodenkultur, Abteilung für Hydrobiologie, Fischereiwirtschaft und Aquakultur, 104 pp. (in German).

Koller R. (1907) Die geographische Verbreitung des Huchens. *Österreichische Fischerei-Zeitung* **IV/9**, 137–142 (in German).

Kollmann J. (1898) Karten der Fischarten vom Land Salzburg. Fischartenspektrum in der Salzach, 1 pp. (in German).

Krafft C. (1874) Die neuesten Erhebungen über die Zustände der Fischerei in den im Reichrathe vertretenen Königreichen und Ländern an den österreichisch-ungarischen Meerküsten. Wien, K.k. statistischer Central-Commission, 145 pp. (in German).

Kreissl E. (1991) Erläuterungen zur Roten Liste gefährdeter Fische and Neunaugen der Steiermark. *Mitteilungen der Abteilung Zoologie, Landesmuseum Joanneum* **44**, 17–32.

Mader H., Unfer G. & Schmutz S. (1998) The effectiveness of nature-like bypass channels in a lowland river, the Marchfeldkanal. In M. Jungwirth, S. Schmutz & S. Weiss (eds) *Fish Migration and Fish Bypasses*. Oxford: Fishing News Books, Blackwell Science, pp. 384–402.

Michor K. (1990) Gewässerbetreuungskonzept für die Isel in Osttirol. Studie im Auftrag des Baubezirkamtes Lienz, Abt. Wasserbau, Lienz, 68 pp. (in German).

Muhar S. (1994) Stellung und Funktion des Leitbildes im Rahmen von Gewässerbetreuungskonzepten. *Gewässerbetreuungskonzepte – Stand und Perspektiven, Wiener Mitteilungen* **120**, 135–158.

Muhar S., Kainz M. & Schwarz M. (1998) Ausweisung flußtypspezifisch erhaltener Fließgewässerabschnitte in Österreich, Teil 2. Bundesministerium für Land- und Forstwirtschaft, 177 pp. (in German).

ÖNORM/M6232 (1995) Richtlinien für die ökologische Untersuchung und Bewertung von Fließgewässern. *Österreichisches Normungsinstitut*, 81 pp. (in German).

Parasiewicz P., Eberstaller J., Weiss S. & Schmutz S. (1998) Conceptual guidelines for nature-like bypass channels. In M. Jungwirth, S. Schmutz & S. Weiss (eds) *Fish Migration and Fish Bypasses*. Oxford: Fishing News Books, Blackwell Science, pp. 348–362.

Schmutz S. & Jungwirth M. (1999) Fish as indicators for large river connectivity: The Danube and its tributaries. *Archiv für Hydrobiologie* **115**(Suppl.), 329–348.

Schmutz S., Giefing C. & Wiesner C. (1998) The efficiency of a nature-like bypass channel for pike-perch (*Stizostedion lucioperca*) in the Marchfeldkanalsystem. *Hydrobiologia* **371/372**, 355–360.

Spindler T., Zauner G., Mikschi E., Kummer H., Wais A. & Spolwind R. (1997) Gefährdung der heimischen Fischfauna. In T. Spindler (ed.) Fischfauna in Österreich: Ökologie-Gefährdung-Bioindikation-Fischerei-Gesetzgebung. *Umweltbundesamt* **87**, 54–75 (in German).

Starmühlner F. (1969) *Die Schwechat. Ein Beitrag zur Kenntnis der Fließgewässer der Wiener Umgebung*. Wien, Verlag Notring, 226 pp. (in German).

Wallner J. (1917) Beiträge zur Geschichte des Fischereiwesens in der Steiermark. *Archiv für Fischereigeschichte* **9**, 1–12 (in German).

Wedland H. (1965) *So fängt man Huchen*. Hamburg-Berlin: Paul Parey, 141 pp. (in German).

Wiesbauer H., Bauer T., Jagsch A., Jungwirth M. & Uiblein F. (1991) Fischökologische Studie – Mittlere Salzach, Abteilung für Hydrobiologie, Fischereiwirtschaft und Aquakultur, Wien, 170 pp. (in German).

Woschitz G. (1991) Die autochthone Fischfauna der Steiermark. Unveröffentlichtes Manuskript, Wien, 87 pp. (in German).

Zitek A. Zobl S., Schmutz S. & Jungwirth M. (2000) LIFE-Natur Projekt 'Lebensraum Huchen' – Istbestandsaufnahme 1999–2000. HFA-BOKU, Wien, 234 pp. (in German).

Chapter 17
Romanichthys valsanicola, a highly threatened species from the Danube River Basin in Romania

P. BANARESCU*
Institute of Biology, Bucharest, Romania

Romanichthys valsanicola, the single member of the genus *Romanichthys* (family Percidae), was discovered in 1956 in the River Vâlsan, a left-hand tributary of the River Arges (Danube drainage) in southern Romania (Fig. 17.1) and described by Dumitrescu, Banarescu and Stoica (1957). It was subsequently found in greater abundance in the River Arges.

Ichthyologists from Europe, and especially from North America, were interested in this genus, since it has great similarities with American genera of Percidae. Its discovery has renewed interest in the systematics of the family Percidae and various authors have proposed four alternative phyletic schemes for members of this family (e.g. Dumitrescu *et al.* 1957; Collette 1963; Coburn & Caglione 1992; Wiley 1992).

The species lives in mountain rivers and is apparently strictly territorial and nocturnal. Each individual is hidden under a stone during daytime and is active in the night when searching for food, returning in the morning to the same stone (P. Banarescu & R. Bless, unpublished data). Its spawning period probably extends from May to the beginning of July. Gâldean, Nalbant and Dragomirescu (1997) found that ephemeropteran nymphs are the main component of the diet. Eleven species of Ephemeroptera were identified, with *Rhithrogena semicolorata* representing 54.4% of the diet, other Ephemeroptera species 13.2%, followed by plecopetran nymphs (10.9%), blepharicerid midges (9.5%), trichopteran larvae (6.5%) and chironomid larvae (4.7%). In 1990 the benthic macroinvertebrate fauna of the River Vâlsan at Bradet was found to be much poorer than formerly, and several of the species consumed by *Romanichthys* were no longer present (mainly stoneflies) and others, including *Rhithrogena semicolorata*, were very rare (Tatole 1993). Moreover two fish species which formerly were more abundant than *Romanichthys* became extinct from the Bradet–Musatesti stretch of the River Vâlsan: *Alburnoides bipunctatus* (Bloch) and *Gobio uranoscopus* (Agassiz), although *R. valsanicola* survived, suggesting that the species is capable of resisting unfavourable conditions.

Romanichthys valsanicola is the fish genus having the smallest range in Eurasia and was considered by Maitland (1991) the most endangered fish species in Europe. The species became extinct in the River Arges because of the construction of a reservoir in

*Correspondence: Institute of Biology, Splaiul Independentei 296, 69751 Bucharest, Romania (email: pbana@ibiol.ro).

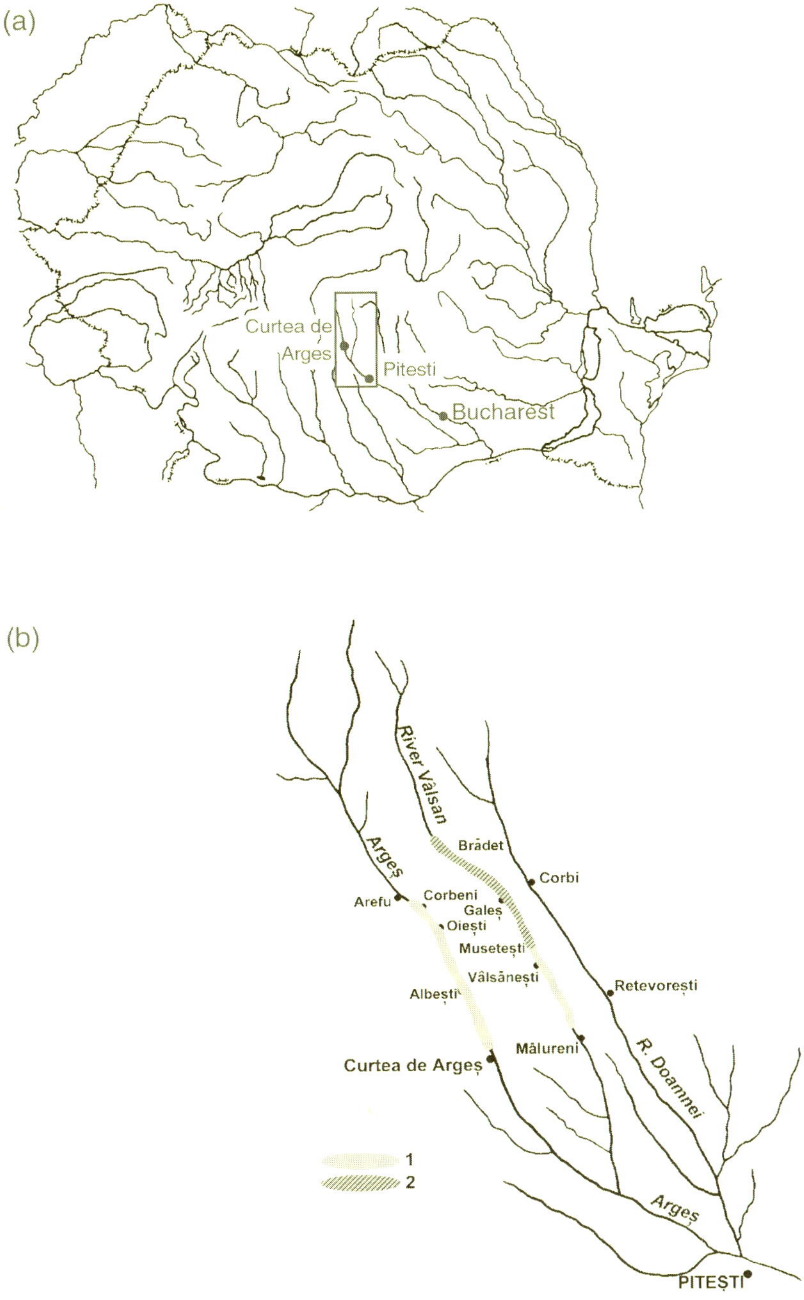

Figure 17.1 (a) Distribution of *R. valsanicola* in Romania. (b) Detailed distribution of *R. valsanicola* before 1967 (1) and 2001 (2)

1967 and the removal of gravel and vegetation from the riverbed. A small reservoir was constructed in the same year on the River Vâlsan, far upstream of the area inhabited by *R. valsanicola,* but the bed of the river was not disturbed. The construction of this reservoir has, however, resulted in a strong reduction in discharge, from 1.3–4.0 $m^3 s^{-1}$ between 1944 and 1966 to 0.6–1.2 $m^3 s^{-1}$ between 1967 and 2000. This caused a decline in abundance of *R. valsanicola* and of the benthic fauna on which it feeds. Another factor which contributed to the decline was the extraction of aggregates from the river bed by local people. Gravel has always been exploited, but before the building of the dam lake large quantities were continuously deposited by the river, from upstream, especially during floods. Some evidence of pollution is also present: represented by debris disposed of by local people and the wastewater from a hospital. When the water discharge was greater these sources of pollution had no influence on the water quality of the river.

Specimens of *R. valsanicola* were taken in 1992 to the laboratory of the Institute for Nature Protection in Bonn, Germany, to carry out captive breeding experiments. This action was unsuccessful, because the males of the species failed to mature in captivity, but further attempts will be made in 2001.

Now the main threats to the survival of *R. valsanicola* are known, a 2-year EU LIFE-NATURE programme was started in late 1999, to take measures to prevent the extinction of the species. The objectives of this project were:

(1) stopping of gravel extraction;
(2) negotiate a better discharge regime from upstream reservoir;
(3) eliminate the small sources of water pollution;
(4) rehabilitate the river by reintroduction of gravels and cleaning of the water;
(5) a permanent stewardship of the River Vâlsan and of its valley.

Under the EU project, the following activities have been performed:

(1) chemical and microbiological analysis of the water has shown it is of good quality; plus measurement of radioactivity levels has demonstrated that it is non-existent;
(2) studies on the benthic fauna on which *R. valsanicola* feeds has shown that this is richer in the stretch inhabited by the species than it was some 5 years ago, but much poorer than upstream of the reservoir;
(3) long stretches of the river valley have been cleaned;
(4) extraction of gravel from the river and fishing (except angling of trout) has been stopped;
(5) signs mentioning that the river is protected, and disposal of debris and exploitation of gravels are prohibited have been placed along the river valley;
(6) collaboration with local authorities for the protection of the river has been established;
(7) four workshops with participants from various institutions and the local authorities have taken place;
(8) interviews with newspapers and the radio have been carried out;
(9) children from schools in villages near the River Vâlsan have been instructed on the importance of *R. valsanicola* and the necessity for its protection;

(10) a permanent stewardship of the River Vâlsan and of its valley has been enacted through two men (one is university educated and the other experienced in nature protection);
(11) small sources of pollution in the villages of Bradulet and Bradet have been identified;
(12) a small supplementary discharge of water from the reservoir has been secured.

References

Coburn M.M. & Caglione J.I. (1992) A comparative study of percid scales (Perciformes: Percidae). *Copeia* **1992**, 986–1001.

Collette B.B. (1963) The subfamilies, tribes and genera of the Percidae (Teleostei). *Copeia* **1963**, 615–623.

Dumitrescu M., Banarescu P. & Stoica N. (1957) *Romanichthys valsanicola* nov. gen. Nov. sp. (Pisces: Percidae). *Travaux du Muséum d'Histoire Naturelle „Grigore Antipa" Bucuresti* **1**, 225–244.

Gâldean N., Nalbant T. & Dragomirescu L. (1997) The food and feeding habits of the sculpin-perch or Romanian darter *Romanichthys valsanicola* (Pisces: Perciformes, Percidae). *Travaux du Muséum d'Histoire Naturelle „Grigore Antipa" Bucuresti* **27**, 287–295.

Maitland P.S. (1991) *Conservation of the Threatened Freshwater Fish in Europe*. Council of Europe Press, Strasbourg Nature and Environment Series, **46**, pp. 1–67.

Tatole V. (1993) Noi consideratii asupra situatiei critice a endemitului *Romanichthys valsanicola* (Nouvelles considerations sur la situation critique de l'endémique *Romanichthys valsanicola). Ocrotirea naturii si a mediului inconjurator, Bucuresti* **37**(2), 25–27 (in Romanian).

Wiley E.O. (1992) Phylogenetic relationships of the Percidae (Teleostei: Perciformes): a preliminary hypothesis. In R. Mayden (ed.) *Systematics, Historical Ecology and North American Freshwater Fishes*. Stanford: Stanford Univeristy Press, pp. 247–267.

Chapter 18
Aspects of the reproduction and early life history of *Ladigesocypris ghigii*, a freshwater fish species endemic to Rhodes Island (Greece): implementation to conservation

M.Th. STOUMBOUDI,* R. BARBIERI
National Centre for Marine Research, Agios Kosmas, Athens, Greece

M. CORSINI-FOKA
NCMR, Hydrobiological Station of Rhodes, Rhodes, Greece

A.N. ECONOMOU
National Centre for Marine Research, Agios Kosmas, Athens, Greece

P.S. ECONOMIDIS
Aristotle University, School of Biology, Thessaloniki, Greece

Abstract

In the course of an EU LIFE-NATURE project aimed at implementing management actions for the endangered freshwater fish *Ladigesocypris ghigii*, endemic to Rhodes Island (Greece), primary data were collected on reproduction and early life history. The results of artificial fertilisation and rearing experiments are described. The reproductive period was determined from gonadosomatic indices and recording the presence and abundance of larvae in the field over a 13-month period. Breeding occurred in spring and early summer; however, the exact onset and duration of the reproductive period differed between areas and possibly years. Based on the presence of large numbers of larvae and juveniles found, *L. ghigii* does not seem to face any serious recruitment bottleneck. Thus, extensive captive breeding and subsequent stocking actions are not recommended as a conservation tool for this species.

Keywords: conservation, *Ladigesocypris ghigii*, life history, recruitment, reproduction.

18.1 Introduction

Ladigesocypris ghigii (Gianferrari) is a small-bodied cyprinid fish, locally known as 'gizani', endemic to Rhodes Island, Greece. It inhabits exclusively the streams, springs and some small reservoirs of the island. Gizani is considered one of the most threatened freshwater fish of Europe (Tortonese 1947; Economidis 1988, 1991, 1995). It is also listed as an endangered species of top priority for protection, in ANNEX II of the

*Correspondence: National Centre for Marine Research, Institute of Inland Waters, Agios Kosmas, Hellinikon, 16604 Athens, Greece (email: mstoum@ncmr.gr).

Habitats Directive (92/43/EEC) of the European Union that concerns the protection of natural habitats and wild fauna and flora.

The present study was performed in the frame of an EU LIFE-NATURE project, entitled 'Conservation measures for the endangered fish *Ladigesocypris ghigii*'. It aims to elucidate aspects of the species reproductive biology and early life stages, to create the basis for the formulation of management and conservation actions that concern captive breeding and subsequent enhancement stocking of different *L. ghigii* populations.

18.2 Materials and methods

Samples were collected monthly at three localities within the EU NATURA 2000 sites of Rhodes Island between February 1999 and February 2000. These sites were a stream flowing into the northern part of the artificial lake Apolakkia, Loutanis stream and Epta Piges springs (Fig. 18.1). In addition, samples were obtained from the

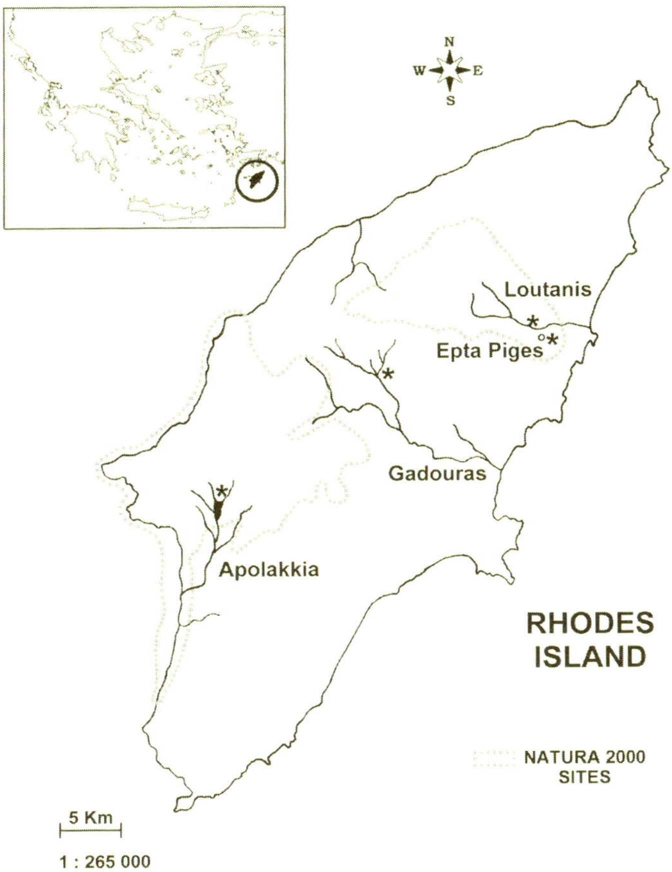

Figure 18.1 Map of Rhodes Island where the NATURA 2000 sites and the aquatic systems where gizani was sampled are shown

population of *L. ghigii* inhabiting Gadouras stream (Fig. 18.1), a location considered the most stable lotic system on Rhodes (Economidis 1988, 1991; Corsini 1994).

Larvae were sampled with beakers and small scoop nets, while juveniles and adults were caught with large scoop nets and electric fishing. Water temperature (°C) was recorded with a Horiba mobile polymeter (at the same time of the day for each site). All fish sampled (3459 in total) were measured in the field (standard length to nearest mm). To avoid depletion of populations, only a small portion of the sampled fish (10–20 adults and 0–15 larvae) was preserved (4% formaldehyde, neutralised with sodium phosphate) each month from each location, for further analyses in the laboratory.

The preserved fish were measured in the laboratory to the nearest 0.1 mm for standard length (SL), to the nearest 0.001 g for net weight (NW) and to the nearest 0.01 mg for gonad weight (gW). The sex of adult fish was also recorded. Gonadosomatic indices (GSI) were calculated as the percentage of gW to NW.

Artificial fertilisation of wild spawners was performed in April 1999, and again in April 2000. Five females and 12 males from the Epta Piges springs population were used on the first occasion, while 10 females and 25 males from Loutanis stream on the second. Gametes were obtained by stripping the mature fish and subsequently eggs and sperm were mixed in glass Petri dishes. The ova were adhesive and negatively buoyant, and stuck to the bottom of the dish, occasionally in layers. Some water was added and the mixture was stirred by moving the dish gently. After about 10 min, the fertilised ova were rinsed with freshwater to remove the unfertilised eggs and the debris. Subsequently, the Petri dishes were placed in portable aquaria in the shade. After a couple of hours the fertilised eggs were bathed in a disinfecting medium and transported to the laboratory (at the Hydrobiological Station of Rhodes). The incubation of eggs took place in 25-L aquaria. When the embryos hatched, they were collected with beakers and pipettes and transferred to aerated rearing aquaria of appropriate size (25–50 L), depending on the density and size of free embryos and larvae. Both rearing experiments took place at room temperature (18–22°C) with a continuous air supply. Free embryos and larvae were fed *ad libitum* on artificial commercial diets for fish fry. Up to the end of the rearing experiment, samples of free embryos and later larvae were taken systematically at 24, 48 or 96 h intervals and were preserved in 4% buffered formaldehyde. Embryos and larvae, produced by artificial fertilisation, were measured by means of a calibrated eyepiece to the nearest 0.1 mm. The embryonic and developmental stages were drawn using a binocular microscope fitted to a camera lucida. The descriptions of the egg, embryo and larva were based on the results of the artificial fertilisation and rearing experiments.

18.3 Results

18.3.1 *Reproductive period*

Reproduction of gizani occurs in spring and early summer (Fig. 18.2), but the exact onset, peak and duration of the breeding season may differ between areas and possibly

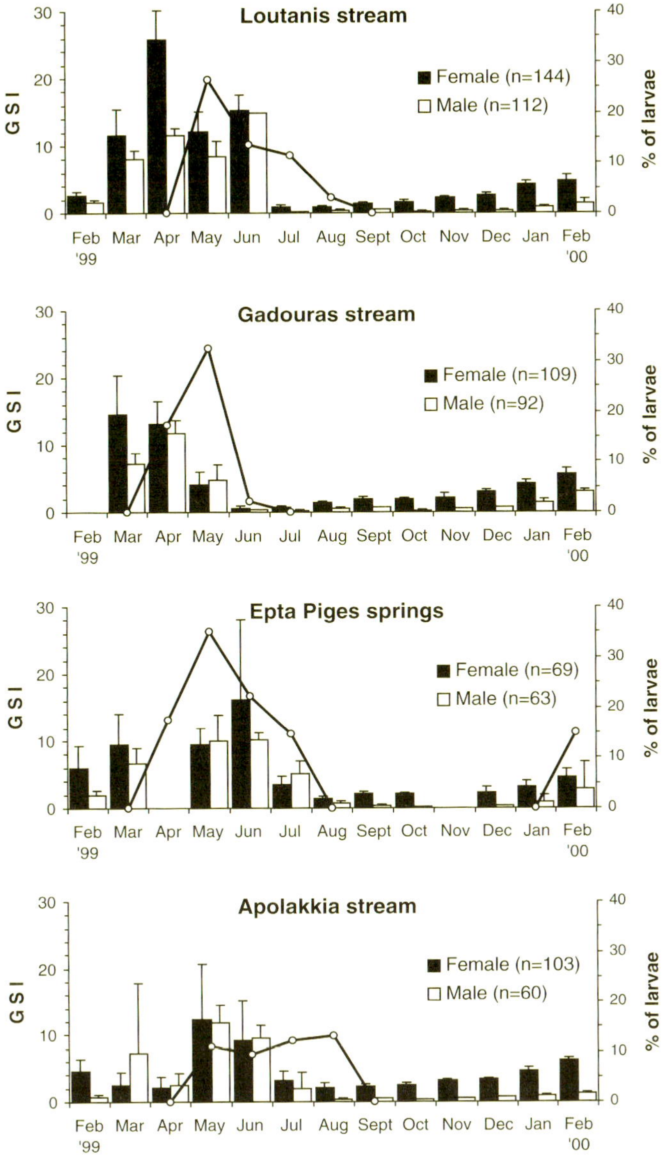

Figure 18.2 Monthly gonadosomatic indices (error bars: 95% confidence limits) of adult *L. ghigii* and monthly larvae abundance (percentage of larvae with SL < 10 mm, to the total sample of fish)

years. Generally, in both sexes gonad recrudescence starts in late autumn and continues during winter, maturation occurs in spring and lasts until early summer, while in late summer and early autumn gonads are in the resting phase.

The populations of *L. ghigii* in Loutanis and Gadouras streams exhibited a well defined reproductive cycle. Nevertheless, in Loutanis reproduction begins a month

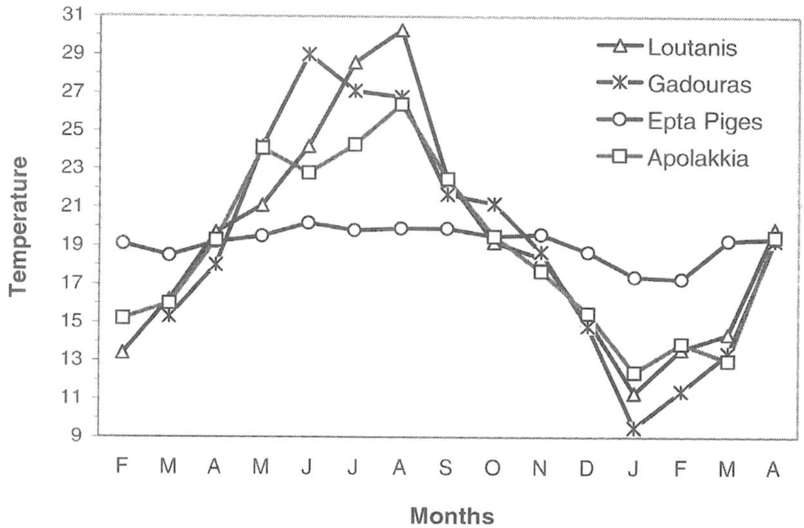

Figure 18.3 Monthly variation of water temperature (from February 1999 to April 2000) in the streams of Apolakkia, Loutanis and Gadouras and the Epta Piges springs

later and lasts longer than in Gadouras (since larvae were observed in Gadouras between April and June, but in Loutanis between May and August).

From the GSI analysis, the population of gizani in the Epta Piges springs exhibited a prolonged breeding period. Additionally, in the year 2000 first larvae appeared in this system in February. This was attributed to the specific environmental conditions occurring in Epta Piges, where temperature is relatively stable throughout the year (Fig. 18.3), but water quantity fluctuates markedly, mainly due to the operation of a small dam that collects most of the water from the springs.

In Apolakkia stream, although the temperature increases in spring in a manner similar to the other two streams (Fig. 18.3), the population of *L. ghigii* attained a lower GSI than those of Loutanis and Gadouras. Larval abundance was also lower than in the other streams. This situation possibly reflects the adverse environmental conditions occurring in this stream, which is much smaller in size as well as in water discharge than the Loutanis and Gadouras streams. In Apolakkia stream, water quantity is extremely limited during summer, resulting to low density and poor longevity of the local population (unpublished data).

18.3.2 *Egg characteristics*

In the wild, gizani deposits its eggs on aquatic vegetation, with a preference for filamentous algae (*Cladophora*). Fertilised eggs are adhesive, semi-transparent, with a narrow perivitelline space. The principal egg characteristics, after water hardening and fertilisation, are summarised in Table 18.1.

Table 18.1 Main characteristics of the *L. ghigii* eggs after hardening and fertilisation

Distribution: demersal	Oil globules: absent
Shape: spherical	Adhesive: yes
Dimensions: 1.33 ± 0.05 mm	Texture of the egg envelope: smooth
Colour: yellowish	Adhesive filaments: absent
Yolk: homogenous	Pigments on embryos: yes

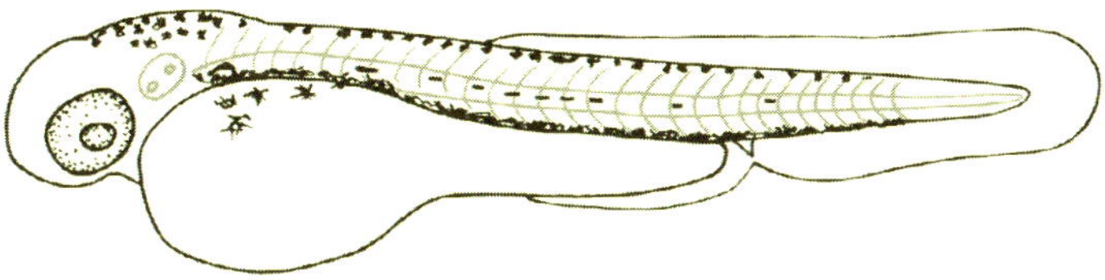

Figure 18.4 Free embryo just after hatching. SL: 4.5 mm

18.3.3 *Yolk-sac larval (free embryo) characteristics*

Fertilised eggs were incubated in the laboratory at room temperature (18–22°C). Hatching occurred four days after fertilisation and continued for approximately 24 h. The embryos hatch at approximately 4.4 mm SL (± 0.09, $n = 8$), with no functional mouth, anus and swimbladder (Fig. 18.4). The yolk sac has a pyriniform shape, the primordial fin is relatively narrow, the anus is behind the midpoint of the body and the pectoral fins are present as small buds (Fig. 18.4). Eyes are quite large, dark brown and elliptical (Fig. 18.4).

The body is also covered by conspicuous melanophores (Figs 18.4 and 18.5). The pigmentation pattern is similar to most other cyprinid larvae. It consists of many dorsal and ventral postanal melanophores arranged in double rows (Fig. 18.5). Melanophores are also present in a single mediolateral and a ventral preanal row, an array on the head and scattered on the yolk sac (Figs 18.4 and 18.5).

Two days after hatching the characteristic jerking respiratory movements for cyprinids began, and within two more days the transition to exogenous feeding was completed with the contemporary absorption of the yolk.

Newly hatched larvae were never observed in the field. This was attributed to the preference of fish of this developmental stage to remain hidden and motionless, clinging to the aquatic vegetation with their cephalic adhesive glands. In the aquaria, free embryos lay motionless on the bottom and moved only if disturbed.

Larvae of later developmental stages were found close to the banks among vegetation or in segments of the stream with minimal water flow.

18.4 Discussion

Ladigesocypris ghigii is a small-sized (maximal standard length 90 mm for females and 65.5 mm for males), short-lived species (up to 3 years in the wild) that matures at

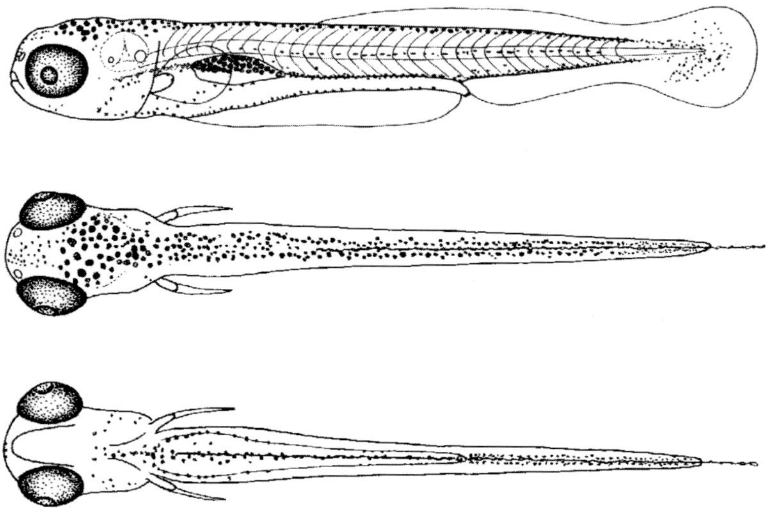

Figure 18.5 *Ladigesocypris ghigii* larva, after yolk sac absorption, at lateral, dorsal and ventral view, where the pigmentation pattern is shown

age 1+ (unpublished data). The reproductive period of *L. ghigii* is concomitant to the season of rising temperature and photoperiod and of maximal biological activity. The peak spawning and the duration of a prolonged breeding season differ between aquatic systems and/or years. These differences were attributed to:

(1) divergent ecological characteristics of the habitats *per se* and in the course of time (in both cases deriving from water quantity fluctuations mostly);
(2) dissimilar energetic conditions of the spawners, attributable to the instability of the environment.

In recent years, captive breeding has been used extensively for the preservation of endangered endemic fish species (Philippart 1995). Captive-bred fish are later stocked, either to replace extinct populations or to enhance existing wild ones. Stocking is performed to overcome intrinsic factors threatening the survival of a species, such as, poor distribution due to the extinction of populations, poor reproduction and/or recruitment, high juvenile mortality and inbreeding that is caused by drastic reduction of the effective population size.

Based on the high abundance of larvae and juvenile fish observed during the present study, the reproductive success and subsequent recruitment of *L. ghigii* does not seem to be weak or problematic, at least in the cases of Loutanis and Gadouras streams, as well as the Epta Piges springs. However, the survival of juvenile and adult fish is jeopardised by the seasonal drying of the streams, resulting from summer to autumn drought and synchronous overpumping of water for irrigation and urban use (Corsini 1994; Economidis 1995; M. Th. Stoumboudi, M. Corsini & R. Barbieri, unpublished data). Rainfall in 1999 was substantial for Rhodes and the reproduction of gizani during the spring was prolific. During late summer, there were considerable numbers of young fish in the main course of the streams, while the best niches near the banks (among roots of trees, bank cavities, vegetation or dead segments of the stream) were occupied by larger specimens. This period was followed by a completely dry autumn and in winter 2000 rainfall was very limited. In the following spring the populations in the different aquatic systems were already diminished, especially in the Apolakkia stream.

Thus, water quantity seems to be the key issue for the survival of the species. If water is further reduced by a series of poor dry years combined with overabstraction, different populations of *L. ghigii* will probably collapse due to habitat loss, and the species will become in danger of extinction. To eliminate or reduce this threat, immediate measures for the sustainable management of the island's water resources have to be taken and applied, aiming at rehabilitating the original habitats, so that they will be appropriate to support larger number of individuals in each population. However, until habitat rehabilitation is realised, extensive captive breeding and subsequent enhancement stocking actions are not recommended as a conservation tool for this species. Artificial fertilisation of wild spawners as well as collection of wild adult specimens will be used, to produce and establish a gene bank of captive breeding stocks for each *L. ghigii* population separately (with priority given to the Apolakkia population). As in other endangered fishes (Philippart 1995), these stocks will be used to re-establish a population, if it becomes extinct in the wild.

Acknowledgements

The authors are indebted to Ian Cowx for useful suggestions and reviewing of the manuscript. The project was financed under the EU-LIFE project B4-3200/98/EEC.

References

Corsini M. (1994) Distribution of the populations of an endemic cyprinid *Ladigesocypris ghigii* in the island of Rhodes (Greece) and efforts for its conservation. *Proceedings of the European Union of Aquarium Curators E.U.A.C.*, Leipzig, pp. 63–67.

Economidis P.S. (1988) The state of the populations of the endemic fish 'ghizani' in the freshwater systems of Rhodes. *Technical Report*, 11 pp. (in Greek).

Economidis P.S. (1991) *Check List of Freshwater Fishes of Greece*. Athens: Hellenic Society for the Protection of Nature, Special Publication, 47 pp.

Economidis P.S. (1995) Endangered freshwater fishes of Greece. *Biological Conservation* **72**, 201–211.

Philippart J.C. (1995) Is captive breeding an effective solution for the preservation of endemic species? *Biological Conservation* **72**, 281–295.

Stoumboudi M.Th., Barbieri R., Psarras Th., Giakoumi S. & Economou A.N. (2000) First results on the distribution and life span of *Ladigesocypris ghigii*. *Proceedings, 9th Panhellenic Ichthyological Congress, Messolonghi, Greece*, pp. 193–196 (in Greek, English abstract).

Tortonese E. (1947) Ricerche zoologiche nell'isola di Rodi (Mar Egeo). *Pesci, Bollettino Pesca, Piscicultura e Idrobiologia* **23**, 155–156 (in Italian).

Chapter 19
A conservation strategy for *Anaecypris hispanica*: a picture of LIFE for a highly endangered Iberian fish

M.J. COLLARES-PEREIRA*
Centro de Biologia Ambiental, Faculdade de Ciências, Lisboa, Portugal

I.G. COWX
Hull International Fisheries Institute, University of Hull, Hull, UK

J.A. RODRIGUES
Centro de Biologia Ambiental, Faculdade de Ciências, Lisboa, Portugal

L. ROGADO
Instituto da Conservação da Natureza, Lisboa, Portugal

Abstract

Anaecypris hispanica is one of the most endangered freshwater fish species in Europe. In Portugal, the distribution of the species was originally throughout the Portuguese sector of the Guadiana catchment, but its distribution range and abundance have contracted considerably over the past 20 years and if concerted action is not forthcoming the species is likely to become extinct in the near future. Information collected between 1997 and 2000 was used to develop a management plan to conserve the species. Key areas for action to prevent further deterioration of the stocks and enhance their status, include: (1) designation of the rivers which still support *Anaecypris* as special protected areas; (2) measures to improve water quality and restore degraded areas; (3) control of the dispersal and stocking of exotic species; (4) international collaboration; (5) stocking of recovered habitats, based on the results of the genetic screening of the extant populations. The key options for each river sub-basin which supports or has the potential to support *Anaecypris* were prioritised. The management plan provides all the necessary information to formulate projects to target the various issues affecting the species. The future of the species needs to be taken up by the responsible bodies, and consultation made between stakeholders to optimise development, but ensure that conservation of the endemic fish species in the region is not compromised.

Keywords: *Anaecypris hispanica*, conservation, Cyprinidae, endangered species, management plan.

19.1 Introduction

Anaecypris hispanica (Steindachner), is a small Iberian cyprinid locally known as saramugo (Portugal) or jarabujo (Spain). It is presently limited to the Guadiana River

*Correspondence: Faculdade de Ciências de Lisboa, Campo Grande, C2, 3° Piso, 1749-016 Lisboa, Portugal (email: mcolares@fc.ul.pt).

Basin, and was classified as 'endangered' in both the Portuguese and Spanish Red Data Books (SNPRCN 1991; Blanco & Gonzalez 1992), and in the IUCN Red List (Baillie & Groombridge 1996). The species is listed in the Bern Convention (Annex III) and in Annexes II and IV of the EC Habitats Directive (92/43/EEC).

The distribution and rarity of the species in Spain, was only recently confirmed by Doadrio and Elvira (personal communication) and Doadrio (2001) in 17 rivers – Albarregas, Albuera, Alcarrache, Alcorneo, Aljucén, Ardila, Estena, Estenilla, Fresnedoso, Gévora, Guadámez, Matachel, Murtigas, Quejigares, Retín, Sansustre and Sillo. In Portugal, the distribution of the species was originally throughout most of the Portuguese sector of the Guadiana, but due to its marked decline (see Collares-Pereira 1990), the species was the subject of an intensive study between 1997 and 2000, under the auspices of an European Union LIFE-NATURE project, to investigate the distribution, biology and genetic structure of the species, the factors responsible for its demise in Portugal, and finally to develop a conservation strategy for the species.

This chapter presents an overview of (1) the main results (with the exception of genetic data, see Coelho, Salgueiro, Alves & Collares-Pereira, Chapter 30), and (2) of the proposed management plan for *A. hispanica* in Portugal, which was developed to address all the issues known to impact on the extant stocks. It was anticipated that such action would also help conserve the remaining sympatric fish species, seven of which are listed in the National Red Book, viz. *Chondrostoma willkommii* Steindachner, *Ch. lemmingii* (Steindachner), *Barbus comizo* Steindachner, *B. steindachneri* Almaça, *B. microcephalus* Almaça, *B. sclateri* Gunther and *Salaria fluviatilis* (Asso) (SNPRCN 1991) – the majority are also included in the Habitats Directive.

19.2 Strategy

Field work comprised 31 visits of 1 week duration performed monthly from October 1997 to June 2000, corresponding to 958 samplings at 221 sites covering all the catchment. During the first year, sampling sites were distributed throughout the 32 main tributaries in Portugal to improve knowledge on the current distribution of the species. In the following years, the strategy changed to allow a better understanding of the life history and annual variations in distribution and abundance of saramugo. To achieve this, two river sub-basins with the highest densities of *A. hispanica* were selected – the Caia and Vascão – and they were intensively sampled, especially the latter.

Fish were sampled by electric fishing (DC at 300/600 V and 4–6A), measured (fork length, nearest mm) and weighed (nearest 0.1 g) and returned live to the river after removal of some scales for growth studies and, whenever necessary, of fin clips for genetic studies. Assessment of the reproductive traits was based on material from the River Degebe, held in the collections of the Museu Bocage (Lisbon).

Details of the methods used for field work regarding both spatial and temporal monitoring of fish communities and sampled sites, plus subsequent analyses, are described in Collares-Pereira, Cowx, Rodrigues, Rogado, Ribeiro, Mendes, Picciochi, Salgueiro, Alves and Coelho (2000a). Water quality was monitored for phosphates (PO_4^{3-}), nitrates (NO_3^-) and sulphates (SO_4^{2-}) at some key sites from May 1998 until December

1999, and the BMWP index (Biological Monitoring Working Party) and the ASPT index (Average Score Per Taxon) for macroinvertebrate communities, were determined between October 1997 and June 1998.

19.3 Main outputs of the project

19.3.1 *Distribution and abundance of* Anaecypris hispanica

The species, which was once prevalent across the catchment, now exhibits a highly fragmented distribution pattern, apparently based on 10 sub-basins – the Xévora, Caia (upstream to the dam), Álamo (upper reaches), Degebe (rivers Pardiela and Pecena), Ardila (rivers Múrtega, Safareja and Murtigão), Chança (upstream do the dam), Carreiras, Vascão, Foupana and Odeleite (upstream to the dam) (Fig. 19.1).

A total of 5967 specimens of *A. hispanica* were collected. Table 19.1 presents the total number of specimens captured in each sub-basin *v.* the number of samplings performed, and Figure 19.1 the relative abundance of the species per 100 m^2 in each tributary.

The populations from the most southern and northern rivers had comparatively higher densities, in particular the Vascão, Foupana and Odeleite, and the Caia, respectively. In the middle reaches of the Guadiana in Portugal, the species was most abundant in the River Álamo, although confined to the upper zone, due to the heavy pollution downstream of Reguengos de Monsaraz municipality. The River Carreiras had the smallest population, being highly threatened by the presence of exotic species. Thus, the northern and central region populations, under greater pressure from human presence, showed greater evidence of serious decline. In general, there was a reduction in the number of captures from April to July, probably due to the local Mediterranean climatic conditions which causes a progressive drying up of rivers in the post spring period, resulting in some tributaries drying up or becoming series of isolated pools. Water quality conditions in the Caia, Chança, Foupana and Odeleite sub-basins were generally good and reflected in the chemical and biological indexes determined, although evidence of pollution and eutrophication was found where the abundance of fish was more reduced (Collares-Pereira *et al.* 2000a).

19.3.2 *Biology of* Anaecypris hispanica

The species reaches a maximum length of 7 cm and has a short life span of <3 years, achieving a small size (<60 mm fork length) with rapid growth (≈40 mm) to maturity in its first year of life, and little growth thereafter (10–15 mm $year^{-1}$). Growth of fish from the River Caia was marginally faster than that of those of the middle reaches (Degebe), although the difference was not significant. However, the growth of *A. hispanica* from the middle and lower reaches of Vascão and Foupana, two of the three most southerly rivers where the species was caught, was significantly slower than that of the other populations examined (Ribeiro, Cowx & Collares-Pereira 2000). The species spawns for the first time early in its second year of life (April–May

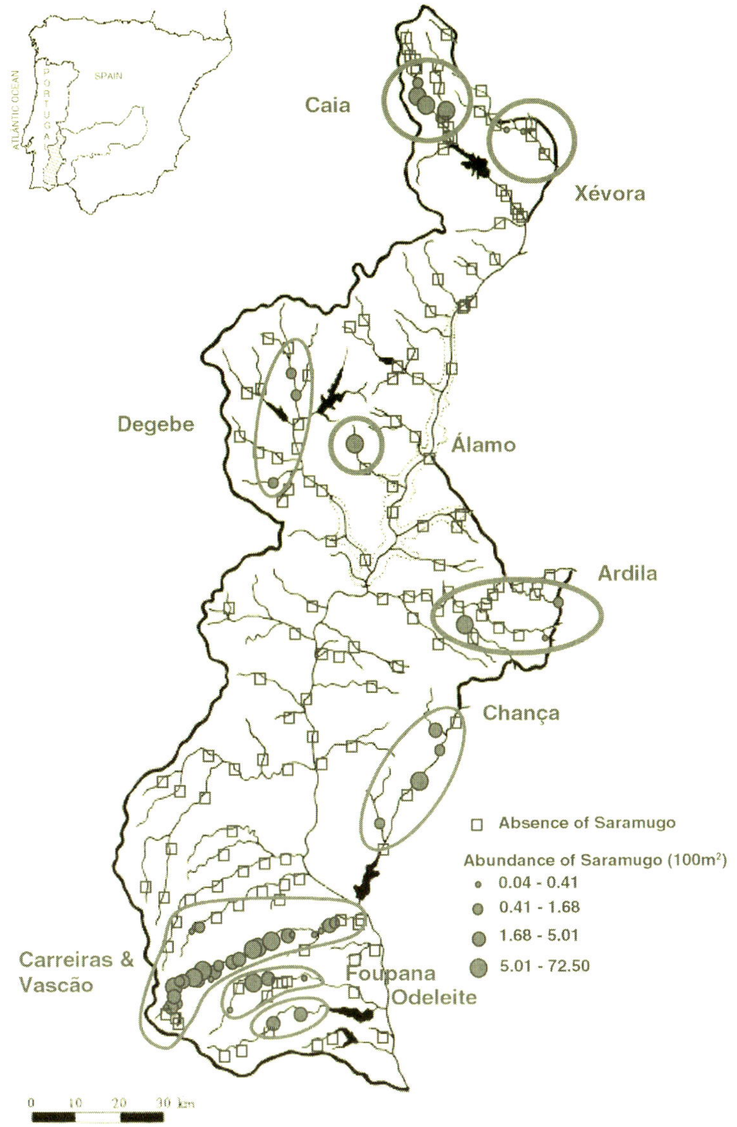

Figure 19.1 Graphic representation of the relative abundance per 100 m^2 of *A. hispanica* in the sampling sites of the Guadiana tributaries where the species still occurs in Portugal

period). In view of several different sized maturing oocytes in the ovaries, the species was considered to be a fractional spawner, laying a small number of oocytes (<100) in each batch (Ribeiro *et al.* 2000).

Anaecypris hispanica appears to occur in specific habitats which vary according to season, mainly as a result of changes in environmental characteristics caused by variation in discharge. The species mainly occurs in waters with high dissolved oxygen concentrations (>9 mg L^{-1}), but is able to tolerate lower concentrations (down to

Table 19.1 Total number of *A. hispanica* collected during the project in the distinct sub-basins and sampling effort[a]

Sub-basins	Number of samplings	Number of samplings at sites with saramugo (a)	Number of fish collected (N)	Capture index (N/a)
Xévora	29	20	28	1.4
Caia	139	95	1321	13.9
Álamo	3	1	13	13.0
Degebe	46	18	49	2.7
Ardila	62	29	167	6.2
Chança	40	37	306	8.3
Carreiras	26	7	5	0.7
Vascão	345	316	3509	11.1
Foupana	33	27	262	9.7
Odeleite	31	28	307	11.0
Total	754	578	5967	10.3

[a]Total number of samplings performed and number of samplings at sites where the species occurs.

3 mg L^{-1}) during the summer when it is found in the shallower permanent pools. Fish are mainly found associated with coarser substratum, especially gravels, cobbles and large rocks, usually with little submerged vegetation, although they are sometimes captured around instream emergent vegetation. They are able to tolerate temperatures up to 30°C, but seem to prefer temperatures below 25°C. The species is more commonly found in shallower (<60 cm deep) and smaller streams (<10 m wide), but the latter could reflect poor sampling efficiency in larger rivers. Although the species is frequently found in areas with very low current velocity, especially in the summer, it is a rheophilic fish, captured across a range of flow velocities up to about 80 cm s^{-1} (see Collares-Pereira, Pires, Coelho & Cowx 1998; Ribeiro *et al.* 2000).

Data collated on fish community structure and biology of other endemic species provided valuable information on the status and habitat preferences of the remaining threatened species and of the most important exotic species, which were found to have a marked impact on fish assemblages by significantly changing their species composition and relative abundance of endemics (see Collares-Pereira *et al.* 2000a).

In addition, *A. hispanica* exhibits migratory tendencies (Collares-Pereira, Cowx, Rodrigues, Rogado & Moreira da Costa 1999; Ribeiro *et al.* 2000). Mature fish appear to migrate upstream to spawn, as evident from the size distribution and greater proportion of mature fish in the upper reaches of rivers when flows are reinstated in the winter, and appearance of juveniles in the summer. The majority of larger specimens migrate downstream as flows recede with the onset of summer, presumably to areas where the flow is maintained throughout the year or where large, deep, permanent pools exist, thus surviving the potentially harsh summer drought conditions. Such a tactic should enhance the survival of the species by maintaining a population of mature fish as a basis for future recruitment (Ribeiro *et al.* 2000). However, the species was never found in the River Guadiana or the main courses of other larger rivers with permanent flows, such as the Degebe and the Ardila.

19.3.3 *Management plan for the conservation of* **Anaecypris hispanica**

The main threats to *A. hispanica* are human activities in the region, in particular water resource development schemes, pollution, overabstraction of water especially in summer refuges, sand extraction and the spread of exotic species. A review of the location and intensity of these threats, coupled with information collated on biological features was used to formulate a management plan to maintain, improve and develop the extant populations in Portugal (Collares-Pereira, Rodrigues, Rogado & Cowx 2000b).

The main actions proposed are:

(1) designation of special protected areas (SCIs) under the auspices of the EU Habitats Directive;
(2) rehabilitation of degraded systems;
(3) establishment of the non-tangible value of the resource and conservation awareness, with special emphasis on public education and extension;
(4) formulation of fiscal measures to regulate the main anthropogenic activities;
(5) integration of fish conservation into the Guadiana River Basin Management Plan (GRBMP);
(6) establishing international collaboration for sustainable use of aquatic resources; and
(7) stocking of recovered habitats (Table 19.2).

The designation of sites was based on three priority and action levels (see Fig. 19.2 for exact location) as follows.

- *Level A*: areas with high priority and immediate action (already proposed or to be designated as SCIs) due to their conservation value for fish (A1–A4).
- *Level B*: areas with high priority (already proposed or to be designated as SCIs) but dependent on bilateral agreements with Spanish authorities, due to their location (B1–B4).
- *Level C*: areas of secondary priority that need habitat rehabilitation actions, where the species still exists or was known to occur in the recent past, and should be after restocked (C1–C6, with the exception of C6).

The plan provided all the necessary information to formulate projects to target the various issues affecting the species. The key options for rehabilitating the environment, enhancing the endemic fish stocks or protecting the extant stocks for each river sub-basin, which supports or has the potential to support the species, were prioritised. This information needs to be taken up by the responsible bodies and consultation made between stakeholders to optimise development but ensure the conservation of the endemic fish in the region is not compromised. However, for the plan to be of value, it needs careful integration into the management of the drainage aquatic resources through the GRBMP, which is in preparation under the auspices of the EU Water Framework Directive, and also an effective collaboration with Spanish authorities. All

Table 19.2 List of actions required for the conservation of *Anaecypris* and the other endangered species in Portuguese Guadiana

Action	Priority[a]	Responsibility[b]	Timescale
Designation of SCIs/SPAs	H	ICN, EU	December 2000
Rehabilitation of sites or rivers where populations of *Anaecypris* have been seriously impacted by human interventions and exhibit potential for reinstatement	H	ICN, DRAOTs, DRAs, DGF, EU	December 2004
Formulate and enact fiscal measures to regulate indiscriminate abstraction of water and prevent localised habitat degradation in sensitive zones	H	INAG, DRAOTs	December 2002
Continued monitoring of the status of the fish stocks at strategic areas in the Guadiana drainage	H	ICN, DGF, EDIA	Ongoing
Full integration of fisheries conservation issues into the Guadiana River Basin Management Plan	H	ICN, INAG	December 2000
Continued education and extension activities to promote and support the importance of conservation activities	M	ICN, EDIA, DGF, IPAMB	Ongoing
Introduce correct EIA procedures for future development projects to prevent further degradation of the aquatic ecosystem	M	ICN, IPAMB, INAG	Ongoing
Initiate mechanism for evaluating the economic value of aquatic resources in the Guadiana catchment, especially those of conservation interest	M	ICN, DGF, DRAOTs, INAG	December 2001
Formulate and enact fiscal measures for protection of fish stocks and enable improvement in environmental/habitat conditions	M	ICN, DRAOTs, DGF	December 2001
Formulate and enact fiscal measures to regulate movement of exotic species, minimise their dispersion and reduce their numbers in sensitive zones	M	ICN, DRAOTs, DGF	December 2002
Establish mechanism for international collaboration over protection and sustainable utilisation of aquatic resources in trans-national rivers	M	ICN, DRAOTs, DGF, EU, S & P Governments	Ongoing
Initiate mechanisms for enhancement of stocks of *Anaecypris* through stocking of level C sites following the recommendations of the genetic study	M/L	EDIA, DRAA, CMM, CREPA	December 2004

[a] H, high; M, medium; L, low.
[b] EU, European Union; ICN, Nature Conservation Institute; INAG, National Water Institute; DGF, Forestry Directorate; EDIA, Multipurpose Alqueva Project Entreprise; IPAMB, Environmental Promotion Institute; DRAs, Regional Directions of Agriculture of Alentejo and Algarve; DRAOTs, Regional Directions of Territorial Management of Alentejo and Algarve; CMM, Moura Municipality; CREPA, Centre for Reproduction of Fish Threatened Species.

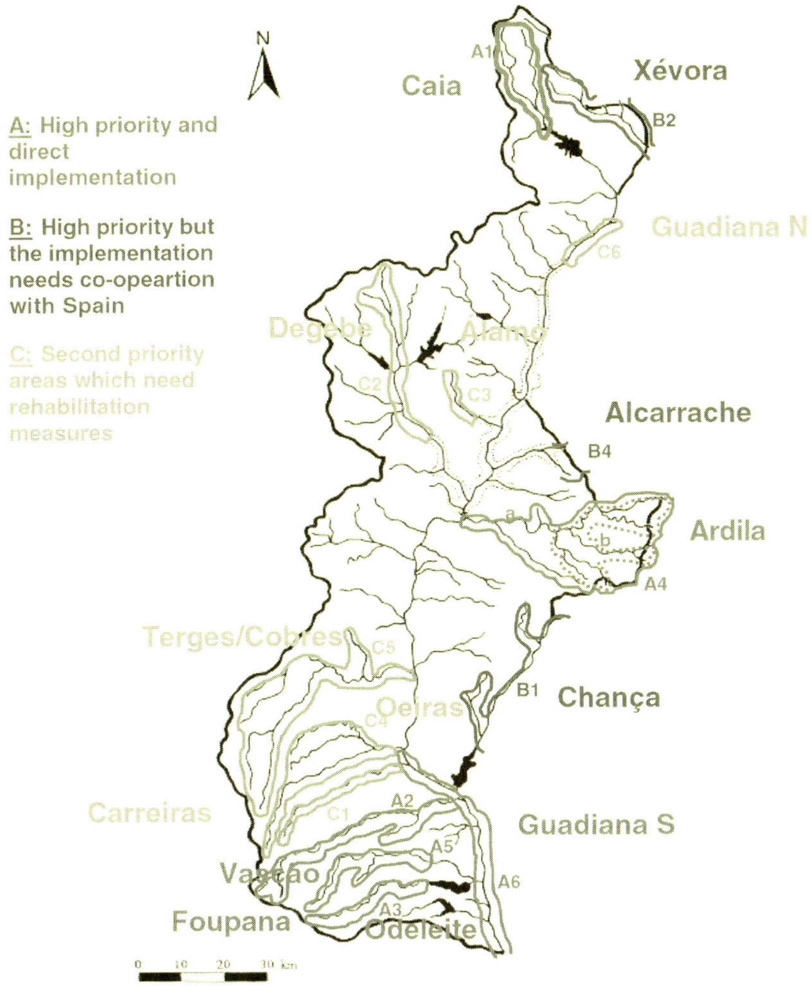

Figure 19.2 Distribution of the sites proposed for the National List of SCIs in the 'Management Plan for the Conservation of Saramugo', according to the A, B and C priority levels (Collares-Pereira *et al.* 2000b) (see also text)

problems affecting saramugo should be included in this plan, as they will help to conserve and enhance the biota and the ecosystem integrity of the river basin.

19.4 Discussion

Anaecypris hispanica has adopted a 'r' selected life-history strategy (Ribeiro *et al.* 2000) to survive the variability of the environmental conditions prevalent in intermittent streams. However, alterations to habitat, flow regime through damming, pollution, overabstraction of water especially in summer pools, sand extraction and the spread of alien fish species have contributed to the demise of the species across its natural range in southern Iberia (see also Doadrio 2001).

Although the extant populations exhibit a high genetic diversity, the low to moderate dispersal ability of *A. hispanica* has important implications regarding its regional persistence. When the gene flow between adjacent populations is low, the chance of recolonisation following an extinction event is correspondingly reduced. Low to moderate dispersal also limits the possibility of 'topping up' vulnerable populations, increasing their probability of extinction (see Coelho *et al.*, Chapter 30).

These issues are particularly significant for the long term persistence of the species, since the semi-arid regime of the Guadiana drainage together with the increasing human pressure, make local bottlenecks and extinctions likely, and it may explain why the species is no longer found in some rivers which apparently have suitable habitats. Therefore, concerted action is needed to conserve the few remaining habitats where the species survives, and to rehabilitate the degraded rivers where the species was found in the recent past. Stocking actions should be only implemented in the recovered habitats (level C) after removing the existing environmental bottlenecks, and be based on the recommendations pointed out by Coelho *et al.* (Chapter 30).

Consequently, key management actions include designation of the tributaries (in their entirety) where the species exist in the National List of Sites under the NATURA 2000, although with distinct priority levels, the planning of a local breeding centre for restocking recovered habitats, and a massive public awareness campaign based on stakeholder meetings (see Portuguese/English web site http://hello.to/saramugo).

Conservation benefits for NATURA 2000

The project showed that the key factor in saramugo recovery was not specifically aimed at the species, which still has considerable genetic diversity, but more the conservation/rehabilitation of natural habitats. Thus, conservation of the endemic fish species demands proactive habitat management and the recovery of degraded zones. Therefore the recognition of SCIs, although based on species value, should give special attention to the intrinsic value of the habitats. This aspect must be considered by funding agencies when supporting projects for conserving individual species listed as threatened under various conventions. The management plan proposed a list of new sites for the National List of SCIs for the NATURA 2000, and the extension of the area of some already listed.

Demonstration and innovation value of the project

Habitat changes/destruction caused by human activities are considered the main threats to freshwater fish species (see Boon, Davies & Petts 2000), and their effects are particularly damaging in rivers of arid and semi-arid regions. In the Guadiana River Basin, the hierarchy of the different interventions affecting the fish is difficult to establish, due to their synergistic action.

Compared to other species/animal groups, when the project started, there was a paucity of information on the target species with respect to its life history, habitat preferences and genetic structure. Such aspects are essential to formulate a management plan for the conservation of any species, and a strategy for the analysis of species conservation status was developed under the frame of this project. Considerable effort was also made to increase public awareness, although this proved initially difficult, but ultimately rewarding, because these stream fishes have little public appeal. This aspect was considered a key action for conserving saramugo, i.e. involving people in the conservation battle needs to be strengthened. Indeed, until conservation/biodiversity management includes major short term incentives and responsibility is devolved to the people, it will be over-ridden by the market tendencies of the modern consumer society. Similarly, contacts were made with other on-going EU LIFE-NATURE projects, to disseminate the strategies developed and to promote a discussion forum on similar actions.

19.5 Final remarks

Despite the actions proposed, the fish species of Guadiana River Basin remain threatened by large scale water resource development schemes, which are driven by economic and political interests, in general with little concern of the existing legislation and specific recommendations of Environmental Impact Assessments (EIAs) carried out by learned, independent entities. The Habitats Directive is the main legal instrument in the European Union, which is now preparing proposals (NATURA 2000) to prevent further degradation of habitats and species. However, it is recognised that the list of species included in the distinct annexes, due to new data, should be revised. Moreover, the measures proposed in the Water Framework Directive, namely the first generation of River Basin Management Plans and the River Basin Councils, may be inefficient if not integrated with other aspects, namely multi-national co-operation for international rivers, and the active participation of the users through specific awareness actions to optimise the benefits and minimise the costs of exploiting aquatic ecosystems (Cowx & Collares-Pereira 2000). The project created a conservation awareness with regards to freshwater fishes, which did not exist previously in Portugal, thus opening new perspectives for future actions.

Appropriate resources to implement the preferred options should be sought and allocated. However, in the meantime, it is essential that the proposed sites are designated in NATURA 2000, so the key rivers are protected for future generations. Therefore, governmental entities must harmonise the conservation principles with human needs and economic demands. Such a process in the short–medium term will be a challenge for scientists, technicians, managers and politicians. Although the project started to create a conservationist attitude in the region, it also needs continuous reinforcement in the future, to ensure sustainable use of aquatic resources.

Finally, the present constraints on the conservation of fish species in the Guadiana drainage, especially those which will be created by Alqueva and Pedrogão reservoirs (Cowx & Collares-Pereira 2000), make the conclusions and recommendations of this project particularly useful in the formulation of regional developmental strategies. Indeed, some of the rivers are still in good environmental condition, i.e. those proposed for NATURA 2000 in the A and B priority levels, and the species still has a high genetic diversity. It is expected that the GRBMP will continue to monitor the status of the threatened species and changes in habitats and land use, as well as to ensure the integrated management of the international rivers. The implementation of specific projects for habitat rehabilitation at level C priority sites, should include population viability analysis on the donor populations before any stocking programme is implemented, and guarantee the involvement of a wider range of institutions, such as INAG (National Water Institute), specialists on environmental engineering and the local authorities, which are now much more aware of their role in conserving fish species and habitat quality.

Acknowledgements

Thanks are especially due to the European Commission and all the institutional partners of the project LIFE-Saramugo (CBA/FCUL, ICN, DGF, MB) for having funded this research under the contract B4-3200/97/280. Also, to all the remaining team of FCUL, the authors are grateful for their enthusiastic support in different phases of the field work – F. Ribeiro, A.F. Filipe, P. Tiago, T. Marques, L. Moreira da Costa, L. Correia, S. Seabra, R. Silva, P. Picciochi, F. Mendes, R. Mendes and J. Reis – and for some data analyses – A. Mendes (FCUL) and A. Nunn (Hull University).

References

Baillie J. & Groombridge B. (1996) *IUCN Red List of Threatened Animals.* Switzerland: IUCN, 368 pp.+annexes.

Blanco J.C. & Gonzalez J.L. (eds) (1992) *Libro Rojo de los Vertebrados de España.* Madrid: Colección Técnica, Icona, 714 pp. (in Spanish).

Boon P.J., Davies B.R. & Petts G.E. (eds) (2000) *Global Perspectives on River Conservation: Science, Policy and Practice.* West Sussex: John Wiley & Sons, 548 pp.

Collares-Pereira M.J. (1990) *Anaecypris hispanica* (Steindachner), a cyprinid fish in danger of extinction. *Journal of Fish Biology* **37**(Suppl. A), 227–229.

Collares-Pereira M.J., Cowx I.G., Rodrigues J.A., Rogado L. & Moreira da Costa L. (1999) The status of *Anaecypris hispanica* in Portugal: problems of conserving a highly endangered Iberian species. *Biological Conservation* **88**, 207–212.

Collares-Pereira M.J., Pires A.M., Coelho M.M. & Cowx I.G. (1998) Towards a conservation strategy for *Anaecypris hispanica*, the most endangered non-migratory fish in Portuguese streams. In I.G. Cowx (ed.) *Stocking and Introduction of Fish.* Oxford: Fishing News Books, Blackwell Science, pp. 437–449.

Collares-Pereira M.J., Cowx I.G., Rodrigues J.A., Rogado L., Ribeiro F., Mendes A., Picciochi P., Salgueiro P., Alves M.J. & Coelho M.M. (2000a) Uma estratégia de conservação para o saramugo (*Anaecypris hispanica*), um endemismo piscícola em extinção. *Final Technical Report, LIFE-NATURE Programme, contract B4-3200/97/280*, Vol. I (121 pp.) and Vol. II (13 Annexes) (in Portuguese, with an English summary).

Collares-Pereira M.J., Rodrigues J.A., Rogado L. & Cowx I.G. (2000b) Plano de Gestão para a Conservação do Saramugo. *Technical Report, LIFE-NATURE Programme, contract B4-3200/97/280*, CBA-FCUL/ICN, 75 pp.+annexes (in Portuguese).

Cowx I.G. & Collares-Pereira M.J. (2000) Conservation of endangered fish species in the face of water resource development schemes in the Guadiana River, Portugal: harmony of the incompatible. In I.G. Cowx (ed.) *Management and Ecology of River Fisheries*. Oxford: Fishing News Books, Blackwell Science, pp. 428–438.

Doadrio I. (2001) *Atlas y Libro Rojo de los Peces Continentales de Espana*. Dirección General de Conservación – Museo Nacional de Ciencias Naturales, Madrid, 364 pp. (in Spanish).

Ribeiro F., Cowx I.G. & Collares-Pereira M.J. (2000) Life history traits of the endangered Iberian cyprinid *Anaecypris hispanica* and their implications for conservation. *Archiv für Hydrobiologie* **149**, 569–586.

SNPRCN (Serviço Nacional de Parques, Reservas e Conservação da Natureza) (ed.) (1991) *Livro Vermelho dos Vertebrados de Portugal. Vol. II – Peixes Dulciaquícolas e Migradores*. Secretaria de Estado do Ambiente, Lisboa, 55 pp. (in Portuguese).

Section III
Analysis of threats and constraints

Chapter 20
Analysis of threats to freshwater fish conservation: past and present challenges

I.G. COWX*
Hull International Fisheries Institute, University of Hull, UK

Abstract

Many of the threats to freshwater fish and fisheries focus on gross changes to the habitat or community structure, e.g. pollution, flow regulation, loss or change of aquatic habitat, overexploitation and introduction of exotic species. These have all contributed towards the critical situation facing many freshwater fish species. Although action has been taken to address this precarious situation, traditional methods of conservation management, e.g. regulation of exploitation, nature reserves, captive breeding programmes, do not seem to be as effective for freshwater fishes as for terrestrial flora and fauna.

This chapter provides examples of the major threats to the conservation of freshwater fisheries and asks whether it is appropriate to treat these problems as single entities as they most frequently are, and whether more subtle environmental, social and economic issues should be given equal weighting in the conservation battle. The chapter then evaluates the efficacy of traditional conservation management mechanisms and concludes that until conservation management includes major short-term economic incentives and responsibility is devolved to the people, it will be overridden by the market tendencies of the modern consumer society.

Keywords: conservation, fisheries value, legislation, pollution, rehabilitation, river regulation, stock enhancement.

20.1 Introduction

Aquatic systems are utilised by a wide variety of users (Fig. 20.1). These activities all have a legacy for impacting on the environment with subsequent knock on effects on the biota, including fisheries (Cowx & Welcomme 1998). However, the scale of the impact varies depending on the dimension of the activity, its location in the aquatic ecosystem and the current status of the environment. Furthermore, the impacts of these activities are not mutually exclusive and can be synergistic in their outcome when combined. These have all contributed towards the critical situation facing many freshwater fish species (Lelek 1987; IUCN 2000), and in the worst case scenario, loss of biodiversity in the widest sense. Although action has been taken to address this precarious situation, traditional methods of conservation management, e.g. regulation of

*Correspondence: Hull International Fisheries Institute, University of Hull, Hull HU6 7RX, UK (email: i.g.cowx @hull.ac.uk).

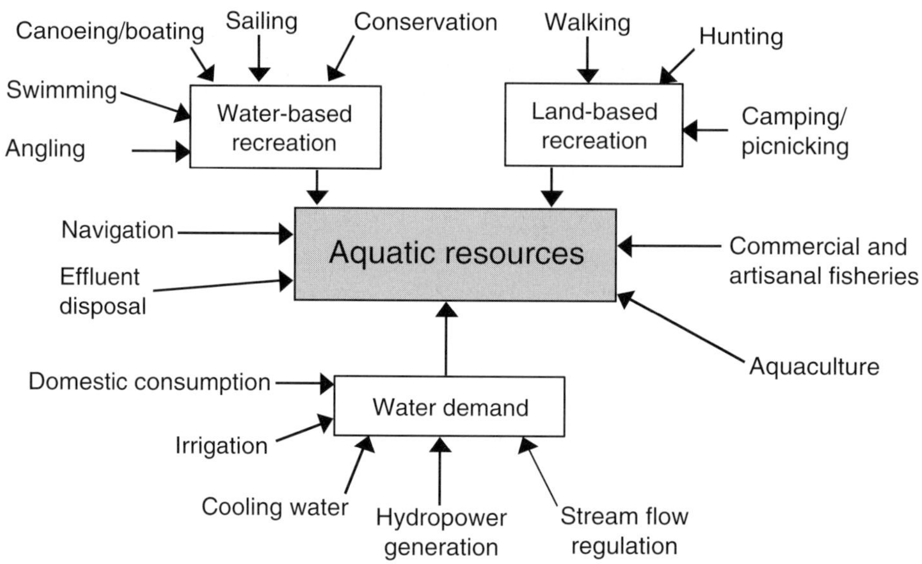

Figure 20.1 Array of aquatic resources users that may impact on fish conservation

exploitation, nature reserves, captive breeding programmes, do not seem to be as effective for freshwater fishes as for terrestrial flora and fauna. Indeed, the number of examples of conservation actions targeting fish in the literature is small in comparison to other animal species (e.g. between 1996 and 1999 only eight EU LIFE projects were on fish whereas 20 projects targeted mammals and 54 targeted birds; http://europa.eu.int/comm/environment/nature/home.htm), and the number of examples where populations have increased and expanded is pitiful (see Kirchhofer & Hefti 1995 and this volume for examples).

This chapter provides examples of the major threats to the conservation of freshwater fisheries and asks whether it is appropriate to treat these problems as single entities as they most frequently are, and whether more subtle environmental, social and economic issues should be given equal weighting in the conservation battle. The chapter then evaluates the efficacy of traditional conservation management mechanisms and asks why they are inadequate for freshwater fisheries and what developments are needed to overcome any recognised deficiencies.

20.2 Analysis of threats

A review of the documentation submitted to the symposium on *Freshwater Fish Conservation: Options for the Future* highlights the perceived main threats to freshwater fish and fisheries (Fig. 20.2). This analysis scored the number of times a particular threat was mentioned and ranked them in order of response. Three main threats

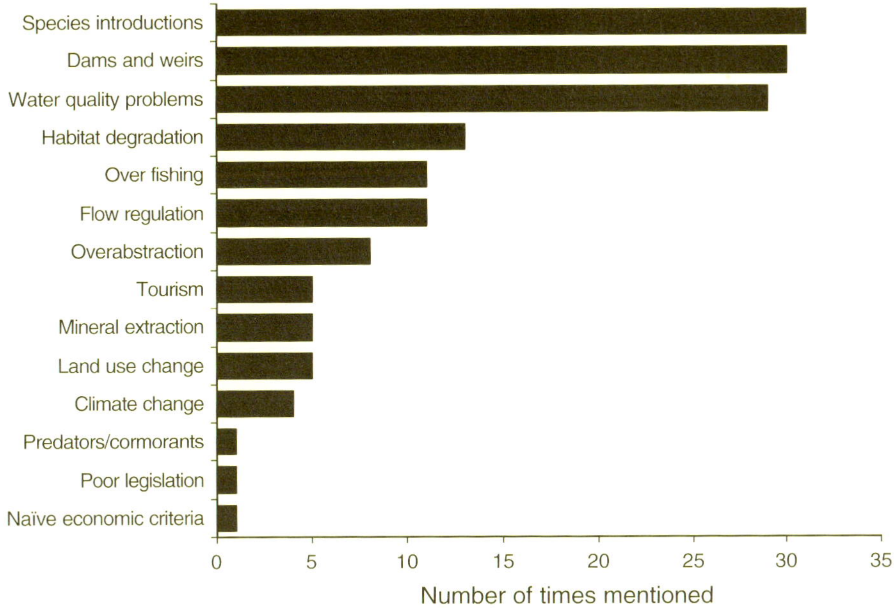

Figure 20.2 Principal threats to freshwater fishes based on assessment of the literature

dominated, viz.: species introduction, impoundments (dams and weirs) and water quality problems. Habitat degradation, overfishing, flow regulation and overabstraction of water were also prevalent but apparently of lesser importance. However, there seems to be a lack of awareness of activities that do not physically impact directly on the aquatic environment, e.g. tourism, land use change and climate change, or institutional and economic arrangements.

Although this analysis provides an indication of the principal threats to freshwater fishes it fails to elucidate the underlying issues that impact on fish populations, which is fundamental for any management regime to function. In the case of freshwater fish and fisheries this can be broken down into categories: fish and fishery-related issues, environment or watershed related issues, and the social and economic dimension.

20.2.1 *Fish and fishery-related issues*

One of the primary issues impacting on freshwater fisheries is the exploitation pattern. Fisheries in developed countries are exploited by commercial and recreational fisheries, usually for pleasure but also for food. Commercial fisheries in this region are now declining largely because the fish stocks have been depleted or altered by other human interventions. In developing countries, exploitation is largely by commercial and artisanal fishermen or subsistence fishing by local residents supplementing their food supply (Welcomme 2000), although recreational fishing is developing as part of the tourist sector (Cowx 2002a). In many cases, fisheries are showing signs of overexploitation typically seen by decline in catch per unit, reduce size of capture, loss of larger-sized species, change in exploitation to smaller, less valuable fish species (Caddy & Mahon 1995).

The general effect of overexploitation of the fisheries are poor catches of desired species, loss of species or disruptions of local stock adaptations, and poor population or community structure (size and species). The typical response of fishery managers is to overcome these limitations by modifying the fish assemblages through stocking and introductions to support exploitation (Petr 1998; Cowx 1999). Stock enhancement activities, which are frequently carried out with no due regard for the environmental or ecological consequences, are known to impact on indigenous fish populations in a multitude of ways. Stocking can be damaging to the native stocks through competition, predation, loss of genetic integrity, or the spread of disease and parasites (Cowx 1994, 1998a, 1999). Unfortunately there is relatively little information about the effects, including successes, of various stocking practices (Cowx 1998a). If the species is being released in high numbers then changes in the ecosystem are likely to occur via fish species interactions and food web dynamics (Persson 2002). Vulnerable fish species, and aquatic flora and fauna can be eliminated through predation, and the stocking, intended to enhance the fishery, can have the opposite effect, including elimination of species (Cowx 1997). The loss of genetic integrity of the native stock, which is thought to have adapted to local environmental conditions over many years, is a major issue and considered to be responsible for a decline of many fisheries, especially salmon, *Salmo salar* L. (Carvalho & Cross 1998), and brown trout, *Salmo* sp. (Persat & Keith, Chapter 32), stocks.

The introduction of new species to promote fishery diversity was common practice worldwide in the 1960s and 1970s (Welcomme 1988), but concerns over the impact of this activity and the implementation of tough regulations has restricted the practise in industrialised countries in recent years. Introductions are considered successful in a number of cases, e.g. rainbow trout, *Oncorhynchus mykiss* (Walbaum), and large mouth bass, *Micropterus salmoides* (Lacépède), but this has usually been at great cost, most often expressed through the demise of indigenous species or spread of diseases (Cowx 1997, 1998a). Other introductions have been disastrous, with wholesale ecosystem change and loss of species (Cowx 1997). With the recent decline in the status of many fisheries, there is renewed interest in species introductions. This should be prevented at all costs, and the causes of the deterioration in extant fisheries should be identified and where possible addressed. If a species is to be considered for introduction, it should be carried out under the auspices of the appropriate governmental organisation and following guidelines such as those promoted by the International Council for the Exploration of the Seas (ICES 1988), the European Inland Fisheries Advisory Committee (EIFAC 1988) or the World Conservation Union (IUCN 1995).

In reality, stock enhancement practises are often carried out in response to pressure from those exploiting the resource and do not address the underlying problems within the fishery, e.g. recruitment bottlenecks or overfishing. In this respect stocking exercises are just short-term solutions and sustaining the exploitable stock requires continuous intervention (Cowx 1994). It is thus not necessarily a desirable solution and alternative mechanisms should always be sought in the first instance.

In addition to planned introductions, there are many examples where fish species have been illegally or accidentally introduced to increase the diversity of the target species for anglers. There are also many cases of fish introductions stemming from the use of live bait for catching predators (Cowx 1997). All these practices are potentially damaging to fish diversity and have rarely proved successful, and most worrying have led to the demise of many fish species (Cowx 1997). Anglers and managers need to be educated in the problems and consequences associated with stocking and introduction of fish, whether deliberate or accidental, and all efforts should be made to minimise this management practise.

It should be recognised that both overfishing and fishery enhancement strategies conflict with the general precept to maintain species diversity elaborated in the Code of Conduct for Responsible Fisheries adopted FAO (1995), and also have damaging effects on conservation efforts for threatened species. In reality, many of the problems that exist with respect to conservation of fish species are the direct consequence of the management of the fishery.

One of the fundamental constraints on fish populations in fresh waters is poor recruitment of fish. This can arise from lack of optimal spawning habitat, inadequate food resources, especially for larvae after they have absorbed the yolk sac, or high mortality of juvenile life stages unable to support maintenance of the stock.

An issue that is prevalent in inland fisheries across Europe and North America is the explosion of avian piscivores (Cowx 2002b). This has arisen because many bird species are protected under environmental and species laws and their numbers, especially cormorants (*Phalacrocorax* spp.), have expanded in inland regions. Many managers perceive this expansion of the bird populations as detrimental to the fish populations through depredation on fish stocks. However, the general concensus is that the problem is specific to individual fisheries, and still waters are more prone to damage than rivers (Feltham, Cowx, Davies, Harvey, Wilson, Holden & Britton 1999).

20.2.2 *Environment or watershed related issues*

It should be recognised that the diversity and scale of anthropogenic perturbations associated with fresh waters are vast and the relative vulnerability of different freshwater ecosystems (Table 20.1) are often underestimated. The following section reviews the impact of various human interventions on the aquatic environment, and consequently fish populations, and highlights where the biggest threats to fish conservation are.

Water quality issues

Pollution is probably one of the biggest threats to the freshwater environment (Moss 1998). This is because discharge into fresh waters is a convenient mechanism for disposal, and the capacity of large water bodies to accommodate such wastes is high. However, with the increasing human population, pressures exerted on fresh waters have become intolerable and many are contaminated, which in turn has resulted in the deterioration of fish stocks and biodiversity.

Table 20.1 Relative vulnerability of ecosystems to user impact

User impact	Streams	Rivers	Lakes	Wetlands
A. Indirect users				
(1) Agriculture and forestry	++	++	++	++++
(2) Civil works	++++	++	+	++++
(3) Extraction industries	++++	++	+	++
(4) Manufacturing industry	++++	++++	++	+++
(5) Urbanisation	++++	+++	++	+
B. Direct users				
(1) Dams and barrages	+	++++	++	++
(2) Land drainage	+++	+	+	++++
(3) Industrial and urban use	+	+	+	+
(4) Recreation and transport	+	+	+	++

NB: This is a subjective evaluation and individual activities will vary greatly in their impact.

Pollutants include, organic wastes, nutrients, pesticides, metals, poisons, suspended solids and cooling water from urban, industrial and agricultural sources. These can act directly on the fish, e.g. toxicity of chemicals, which may have an acute or chronic affect depending on concentrations of chemicals concerned, or indirectly by changing water quality parameters, and consequently the suitability of the water body. Processed waters (i.e. water which is someway used as an integral part of the manufacturing process), if discharged untreated, may contain large concentrations of pollutants. For example the organic materials and suspended inert solids that derive from agricultural or food processing enterprises; toxic heavy metals from tanning and metal plating industries; and fibres discharged by the paper manufacturing industry. Effluents from large chemical manufacturing concerns and industrial complexes may contain a cocktail of unwanted materials, often combining to produce unexpected toxic effects.

Many industries, especially those concerned with the brewing industry, food processing, paper manufacture, produce large quantities of liquid waste, which has a high organic and nutrient content. The discharge of human excretory wastes also has the immediate effect of raising the concentration of organic material in the water. When these effluents are discharged into natural water bodies at relatively small concentrations the input of nutrient material may raise the level of productivity. However, with excess loads of organic materials, either as solids or in solution, the immediate observable effect is the creation of blooms of green algae or massive growths of rooted or floating higher plants (i.e. eutrophication). The water body may suffer further from falls in the levels of dissolved oxygen to the point where the water and its substrate becomes anoxic with the production of hydrogen sulphide and methane. Under these conditions only a few species of organisms are able to survive, although they may do so in populations with very high numbers of individuals. If this species change replaces favoured species with less desirable species the increased levels of production may not be seen to be advantageous.

Acidification, brought about by acid precipitation reaching water courses, has the opposite affect of eutrophication because its final stage is an almost dead lake with no decomposition taking place. In part this low production is brought about by the acid environment creating conditions where the nutrients present are not readily available to the plants. This type of pollution is most notable in northern Europe and Canada where large numbers of lakes are affected and they are virtually fishless. Acidification is caused by sulphur and nitrogen compounds being emitted into the atmosphere from burning fossil fuel then being blown hundreds of kilometers before being deposited later as sulphuric and nitric acid. The rain may have a pH of less than 4.

Problems relating to pollution are in the main being addressed in western industrial countries, and waters that were once fishless or supported poor stocks, are now recovering. Such water quality improvements are, however, rare in the developing world and the scenario remains one moving towards environmental degradation and demise of the fisheries. Notwithstanding, water quality improvements do not always lead to a desirable outcome in terms of fishery status. The cleaning up of water bodies often results in reduced nutrient input and loss of productivity of the water body concerned, which has a knock on effect on the fish community structure and stock size. This results in a change in the catch composition and reduction in catches, which are quickly followed by complaints (Cowx 2002a).

Finally, many industries directly exploit the water resources as a means of cooling. The biological effects of the discharge of cooling waters is similar to the effects of the discharge of organic materials. In relatively small amounts the raising of the temperature of the receiving waters, and especially the maintenance of higher and constant temperatures throughout the annual seasonal cycle, will raise the level of production. However, as temperatures rise above the optimum for the environment they cause a decrease in the amount of available dissolved oxygen and may create localised, lethal heat stress conditions for fish.

Dams and reservoirs

The supposed benefits and high prestige derived from the construction of dams have made them very attractive to funding agencies and national governments. However, dams, and the resultant man-made lakes, generate a complex web of impacts which affect the human, biological and physical components of the environment. They represent the greatest source of hydrological interference by man and are perhaps one of the single most damaging (Table 20.2), in terms of loss of fish diversity, development activities associated with fresh waters.

The most direct impact of dams is that the river continuum is interrupted and longitudinal connectivity is blocked. The barrier across the water course will obstruct the free migration of fish both upstream and downstream (Table 20.2). As many of these migrations are essential for the breeding of the species this constraint may cause catastrophic falls in numbers, or complete absence, of fish in areas where they were once abundant.

The change in the ecosystem, from that of a river to that of a lake, will inevitably create a change in the species dominance of the fish fauna. There is a usual tendency for rheophilic species to disappear, especially anadromous species, unless appropriate arrangements for their passage are put in place. Species extinctions have often been associated with the introduction of exotic species aimed at enhancing the fish potential

Table 20.2 The effects of river impoundment upon anadromous fishes (from Petts 1984)

Structural effects	Physico-chemical effects
Construction of dam	*Discharge, sediment load, water quality, food*
supply,	
(i) Obstruction to upstream migration	*channel form, riparian and aquatic vegetation*
(ii) Loss of fry during passage over or through dams	(i) Unacceptable changes of seasonal discharge pattern
(iii) Gas supersaturation	(ii) Unnatural short-term flow fluctuations
	(iii) Unacceptable changes of seasonal water quality pattern (particularly temperature)
Creation of lake	(iv) Elimination of natural trigger-mechanisms for migration (related to discharge and water quality)
(i) Flooding of spawning grounds	
(ii) Delays to migration of spawners through reservoirs to spawning grounds	(v) Unnatural 'toxic pulses' of poor-quality water
(iii) Delays to migration of fry downstream through reservoirs	(vi) Removal of flushing capacity for dilution of toxic wastes
	(vii) Suitability of substrate for spawning
	(viii) Survival of eggs in gravels
	(ix) Amount of drifting food organisms
	(x) Spatial and territorial requirements
	(xi) Shelter and susceptability of young fish to predation from birds

of the newly established lake environment. In addition, the construction of the reservoir frequently drowns out spawning and nursery areas and results in a loss of recruitment to the catchment.

Regulation of flow below the dam can lead to alteration in the seasonal or diel flow regimes. Evidence exists that alteration in timing and volume of flows interrupts migration of salmonid and cyprinid species on a diel and seasonal basis. The loss of the seasonal cycle of low summer flows and autumn and spring floods can lead to inappropriate stimuli for fish breeding, leading to shifts or failure of seasonal spawners, which include many endangered species, being displaced by those with more flexible spawning habits. In addition, rapid and often extensive changes in discharge regime associated with hydropower releases cause displacement of fish, particularly juveniles and can lead to loss of the fish populations.

Water temperature in rivers characteristically exhibits a high degree of seasonal and diurnal variability (up to 15°C in European rivers). Thermal regulation by reservoirs tends to dampen these natural variations and stream temperatures below dams are usually higher than normal in winter, lower in summer, fluctuate less seasonally, and can exhibit a seasonally displaced maximum. In temperate regions, faunas are adapted to diurnal fluctuations in temperature, so that a more constant regime would be disadvantageous. Seasonal thermal regulation can have similar consequences, because different physiological and behavioural components have different temperature optima. Thermal influences of impoundments have made significant contributions to the elimination of many temperature-specific fish species. Reduced spawning success resulting from temperature changes is a key factor causing reduction in native species diversity.

Water transfers and abstractions from donor waters occur for a variety of reasons, such as cooling water for fuel fired electrical generation, industry, potable water supplies, irrigation, and fish farming. These can also lead to an increased volume in the recipient watercourse and introduces the possibility of disease organisms, planktonic species and fish being transferred. Flow patterns can be changed in both the donor and recipient rivers, especially during low flow periods, despite consented abstraction controls.

Land drainage and channelisation

Channelisation is a widespread activity in Europe and North America for a variety of reasons, including: flood control, navigation, urbanisation, and drainage for agricultural purposes. The primary mechanism is to simplify the channel topography by widening and deepening the channel, increasing the size of banks (levees), removing meanders, revetment of banks, and eliminating instream and riparian habitat features. This leads to a reduction in habitat diversity, causing loss of habitats such as secondary channels, i.e. braiding, and prevention of flooding and isolation of floodplain lakes, old reaches, cutoffs, etc. Access to dead arms is denied due to confinement of higher flows within channel. Channelisation is usually associated with downgrading of riparian vegetation and nutrient dynamics, allochthonous inputs and refuges for feeding and breeding are affected causing reduction in fish numbers and biomass of food organisms. There is also a loss of niche diversity with sheltered, shallow, warm areas disappearing, reducing optimal refuges for small fish species and young of the year for larger species. Ultimately this leads to a reduction in fish diversity and can cause selective disappearance of obligate floodplain and opportunistic spawning species of fish.

Agriculture and forestry

The major impact of agricultural activities comes from so-called 'non-point source' pollution. This is the diffuse discharge of dissolved and suspended materials in run-off or seepage drainage waters. Such discharges occur with drainage from unmodified catchments but may be greatly increased where the land is cultivated or is exploited in other ways (e.g. forestry). Typically this input is intermittent in its intensity and will depend particularly upon the rainfall pattern and the seasonal cycle of agricultural inputs. The effects may be reduced by efficient land management and conservation measures.

Deforestation and poor forestry practice have been shown to lead to habitat destruction and modifications in discharge regimes and seston flows. This causes increases in silt loading from erosion due to felling, timber extraction, logging-road construction, particularly where 'clear-cut' techniques are employed, and cause changes in channel and floodplain morphology. Habitat and community diversity decreases, log and debris dams proliferate, altering flow regimes in some instances with loss of food organisms and choking of spawning substrates for psammophilic and lithophilic fish species.

On catchment slopes deforestation affects water, topsoil and nutrient conservation; run-off is decanted straight into stream channels and flood peaks become higher and shorter. Water conservation and slower release of water to streams is lost and lower dry season flows combine to produce adverse conditions for some fish species.

Climate change

Climatic change or global warming is a problem that has received little attention with respect to fish conservation. Much of the literature (reviewed by Lehtonen 1996) examines the effects on the physical and chemical environment with little attention focussed on the biological impacts. Climatic change is predicted to lead to increased water temperatures, and increased prevalence to floods and droughts, but the influence of environmental change on ecological processes, including physiological tolerances and ecosystem dynamics, seems to have not been considered in detail. In Europe the general prediction is that fish communities will shift towards cyprinid and percid fish assemblages coupled with a decrease or collapse of salmonid and other cold water fish populations. Beyond these changes in the natural assemblages, there could be a spread and proliferation of alien species with impacts similar to those observed for species introductions. Thus it is highly likely that many more species, especially stocks with genetic or phenotypic adaptations at the limits of their species range (Persat & Keith, Chapter 32), could be lost.

20.2.3 *Social and economic dimension*

Many of the problems that exist with respect to conservation of fish species are the direct consequence of the management or objective of the fishery (Table 20.3). There is a dichotomy in fisheries management strategies between developed and developing

Table 20.3 Strategies for management of inland waters for fisheries in developed and developing countries (after Welcomme 2000)

Conservation (developed)	Production (developing)
Objectives	
• Conservation	• Provision of food
• Recreation	• Income
Mechanisms	
• Sport fisheries	• Food fisheries
• Habitat restoration	• Habitat modification
• Environmentally sound stocking	• Enhancement through intensive stocking and management of ecosystem
• Intensive, discrete, industrialised aquaculture	• Extensive, integrated, rural aquaculture
Economic	
• Capital intensive	• Labour intensive

countries. In the former, freshwater fisheries are not typically heavily exploited for commercial purposes and management is orientated towards recreation and conservation. This is often achieved through rehabilitation of habitat, environmentally sensitive stocking and protection of the resources. In developed countries the fisheries are managed for provision of food and provision of income. The latter regime offers no incentive to conserve fish species. Again the typical response of managers is to compensate for the poor production through stocking to bypass the recruitment bottleneck and utilise the potential production of the water body. This action is often in response to pressure from fishermen and requires continuous intervention, often at high cost, with marginal returns (Cowx 1998a, b). A more appropriate strategy would be to address, if possible, the recruitment bottleneck either through more traditional fishery management regulation or rehabilitation of the water body (see Section 20.4).

Aquatic resources are subject to numerous anthropogenic perturbations (Fig. 20.1), which have resulted in a shift in the status of freshwater fish communities and a general decline in biodiversity. In these circumstances fish and fisheries are not considered of sufficiently high priority or value, and thus suffer in the face of economically and socially higher priorities, e.g. agriculture, hydroelectric power production or other water sports. The underlying tenet here is that the ever-expanding human population, which is imposing greater demands for a higher quality of life, are putting unprecedented pressures on the environment. This can only be resolved by increasing public awareness of the problems and internalise the costs of minimising the environmental damage or rehabilitation against development, i.e. the polluter-pays principle.

From the previous section, it should be recognised that fish conservation and maintenance of fish biodiversity in a multiple-user environment are fraught with cross-sectoral problems. Fisheries and conservation are often considered marginal activities because the value of the resource is usually ill-defined and poorly represented from an economic and social perspective. Conservation is traditionally based on trying to avoid further deterioration of the diversity but it is rarely argued from an economic perspective. This is born out by the paucity of information on the economic value of biodiversity or fish stocks *per se* (e.g. Kennedy & Crozier 1997; Peirson, Tingley, Spurgeon & Radford 2001). Consequently, conservation is given low priority in any consultation process and it is difficult to attract investment for protection of biodiversity, despite obligations under Agenda 21 of the Rio Convention.

To overcome this problem there is an urgent need to provide accurate social and economic valuation of biodiversity for justification of its position in multiple-user conflict situations. There are various direct (financial and cost benefit analysis; e.g. Kennedy & Crozier, 1997) and indirect (contingency valuation analysis or bequest and existence value; e.g. Baker & Pierce 1996; Gautam & Steinback 1998; Postle & Moore 1998; Peirson *et al.* 2001) methods of undertaking this appraisal which have been applied to fisheries, but it must be recognised that they rarely give a true economic value of the fishery. If freshwater fish diversity is to be preserved in the future, it is essential that appropriate techniques to provide sound economic valuation are developed.

Conservation is an emotive science, driven by a network of dedicated, enthusiastic persons working towards a common goal. However, following recent recognition about the deteriorating state of the environment and increasing loss of biodiversity it has reached the public imagination much more. Despite this general awareness, the public is poorly informed of the underlying issues concerning specific species or habitats. As a result, they often, unknowingly, contravene legislation in the normal routines of life and unwittingly damage fragile environments or habitats, e.g. surface water abstraction from permanent pools supporting *Anaecypris hispancia* by subsistence farmers in Portgual (Cowx & Collares-Pereira 2000). Meffe (Chapter 33) considered this was because scientists, managers and policy makers each have their own interests and modes of operation, and they respond to different motivations, constituencies and reward systems. However, as Meffe (*op. cit.*) pointed out, if conservation is to succeed, each must be engaged and work effectively with the others as a unit, and more importantly engage the public in the debate. Consequently one of the mechanisms for improving elevating the role of conservation is to improve communication between all stakeholders and inform them of the problems and how they can become involved.

20.3 Problems relating to current conservation practices

It is evident from the previous section that inland waters have suffered from an array of activities, which have interrupted, degraded or destroyed the functioning of aquatic ecosystems. The net result has been depreciation in the resource value for other users, and demise of fish biodiversity and fisheries *per se*. To identify the conservation strategies that have been used in an attempt to protect extant fish populations that are threatened a similar analysis to that carried out to assess the major threats to freshwater fishes was undertaken. Again the literature submitted to the symposium was reviewed and the key actions for fish conservation were prioritised (Fig. 20.3).

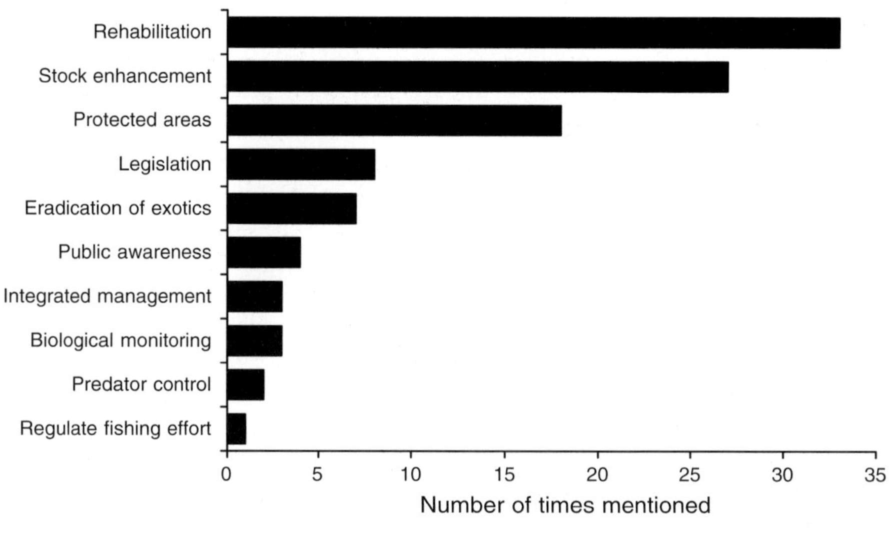

Figure 20.3 Principal actions used in the conservation of freshwater fish

The three main actions, in rank order, were habitat restoration, stock enhancement and protected areas. The dominance of these actions is alarming for a number of reasons, not least because habitat restoration and stock enhancement are reactive responses to problems when they are usually deeply entrenched, and protected areas are rarely set up with freshwater fish as the target organism for conservation (Crivelli, Chapter 34). However, there are more fundamental concerns about the contribution of these management initiatives in the conservation battle.

An underlying problem is a lack of basic information about the distribution and abundance of fish species in general, and threatened species, in particular. Assessment of fish populations is notoriously difficult, even with modern equipment (Cowx 1996). Consequently, knowledge of the status of stocks is often weak (Hefti, Chapter 38) and comparison with the historical, pre-perturbation, situation is highly subjective. Thus, identifying why species' distribution and abundance have retracted is a precarious undertaking, fraught with assumptions. Although anecdotal and indigenous knowledge is valuable to address this issue it does not necessarily help identify the factors contributing towards the demise of the species.

Restoration of freshwater habitats towards pristine conditions, which is the objective of most managers, is a utopian view. In most cases, freshwater habitats have experienced extensive changes because of human activity. The most dramatic impacts resulted from deforestation, land use changes, intensification of agricultural and industrial activity, and the modification of river channels to control floods or provide for navigation. More recently, demands for water resources and hydropower have created new impacts. These changes have been superimposed upon environmental changes caused by recent global climatic variations, and many are now irreversible.

Furthermore, many of the rehabilitation schemes are localised dealing with only small sections or reaches of the water body and ignore the problems associated with adjacent sections. For example water quality problems upstream inevitably will have an impact downstream which will compromise any physical habitat improvement measures. Similarly, construction of barriers (dams and weirs) in downstream reaches will impede upstream migration of fish and impose a serious bottleneck to recruitment, if, as is often the case, they are not addressed. In addition, the cost of improvement schemes is high and financial resources necessary to undertake the job in a thorough manner are rarely available. Also, many of the so-called improvement schemes conflict with the water resource activities in the catchment, especially hydropower generation and flood alleviation. In these scenarios, fish conservation and fisheries are not considered of sufficiently high priority or value and thus suffer in the face of economically and socially higher priorities. Consequently, most schemes are unable to provide all the features of the habitat predicted as necessary to ensure a return to the former status.

Notwithstanding this, manipulation of the habitat should be based on ecologically sound principles where relations between fish populations and habitat characteristics have been demonstrated and knowledge of the potential capacity of a system to produce fish exists. Unfortunately, insufficient is known about the interrelationships between fish and their habitat because information on pre-perturbation habitat is rarely available, thus it is often difficult to identify the bottleneck in the life cycle of the threatened species. In addition, there is naïvity about the response of fisheries to habitat improvement measures because of lack of information from existing schemes. It must also be recognised that rehabilitation (or restoration) is not a reversal of the degradation, because ecosystem dynamics are far more complex than merely reinstating the habitat for a particular species or fish community. This issue is illustrated in Fig. 20.4, which shows how fish communities typically respond to habitat manipulations either by (A) increasing overall standing stock at the expense of biodiversity or (B) increase species diversity but at much lower standing abundance than expected.

Stock enhancement for conservation purposes also has a legacy of issues that need attention. In reality, this option should only be considered when limiting factors to stock recovery or improvement have been removed or reduced, e.g. water quality improvement, habitat rehabilitation, the easing of passage for migratory fish or management of fishing pressure. Stock enhancement in the case for restoration is justifiable because the underlying problems limiting production should have been tackled and long-term benefits are likely to accrue. Stocking programmes of this type should be temporary measures and require a more active management strategy for the aquatic ecosystem and the fish populations. Management can be in the form of control over the exploitation of the fishery or habitat manipulation, or both. The ultimate objective for conservation in this scenario is to create a fish stock that is self-sustaining. However, many projects are ill conceived and do not fully address the issues that have led to the requirement to enhance the fish stocks and possible constraints on the enhancement procedures adopted. Thus before any resources are invested, a full prefeasibility study should be carried out and the stocking programme should be linked to the wider development objectives, not just the fish conservation objectives.

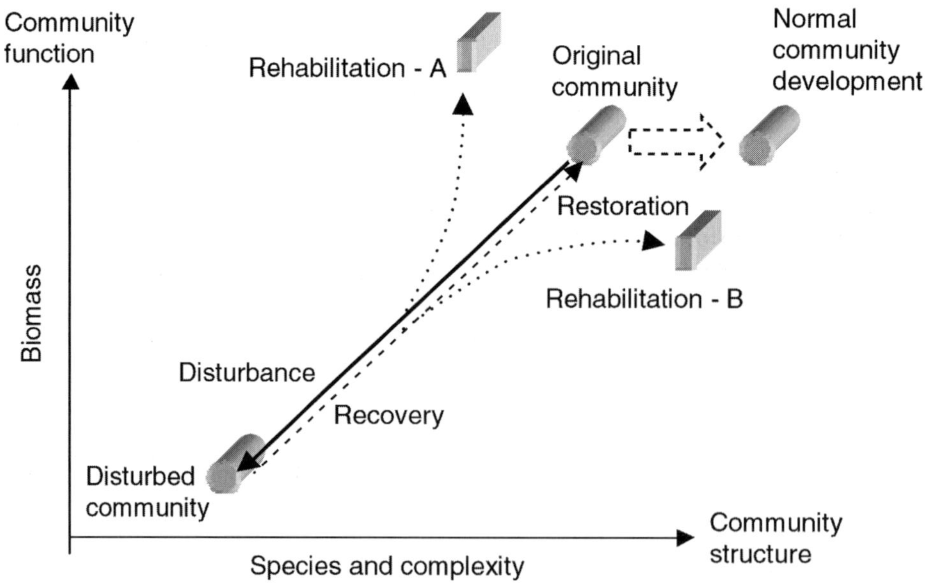

Figure 20.4 The process of rehabilitation of ecosystems whereby the response is rarely towards returning to pristine conditions but an increased biomass (A) or species complexity (B) (modified from Bradshaw 1996)

It should also be recognised that the enhancement of the stocks of fish species that have been designated endangered is a complex issue and the strategy must take due account of the potential problems that may occur as a direct result of the management action(s). Strategies that have proved successful to increase fish yield or for mitigation purposes (Cowx 1994) may not necessarily be suitable for species that are classified as endangered because they may adversely affect the residual stocks.

One of the most common strategies for enhancing fish populations is through stocking of artificially-reared fish or fish from other regions (Cowx 1994; Poncin & Philippart, Chapter 36). Unfortunately, the prospects for artificial propagation and restocking of many endangered species are limited. Paucity of information on the general biology, and in particular the reproductive characteristics, and the small size and low fecundity of many species (Economidis, Chapter 8) make them difficult candidates for developing artificial rearing techniques. Development of artificial rearing strategies may also put increasing pressure on the already stressed residual stocks. Similarly, translocation of fish from other, already stressed, populations may further reduce the spawning and recruitment potential and put extant donor stocks under further pressure.

In addition to the issues associated with whether stock enhancement should take place, due attention should also be given to the issue of genetic integrity of extant stocks and potential dilution/erosion of the gene pool. Molecular studies have demonstrated the impact of stocking on the loss of genetic diversity (Weiss, Uiblein & Jagsch, Chapter 7; Almodóvar, Nicola & Suárez, Chapter 31; Persat & Keith, Chapter 32), but Moritz, McGuigan and Bernatchez (Chapter 27) argued that freshwater fishes have the ability to phenotypically evolve and reset populations; therefore the question should be focussed on protection of the species or the habitat. Irrespective, the use of molecular markers to discriminate evolutionary diversity within species, linked to the concept of evolutionarily significant units (ESUs) (Moritz *et al.*, Chapter 27) offers an opportunity to protect conservation units within species. This information should be used to both restrict and direct stock enhancement programmes more effectively.

Despite the benefits that the use of molecular markers can bring, strong arguments should be formulated, based on the deleterious impacts of stocking on extant populations and loss of fitness (Carvalho & Cross 1998), to drive managers towards rehabilitation actions, which remove bottlenecks in the fish's life cycle, an issue supported by Moritz *et al.*'s (Chapter 27) arguments. This would not only be a more acceptable long-term solution to the problem because the environment will be improved and natural recruitment processes reinstated, but probably also benefit other stressed and threatened species in the target water body.

There are, however, possible exceptions to the arguments against conservation-driven, stock enhancement as a conservation tool. Where the species is on the verge of extinction or where bottlenecks to natural recruitment cannot be eliminated, continuous stocking is appropriate. This is particularly relevant where added benefits can accrue from the enhancement programme, e.g. River Thames (UK) Salmon Restoration Scheme which has considerable environmental spin offs, particularly the public perception that the river is now clean and the increased responsibility of water users to ensure this status is maintained (Mills 1990).

The use of protected areas (nature reserves or sanctuaries) was also prominent in the analysis of conservation measures, yet Crivelli (Chapter 34) indicated few have been established specifically for freshwater fishes. As knowledge about the status and distribution of fish species improves this should change, and better use can be made of this tool. Appropriate authorities and institutions should extend the existing network of nature reserves (or NATURA 2000 sites in Europe) to offer more protection for freshwater fishes, as was proposed by Collares-Pereira, Cowx, Rodrigues and Rogado (Chapter 19) for *Anaecypris hispanica* (Steindachner) in Portugal. Care must, however, be taken not to designate small sections of the watershed where the species persists but whole catchments, including adjacent land, to protect the populations from further deterioration (Collares-Pereira *et al.*, Chapter 19; Persat & Keith, Chapter 32; Crivelli, Chapter 34).

Other actions proposed for conservation of freshwater fishes (Fig. 20.3) included eradication of alien species, predator control, legislation and improved public awareness. Eradication of alien species is virtually impossible without poisoning the water body, and this could do irreparable damage to the ecosystem integrity. The use of rotenone to remove fish predators is feasible in small isolated areas (where barriers to prevent dispersion of fish exist), although this has rarely proved totally effective because poisoning is not 100% efficient and residual populations often remain. Predator control, particularly of birds, is equally problematic because the predators rapidly recolonise and many species themselves have protected status (e.g. cormorants under the EU Habitats Directive 92/43/EEC, Annex B).

With regards to the legal framework, appropriate legislation exists in many countries, the problem is one of enforcement. Rarely are the resources made available for implementation and major economic and political arguments are frequently used to denigrate legislation. Greater power and resources should be given to institutions to enforce legislation and oppose such arguments. To support this strategy, public awareness must be improved. As mentioned earlier, fish conservation projects generally do not have high profile with the general public. This is partly because fish are not charismatic like bears, elephants or tigers, but there is no reason why such a profile cannot be raised. There are good examples where public awareness has benefited conservation of fish, e.g. the reintroduction of salmon in the rivers Thames and Rhine (Mills 1990; Raat 2001), protection of *A. hispanica* in Portugal (Collares-Pereira *et al.*, Chapter 19) or *Sandelia bainsii* in South Africa (Cambray, Chapter 9), and lessons should be learnt from these projects.

20.4 Options for future conservation management

The previous sections highlighted many of the threats to freshwater fishes and weaknesses in conservation management of freshwater fishes. Although it is desirable to understand how ecosystems function, the state of many inland fish communities and populations are now so precarious that unless concerted action is taken in the near future many fish species and resources may be irretrievably lost. By undertaking further research to identify the problems and solutions the inevitable is probably being delayed, i.e. the time when harsh decisions will have to be made to manage the resources on a rational, sustainable basis is being put off. There is an urgent need to evaluate existing information and use it to formulate rational management policy. Unfortunately, mechanisms for implementing fish conservation measures in a multiple resource use environment are not well developed. However, the weaknesses in fish conservation can be turned to strengths given the appropriate action.

It is recommended that a multi-dimensional approach to fish conservation is adopted. This approach is based on three levels, all of which run in parallel but are not mutually exclusive.

Dimension one is orientated towards research to underpin the decisions to be made. Research should provide additional information to improve management decision-making and remove uncertainty as more appropriate information comes on stream, but should not be used as an excuse to delay making management decisions where a species is threatened. The type of research to be carried out should include:

- assessment of species distribution and abundance, both temporally and spatially;
- comparison of species current distribution with historical information;
- improved understanding of biology of species, especially habitat requirements to support rehabilitation activities; and
- identifying bottlenecks acting on recruitment and survival of target species using Environmental Impact Assessment methodologies and prioritising the problems for amelioration. It is essential that the assessment is based on watersheds and not from a localised perspective.

The second dimension is orientated towards promoting stability and enhancement of the fish populations. For stability, traditional conservation measures such as protected areas, legislation and fiscal measures should be utilised and promoted wherever possible. This action should involve all stakeholders thus increasing public awareness, an aspect that cannot be overemphasised. Developing partnerships and influencing decision-making is paramount to successful freshwater fish conservation.

To enhance the status of threatened fish populations and communities there is a need to identify indicator species, such as *A. hispanica* (Collares-Pereira *et al.*, Chapter 19) or *S. bainsii* (Cambray, Chapter 9), which can act as the target of enhancement actions that are likely to benefit other members of the community. Where possible rehabilitation actions should be formulated and implemented to address bottlenecks in the life cycle of the species. This should be based on appropriate scientific information, if available, but if not the decisions should be based on expert opinion. In the latter case, low risk rehabilitation techniques in degraded habitats should be promoted to minimise any unforeseen problems. If necessary, especially where natural dispersal of extant populations is not likely, reintroduction of a species should be considered where the habitat has been improved and potential bottlenecks to recruitment success have been removed. All the issues concerning genetic integrity should be addressed before such an action is sanctioned.

The third dimension is managing freshwater fish conservation in a wider multiple resource use framework. Fish biodiversity represents an extremely important commodity which is under threat from many sources. It is naïve to believe that fish conservation issues will stem development, thus fish conservation must be integrated into an overall framework for the optimal use of the aquatic resource base. To overcome the problems there are a number of tactics that can be employed to support fish conservation interests.

Initially, an evaluation of the current and future conflicts, both real and perceived, between fisheries and other user groups is needed. This can most easily be achieved using matrix analyses such as those used in environmental impact assessments (Cowx 1998b). This will identify the uses which are creating the greatest impact or conflict. It is these uses that will require the greatest focus in any conflict resolution. This should be coupled with an economic assessment of the fish diversity. However, it should be recognised that the output may fall well short of the true value because the methods used are often fisheries specific and do not consider the upstream of economic value in terms of aesthetic and conservation value or the downstream value associated with the service sectors. Once this information is available, value is a useful tool for arguing the case of fish conservation. However, it must be recognised that it is not the only tool to be used because the economic value of a major water resource scheme will far out weigh conservation value.

Harmonisation of the various user groups requires careful orchestration. This can be best achieved through an integrated management whereby all the stakeholders are involved in the decision-making process. The objective of this strategy must be to promote the sustainable use of the water body to yield the greatest benefit to the present population whilst maintaining the potential of the water body to meet the needs and aspirations of future generations, in a manner compatible with the maintenance of the natural properties of the water body and their value for wildlife. That is, it should be akin to the FAO Code of Conduct for Responsible Fisheries which targets sustainable development (FAO 1995).

The approach must be interactive, thus allowing wider issues than those related to a single activity, in this case fish conservation, to be taken into account during the decision-making process. It should also evaluate the likely effect of a particular development activity upon the environment and other users. In this process fish conservation should be well represented and armed with all the appropriate information to quantify potential impact.

Finally, there is a need to promote conservation issues. This can be achieved through education and extension methodologies to disseminate information and provide feedback to the wider public, who will influence political and social processes. Without their support, conservation will remain a weak player in the wider arena.

To summarise, there is an immediate need to protect threatened species from further loss through sanctuaries and enforcement of legislation. Overcoming the problems that persist can only be achieved by forming partnerships and increasing public awareness to influence the policy makers.

References

Baker D.L. & Pierce B.E. (1997) Does fisheries management reflect societal values? Contingent valuation evidence for the River Murray. *Fisheries Management and Ecology* **4**, 1–16.

Bradshaw A.D. (1996) Underlying principles of restoration. *Canadian Journal of Fisheries and Aquatic Sciences*, **53**(Suppl. 1), 3–9.

Caddy J.F. & Mahon R. (1995) Reference points for fisheries management. *FAO Fisheries Technical Paper* **347**, 83 pp.

Carvalho G.R. & Cross T.F. (1998) Enhancing fish production through introductions and stocking: genetic perspectives. In I.G. Cowx (ed.) *Stocking and Introduction of Fish.* Oxford: Fishing News Books, Blackwell Science, pp. 329–337.

Cowx I.G. (1994) Stocking strategies. *Fisheries Management and Ecology* **1**, 15–31.

Cowx I.G. (1996) The integration of fish stock assessment into fisheries management. In I.G. Cowx (ed.) *Stock Assessment in Inland Fisheries.* Oxford: Fishing News Books, Blackwell Science, pp. 495–506.

Cowx I.G. (1997) L'introduction d'espèces éstrangères de poissons dans les eaux douces européennes: succès économiques ou désastre écologique? *Bulletin Français de la Pêche et de la Pisciculture* **344/345**, 57–78.

Cowx I.G. (ed.) (1998a) *Stocking and Introduction of Fish.* Oxford: Fishing News Books, Blackwell Science, 456 pp.

Cowx I.G. (1998b) Aquatic resource management planning for resolution of fisheries management issues. In P. Hickley & H. Tompkins (eds) *Recreational Fisheries: Social, Economic and Management Aspects.* Oxford: Fishing News Books, Blackwell Science, pp. 97–105.

Cowx I.G. (1999) An appraisal of stocking strategies in the light of developing country constraints. *Fisheries Management and Ecology* **6**, 21–34.

Cowx I.G. (2002a) Recreational fisheries. In P.J.B. Hart & J. Reynolds (eds) *The Fisheries Handbook, Vol. 2*, Oxford: Blackwell Science (in press).

Cowx I.G. (2002b) *Interactions between Fish and Birds: Implications for Management.* Oxford: Fishing News Books, Blackwell Science (in press).

Cowx I.G. & Collares-Pereira M.J. (2000) Conservation of endangered fish species in the face of water resource development schemes in the Guadiana River, Portugal: harmony of the incompatible. In I.G. Cowx (ed.) *Management and Ecology of River Fisheries* Oxford: Fishing News Books, Blackwell Science, pp. 428–438.

Cowx I.G. & Welcomme R.L. (eds) (1998) *Rehabilitation of Rivers for Fish.* Oxford: Fishing News Books, Blackwell Science, 260 pp.

EIFAC (1988) Codes of practice and manual of procedures for consideration of introductions and transfers of marine and freshwater organisms. FAO/EIFAC Occasional Paper No. 23, 44 pp.

FAO (1995) *Code of Conduct for Responsible Fisheries.* Rome: FAO, 41 pp.

Feltham M.J., Cowx I.G., Davies J.M., Harvey J.P., Wilson B.R., Holden T. & Britton R.J. (1999) Case studies of the impact of fish-eating birds on inland fisheries in England and Wales. Report to Ministry of Agriculture, Fisheries and Food, London, 234 pp.

Gautam A. & Steinback S (1998) Valuation of recreational fisheries in the north-east US. Striped bas: a case study. In P. Hickley & H. Tompkins (eds) *Recreational Fisheries: Social, Economic and Management Aspects.* Oxford: Fishing News Books, Blackwell Science, pp. 97–105.

ICES (1988) Codes of practice and manual of procedures for consideration of introductions and transfers of marine and freshwater organisms. ICES Cooperative Research Report No. 159, Copenhagen 44 pp.

IUCN (1995) The IUCN guidelines for reintroductions. *www.rbgkew.org.uk/conservation/RSGguidelines.html*

IUCN (2000) The 2000 IUCN Red List of threatened species. *www.redlist.cymbiont.ca/resultslist.asp*

Kennedy G.J.A. & Crozier W.W. (1997) What is the value of a wild salmon, *Salmo salar* L., smolt? *Fisheries Management and Ecology* **4**, 103–110.

Kirchhofer A. & Hefti D. (eds) (1995) *Conservation of Endangered Freshwater Fish in Europe.* Basel: Birkhäuser Verlag, 341 pp.

Lehtonen H. (1996) Potential effects of global warming on northern European freshwater fish and fisheries. *Fisheries Management and Ecology* **3**, 59–71.

Lelek A. (1987) *Threatened Fishes of Europe* (The Freshwater Fishes of Europe. Vol. 9). Wiesbaden: AULA-Verlag, 343 pp.

Mills D.H. (ed.) (1990) *Strategies for the Rehabilitation of Salmonid Rivers*. London: Linnean Society, 208 pp.

Moss B. (1998) *Ecology of Fresh Waters*, third edition. Oxford: Blackwell Science, 560 pp.

Petts G.E. (1984) *Impounded Rivers*. Chichester: Wiley and Sons, 326 pp.

Peirson G., Tingley D., Spurgeon J. & Radford A. (2001) Economic valuation of inland fisheries in England and Wales. *Fisheries Management and Ecology* **8**, 415–424.

Persson L. (2002) Community Ecology of freshwater fishes. In P.J.B. Hart & J. Reynolds (eds) *The Fisheries Handbook, Vol. 1*, Oxford: Blackwell Science (in press).

Petr T. (ed.) (1998) Inland fishery enhancements. Rome: *FAO Technical Paper* **374**, 463 pp.

Postle M. & Moore L. (1998) Economic valuation of recreational fisheries in the UK. *Recreational Fisheries: Social, Economic and Management Aspects*. In P. Hickley & H. Tompkins (eds). Oxford: Fishing News Books, Blackwell Science, pp. 184–199.

Raat A (2001) Ecological rehabilitation of the Dutch part of the River Rhine with special attention to the fish. *Regulated Rivers: Research and Management* **17**, 131–144.

Welcomme R.L. (1988) International introductions of inland aquatic species. *FAO Fisheries Technical Paper* **294**, 318 pp.

Welcomme R.L. (2000) Principles and approaches for river fisheries management. In I.G. Cowx (ed.) *Management and Ecology of River Fisheries*. Oxford: Fishing News Books, Blackwell Science, pp. 331–345.

Chapter 21
An overview of the challenges of conserving freshwater fishes in South Africa

P.H. SKELTON*
JLB Smith Institute of Ichthyology, Grahamstown, South Africa

Abstract

Freshwater fishes evolved in isolation in the far south of Africa since the breakup of Gondwanaland. Legacies of isolation are ecological fragility and restricted distribution ranges. South African freshwater ecosystems have been seriously impacted by human activities, especially over the past century. The present conservation status of freshwater fishes in South Africa is reviewed to indicate a trend of increasingly threatened species. The socio-political change in South Africa has introduced constitutional momentum in support of sustainable use of the environment and the conservation of biodiversity. The more stringent task of directly preserving biodiversity is being addressed through small-scale independent initiatives. The question of defining effective conservation units is only just beginning to be tackled. The development of modern electronic databases is assisting to focus attention on priority areas of concern. Critical challenges facing freshwater fish conservation in South Africa include the severe lack of skilled human resource capacity in systematics and conservation biology, management and law enforcement.

Keywords: action plans, alien invasives, biodiversity, ecosystems, ethics, river health.

21.1 Introduction

Conservation is all about human interactions with the natural environment. Among the many factors to be considered, two important ones are the socio-political system prevailing in a country and the philosophical approach to the problem. At the present time the conservation of freshwater fishes in South Africa has to be seen against a dual background of major fundamental changes to the socio-political framework of the nation in the past decade and the natural fragility of the fauna. There are two prevailing conservation paradigms at present – an ecosystem view and an approach based on biodiversity considerations (Callicott 1995). Skelton (2000), however, argued that both ecosystem and systematic or biodiversity conservation ethics have a role to play in conserving the freshwater fish fauna in southern Africa.

South Africa has undergone dramatic political revolution in the 1990s with major impacts on every aspect of life in the country, including the conservation of biodiversity and the environment. The new South African Constitution provides protection for

*Correspondence: JLB Smith Institute of Ichthyology, Private bag 1015, Grahamstown 6140, South Africa (email: P.Skelton@ru.ac.za).

the environment as a fundamental right for all South Africans. These rights are now being incorporated extensively into policy and legislation in every government sector. South Africa has embraced important international conventions such as the Convention on Biological Diversity (the Rio Convention). As a member of the Southern African Development Community (SADC) South Africa signed a protocol on shared watercourse systems within the region.

Provincial restructuring has severely disrupted conservation services in the country. At the same time there is a general trend towards accountability for the environment and conservation in South African society that is in tune with global trends of concern for environmental issues.

The natural fragility of freshwater fishes in South Africa relates to a long period of faunal evolution in isolation under harsh environmental conditions. Southern Africa juts into the south-eastern Atlantic and western Indian oceans and has been isolated since the breakup of Gondwanaland beginning in the late Jurassic and early Cretaceous some 160–120 million years ago (Eriksson 2000). Recent studies on biogeography of the freshwater fishes of the sub-continent indicate that the Southern or Cape fauna has evolved in isolation within an arena defined by the basin of the Orange River and surrounding coastal catchments (Skelton 1994). To the north, the tropical (or Zambezian) fauna also evolved for a long period of time within two historical arenas that extend to and partly overlap the southern arena.

There are fewer than 100 indigenous species in South Africa of which about half are endemic (Skelton, Cambray, Lombard & Benn 1995). The Zambezian fauna includes 61 species within South Africa of which 13 (21%) are endemic to the south-eastern coastal drainages (Limpopo, Incomati and Phongola). Many tropical species have fairly broad distribution ranges with marginal populations within South Africa. An ecotonal boundary runs along the escarpment (below 300 m.a.s.l.) and across to the coast south of the St Lucia estuary.

The Cape fauna is small (33 species in four families) and dominated by cyprinids (27 species, 81%), mainly barbines (70%, 23 species). Of these barbines, 18 (78%) are polyploids – 12 (52%) tetraploids and 6 (26%) hexaploids; the balance (5 species, 22%) are diploids. This is in contrast to the dominant global cyprinid ratio of approximately (85%) diploid, (8–10%) tetraploid and (4%) hexaploid (Klinkhardt 1998). The balance of the Cape fauna comprises three austroglanid catfishes, two anabantids and a galaxiid. Local endemicity (single river basin) is high (45%, 15 species) with 27% (9 species) in only two river basins and a further nine species (27%) in fewer than five river basins (Skelton *et al.* 1995). Most (19 species, 58%) of the species are relatively small-sized, generally less than 150 mm SL when adult. None are entirely piscivorous, although the anabantids and some of the larger barbines are partly piscivorous as adults.

21.2 Conservation status and hotspots of threatened fishes

The most recent evaluation of the conservation status of threatened fishes in South Africa (Tables 21.1 and 21.2; Baillie & Groombridge 1996) indicates that there has

Table 21.1 Numbers of IUCN red list threatened freshwater fishes from South Africa (and Lesotho) since 1977

	Old IUCN categories			1996 IUCN categories	
	1977	1987	1993		1996
En	5	4	8	CR	7
Vu	2	7	5	En	6
Rare	15	10	10	Vu	9
I			3	LR	7
				DD	(1)
Total	22	21	26		30

Threatened categories for 1977–1993 as defined by Groombridge (1993) and for 1996 by Baillie and Groombridge (1996). Abbreviations: CR, critically endangered; En, endangered; Vu, vulnerable; I, indeterminate; LR, lower risk; DD, data deficient.

been deterioration in the conservation status of freshwater fishes since 1987. This deterioration is born out by studies on two threatened river cyprinids, the Twee River redfin, *Barbus erubescens* Skelton, and the Maloti minnow, *Pseudobarbus quathlambae* (Barnard).

The Twee River redfin is restricted by a 12-m high vertical waterfall to a third-order tributary system of the Olifants River in the Western Cape (Fig. 21.1). Marriott (1998) showed that citrus, deciduous fruit and vegetable farming has increased considerably in the catchment and has raised the threat from water abstraction and pesticides. Up to 15% of the mean annual runoff is extracted during the summer months when stream flow is at a minimum. Marriott also found four alien fish species are established in the tributary system including the Cape kurper, *Sandelia capensis* (Cuvier), bluegill sunfish, *Lepomis macrochirus* Rafinesque, rainbow trout, *Oncorhynchus mykiss* (Walbaum), and the Clanwilliam yellowfish, *Barbus capensis* A. Smith. The range of the Twee River redfin has been reduced by 40% and the population fragmented into three clusters (Fig. 21.1). The Twee River redfin was rated vulnerable in 1987 and has been re-assessed as critically endangered.

The Maloti minnow was already threatened when first described in the Umkomazana River in Natal (Barnard 1938). Subsequently it was extirpated from this river and only rediscovered in Lesotho in 1971 (Pike & Tedder 1973). Six isolated river populations have been discovered in tributaries of the Orange River system in Lesotho (Fig. 21.2). Three populations are protected from predation by trout (mostly *Oncorhynchus mykiss* but also *Salmo trutta* L.) by barrier waterfalls. In three instances where trout occur together with the minnow the habitat is not optimal for trout and the populations of both species are small and marginal. The development of the Lesotho Highlands Water Project (LHWP) directly threatens the largest population of the minnow in the Senqunyane tributary through flooding of the habitat and almost certain invasion by trout (Rall 1993, 2000). The infrastructure developed for the LHWP has opened up the highlands for other developments and has stimulated human settlements and increased animal grazing and crop production pressures. The Alpine streams in

Table 21.2 Criteria for listing and major threats for IUCN red listed threatened freshwater fishes from South Africa (and Lesotho)[a]

Status	Species	Criteria	Primary threat	Secondary threat	Remarks
CR	*Pseudobarbus burgi* (Boulenger)	A1c + e, A2c + e, B1, B2a + b + c + e, B3a + c + d	A	HD	Taxonomy uncertain; divided ESUs identified by genetic analysis
	Pseudobarbus quathlambae (Barnard)	B1, B2a + b + c + d	A	D	Assessed by detailed study. Major development project poses new threats
	Barbus erubescens Skelton	B2a + e, B3a,	HD	A	Assessed by detailed field study
	Barbus trevelyani Günther	A1e	A	HD	Surveys made
	Labeo seeberi Gilchrist & Thompson	A1a + c + e	HD	D, A	Surveys made
	Austroglanis barnardi (Skelton)	B1, B2b + c + d	A	HD	Detailed ecological and population genetics study in progress
	Chiloglanis bifurcus Jubb & Le Roux	A1a + e	HD		Surveys made
EN	*Pseudobarbus burchelli* Smith	A1c + e, A2c + e, C1	A	HD	Surveys and population genetics being studied
	Pseudobarbus phlegethon (Barnard)	B1, B2a + c	A	HD	Surveys and population genetics studied
	Pseudobarbus tenuis (Barnard)	A1c + e, A2c + e, C1	A	HD	Surveys and population genetics being made
	Barbus calidus Barnard	B1, B2a + c	A	HD	Surveys made
	Barbus serra Peters	B1, B2a + b + d + e, C1	D	HD, A	Surveys made
	Sandelia bainsii Castelnau	A1e, B1, B2a + b + c + d	HD	A	Detailed studies made

VU	*Opsaridium peringueyi* (Gilchrist & Thompson)	B1, B2a + b + c, D2	HD	D	Revised taxonomy affects status
	Pseudobarbus asper (Boulenger)	A1c + e	A	HD	Biology studied
	Barbus motebensis Steindachner	A1c	A	HD	Genetics studied
	Barbus brevipinnis Jubb	A1c	HD	A	Taxonomy uncertain, genetic and ecological studies made
	Barbus andrewi Barnard	A1c + d + e	A	HD, D	Detailed studies made
	Barbus capensis A. Smith	A1c + e	D	HD, A	Detailed surveys made
	Barbus kimberleyensis Gilchrist & Thompson	A1c	D	HD	
	Austroglanis gilli (Barnard)	A1c + e, B1, B2c	A	HD	Detailed studies being made
	Chetia brevis Jubb	A1e	HD		
LR	*Pseudobarbus afer* (Peters)	nt	A	HD	Taxonomy uncertain; biology studied
	Barbus treurensis Groenewald	cd	A	HD	Surveys made; translocated
	Barbus hospes Barnard	nt	HD	HD	Detailed studies made
	Chiloglanis swierstrai Van der Horst	1c	HD	HD	
	Galaxias zebratus Castelnau	nt	A	HD	Taxonomy uncertain; population genetics being studied
	Serranochromis meridianus Jubb	cd	HD		
	Sandelia capensis (Cuvier)	1c	A	HD	Population genetics being studied
DD	*Austroglanis sclateri* (Boulenger)		HD		Detailed study made

[a]Status abbreviations as follows: CR, critically endangered; EN, endangered; VU, vulnerable; LR, lower risk; DD, data deficient. Abbreviations of criteria are as given by Groombridge (1993). Threat abbreviations as follows: HD, habitat destruction; D, dams; A, aliens. IUCN criteria: A, Population reduction quantified or estimated; B, Extent of occurrence; C, Population estimation (mature individuals) with continuing decline; D, Population estimation (mature individuals); E, Quantitative analysis of probability of extinction in the wild.

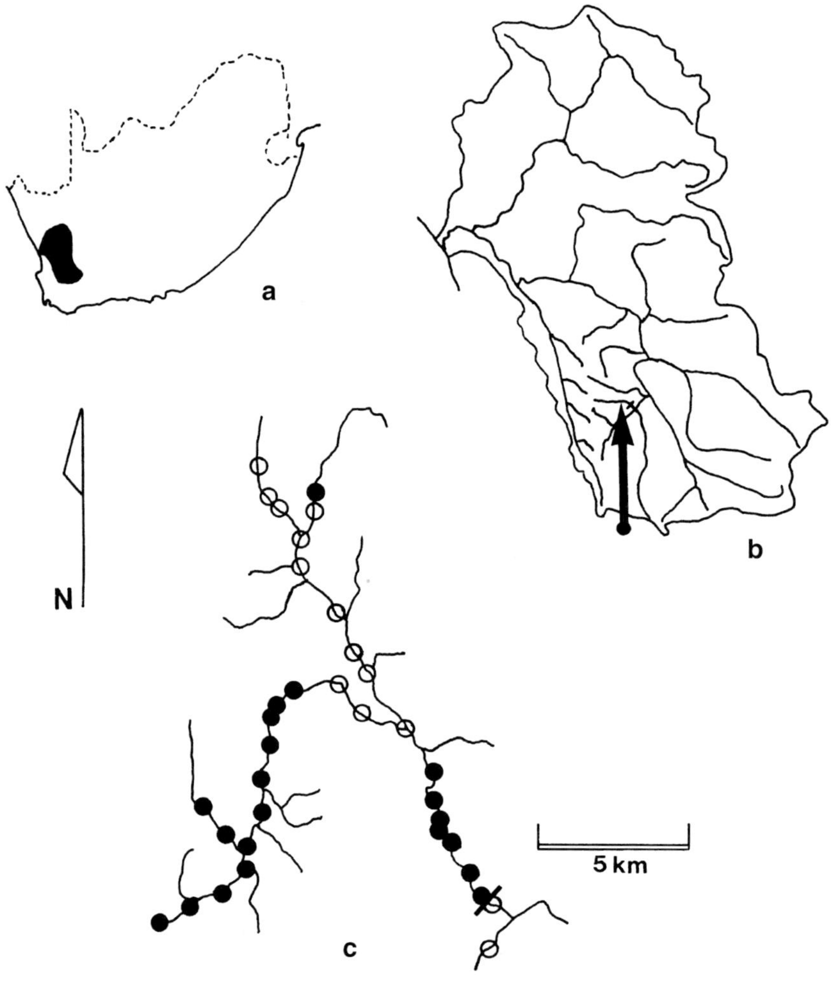

Figure 21.1 Distribution of *Barbus erubescens* in the Twee River, Olifants River System, Western Cape, South Africa – results of a survey conducted in 1997 (Marriott 1998). (a) South Africa showing Olifants River catchment (shaded); (b) Olifants River catchment with Twee River tributary pointed out; (c) Twee river tributary system showing survey sites (all circles) and localities where *Barbus erubescens* was observed or collected (full circles). Bar indicates barrier waterfall

which the minnow survives are sensitive to climatic fluctuations and, with downstream prospects for all the populations now eliminated by the presence of alien predators, the long-term prospect for this critically endangered species is poor.

Skelton *et al.* (1995) analysed the distribution patterns of South African freshwater fishes within a conservation context. Hotspots of threatened fish diversity were identified in the western, south-western and southern Fold Mountain catchments of the Western Cape, the Amatola-Winterberg catchments (Eastern Cape), coastal reaches in north-eastern Kwazulu-Natal and the eastern escarpment rivers of Mpumalanga

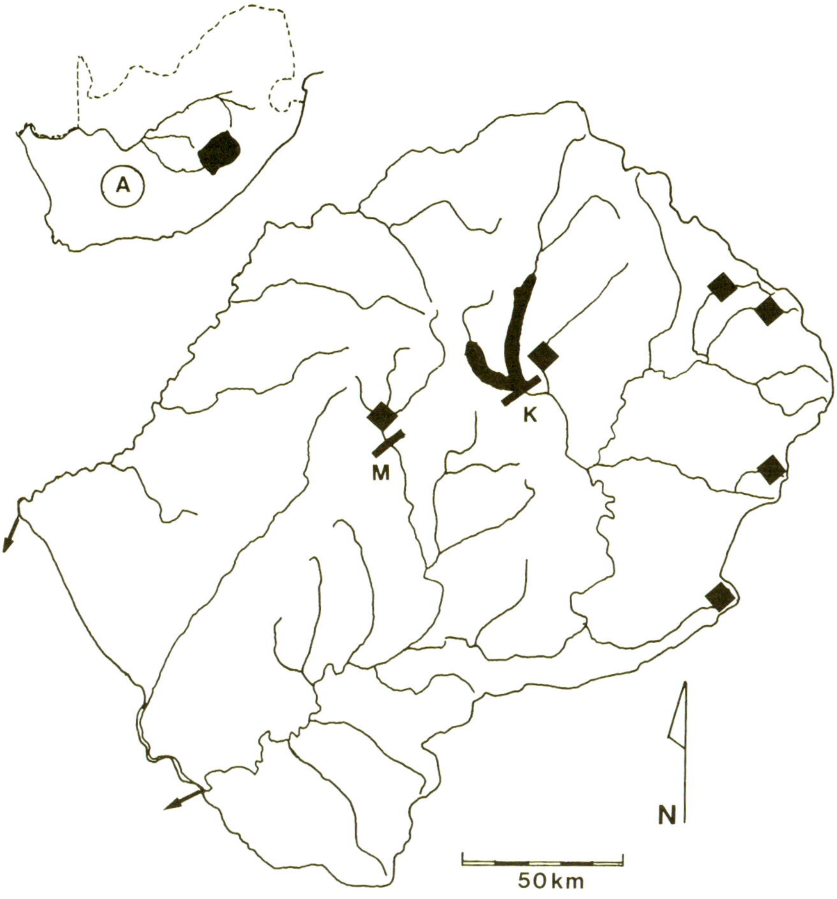

Figure 21.2 Distribution of *Pseudobarbus quathlambae* in Lesotho (after Rall 2000). Populations confirmed by survey conducted in September 2000. A, South Africa showing Lesotho (shaded) in relation to Orange River system. Diamonds indicate recorded population of Maloti minnow. K, Katse Dam; M, Mohale Dam

(Fig. 21.3). These hotspots reflect the pattern of endemic indigenous freshwater fish species richness (Fig. 21.4).

21.3 Threats

Major threats to African freshwater fishes are habitat degradation and fragmentation, disruption of essential ecological processes, hydrological manipulations such as dams and channelisation, pollution, overabstraction of water, impacts of alien organisms, breakdown of natural barriers through inter-basin transfer, genetic contamination, global effects (global warming, increased UV radiation) and policy deficiencies (Davies, O'Keeffe & Snaddon 1993; Lévêque 1997; Shumway 1999). All these factors

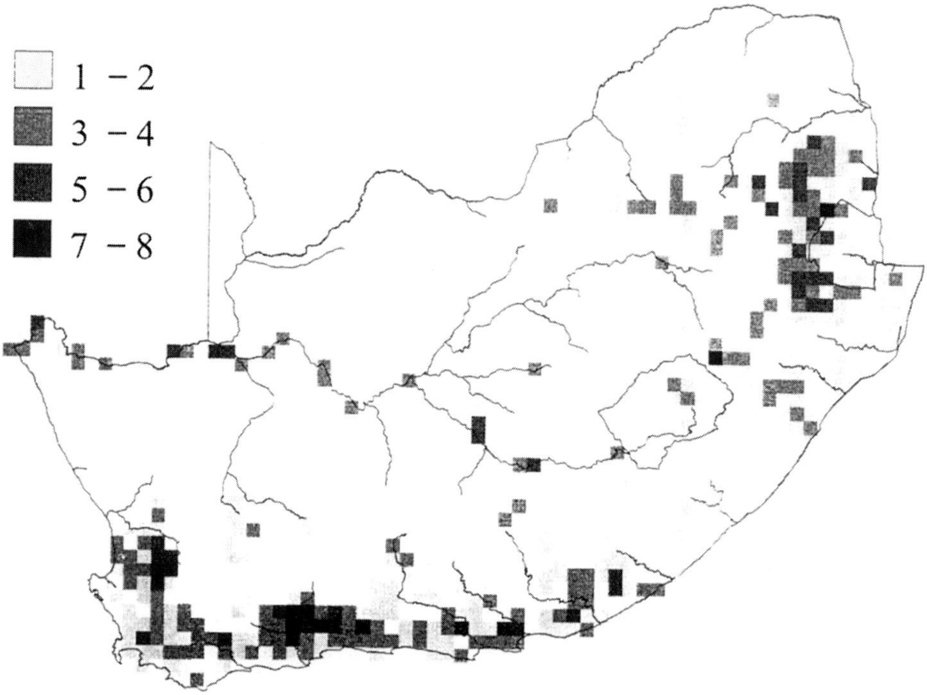

Figure 21.3 Endemic indigenous freshwater fish species richness in South Africa. Plots from records in the JLB Smith Institute of Ichthyology and the Albany Museum (after Skelton *et al.* 1995)

threaten the freshwater fishes of South Africa (Skelton 1987, 1990; Skelton *et al.* 1995). Generally a combination or synergistic array of factors threaten species.

The impact of alien organisms and the fragmentation of populations through various means are serious problems (Table 21.2). Alien predators is especially serious for both directly eliminating and fragmenting populations of indigenous species. Alien predators, notably rainbow and brown trout and the centrarchid basses (largemouth *Micropterus salmoides* (Lacépède), smallmouth *Micropterus dolomieu* (Lacépède), and spotted *Micropterus punctulatus* (Rafinesque)) are the species causing the greatest problems in South Africa (De Moor & Bruton 1988). Other predators such as *Clarias gariepinus* (Burchell) and species of African largemouth breams (*Serranochromis* species) have been translocated within southern Africa, but their impact is less well studied or understood (De Moor & Bruton 1988).

Why should these species have had so much impact? The habitat of low-order mountain streams is usually restricted in size (width and depth), and the water is generally clear and cool. Depending on the streambed, cover can be limited or restricted to certain reaches such as in riffles or rocky runs. Productivity is relatively low, and food chains are short and direct from primary producers through one or two consumer links to the top predators. Fish communities in the Cape fauna are small, usually consisting of no more than four or five species even in mainstream habitats. The life-history

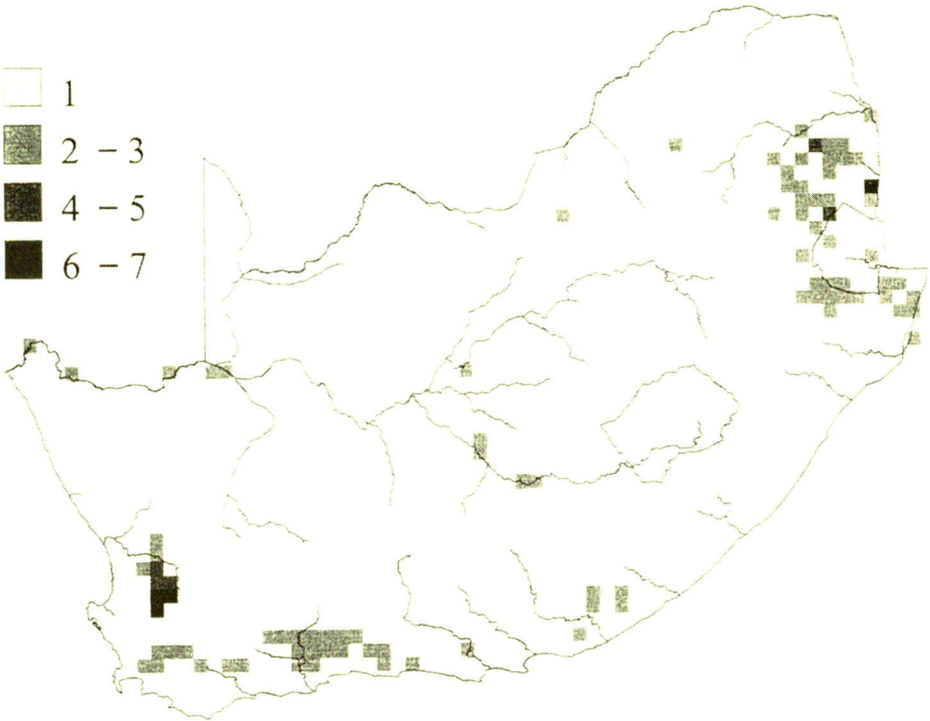

Figure 21.4 Endemic threatened freshwater fish species richness in South Africa plotted from records in the JLB Smith Institute of Ichthyology and the Albany Museum (after Skelton *et al.* 1995)

styles of both predator and prey are important in the extent to which native fishes resist an introduced alien predator. Trout breed in winter and their offspring are therefore well geared to prey on the larvae or small juveniles of Cape species when they breed in spring and early summer. Centrarchid bass are brood guarders, which potentially enhances survival chances of their offspring. Indigenous stream-dwelling minnows are adapted to conditions of stable food supply and consistent annual mortality rates. They are relatively short-lived and do not reach a size large enough to resist predation from either trout or bass. They are opportunist open-water foragers feeding from either the substratum or the water column and surface. They breed territorially with males displaying bright colours, but they do not defend the fertilised eggs, larvae or juveniles (Cambray 1994).

In the case of long-isolated communities the lack of integrated ecology between indigenous species and an introduced predator is a crucial factor. In North American streams predator–prey and other community interactions between suckers, minnows and centrarchids are old and balanced (Matthews 1998). Redfins and other stream fishes of the Cape are notoriously naïve in the presence of predatory species and were rapidly eliminated following the introduction of an alien top predator. Similar destructive contact between predator and prey species does not necessarily occur in more diverse communities or where the habitat is less restricted in size or cover. In Zimbabwe, for example, introduced largemouth bass appeared not to have seriously affected indigenous fishes, although the later introduction of a faster growing 'Florida' strain might have changed this situation (Marshall 1999; Gratwick 2000).

Genetic pollution is a rising threat to indigenous fishes. The Nile tilapia *Oreochromis niloticus* (L.) was introduced to Zimbabwe and Zambia in the 1980s for aquaculture and has invaded several river systems, including the Limpopo River where it was found to hybridise with the indigenous *Oreochromis mossambicus* (Peters) (Moralee, van der Bank & van der Waal 2000). Minnow species have been stocked into certain rivers as bait for sport fishes or accidentally from state hatcheries (Barnard 1943). Engelbrecht & van der Bank (1996) reported that two goldie barb populations from different river systems are genetically identical, possibly as a result of accidental introduction of one population by the fisheries officials.

South Africa is essentially a dry country and its water resources are under increasing pressure to service human needs in a rapidly increasing population (Allanson, Hart, O'Keeffe & Robarts 1990). Agriculture, industry and urban domestic users consume 66% of available surface resources (DWAF 1986). The insidious depletion of instream water resources through the widespread development of forest plantations has reduced stream flows by up to 27% in naturally grassland catchments and 3.5% in South Africa as a whole (Pallet 1997). This consumption is exacerbated by the extensive invasion of riparian zones by alien plants such as the black wattle *Acacia mearnsii* De Wild. Invading alien plants consume about 6.7% of the total surface runoff (Le Maitre, Versveld & Chapman 2000). The invasion and impact is most serious in the Western Cape where alien plants account for up to a third of the estimated total water use.

The extensive building of dams and weirs has undoubtedly impacted on indigenous fish populations (Davies *et al.* 1993). The impact of channel obstruction is most evident on threatened large mainstream species such as the Clanwilliam yellowfish, *Barbus capensis* Smith, the sawfin, *Barbus serra* Peters, and the Clanwilliam sandfish, *Labeo seeberi* Gilchrist & Thompson. River regulation affects the down stream hydrological and biological dynamics with differential effects on fish species. The Namaqua barb, *Barbus hospes* Barnard, is confined to the channel of the lower Orange River and a single tributary in Namibia. Mainstream regulation has resulted in a steady all year round flow and has favoured the species. In contrast, the reduced flows and lentic conditions in one of its tributaries, the Fish River, has harmed the species such that its survival is dependent on recruitment from the mainstream (Hay, van der Bank & Ferreira 1997).

Inter-basin transfer systems and direct human intervention have been responsible for the transfer of various fish species between the different basins (Davies *et al.* 1993). Where non-predatory species are involved the impact of such transfers appears to be benign, but in the case of an ecologically dominant and aggressive invader species like the sharptooth catfish *Clarias gariepinus* (Burchell) the impact on native fauna may be serious. The sharptooth catfish has been transferred extensively beyond its native range to river systems in the Eastern Cape and the Western Cape where it poses a direct threat to endangered species like *Sandelia bainsii* (Castelnau) and the witvis *Barbus andrewi* Barnard.

Soil erosion is widespread in areas where rural populations are high. It has been estimated that modern rates of geological erosion ($322\,t\,km^{-2}year^{-1}$) in KwaZulu-Natal where high rural populations exist, are as much as 12–22 times the geological average (Martin 1987). The impact of high turbidity and sedimentation in rivers on fish life is well appreciated (Bruton 1985), but there are no specific studies in southern Africa to quantify the problem.

Soil degradation through physical and chemical means is a widespread problem in South Africa where crop cultivation and water irrigation schemes are active (Laker 2000). Salinisation of the Berg River in the south-west Cape is likely to have been one of the factors impacting on the critically endangered population of *Barbus andrewi* (Impson & Bloomer 1998/1999). Pollution from mining activities is another widespread phenomenon responsible for general riverine ecosystem degradation in South Africa.

21.4 Indirect threats

Environmental conservation depends upon skills and knowledge of the people involved. Greenwood (1992) argued that the decline of taxonomy and of the number of taxonomists itself is a threat to the conservation of fishes. Taxonomists generate primary knowledge of biodiversity and alert society to situations of fragile and threatened biodiversity. A loss of unique expertise sets back the clock on knowledge generation and strangles the flow of knowledge to society (D. Pauly, personal communication). The loss of personnel in various nature conservation agencies is a threat to conservation (Impson & Hamman 2000). Nature conservation scientists identify priority issues, communicate with funding agencies and mobilise researchers (e.g. Impson, Bills, Cambray & le Roux 1999).

21.5 Conservation options

Conservation of freshwater environments and fishes can focus on either ecosystem processes or maintenance of biodiversity. In South Africa the movement towards ecosystem management gained momentum in the 1980s (O'Keeffe 1989) and matured in the 1990s to become a dominant paradigm driving the Water Policy and new legislation of the country (Davies *et al.* 1993; DWAF 1997). Steps taken towards the ideal of national riverine ecosystem management included developing a protocol for the definition of desired states of riverine systems (Rogers & Bestbier 1997) and of establishing a national river classification (Eekhout, King & Wackernagel 1997).

The national classification of rivers was based on riverine fauna and flora (Eekhout *et al.* 1997). Ten biotic regions provide a framework for establishing the site selection for a national biomonitoring network known as the National River Health Programme (RHP). The RHP uses various indices of the integrity of the instream and riparian biological communities, including fish, as a measure of the ecosystem health (Kleynhans 1999). The indices establish the present ecological state of the different river segments studied against which desirable management state goals can be set. Although the RHP is currently in a development phase it is set to become the basis of a national river environmental management system (Roux, Kleynhans, Thirion, Hill, Engelbrecht, Deacon & Kemper 1999).

21.6 Biodiversity conservation

Ecosystem management alone will not ensure the survival of threatened fish or other aquatic species. Certain major threats to species such as invading alien predators do not necessarily threaten the overall health of the system sufficiently to concern water resource managers.

There are several examples of *in situ* conservation actions that have been taken to conserve particular species or communities. Skelton *et al.* (1995) discussed the difficulties of creating reserves for freshwater fishes that under certain conditions are necessary to save a species from becoming extinct in the wild. Although translocating organisms poses ethical and practical problems (Tharme & Anderson 1999), the positive value of local translocations of threatened species to safe areas was demonstrated in the case of *Barbus treurensis* Groenewald (Engelbrecht 1998/1999). In the 1970s a population of the Maloti minnow was translocated to a safe stretch of river protected by a barrier waterfall, all within the boundaries of the Sehlabathebe National Park. A recent survey of the species found that the original (downstream) population was practically extinct but the translocated population was secure. Impson and Tharme (1998) reported on a conservation-sanctioned translocation of several threatened fish species to above a waterfall protected river stretch in the Olifants River system. This translocation was made within a private nature conservation reserve and was partly motivated to develop an angling resource on the reserve. A careless translocation by the conservation authorities of one threatened species, the Clanwilliam yellowfish, *Barbus capensis*, into the restricted range of another threatened species, the Twee River redfin, *Barbus erubescens* Skelton, illustrates how carefully translocations need to be researched before they are made.

Few attempts have been made in South Africa to research the ecology of invasive species or to develop effective means to control or eradicate invasive populations from river systems. An example where largemouth bass were successfully eliminated to allow the indigenous fishes to recover was in the Blindekloof River within the Groendal Wilderness Area (Skelton 1993). In this case the stochastic nature of the invaded stream allowed for the elimination of largemouth bass through angling and electric fishing.

Introduced sport fishes raise a conflict of interest between conservation and the influential angling community. Control of alien sport fishes is complicated by a maverick attitude in some quarters for introducing such fishes and a low law-enforcement capacity. Conservation authorities often lack clear policy priorities, and the will to uphold such policy in the face of such public conflicts of interest.

In South Africa there has been a welcome trend towards research into conservation genetics and the development of databases to assess and analyse patterns of diversity and priority areas of concern. Several recent genetic studies exposing conservation issues include Engelbrecht and van der Bank (1996), Engelbrecht and Mulder (1999),

Waters and Cambray (1997), Impson and Bloomer (1998/1999) and Bloomer and Impson (2000). The latter authors conducted a mitochondrial DNA investigation of the critically endangered Berg River redfin, *Pseudobarbus burgi* (Boulenger) to reveal that populations from the adjacent Berg and Verlorenvlei river systems are genetically and morphologically distinct. As distinct evolutionarily significant units (ESUs) of a species with a critically endangered status they require independent management attention.

Establishing a river classification in South Africa and the implementation of the River Health Programme have both required the development of databases (Eekhout *et al.* 1997; DWAF 2000). Museum collections are one of the most valuable long-term sources of conservation data and are scientifically robust (Cambray 1990; Skelton *et al.* 1995). The advent of modern computing technology and the Internet have greatly facilitated database development. Skelton, Verheust and Scott (2000) described the development of a GIS Atlas for freshwater fishes for southern Africa based on a database assembled from a variety of international museum collections. The database has been used for conservation purposes to identify hotspots of threatened fish diversity in South Africa (Skelton *et al.* 1995).

21.7 Conclusions

Aquatic environmental degradation is a major problem for water resource managers and for conservation of aquatic life in South Africa (and elsewhere in Africa) (Davies *et al.* 1993). In circumstances where poverty and ignorance prevail, a Constitution that defends the environment is fundamental for conservation to take root and be embraced by society. Although the status of fish species continues to decline in the face of rising population growth and development in South Africa, advances have been made in terms of aquatic conservation and management. A tension persists between conserving ecosystems and focusing on biodiversity (Callicott 1995). Depleted provincial conservation agencies focus efforts on biodiversity conservation while national resource management agencies focus on conserving ecosystems.

In an open and enlightened society, conservation values must move beyond economic justification. Recent interactions between conservation agencies and the angling fraternity suggest that this may be happening in South Africa. About 100 years ago, state and private sector interests converged to introduce alien predatory fish species for inland recreational fisheries. This collusion persisted for most of the 20th century under an ethical framework of autocratically administered preservationist conservation. The situation collapsed when the extent of impact by the introduced aliens was too large to be ignored by the conservation agencies. By the 1990s most conservation agencies had divested their interests in breeding and stocking out alien species (Skelton 2000). Subsequently, certain responsible sport fishing bodies are beginning to work with conservation interests to promote indigenous species and support conservation projects (Impson 1999). Such convergence of values and ethics are essential if conservation is to progress in an enlightened, open society. The challenge for conservation is to achieve a balance that the majority, at least, can buy into.

Acknowledgements

I am grateful to Roger Bills, Andrew Deacon, Lil Haigh, Dean Impson, Neels Kleynhans, Tally Palmer, Ian Russell, and Alan Whitfield for information or advice in preparing this paper. Jane Stockwell provided much assistance under tight time-frames. The National Research Foundation provided funds for my research programme and to attend the conference for which the work was prepared and presented.

References

Allanson B.R., Hart R.C., O'Keeffe J.H. & Robarts R.D. (1990) *Inland Waters of Southern Africa: An Ecological Perspective.* Dordrecht: Kluwer, 458 pp.

Baillie J. & Groombridge B. (1996) *IUCN Red List of Threatened Animals*. Gland: IUCN, 368 pp.

Barnard K.H. (1938) A new species of freshwater fish from Natal. *Annals of the Natal Museum* **8**, 525–528.

Barnard K.H. (1943) Revision of the indigenous fresh-water fish fauna of the S-W Cape region. *Annals of the South African Museum* **36**, 101–262.

Bloomer P. & Impson N.D. (2000) Mitochondrial DNA differentiation in the Critically Endangered Berg River redfin (*Pseudobarbus burgi*). *The Journal of Heredity* **91**, 122–127.

Bruton M.N. (1985) The effects of suspensoids on fish. In B.R. Davies & R.D. Walmsley (eds) *Perspectives in Southern Hemisphere Limnology. Developments in Hydrobiology* **28**, 221–242.

Callicott J.B. (1995) Conservation ethics at the crossroads. *American Fisheries Society Symposium* **17**, 3–7.

Cambray J.A. (1990) The role of natural history museums in aquatic conservation. In E.M. Herholdt (ed.) *Natural History Collections: Their Management and Value. Special Publication* No. 1, Pretoria: Transvaal Museum, pp. 129–137.

Cambray J.A. (1994) The comparative reproductive styles of two closely related African minnows (*Pseudobarbus afer* and *P. asper*) inhabiting two different sections of the Gamtoos River system. *Environmental Biology of Fishes* **41**, 247–268.

Davies B.R., O'Keeffe J.H. & Snaddon C.D. (1993) A synthesis of the ecological functioning, conservation and management of South Africa river ecosystems. *WRC Report No. TT 62/93*, 232 pp.

De Moor I.J. & Bruton M.N. (1988) Atlas of alien and translocated indigenous aquatic animals in southern Africa. *South African National Scientific Programmes Report No. 144*, 310 pp.

DWAF (Department of Water Affairs) (1986) *Management of the Water Resources of the Republic of South Africa*. Pretoria: Department of Water Affairs, 420 pp.

DWAF (Department of Water Affairs) (1997) *White Paper on a National Water Policy for South Africa*. Pretoria: Department of Water Affairs and Forestry, 37 pp.

DWAF (Department of Water Affairs) (2000) The River Database (version 1.03). *River Health* **7**, 3–4.

Eekhout S., King J.M. & Wackernagel A. (1997) *Classification of South African Rivers*, Vol. 1. Pretoria: Department of Environmental Affairs and Tourism, 125 pp.

Engelbrecht J.S. (1998/1999) A note on the reintroduction of *Barbus treurensis* Groenewald 1958 (Pisces: Cyprinidae) into the Treur River, Mpumalanga, South Africa. *Southern African Journal of Aquatic Sciences* **24**, 153–156.

Engelbrecht G.D & Mulder P.F.S. (1999) Allozyme variation in the river sardine, *Mesobola brevianalis* (Pisces: Cyprinidae). *Water SA* **25**, 293–295.

Engelbrecht J.S. & van der Bank H. (1996) Genetic relationships between seven species within the chubbyhead and goldie barb groups of minnows (Pisces: Cyprinidae). *Journal of African Zoology* **110**, 381–396.

Eriksson P. (2000) The geological template. In R. Fox & K. Rowntree (eds) *The Geography of South Africa in a Changing World.* Cape Town: Oxford University Press, pp. 257–283.

Gratwick B. (2000) The impact of bass and robustus in streams. *Zimbabwe Fisherman* **12**, 30–31.

Greenwood P.H. (1992) Are the major fish faunas well-known? *Netherlands Journal of Zoology* **42**, 131–138.

Hay C.J., van der Bank F.H. & Ferreira J.T. (1997) Aspects of the ecology of *Barbus hospes* from the Fish River, Namibia. *Madoqua* **19**, 95–97.

Impson D. (1999) Fish distribution in the Western Cape's upper Doring River and its implications for river management and flyfishing. *Piscator* **131**, 86–92.

Impson N.D. & Bloomer P. (1998/1999) Genetic variation between four whitefish *Barbus andrewi* (Pisces: Cyprinidae) populations and possible implications for conservation management. *Southern African Journal of Aquatic Sciences* **24**, 2–10.

Impson N.D. & Hamman K.C.D. (2000) Conservation status, threats and future prospects for the survival of freshwater fishes of the Western Cape Province, South Africa. In I.G. Cowx (ed.) *Management and Ecology of River Fisheries.* Oxford: Fishing News Books, Blackwell Science, pp. 418–427.

Impson N.D. & Tharme R. (1998) Translocation of indigenous freshwater fishes for conservation and flyfishing: the Boontjies River, Western Cape as a case study. *Piscator* **130**, 87–92.

Impson N.D., Bills I.R., Cambray J.A. & le Roux A. (1999) The primary freshwater fishes of the Cape Floristic Region: conservation needs for a unique and highly threatened fauna. *Report, Cape Nature Conservation*, Cape Town, 26 pp.

Kleynhans C.J. (1999) The development of a fish index to assess the biological integrity of South African rivers. *Water SA* **25**, 265–278.

Klinkhardt M.B. (1998) Some aspects of karyological evolution in fishes. *Animal Research and Development* **47**, 7–36.

Laker M.C. (2000) Soil resources: distribution, utilization, and degradation. In R. Fox & K. Rowntree (eds) *The Geography of South Africa in a Changing World.* Cape Town: Oxford University Press, pp. 326–360.

Le Maitre D.C., Versveld D.B. & Chapman R.A. (2000) The impact of invading alien plants on surface water resources in South Africa: A preliminary assessment. *Water SA* **26**, 397–408.

Lévêque C. (1997) *Biodiversity Dynamics and Conservation. The Freshwater Fish of Tropical Africa.* Cambridge: Cambridge University Press, 438 pp.

Marriott M.S. (1998) Conservation biology and management of the Twee River redfin, *Barbus erubescens* (Pisces: Cyprinidae). Unpublished MSc Thesis, Rhodes University, Grahamstown, 104 pp.

Marshall B. (1999) The fishes of Zimbabwe: A century of change. *The Zimbabwe Science News* **33**, 45–52.

Martin A.K. (1987) Comparison of sedimentation rates in the Natal Valley, south-west Indian Ocean, with modern sediment yields in east coast rivers of Southern Africa. *South African Journal of Science* **83**, 716–724.

Matthews W.J. (1998) *Patterns in Freshwater Fish Ecology*. London: Chapman & Hall, 756 pp.

Moralee R.D., van der Bank F.H. & van der Waal B.C.W. (2000) Biochemical genetic markers to identify hybrids between the endemic *Oreochromis mossambicus* and the alien species *O. niloticus* (Pisces: Cichlidae). *Water SA* **26**, 263–268.

O'Keeffe J.H. (1989) Conserving rivers in Southern Africa. *Biological Conservation* **49**, 255–274.

Pallett J. (ed.) (1997) *Sharing Water in Southern Africa.* Windhoek: Desert Research Foundation of Namibia, 121 pp.

Pike T. & Tedder A.J. (1973) Rediscovery of *Oreodaimon quathlambae*. *The Lammergeyer* **19**, 9–15.

Rall J.L. (1993) An ecological study on the Maloti minnow (*Pseudobarbus quathlambae*) in the catchment area of Phase 1B of the Lesotho Highland Water Scheme. Unpublished PhD Thesis, Rand Afrikaans University, Johannesburg, 99 pp.

Rall J.L. (2000) Development of a conservation program on *Pseudobarbus quathlambae* in the catchment area of phase 1B of the LHWP. *Final Report LHDA Contract No. 170*. Johannesburg: Rand Afrikaans University, 123 pp.

Rogers K. & Bestbier R. (1997) *Development of a Protocol for the Definition of the Desired State of Riverine Systems in South Africa.* Pretoria: Department of Environmental Affairs and Tourism, 100 pp.

Roux D.J., Kleynhans C.J., Thirion C., Hill L., Engelbrecht J.S., Deacon A.R. & Kemper N.P. (1999) Adaptive assessment and management of riverine ecosystems: The Crocodile/Elands River case study. *Water SA* **25**, 501–518.

Shumway C.A. (1999) *Forgotten Waters: Freshwater and Marine Ecosystems in Africa. Strategies for Biodiversity Conservation and Sustainable Development.* Boston: Boston University, 167 pp.

Skelton P.H. (1987) South African Red data Book – Fishes. *South African National Scientific Programmes Report No. 137*, 199 pp.

Skelton P.H. (1990) The conservation and status of threatened fishes in southern Africa. *Journal of Fish Biology* **37**(Suppl. A), 87–95.

Skelton P.H. (1993) On bass in Blindekloof – the impact of an alien predator on a wilderness stream. *The Naturalist* **37**, 21–27.

Skelton P.H. (1994) Diversity and distribution of freshwater fishes in East and Southern Africa. *Annals Museum Royale de l'Afrique Centrale* **275**, 95–131.

Skelton P.H. (2000) Flagships and fragments – perspectives on the conservation of freshwater fishes in southern Africa. *Southern African Society of Aquatic Scientists Journal* **25**, 37–42.

Skelton P.H., Cambray J.A., Lombard A. & Benn G.A. (1995) Patterns of distribution and conservation status of fishes in South Africa. *South African Journal of Zoology* **30**, 71–81.

Skelton P.H., Verheust L. & Scott L.E.P. (2000) A collection-based GIS atlas of fish distribution in South Africa. *Samab* **25**, 11–15.

Tharme R & Anderson A. (1999) A real dilemma: Translocating indigenous fish in southern Africa - some scientific and ethical considerations. *African Wildlife* **53**, 20–21.

Waters J.M. & Cambray J.A. (1997) Intraspecific phylogeography of the Cape galaxies from South Africa: evidence from mitochondrial DNA sequences. *Journal of Fish Biology* **50**, 1329–1338.

Chapter 22
Tropical forest degradation and aquatic ecosystems: our current state of knowledge

L.J. CHAPMAN and C.A. CHAPMAN*
Department of Zoology, University of Florida, Gainesville, and Wildlife Conservation Society, Bronx, USA

Abstract

Recently available data illustrate that tropical forests are being degraded over very large spatial scales. Estimates indicate that tropical countries are losing 127 300 km^2 of forest annually, a further 55 000 km^2 of tropical forests are logged each year, and while it is difficult to quantify the amount of tropical forest that is burnt each year, it may be as much as 30 000 km^2. Conservation agencies have raised interest in addressing deforestation impacts; however, the vast majority of the efforts have been directed towards understanding and mitigating impacts to terrestrial habitats. In the tropics, the impacts of forest degradation on aquatic systems have received little attention. Available evidence on the impacts of tropical forest degradation on aquatic systems, including changes in flood regimes, siltation, sedimentation, removal of overhanging vegetation, and changes in invertebrate and fish communities, are reviewed.

Keywords: deforestation, fire, fish, logging, siltation, tropical aquatic systems.

22.1 Introduction

Accelerating deforestation and forest degradation in the tropics is generating widespread concern throughout the world. Conservation agencies have raised awareness and interest in addressing deforestation impacts; however, the vast majority of the efforts have been directed towards understanding and mitigating impacts to terrestrial habitats. In the tropics, impacts of deforestation and forest degradation on aquatic systems are largely unknown. Yet, such ecosystems are extensive and support an extremely high richness of freshwater fishes. For example, the Congo River is 4650 km long, its drainage basin encompasses 3.8 million km^2, and it supports more than 560 species of fishes. Similarly, the Amazon River is 5500 km long with a drainage of 5.6 million km^2 and supports over 1300 fish species (Chapman 2001). This biodiversity is easily jeopardised by severe perturbation (Chapman, Chapman, Ogutu-Ohwayo, Chandler, Kaufman & Keiter 1996). For example, Mohd (1994) documented that 41% of the native fish species in the Gambak River of Southeast Asia were lost between 1969 and 1990 and attributed this to logging, highway construction, and land clearing for agriculture. Millions of people depend on these aquatic systems for food,

*Correspondence: Department of Zoology, University of Florida, 223 Bartram Hall, PO Box 118525, Gainesville, FL 32611-8525, USA (email: ljchapman@zoo.ufl.edu).

drinking water, transport, etc. For example, it is estimated that 30 million people depend on the fisheries of Lake Victoria in some capacity (Kaufman 1992; Kaufman, Chapman & Chapman 1997).

Here the patterns of deforestation and forest degradation in the tropics are reviewed and the potential consequences of these events on inland waters are examined. Wherever possible case studies conducted in the tropics are presented but, due to the paucity of information on tropical systems, potential consequences of forest degradation are also illustrated by reference to studies conducted in temperate regions or through speculation. The process of conducting this review illustrates the current state of knowledge of the impacts of tropical forest degradation on tropical aquatic systems, and suggests a number of areas where further study is required.

22.2 Tropical forest degradation

Current data on worldwide forest cover are provided by FAO (1999), making it possible to estimate the magnitude of impact that deforestation and forest degradation will have on aquatic systems. For developing countries the FAO defines deforestation as the depletion of tree cover in closed-canopy forests to less than 10%, a canopy thinning threshold that will likely have significant impacts on neighbouring water bodies. In 1995 there were 19 555 700 km^2 of forest remaining in tropical countries (FAO 1999). Forest loss between 1980 and 1995 was 10.5% for Africa, 9.7% for Latin America and the Caribbean, and 6.4% for Asia and Oceania. Tropical countries lose 127 300 km^2 of forest annually; an area greater than Mississippi (122 335 km^2) or just smaller than Greece (131 985 km^2). The highest losses occurred in countries with large expanses of tropical forest, including average annual conversions of 25 540 km^2 for Brazil, 10 840 km^2 for Indonesia, and 7400 km^2 for the Democratic Republic of Congo (Fig. 22.1). The top four countries losing the greatest proportion of their remaining forest cover are the Philippines (annual deforestation rate of 3.87%), El Salvador (3.81%), Costa Rica (3.29%) and Sierra Leone (3.28%).

Tropical deforestation appears to be driven primarily by frontier expansion of subsistence agriculture and large economic development programmes involving resettlement, agriculture and infrastructure (FAO 1999). Growing external debts place strong pressures on governments to encourage timber harvesting and increased agricultural activity. For example, each year the countries of sub-Saharan Africa return a mean of 58% of their gross national product (GNP) in repayments of foreign debts and the repayment can be as high as 241% of the GNP (Stuart, Adams & Jenkins 1990).

According to the FAO definition, selective logging is not considered deforestation, since it does not decrease forest cover to less than 10% of its original level. Yet, selective logging operations can dramatically impact aquatic systems (Pringle & Benstead, 2001). Currently, it is estimated that between 50 000 and 60 000 km^2 of tropical forests are logged each year, approximately a third of the area that is completely deforested (FAO 1993).

With the proliferation of forest fires throughout Southeast Asia (Kinnaird & O'Brien 1999) and South America (Nepstad, Veríssimo, Alencar, Nobre, Lima,

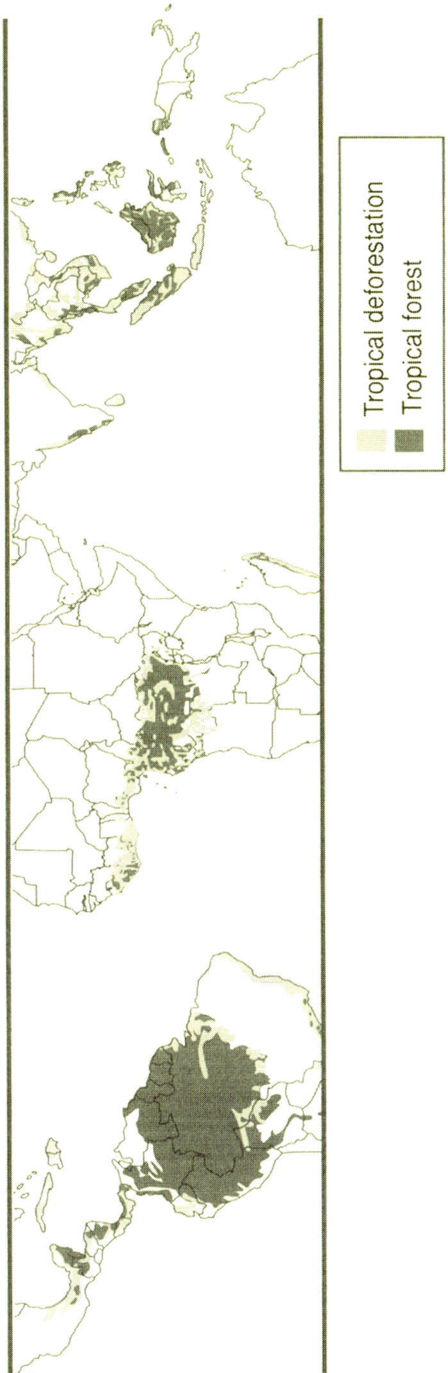

Figure 22.1 Map of the world illustrating the major regions of moist and wet forest and the extent of deforestation in these areas (adapted from National Geographic Atlas 1992)

Lefebvre, Schlesinger, Potter, Moutinho, Mendoza, Cochrane & Brooks 1999), and the media coverage that they have incited, it is increasingly evident that wildfires are having significant impacts on tropical ecosystems. The prevailing opinion concerning fire ecology in tropical forests is that natural fires are relatively rare, and that the majority of fires seen today are either induced or aggravated by human activity (Janzen 1986; Tutin, White & Missandzouo 1996). For example, when logging operations leave an area, large amounts of deadwood are left behind. Since, the canopy has been opened, this wood dries and becomes a large source of flammable material.

Determining the amount of tropical forest recently burned using conventional satellite imagery is at best difficult (FAO 1999) because many fires are restricted to the understory, and leave much of the canopy relatively intact (Nepstad *et al.* 1999; Peres 1999). Obtaining representative figures for the amount of tropical forest that burns annually is further complicated because there is large year-to-year variability in the extent of fires, which are primarily mediated by supra-annual El Niño events. Despite these difficulties in quantifying the amount of tropical forests burned each year, the magnitude of forest fires can be illustrated by presenting a couple of examples. FAO (1999) estimated that in 1997 and 1998 an area of 2 million ha of forest burned in Brazil and 4 million ha burned in Indonesia. Between December 1997 and April 1998, more than 13 000 fires burned in Nicaragua destroying vegetation on more than 800 000 ha of land. These estimates appear to be conservative since at least 1 million ha of intact forests burned in the State of Roraima (Brazil) alone following the 1997/1998 El Niño dry season (Shimabukuro, Krug, Santos, Novo & Yi 2000). At this time, almost half of the forest cover in the entire Brazilian Amazon (1 550 000 km^2) had already completely exhausted its ground-water supply to a depth of at least 10 m, and were therefore highly inflammable (Nepstad *et al.* 1999).

If one made a conservative estimate that 30 000 km^2 of forest burns each year, then the amount of forest deforested, logged, or burned is approximately 212 000 km^2. While conservation agencies have raised widespread concern about this extensive degradation of tropical forests, the impacts of this degradation on aquatic systems has been largely ignored.

22.3 Impacts of deforestation on aquatic systems

22.3.1 *Siltation and sedimentation*

An effect of deforestation readily apparent is increased siltation and sedimentation. There are many ways to remove logs from a logging site. Ground-based timber yarding is the predominate approach in the tropics, and involves felling widely separated trees and hauling out the sawn timber or logs. When trees are cut on site (e.g. pitsawing) and the timber is hauled out manually, there is little damage to neighbouring trees or to the soil (Chapman & Chapman 1996). However, more typically, tractors or bulldozers are used to haul out whole logs. When this is done roads must be constructed to allow machinery access. Logging with bulldozers in Brazil that removed only 2% of the trees greater than 10 cm diameter at breast height (DBH), resulted in damage to

26% of the remaining trees, canopy opening of 50%, and the development of logging roads covering 8% of the forest floor (Frumhoff 1995; Johns, Barreto & Uhl 1997). Uprooted trees and logging roads expose the soil to the rains and create situations where sediment readily runs off of the land into adjacent aquatic systems.

Deforestation of the watershed leads to changes in the seasonal flood regime. In forested watersheds, vegetation and topsoil help retain water. With their removal, flood peaks tend to become higher and shorter, because run-off is not delayed by the holding capacities of the forest. Dramatic changes in the flood regime can negatively impact fish populations that require a smoother seasonal transition (Welcomme 1985). The felling or poisoning of 40% of the upperstory trees in South America corresponded to a 55–70% increase in the water yield (C. Bruijnzeel, unpublished data). This increase can last for several years, since the regrowth that comes after logging possesses a smaller total leaf area that does not intercept rainfall to the same extent of mature vegetation (Uhl & Jordan 1984), and this secondary vegetation has a less-developed root system that cannot exploit soil moisture levels to the extent of mature forest trees (C. Bruijnzeel, unpublished data).

Removal of vegetation decreases the evapotranspiration rate and the interception of precipitation, and therefore increases runoff and sediment yield. In many tropical forested and savanna rivers, annual sediment yield is low. Sedimentation increases dramatically in the forested rivers following deforestation (Fig. 22.2). Increased turbidity

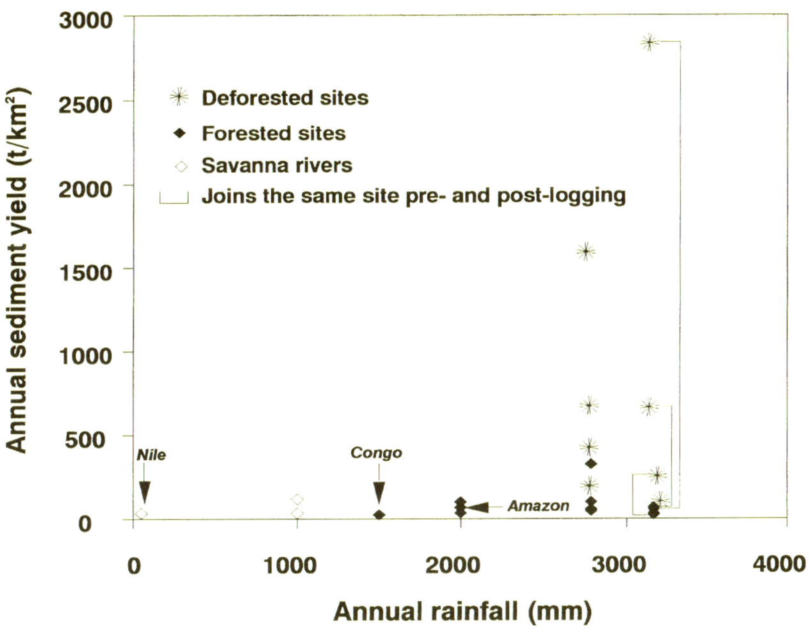

Figure 22.2 Annual sediment yield (t km^{-2}) for forested, savanna, and deforested catchments in the tropics *v.* annual precipitation for the region. Data abridged from Holeman (1962), Welcomme (1985) and Douglas *et al.* (1993). A vertical line joins sites for which data were gathered pre- and post-disturbance. All deforested sites are from Peninsular Malaysia

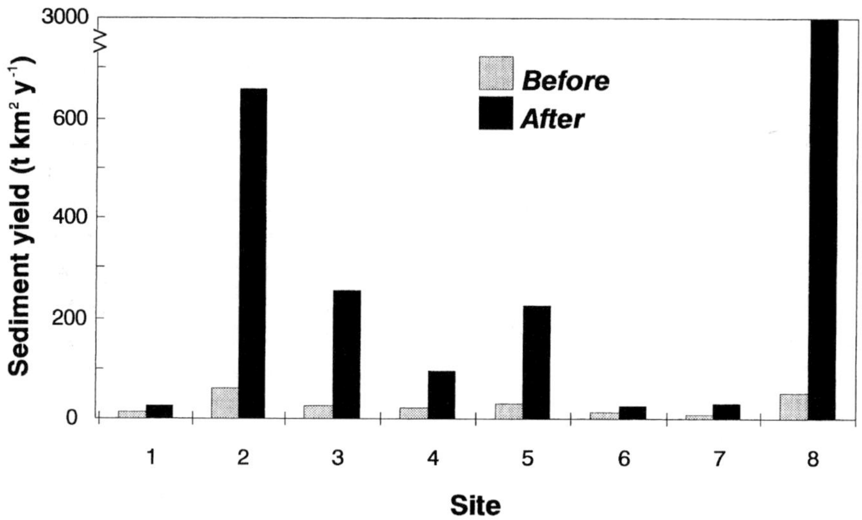

Figure 22.3 Sediments yields from rainforests in Malaysia and Kalimantan before and after logging (data from Kasran 1988; Douglas *et al.* 1993; Kasran & Nik 1994)

is associated with increased sedimentation and siltation, and much of the increased silt load of tropical rivers in recent history has been associated with the deforestation in the upper regions of the watershed (Marlier 1973; Eckholm 1976).

The most detailed tropical studies of how logging effects sediment loads in rivers have been conducted in Malaysia and Kalimantan (Fig. 22.3). A number of these studies also evaluated the degree to which reduced-impact logging decreased the amount of sediment coming off of logged areas. Reduced-impact logging typically involves removing the same amount of timber from an area as conventional logging, but it decreases forest damage by closely supervising tree felling, cutting vines before felling, by minimising road construction, and occasionally by cutting cross drains on steep logging roads. For a dipterocarp forest in Peninsular Malaysia, sediment yield increased from pre-logging levels by 97% for conventional logging and 70% for reduced-impact logging Kasran (1988). Yusop and Suki (1994) showed that relative to a control site in the Berembun Forest Reserve, Peninsular Malaysia, conventional logging increased turbidity by 9 times and suspended soil by 12 times. Reduced-impact logging only resulted in a two-fold increase in these characters. At a different study site, Kasran and Nik (1994) found suspended sediment yield increased 177% the year following logging and further increased 297% in the following year. Four years after, with the growth of secondary vegetation, the yield returned to the pre-logging level. Douglas, Greer, Bidin and Silsbury (1993) documented that logging or ground clearance increased river sediment yields by 2–50 times in Malaysia.

These studies in Malaysia are the type of research that is needed, because they not only assess the nature of perturbation that logging causes to aquatic systems, but also evaluate possible management options. However, it is questionable just how appropriate it is to use sediment-loading studies conducted in Southeast Asia as models for activities in other parts of the tropics. First, studies have shown a high degree of variation in the amount of sediments coming off of logged areas depending on the slope, the number and extent of road construction, and the underlying geology (Douglas *et al.* 1993). Second, much of the sediment released from logged areas can be deposited during very brief storm events, illustrating the need for intensive monitoring over a number of years. Unfortunately, long-term studies of this nature are rare. Finally, and possibly most importantly, the nature of the extraction differs among tropical regions. For example, African forests are very heterogeneous in their species composition. Consequently, harvesting activities produce a comparatively low volume per hectare compared to Southeast Asia. Yields from selectively logged forests in Africa are generally not more than $15\,m^3\,ha^{-1}$, while $60\,m^3\,ha^{-1}$ are commonly extracted from Southeast Asian forests (Bakouma & Buttoud 1996).

In a general survey of watersheds in Kenya, Dunne (1979) demonstrated that forested catchments lose 20–30 $t\,km^{-2}\,year^{-1}$. Sediment yields of agricultural lands vary enormously with runoff. In the wettest, steepest cultivated catchments, the soil loss exceeds 4000 $t\,km^{-2}\,year^{-1}$. The sediment yields from rangeland catchments are also variable, the driest catchments lose less than 100 $t\,km^{-2}\,year^{-1}$, while 20 000 $t\,km^{-2}\,year^{-1}$ are exported from the wettest steepest grazed catchment.

22.3.2 *Changes in the biota and cascading effects*

Surveys of aquatic biota in tropical waters affected by different types of human habitat modification are few; however, researchers are becoming increasingly interested in using macroinvertebrates as indicators of stream 'health' (Fore, Karr & Wisseman 1996; Richards, Johnson & Host 1996; Stone & Wallace 1998). Changes in the physical environment associated with human modifications to the landscape, such as changing the amount of silt in the water, or changing light level, will impact the aquatic biota. These changes are complex, difficult to predict, and there are too few studies available from tropical areas to make generalisations that are sufficiently reliable to use in management. In Oregon, USA, Carlson, Andrus and Froelich (1990) documented that macroinvertebrate density was 20–113% greater at logged sites relative to unlogged sites, despite similar diversity between the logged and undisturbed sites. In contrast, sand deposition resulting from stream diversion in Cornwall, UK, was associated with a dramatic reduction in aquatic plants, the elimination of several species, and a lower diversity of macroinvertebrates (Nuttall 1972). Stone and Wallace (1998) found that 16 years after logging in the southern Appalachian mountains of the USA, benthic invertebrate abundance was three times higher in the disturbed stream.

Increased sedimentation and higher turbidity can lead to the decline of plankton through a reduction in light penetration, and the disappearance of many benthic, rheophilic animals that are sensitive to mud on their integument and gills or lose their interstitial habitats to clogging by silt (Marlier 1973; Welcomme 1983). These include insect groups like the Ephemeroptera, Plecoptera, and Trichoptera. Severe forest modification can result in shifts to invertebrate communities dominated by small burrowing forms, such as larval chironomids. Such burrowing forms may be less available to foraging fishes (Pringle & Benstead 2001). The silt can also provide anchorage for vegetation that can block low order streams (Welcomme 1983). On the floodplain, excessive siltation can choke the standing waters that then disappear at a rate faster than they are created.

In their review of the effects of logging on tropical riverine systems, Pringle and Benstead (2001) found no studies that dealt specifically with effects of increased sedimentation on tropical fishes. However, in temperate systems (focusing primarily on salmonids) negative effects include: smothering of eggs and entrapment of fry by consolidated sediments; mortality of juvenile and adult fish due to clogging of opercular cavities and gill filaments; and sub-lethal effects such as reduced feeding and growth, respiratory impairment, reduced tolerance to disease and toxicants, and physiological stress (Lloyd 1987; Waters 1995). A serious impact is also the loss of habitat for fry that require interstitial space in riffles, and the filling of pools and blanketing of structural cover (Waters 1995). Garman and Moring (1993) documented a decline in abundance of one fish species following clearcut logging and a decline in aquatic prey resources, while a second species increased in abundance because it switched to eating terrestrial arthropods. In a logged Eucalyptus forest in Tasmania, Davies & Nelson (1994) found that a significant decrease in riffle macroinvertebrates, particularly stoneflies and mayflies, corresponded to a decrease in brown trout, *Salmo trutta* L., abundance. Goulding, Smith and Mahar (1996) commented that floodplain deforestation in the Amazon resulted in a dwindling amount of wood that is naturally carried downstream.

Given that many rain forest rivers have relatively low primary productivity and rely largely on allochthonous input, destruction of the forest may result in a decline in aquatic productivity (Davies & Nelson 1994). In areas of the central Congo basin and extensive areas of the Amazon basin where flooding allows fishes to find food, refuge and breeding areas in the inundated forest, the consequences of deforestation may be particularly severe. Although data are few on the impact of flooded forest conversion in tropical rain forests, the recent history of the Mekong system provides a basis on which to speculate. In Cambodia, the Mekong River floods into the Grand Lac, a seasonally fluctuating lake, and surrounding floodplain (formerly forest). Many fishes move into the Grand Lac as it fills and then move out into the inundated zone, returning to the lake and migrating down river as waters fall in the dry season (Lowe-McConnell 1975). Chevey and Le Poulain (1940) found that growth of some cyprinids was faster in the flooded forest than in the open lake and the river that the authors attributed to abundant fish food in the forest. Subsequently, the forest surrounding the Grand Lac area of the Mekong was cleared for agriculture and firewood. This clearance was accompanied by a decline of about 50% in the fishery in 25 years (Welcomme 1985). The decline was attributed to erosion and siltation in the Mekong basin and the dramatic decline of allochthonous food arising from the deforestation. Siltation led to increased turbidity of waters, which in the case of the Grand Lac coincided with a drop in primary production reflected in decreased fish production. Welcomme (1985) reported that the process was reversed during the late 1970s when political troubles led to a collapse of the human population in the region, which permitted the forest around the Grand Lac to regenerate. Forest regeneration coincided with an increase in fish populations, which may have related to the renewal of forest resources and reduced fishing pressure. Similar effects have been observed in the Amazon basin, where the economically important characid, *Colossoma macropomum* Cuvier (the tambaqui) declined near the major city of Manaus. The decline of the frugivorous tambaqui was attributed to overfishing and the disappearance of the inundated forest feeding grounds (Marshall 1995). Welcomme (1983) noted that in heavily forested systems, such as the Amazon, Congo and Mekong, trees appear to act as a nutrient sink and their disappearance from the fundamentally poor systems may have long-term impacts on nutrient balance.

Cohen, Bills, Cocquyt and Caljon (1993) studied the effects of sedimentation on the species richness of three taxonomic groups in Lake Tanganyika: ostracods, fish and diatoms, and by contrasting undisturbed, moderately disturbed, and highly disturbed areas. The most serious threat facing Lake Tanganyika is viewed to be deforestation and subsequent erosion (Cohen *et al.* 1993). In the north end of the lake where deforestation approaches 100% of the forested land, rates of soil erosion range between 28 and 100 t ha^{-1} $year^{-1}$ depending on the slope (Cohen *et al.* 1993). The species richness of ostracods and fish appeared to be heavily affected by sediment, while diatoms showed little change in species richness (Cohen *et al.* 1993). In a later study of the same region, Alin, Cohen, Bils, Gashagaza, Michel, Tiercelin, Martens, Coveliers, Mboko, West, Soreghan, Kimbadi and Ntakimazi (1999) showed that the species richness and density of fishes, molluscs and ostracods were negatively correlated with sedimentation.

22.3.3 *Effects of removing overhanging trees*

When one thinks of tropical riverine systems, one typically thinks of rivers like the Amazon, Congo, Mekong or Nile. However, many rivers are much smaller. For example, more than 90% of Africa's rivers are less than 9 km long (Stiassny 1996). As a result, many tropical river systems are strongly influenced by overhanging terrestrial vegetation. Removing trees bordering rivers causes a dramatic shift in sources of production, reduces the input of terrestrial elements (e.g. leaves, terrestrial invertebrates), increases the light and water temperature, and is associated with changes in the chemical composition of the water. Increased light resulting from the removal or loss of the canopy leads to higher standing stocks of algae (Ulrich, Burton & Oemke 1993). Such changes can restructure the macroinvertebrate communities. For example, in the southern Appalachian mountains, Wallace and Gurtz (1986) found that Ephemeroptera production was 28 times higher in a logged catchment in comparison to an unlogged steam. Presumably, this was the result of an increase in primary productivity associated with the opening of the canopy and increased light levels. Marlier (1973) reported changes in chemical composition associated with increased solar insolation for small streams in the eastern Congo. Waters from deforested areas had higher pH and higher concentrations of salts than protected waters.

The importance of maintaining riparian vegetation is highlighted in studies that have examined the value of maintaining a riparian buffer zone as a strategy to decrease effects of logging on stream habitats. Growns and Davies (1991) contrasted the lotic macroinvertebrate community of logged streams, a stream with a 100-m buffer zone, and control streams. Macroinvertebrate community structure in the buffered stream differed from that in the undisturbed stream, but it was more similar to the latter than to the logged stream. These results suggest that a 100 m wide buffer zone is effective at ameliorating disturbance due to clearfelling (see also Richards *et al.* 1996). Davies and Nelson (1994) documented the effects of logging for both cable and conventional logging in areas with a range of riparian buffer strip widths (0–50 m) and found that all impacts were significant only at buffer widths of less than 30 m.

22.4 Conclusions

The recently available information on deforestation, logging, and fire illustrates that tropical forests have been seriously impacted over very large spatial scales. The new millennium brings an even greater potential for change. The severity of this situation has been widely recognised for the last three decades. However, recognition of impacts to tropical terrestrial systems has not lead to sufficient effort to quantify the impacts of tropical forest degradation on aquatic systems, nor to many major conservation efforts to protect fresh waters. The future offers a great opportunity for research to contribute to freshwater aquatic conservation and management.

Research priorities should include:

(1) documenting patterns of change to terrestrial systems and associated responses of aquatic systems;
(2) understanding cascading effects of forest removal;
(3) predicting how different functional guilds (fishes, benthic invertebrates) will be affected by different types of human activities;
(4) identifying mechanisms determining the pattern of change.

Studies that attempt to address these research priorities should be set in a framework such that concrete management recommendations can result from the research (e.g. document patterns of change that evaluates different conservation approaches). Since the protection of entire aquatic systems is so difficult (e.g. protection of an entire drainage), one of the most rewarding avenues for future tropical research will be to determine simple mechanisms to minimise impacts that degradation of forests will have on neighbouring aquatic systems.

Acknowledgements

Funding for our research in Uganda has been provided by the Wildlife Conservation Society, the National Geographic Society, and the National Science Foundation. Our field programs have greatly benefited from the assistance of numerous collaborators.

We thank Jennifer Piascik for their assistance with the figures. We would like to thank Ian Cowx and Maria Joao Collares-Pereira for encouraging us to participate in the symposium.

References

Alin S.R., Cohen A.S., Bils R., Gashagaza M.M., Michel E., Tiercelin J.-J., Martens K., Coveliers P., Mboko S.K., West K., Soreghan M., Kimbadi S. & Ntakimazi G. (1999) Effects of landscape disturbance on animal communities in Lake Tanganyika, East Africa. *Conservation Biology* **13**, 1017–1033.

Bakouma J. & Buttoud G. (1996) African markets: a future for African sawnwood? *Tropical Forest Update* **6**, 17.

Carlson J.Y., Andrus C.W. & Froelich H.A. (1990) Woody debris, channel features, and macroinvertebrates of streams with logged and undisturbed riparian timber in Northeastern Oregon, U.S.A. *Canadian Journal of Fisheries and Aquatic Science* **47**, 1103–1111.

Chapman L.J. (2001) Fishes of African rain forests: diverse adaptations to environmental challenges. In W. Weber, A. Vedder, H. Simons Morland, L. White & T. Hart (eds) *African Rain Forest Ecology and Conservation*. New Haven: Yale University Press, pp. 263–390.

Chapman C.A. & Chapman L.J. (1996) Exotic tree plantation and the regeneration of natural forests in Kibale National Park, Uganda. *Biological Conservation* **76**, 253–257.

Chapman L.J., Chapman C.A., Ogutu-Ohwayo R., Chandler M., Kaufman L. & Keiter A. (1996) Refugia for endangered fishes from an introduced predator in Lake Nabugabo, Uganda. *Conservation Biology* **10**, 554–561.

Chevey P. & Le Poulain F. (1940) La pêche dans les eaux douces du Cambodge. *Memoire l'Institut Oceanographique de l'Indochine* **5**, 193 pp. (in French).

Cohen A., Bills R., Cocquyt C.Z. & Caljon A.G. (1993) The impact of sediment pollution on biodiversity in Lake Tanganyika. *Conservation Biology* **7**, 667–677.

Davies P.E. & Nelson M. (1994) Relationships between riparian buffer widths and the effects of logging on stream habitat, invertebrate community composition and fish abundance. *Australian Journal of Marine and Freshwater Research* **45**, 1289–1305.

Douglas I., Greer T., Bidin K. & Silsbury M. (1993) Impacts of rainforest logging on river systems and communities in Malaysia and Kalimantan. *Global Ecology and Biogeography Letters* **3**, 245–252.

Dunne T. (1979) Sediment yield and land use in tropical catchments. *Journal of Hydrology* **42**, 281–300.

Eckholm E.P. (1976) *Losing Ground: Environmental Stress and World Food Prospects*. New York: W.W. Norton, 223 pp.

FAO (Food and Agriculture Organization of the United Nations) (1993) Forest resources assessment 1990 – Tropical Countries. *FAO Forestry Paper* **112**, Rome: Italy, 112 pp.

FAO (Food and Agriculture Organization of the United Nations) (1999) *State of the World's Forests*. Rome: Italy, 154 pp.

Fore L.S., Karr J.R. & Wisseman W.R. (1996) Assessing invertebrate responses to human activities: evaluating alternative approaches. *Journal of the North American Benthological Society* **15**, 212–231.

Frumhoff P.C. (1995) Conserving wildlife in tropical forests managed for timber. *Bioscience* **45**, 456–464.

Garman G.C. & Moring J.R. (1993) Diet and annual production of two boreal river fishes following clear-cut logging. *Environmental Biology of Fishes* **36**, 301–311.

Goulding M., Smith N.J. & Mahar D.J. (1996) *Floods of Fortune: Ecology and Economy along the Amazon*. New York: Columbia University Press, 193 pp.

Growns I.O. & Davis J.A. (1991) Comparison of the macroinvertebrate communities in streams in logged and undisturbed catchments eight years after harvesting. *Australian Journal of Marine and Freshwater Research* **42**, 689–706.

Holeman J.N. (1962) The sediment yield of major rivers of the world. *Water Resources Research* **4**, 737–747.

Janzen D.H. (1986) *Guanacaste National Park: Tropical Ecological and Biocultural Restoration.* San Jose: Editorial Universidad Estatal A distancia, Costa Rica, 103 pp.

Johns J.S., Barreto P. & Uhl C. (1997) Logging damage in planned and unplanned logging operations in the eastern Amazon. *Forest Ecology and Management* **89**, 59–77.

Kasran B. (1988) Effect of logging on sediment yield in a hill dipterocarp forest in Peninsular Malaysia. *Journal of Tropical Forest Science* **1**, 56–66.

Kasran B. & Nik A.R. (1994) Suspended sediment yield resulting from selective logging practices in a small watershed in Peninsular Malaysia. *Journal of Tropical Forest Science* **7**, 286–295.

Kaufman L.S. (1992) Catastrophic change in species rich freshwater ecosystems; the lessons of Lake Victoria. *Bioscience* **42**, 846–858.

Kaufman L.S., Chapman L.J. & Chapman C.A. (1997) Evolution in fast forward: haplochromine fishes of the Lake Victoria region. *Endeavour* **21**, 23–30.

Kinnaird M.F. & O'Brien T. (1999) Ecological effects of wildfire on lowland rainforest in Sumatra. *Conservation Biology* **12**, 954–956.

Lloyd D.S. (1987) Turbidity as a water quality standard for salmonid habitats in Alaska. *North American Journal of Fish Management* **7**, 34–45.

Lowe-McConnell R.H. (1975) *Fish Communities in Tropical Freshwaters.* London: Longman, 337 pp.

Marlier G. (1973) Limnology of the Congo and Amazon Rivers. In B.J. Meggers, E.S. Ayensu & W.D. Duckworth (eds) *Tropical Forest Ecosystems in Africa and South America: A Comparative Review.* Washington, D.C.: Smithsonian Institution Press, pp. 223–238.

Marshall E. (1995) Homely fish draws attention to Amazon deforestation. *Science* **267**, 814.

Mohd Z.-I. (1994) Zoogeography and biodiversity of the freshwater fishes of Southeast Asia. *Hydrobiologia* **285**, 41–48.

Nepstad D.C., Veríssimo A., Alencar A., Nobre C., Lima E., Lefebvre P., Schlesinger P., Potter C., Moutinho P., Mendoza E., Cochrane M. & Brooks V. (1999) Large-scale impoverishment of Amazonian forests by logging and fire. *Nature* **398**, 505–508.

Nuttall P.M. (1972) The effects of sand deposition upon the macroinvertebrate fauna of the River Camel, Cornwall. *Freshwater Biology* **2**, 181–186.

Peres C.A. (1999) Ground fires as agents of mortality in a central Amazonian forest. *Journal of Tropical Ecology* **15**, 535–541.

Pringle C.M. & Benstead J.P. (2001) Effects of logging on tropical river ecosystems. In R. Fimbel, A. Grajal, & J. Robinson (eds) *The Cutting Edge.* New York: Columbia University Press, pp. 467–504.

Richards C., Johnson L.B. & Host G.E. (1996) Landscape-scale influences on stream habitats and biota. *Canadian Journal of Fisheries and Aquatic Sciences* **53**, 295–311.

Shimabukuro Y.E., Krug T., Santos J.R., Novo E.M. & Yi, J.L.R. (2000) Roraima: o incêndio visto do espaço. *Ciência Hoje* **157**, 32–34 (in Portuguese).

Stiassny M.L.J. (1996) An overview of freshwater biodiversity: with some lessons from African fishes. *Fisheries* **21**, 7–13.

Stone M.K. & Wallace J.B. (1998) Long-term recovery of a mountain stream from clearcut logging: the effects of forest succession on benthic invertebrate community structure. *Freshwater Biology* **39**, 151–169.

Stuart S.N., Adams R.J. & Jenkins M.D. (1990) *Biodiversity in Sub-Saharan Africa and its Island: Conservation, Management, and Sustainable Use.* Gland Switzerland: IUCN, 242 pp.

Tutin C.E.G., White L.J.T. & Missandzouo A.M. (1996) Lightning strike burns large forest tree in the Lopé Reserve, Gabon. *Global Ecology and Biogeography Letters* **5**, 36–41.

Uhl C. & Jordan C.F. (1984) Succession and nutrient dynamics following forest cutting and burning in Amazonia. *Ecology* **65**, 1476–1490.

Ulrich K.E., Burton T.M. & Oemke M.P. (1993) Effects of whole-tree harvest on epilithic algal communities in headwater streams. *Journal of Freshwater Ecology* **8**, 83–92.

Wallace J.B. & Gurtz M.E. (1986) Response of *Baetis* mayflies (Ephemeroptera) to catchment logging. *American Midland Naturalist* **115**, 25–41.

Waters T.F. (1995) Sediment in streams: sources, biological effects, and control. *American Fisheries Society Monograph* 7, Bethesda MD, American Fisheries Society, 251 pp.

Welcomme R.L. (1983) River basins. *FAO Fisheries Technical Paper* **202**, **60** pp.

Welcomme R.L. (1985) River fisheries. *FAO Fisheries Technical Paper* **262**, 330 pp.

Yusop Z. & Suki A. (1994) Effects of selective logging methods on suspended solids concentration and turbidity level in streamwater. *Journal of Tropical Forest Science* **7**, 199–219.

Chapter 23
Heavy metals as a threat for restored fish populations in a lowland river

L. BERVOETS,* M. LODTS, K. VAN CAMPENHOUT and R. BLUST
Department of Biology, University of Antwerp, Groenenborgerlaan, Antwerp, Belgium

Abstract

Fish are among the top predators in rivers and tend to accumulate metals in their tissues, which can result in harmful effects such as reduced growth and condition. Moreover, metal exposure can induce an increased synthesis of metallothioneins. In this study the effect of metal pollution on a recovered fish population in a natural river, with a pronounced gradient in cadmium and zinc pollution, was investigated. Metal levels were measured in water, sediment, and in tissues of one fish species, the gudgeon *Gobio gobio*. The condition factor (K) of the fish was measured. Accumulated levels were related to environmental levels, to effects, and to induction of metallothionein-like proteins (MTLP). MTLPs were determined using the cadmium saturation assay. Tissue levels of metals were significantly correlated with environmental levels. Metal levels in the liver were significantly correlated with levels of MTLP and with fish condition. However the latter relationship was rather low (18% of the variation described). Probably factors other than metals influence the condition factor.

Keywords: condition, gudgeon, metal pollution, metallothioneins.

23.1 Introduction

As a result of increased purification of municipal waste waters over the last 20 years, water quality in Flanders has improved. Consequently, fish communities have been partially recovered in several watercourses (Vandelannoote, Yseboodt, Bruylants, Verheyen, Coeck, Maes, Belpaire, Van Thuyne, Denayer, Beyens, De Charleroy & Vandenabeele 1998). However, several rivers are still polluted with heavy metals. In 1997, in approximately 15% of river length, Cd and Zn levels were in excess of the Belgian water quality objectives (Geuzens, Nouwen, Cornelis, Craps, De Vlieger, Mensink, Schoeters, Witters, Goyvaerts, Weltens, Van Thuyne, Belpaire, Guns & Brouwers 1999). Nevertheless, in some of those rivers, fish communities are again present.

Fish are among top predators in aquatic ecosystems and tend to accumulate metals in their tissues, which can result in harmful effects. Chronic exposure of fish to sublethal levels of heavy metals causes, among others, reduced growth, disturbed calcium ion balance and altered biochemical or physiological conditions (Bendell-Young, Harvey & Young 1986; Bryan, Atchison & Sandheinrich 1995; Miliou, Zaboukas &

---

*Correspondence: Department of Biology, University of Antwerp, Groenenborgerlaan 171, 2020 Antwerp, Belgium (email: bervoets@ruca.ua.ac.be).

Moraitou-Apostolopoulou 1998). Moreover, metal exposure can induce an increased synthesis of metallothioneins (MT), a specific metal binding protein, whose induction can be seen as a possible detoxification mechanism (De Conto-Cinier, Petit-Ramel, Faure & Garin 1997; Muto, Ren, Hwang, Tominaga, Itoh & Tanaka 1999). All those effects finally could result in a decrease of the fish community and the elimination of the most sensitive species.

Little information is available on the relationship between tissue levels of metals and condition of fish from natural systems. From a conservation point of view we are interested in the metal levels that can be tolerated by a fish community without harmful effects. Therefore the aim of this study was to investigate the effect of metal pollution on a recovering fish community in a natural river, with a pronounced gradient in cadmium and zinc pollution, looking at species composition and fish densities. Further tissue levels of cadmium and zinc in one species, the gudgeon, were related to environmental levels. Finally, the study examined if tissue levels of both metals could be related to fish condition and to the induction of metallothionein-like proteins (MTLP).

23.2 Materials and methods

23.2.1 *Study area*

Ten sampling sites were selected on the Molse Nete–Grote Nete river system, located in Flanders (Belgium) and belonging to the basin of the River Scheldt (Fig. 23.1). In the past 10 years the fish communities have recovered due to improvement of the oxygen concentration in the water. However, elevated concentrations of cadmium and zinc in the water and sediment were reported in the upstream region of the river system (De Cooman, Florus & Devroede 1998; Yseboodt, Clement, Meire & Verheyen 1999). Sampling sites 1, 2, 4–10 were along a cadmium and zinc concentration gradient, the highest metal levels being measured at site 1. A reference site (3) was located on a nearby tributary (Oude Nete). In the study area no fish had been stocked.

23.2.2 *Test species*

Gudgeon, *Gobio gobio* (L.), a benthivore cyprinid, was selected for the metal analysis and condition measurements. This species is common in Europe, both in polluted and non-polluted areas (Flammarion & Garric 1997), and is very abundant in Flanders, including the study area. Moreover, gudgeon has a restricted home range and therefore measured metal body-residues and condition indices of fish should reflect local environmental conditions (Stott, Eldson & Johnston 1963).

23.2.3 *Fish sampling*

Fish were captured at sites 1–6, in November and December 1998, by electric fishing, using an Electra Catch WFC7 generator producing 150 V (smooth DC; max. 4 kW). At

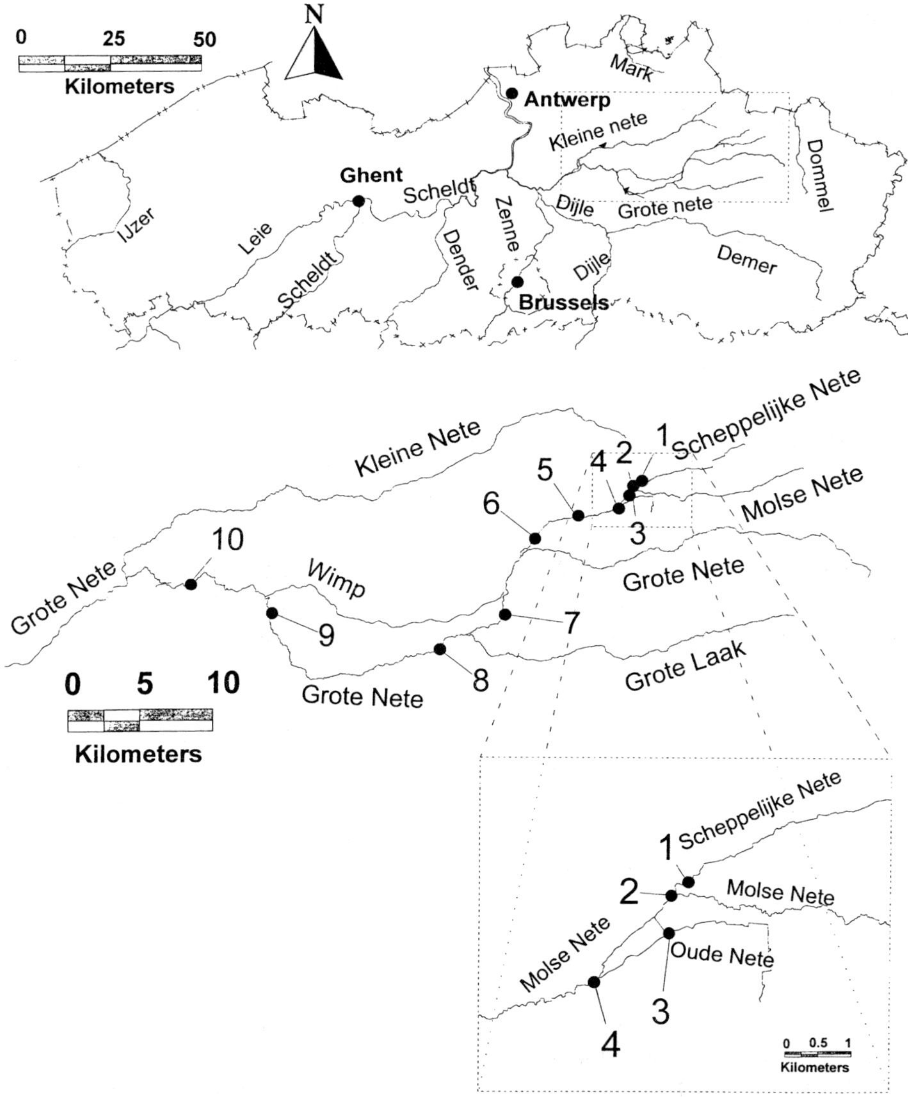

Figure 23.1 Location of the sampling sites. The numbers indicate the sampling sites where the gudgeons were caught and where water and sediment samples were taken

the other sites both electric fishing and fish traps were used. The river sections sampled were about 100 m long. Species composition and densities of the fish communities were assessed. All gudgeon caught were measured (fork length, nearest mm) and weighed (nearest 0.1 g). From each sampling site, 8–10 individuals were killed using the anaesthetic ethyl meta-aminobenzoate methanesulphonic acid (MS222), and tissues (kidney, gill, liver and muscle) were collected in the field. The organs were stored in liquid nitrogen and transported to the laboratory, where they were cut in two parts: one part for metal analysis and one part for the analysis of MTLP. Samples for metal analysis were put in acid-washed polypropylene pre-weighed vials and dried for 24 h at 60°C. Subsequently, the biological material was digested with concentrated ultrapure nitric acid in a microwave oven and stored until analysis (Blust, Van Der Linden, Verheyen & Decleir 1988). Samples for analysis of MTLP were stored at −80°C.

23.2.4 *Metal analysis*

Water was sampled monthly ($n = 8$) between October 1998 and May 1999. The sediment was sampled once in January 1999 with a 'Petit Ponar' grab sampler (Wildco cat. no. 1728; 235 cm^2). Samples were sieved using a 500 μm-mesh sieve and stored in 500 mL polyethylene bags at 4°C. All sediments were sandy. Cadmium, copper, and zinc levels in water, sediments, and biological material were measured by an axial inductively coupled plasma atomic emission spectrometer (ICP-AES, Varian). Sediments were dried at 60°C for 48 h, and a mixture of HNO_3 and HCl (4 : 1) was added; subsequently, samples were put in teflon bombs and digested in a microwave oven (Tessier, Campbell & Bisson 1979; Bervoets, Blust, De Wit & Verheyen 1997). Concentration of the metals in the sediments and tissues was calculated on a dry weight basis and expressed as $\mu g\, g^{-1}$. Analytical accuracy was determined using certified reference material of the Community Bureau of Reference (Belgium) standard for trace elements in river sediment (CRM 320) and muscle tissue (CRM 278). Recoveries were within 10% of the certified values. Total metal load (ML) in the tissues was calculated by dividing the concentration of a single metal by the concentration of that metal measured in tissues of fish from the reference sites. Standardised values were then added using $ML_j = \Sigma_i\, C_{ij}/CR_i$, where ML_j is the metal load in a tissue at site j, C_{ij} the concentration of metal i at site j, and CR_i is the concentration of metal i at the reference site (Meregalli, Vermeulen & Ollevier 2000).

23.2.5 *Condition*

The condition factor (K-factor) was calculated according to Bagenal (1978) – $K = W/L^b \times 100$, where W is the weight (g) and L the fork length (mm). The exponent b is derived from the length–weight relationship at each site which is described by $W = a\, L^b$. K was related to ML in the different organs.

23.2.6 *MTLP-measurement*

Metallothionein-like proteins or (Cd, Zn)-MTLPs in liver, gill and kidney were measured using the Cd-Chelex assay according to Bartsch, Klein and Summer (1989). MTLP present in tissues were saturated with ^{109}Cd-labelled CdCl and analysed with a Minaxi-Autogamma 5530 counter (Canberra Packard). The detection limit of the Cd-Chelex assay was 10 μg MTLP per gram tissue. For (Cd, Zn)-MTLP concentration calculations, a molar ratio of 7 mol Cd per mol MTLP was assumed (Kito, Ose, Mizuhira, Sato, Ishikawa & Tazawa 1982).

23.3 Results

The mean levels of cadmium and zinc in the water declined with increased distance downstream from 23.2 to 1.60 and 2168 to 142 $\mu g\,L^{-1}$ for cadmium and zinc respectively. In the sediment, levels decreased from 47.1 to 11.90 and 2580 to 707 $\mu g\,g^{-1}$ for cadmium and zinc respectively (Fig. 23.2). No significant differences were found among sampling sites for copper.

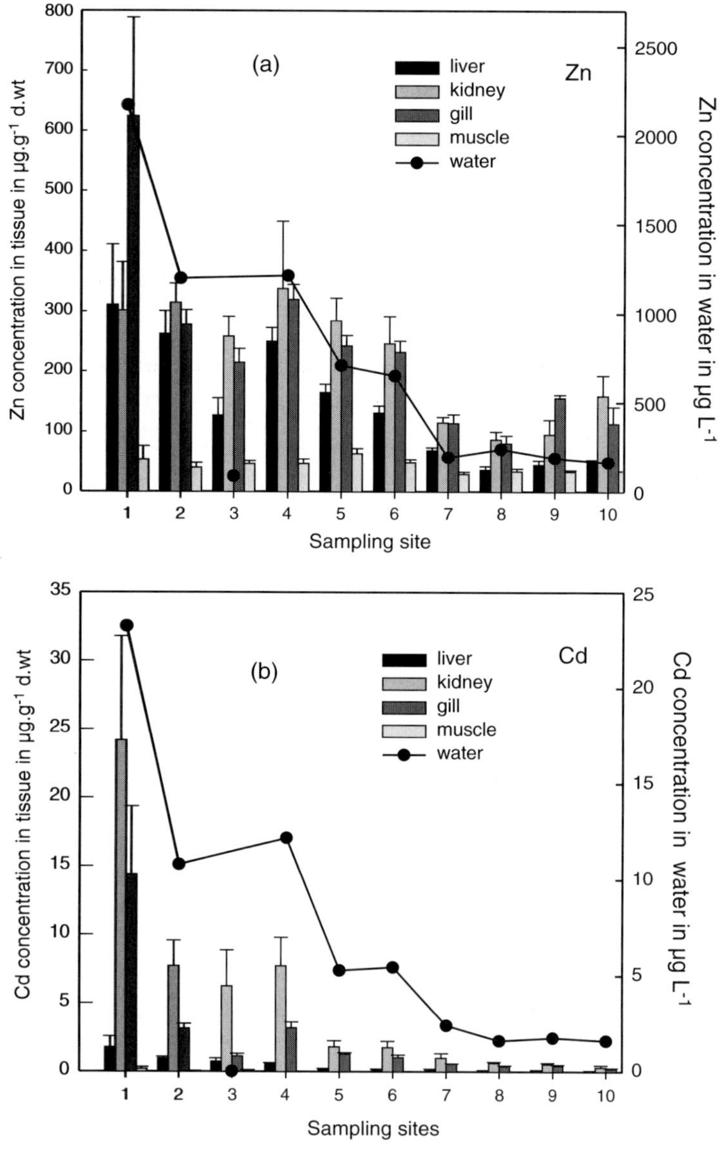

Figure 23.2 Environmental levels of zinc and cadmium at the different sampling sites: (a) in the sediment; (b) in water

Fish species composition was relatively similar along the pollution gradient (Table 23.1; note data on fish density from sites 6 to 10 were obtained from Van Liefferinge, De Cooman, Yseboodt, Bervoets, Schneiders & Verheyen (2000)). However, density and biomass were lowest at the site with the highest environmental concentrations (site 1). Between sites 2 and 10, including control site 3, differences in species composition or density could not be attributed to differences in metal levels.

The cadmium concentration in the tissues decreased in the following order: kidney > gill ⩾ liver > muscle (Fig. 23.3). More or less the same was found for zinc, although differences among tissues were less pronounced. Both metals followed the same gradient in the gill, kidney and liver as the environmental samples, with the highest levels found at the most upstream sampling sites. Significant positive relationships were found between the environmental Cd and Zn concentrations and the liver or gill metal content (results not shown).

The mean condition factor K at the different sites did not reflect the metal gradient in the environment (Fig. 23.4(a)). Although the lowest mean K was measured at the sites with the highest Cd and Zn concentrations (sites 1 and 2), a low K was measured at site 7 with a rather low metal contamination. When the individual K was related to

Table 23.1 Species composition, density and biomass at the different sampling sites

Species	St 1	St 2	St 3	St 4	St 5	St 6	St 7	St 8	St 9	St 10
Anguilla anguilla (L.)		2				5	6	1	3	2
Umbra pygmaea (De Kay)			1	7						
Lepomis gibbosus (L.)	1	3	3	7	5	3	22	5	2	
Abramis brama (L.)			1							1
Blicca bjoerkna (L.)			1						1	8
Carassius auratus gibelio (Bloch)						2	2	1	2	1
Cyprinus carpio L.		1								1
Gobio gobio (L.)	3	26	30	13	16	39	9	31	32	4
Leuciscus idus (L.)		3				1			1	1
Leuciscus cephalus (L.)						1		1		
Pseudorasbora parva (Schlegel)	7	1	1			6				
Rutilus rutilus (L.)	10	8	94	31	35	103	19	19	49	22
Scardinius erythrophtalmus (L.)	6	4	2						1	2
Tinca tinca (L.)						1				
Barbatula barbatula (L.)			15	10	4		2	1		
Ictalurus melas (Rafinesque)	2	2		2		10	2		2	
Perca fluviatilis L.	8	4	6	2	2	2	31	50	13	8
Gasterosteus aculeatus L.	1	12	38	20	5		7		1	3
Pungitius pungitius (L.)	3	3	15	4	5		1	1		
Number of species	9	12	12	9	8	11	10	9	11	11
Total biomass (g)	407	4299	1873	652	516	4420	2900	3086	2786	1682

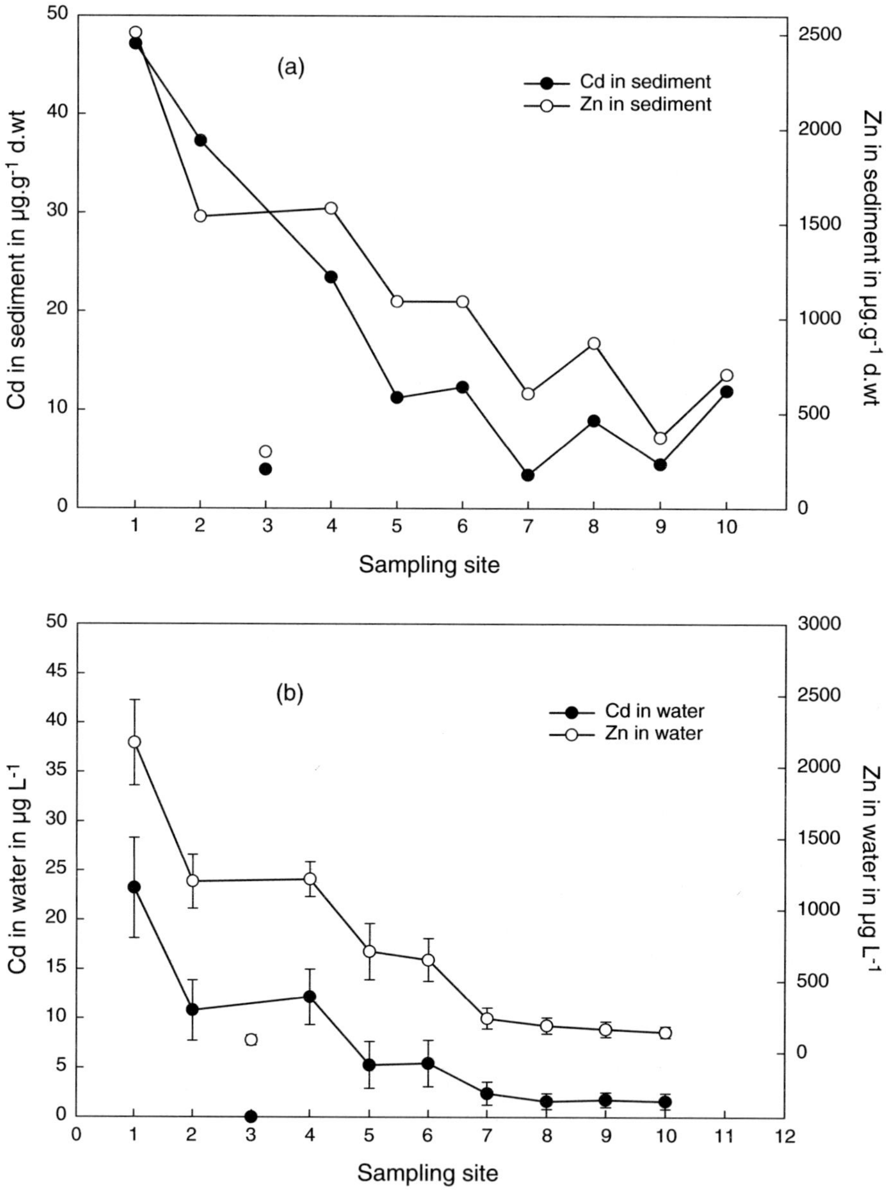

Figure 23.3 Tissue levels of zinc (a) and cadmium (b) at the different sampling sites, together with metal concentrations in the water

ML in the tissues, a significant inverse relationship was measured for liver and muscle ($P < 0.001$), but this only described 18% and 15% of the variation, respectively. Figure 23.4(b) shows the relationship between K and ML in the liver. For the other tissues no significant relationship was found.

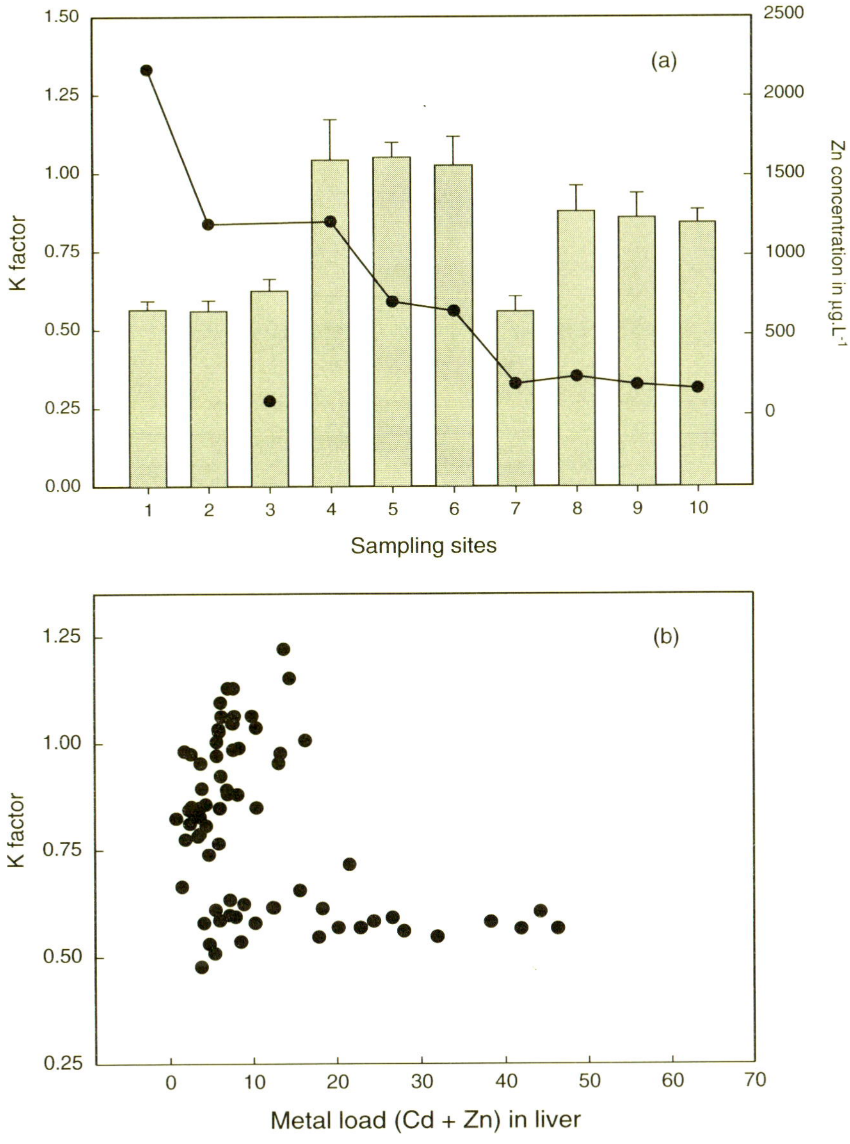

Figure 23.4 (a) Condition factor *K* (columns), mean ± SD, at the different sampling sites, together with concentrations of zinc in the water (dots). (b) Individual condition factors *K v.* individual MLs in the liver

The hepatic (Cd, Zn)-MTLP concentration followed the zinc concentration in the water (Fig. 23.5(a)). A similar relationship was found for Cd (not shown). For the (Cd, Zn)-MTLP contents in the kidney, no significant differences between the sampling sites were found. For the (Cd, Zn)-MTLP in the gills, negative correlations between the metal concentrations in the environment and in the tissue were observed. The hepatic (Cd, Zn)-MTLP concentration was also related to the total metal concentration in the

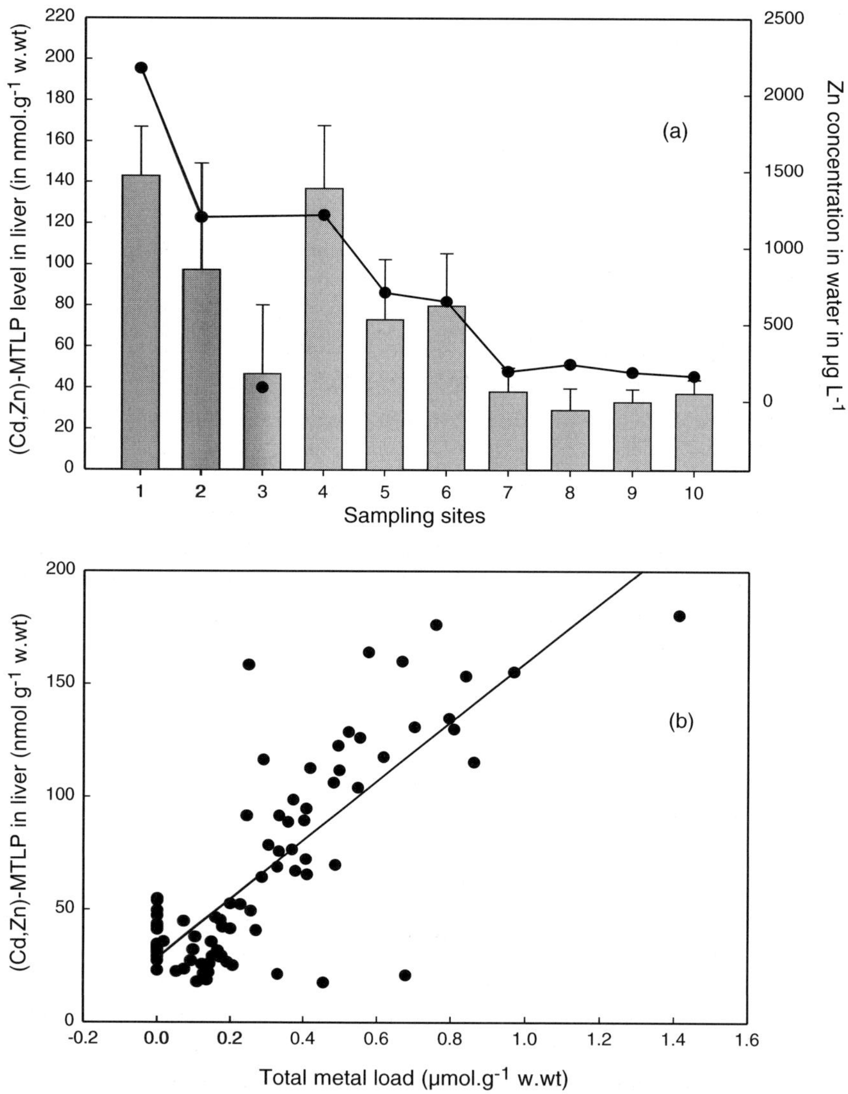

Figure 23.5 Metallothionein-like proteins (MTLP) in the liver: (a) mean values (±SD) at the different sampling sites, together with zinc concentration in the water; (b) individual MTLP levels in the liver related to metal concentration in the liver

liver (Fig. 23.5(b)), describing 68% of the variation. For the other tissues no significant relationship was found.

23.4 Discussion

Concentrations of metals in the water were very high compared to Belgian and European water quality objectives (1 μg L^{-1} Cd; 200 μg L^{-1} Zn) (Vlaamse Regering 2000). Sediment levels were also very high compared to background levels measured in Flanders (De Cooman *et al.* 1998). Similarly, compared to the literature, levels of both metals in organs were high to very high at sites 1–4 (Cheuvreuil, Carru, Chesterikoff, Boët, Tales & Allardi 1995; Allen-Gil, Gubala, Landers, Lasorsa, Crecelius & Curtis 1997; Farag, Woodward, Goldstein, Brumbauch & Meyer 1998).

Despite the very high environmental concentrations of Cd and Zn, up to 12 fish species were captured at mean Cd and Zn levels of 10.8 and 1197 $\mu g\,L^{-1}$, respectively. Species generally considered to be relatively sensitive to water pollution, e.g. stone loach, *Barbatula barbatula* (L.), ide, *Leuciscus idus* (L.), and chub, *Leuciscus cephalus* (L.), were also found at the highly polluted sites. The absence of any relationship between metal pollution and fish community structure might be due to differences in sampling methods and/or differences in structure or type of the river transect, which also influence fish density, species composition and biomass (Huet 1962).

The hepatic (Cd, Zn)-MTLP concentration in the liver varied according to the environmental cadmium and zinc concentrations. It is known, however, that MTLP induction is not mediated through the external metal concentration, but through the internal metal tissue concentration (Hogstrand, Lothner & Haux 1991). It is therefore important to study the relationship between the metal tissue concentrations and the MTLP contents in the different tissues, and a combination of Zn and Cd in the liver described most of the variability of the hepatic (Cd, Zn)-MTLP level. Similar studies by Olsson and Haux (1986) showed a close relationship between the Cd concentration and MT level in the liver of perch, *Perca fluviatilis* L., caught in a cadmium-polluted river, and Hogstrand *et al.* (1991) also found a relationship between the hepatic copper and zinc contents and the liver MT levels in perch. However the latter found (Cd, Zn)-MTLP levels in the gills were negatively correlated with the Cd and Zn concentrations in the gills. This could arise because in the gills the metals might be bound to non-MTLP metal binding proteins (Olsson, Larrson, Maage, Haux & Bonham 1989). Also in the present study the absence of a relationship in the gills could be due to binding to other metal binding proteins.

No evidence of the high metal concentration in either the environment (water and sediment) or fish tissues affecting fish condition was found. This poor relationship between *K* and ML was because environmental factors other than metal pollution, such as habitat quality and food availability, affect condition (Lobon-Cervia, Montaes & De Sostoa 1991). Since the *K* index is not only affected by (metal) pollution but also by other factors, more sensitive endpoints have to be measured. In future experiments physiological condition (respiration and swimming capacity) of gudgeon in the pollution gradient will be measured.

Heavy metal contamination undoubtedly can be an important threat to fish communities. However, from the present study it was not possible to assign environmental levels that negatively affected the present fish communities. Only at the site with the highest metal levels were the number of individuals and the total biomass reduced. Moreover species like eel and stone loach were absent. In gudgeon, metal concentrations in liver of more than 20 times the background levels seemed to have a negative effect on fish condition.

However, to have a better idea of the effects of metal pollution on fish populations and on fish communities diversity, more information is needed on population structure and on acclimation and/or adaptation processes in fish to metal stress.

Acknowledgements

We thank R. Smolders for making the map with the location of the sampling sites. L. Bervoets is a Postdoctoral Fellow and R. Blust is a Research Associate of the Fund for Scientific Research – Flanders (FWO-Vlaanderen).

References

Allen-Gil S.M., Gubala C.P., Landers D.H., Lasorsa B.K., Crecelius E.A. & Curtis L.R. (1997) Heavy metal accumulation in sediment and freshwater fish in US Arctic lakes. *Environmental Toxicology and Chemistry* **16**, 733–774.

Bagenal T.B. (1978) Age and growth. In T.B. Bagenal (ed.) *Methods for Assessment of Fish Production in Fresh Waters*. IBP Handbook No. 3, second edition. Oxford: Blackwell Scientific Publications, pp. 101–136.

Bartsch R., Klein D. & Summer K.H. (1989) The Cd-chelex assay: a new sensitive method to determine metallothionein containing zinc and cadmium. *Archives of Toxicology* **64**, 177–180.

Bendell-Young L.I., Harvey H.H. & Young J.F. (1986) Accumulation of cadmium by white suckers (*Catostomus commersoni*) in relation to fish growth and lake acidification. *Canadian Journal of Fisheries and Aquatic Sciences* **43**, 806–811.

Bervoets L., Blust R., De Wit M. & Verheyen R. (1997) Relationships between river sediment characteristics and trace metal concentrations in tubificid worms and chironomid larvae. *Environmental Pollution* **95**, 345–356.

Blust R., Van Der Linden A., Verheyen E. & Decleir W. (1988) Evaluation of microwave heating digestion and graphite furnace atomic absorption spectrometry with continuum source background correction for the determination of iron, copper and cadmium in brine shrimp. *Journal of Analytical Atomic Spectrometry* **3**, 387–393.

Bryan M.D., Atchison G.J. & Sandheinrich M.B. (1995) Effects of cadmium on the foraging behavior and growth of juvenile bluegill, *Lepomis macrochirus*. *Canadian Journal of Fisheries and Aquatic Sciences* **52**, 1630–1638.

Cheuvreuil M., Carru A.M., Chesterikoff A., Boët P., Tales E. & Allardi J. (1995) Contamination of fish from different areas of the river Seine (France) by organic (PCB and pesticides) and metallic (Cd, Cr, Fe, Mn, Pb, and Zn) micropollutants. *Science of the Total Environment* **162**, 31–42.

De Conto-Cinier C., Petit-Ramel M., Faure R. & Garin D. (1997) Cadmium bioaccumulation in carp (*Cyprinus carpio*) tissues during long-term exposure: analysis by inductively coupled plasma mass spectrometry. *Ecotoxicology and Environmental Safety* **38**, 137–143.

De Cooman W., Florus M. & Devroede M.P. (1998) Karakterisatie van de bodems van de Vlaamse onbevaarbare waterlopen. Report of the Ministry AMINAL, 56 pp. (in Dutch).

Farag A.M., Woodward D.F., Goldstein W., Brumbauch W. & Meyer J.S. (1998) Concentrations of metals associated with mining waste in sediments, biofilm, benthic microinvertebrates, and fish from the Coeur d'Alene River Basin, Idaho. *Archives of Environmental Contamination and Toxicology* **34**, 119–127.

Flammarion P. & Garric J. (1997) Cyprinid EROD activities in low contaminated rivers: A relevant statistical approach to estimate reference levels for EROD biomarker? *Chemosphere* **35**, 2375–2388.

Geuzens P., Nouwen J., Cornelis C., Craps R., De Vlieger I., Mensink C., Schoeters G., Witters H., Goyvaerts M.P., Weltens R., Van Thuyne G., Belpaire C., Guns M. & Brouwers J. (1999) Verspreiding van zware metalen. Garant: Leuven. MIRA-T 1999, pp. 85–109 (in Dutch).

Hogstrand C., Lothner G. & Haux C. (1991) The importance of metallothionein for the accumulation of copper, zinc and cadmium in environmentally exposed perch, *Perca fluviatilis*. *Pharmacology and Toxicology* **68**, 492–501.

Huet M. (1962) Influence du courant sur la distribution des poissons dans les eaux courantes. *Revue suisse d'Hydrologie* **24**, 412–432 (in French).

Kito H., Ose Y., Mizuhira V., Sato T., Ishikawa T. & Tazawa T. (1982) Separation and purification of (Cd, Cu, Zn)-metallothionein in carp hepatopancraes. *Comparative Biochemistry and Physiology* **73C**, 121–127.

Lobon-Cervia J., Montaes C. & De Sostoa A. (1991) Influence of environment upon the life history of gudgeon, *Gobio gobio* (L.): a recent and successful colonizer of the Iberian Peninsula. *Journal of Fish Biology* **39**, 285–300.

Meregalli G., Vermeulen A.C. & Ollevier F. (2000) The use of chironomid deformation in an in situ test for sediment toxicity. *Ecotoxicology and Environmental Safety* **47**, 231–238.

Miliou H., Zaboukas N. & Moraitou-Apostolopoulou M. (1998) Biochemical composition, growth, and survival of the guppy, *Poecilia reticulata*, during chronic sublethal exposure to cadmium. *Archives of Environmental Contamination and Toxicology* **35**(1), 58–63.

Muto N., Ren H.W., Hwang G.S., Tominaga S., Itoh N. & Tanaka K. (1999) Induction of two major isoforms of metallothionein in crucian carp (*Carassius cuvieri*) by air-pumping stress, dexamethasone, and metals. *Comparative Biochemistry and Physiology* **122C**, 75–82.

Olsson P.E. & Haux C. (1986) Increased hepatic metallothionein content correlates to cadmium accumulation in environmentally exposed perch (*Perca fluviatilis*). *Aquatic Toxicology* **9**, 231–242.

Olsson P.E., Larsson A., Maage A., Haux C. & Bonham K. (1989) Induction of metallothionein synthesis in rainbow trout, *Salmo gairdneri*, during long-term exposure to waterborne cadmium. *Fish Physiology and Biochemistry* **6**, 221–229.

Stott B., Eldson J.W.V. & Johnston J.A.A. (1963) Homing behaviour in gudgeon (*Gobio gobio* L.). *Animal Behaviour* **11**, 93–96.

Tessier A., Campbell P.G.C. & Bisson M. (1979) Sequential extraction procedure for the speciation of particulate trace metals. *Analytical Chemistry* **51**, 844–850.

Vandelannoote A., Yseboodt R., Bruylants B., Verheyen R., Coeck J., Maes J., Belpaire C., Van Thuyne G., Denayer B., Beyens J., De Charleroy D. & Vandenabeele P. (1998) Atlas van de Vlaamse beek-en riviervissen. Antwerp: WEL, 303 pp. (in Dutch).

Van Liefferinge C., De Cooman W., Yseboodt R., Bervoets L., Schneiders A. & Verheyen R. (2000) Onderzoek naar het effect van de zoutlozingen van Tessenderlo Chemie op de Grote Nete. Visstandsonderzoek, Waterbodemkwaliteit & Waterkwaliteitsanalyse. Antwerp: Repport UIAntwerp 117 pp. (in Dutch).

Vlaamse Regering (2000) Besluit van de Vlaamse Regering van 1 Juni 1995 houdende vaststelling van het Vlaamse Reglement betreffende de milieuvergunning (Vlarem), zoals gewijzigd bij besluit van 17 Juli 2000. Belgisch Staatsblad, 5 Augustus 2000 (in Dutch).

Yseboodt R., Clement L., Meire P. & Verheyen R.F. (1999) Vergelijking van de waterkwaliteit van de bekkens van de Kleine en de Grote Nete in de periodes 1995–96 en 1997–98. Antwerp: Repport UIAntwerp, 35 pp. (in Dutch).

Chapter 24
A screening of roach, *Rutilus rutilus*, populations for oestrogenic effects in Flanders, Belgium

K. VAN CAMPENHOUT,* W.M. DE COEN
Laboratory for Ecophysiology, Biochemistry and Ecotoxicology, University of Antwerp, Belgium

G. VANDENBERGH
Department of Applied Ecology and Environmental Biology, Ghent University, Belgium

B. DENAYER
Flemish Institute of Forestry and Game Management (IBW), Belgium

C.R. JANSSEN
Department of Applied Ecology and Environmental Biology, Ghent University, Belgium

Abstract

Oestrogenic substances, man-made chemicals that mimic female hormones, may interact with the reproductive function of male fish inducing the synthesis of the female protein vitellogenin (Vtg). Hence Vtg has proven to be a useful indicator of oestrogen exposure in male fish. A field monitoring campaign investigated the occurrence of oestrogenic effects in male roach, *Rutilus rutilus*, populations in Flanders, Belgium. Samples were collected from a number of canals between September and November 1998, to determine condition factor, gonadosomatic index (GSI) and Vtg content in blood. The Vtg content was compared to male roach exposed to EE2 in controlled laboratory conditions using SDS gel electrophoresis. Based on this comparison, one roach population with a markedly higher GSI and blood Vtg content was identified (Stekene canal), strongly suggesting exposure to endocrine disruptors.

Keywords: endocrine disruption, gonadosomatic index, roach, vitellogenin content.

24.1 Introduction

Extensive industrialisation and increased population density have resulted in the release of many man-made chemicals, with a broad diversity of chemical structures, into the environment. A large number of these chemicals are capable of disrupting the endocrine system of animals, including fish (e.g. Bortone & Davis 1994; Arcand-Hoy & Benson 1998). Several of these substances mimic or act like oestrogens, such as synthetic steroids, pesticides, polychlorinated biphenyls (PCBs), surfactants, and plasticisers (Routledge, Sheehan, Desbrow, Brighty, Waldock & Sumpter 1998).

*Correspondence: Department of Biology, University of Antwerp (RUCA), Groenenborgerlaan 171, B-2020 Antwerp, Belgium (email: Kavacaho@ruca.ua.ac.be).

One of the most commonly used biomarkers for oestrogen exposure is vitellogenin (Vtg). This large phospholipoprotein (170–200 kDa) is normally produced only in the liver of female fish as a precursor of egg yolk. Due to the presence of exogenous oestrogens, however, male fish can produce Vtg in quantities approaching those of mature females (Tyler, van der Eerden, Jobling, Panter & Sumpter 1996; Tyler, Van Aerle, Hutchinson, Maddix & Trip 1999). These pseudo-oestrogens can directly disturb hormonal pathways leading to reduced egg production and decreased fertility in females and reduced gonad size and feminisation of genetic male fish (Jobling, Sheahan, Osborne, Matthiessen & Sumpter 1996). The ecological significance of elevated Vtg levels in the blood of male fish, however, remains unclear. Recently it was demonstrated that exposure to oestrogenic substances present in sewage effluents entering rivers can cause reproductive and developmental effects in wild roach populations (Rodgers-Gray, Jobling, Kelly, Morris, Brighty, Waldock, Sumpter & Tyler 2001). This can influence the survival of the fish populations, thus affecting biodiversity. This chapter describes an investigation into the occurrence of endocrine disruption in male roach, *Rutilus rutilus* (L.) in Flanders. This fish species is potentially useful for investigating the effects of oestrogenic chemicals in the aquatic environments, since it is of small size, common and widely distributed.

24.2 Materials and methods

24.2.1 *Sampling procedure*

Sexually mature male roach were sampled between September and November 1998 in canals in three different hydrological basins:

(1) the Ghent–Ostend canal representative of the basin of Ghent and the polders of Bruges;
(2) the Roeselare–Leie canal representative of the basin of the Scheldt River; and
(3) the Stekene canal, also representative of the canals of Ghent.

Both the Ghent–Ostend and Roeselare–Leie canals receive domestic and industrial discharges; whereas the Stekene canal is mainly affected by domestic discharges. An isolated part of the River Leie at Bavikhove was selected as a non-polluted control-sampling site. Fish were sampled by electric fishing, traps and dragnets.

Roach was selected as the test species because of its widespread distribution, easy maintenance in the laboratory and ease of sampling in the field. After anaesthesia with 2-phenoxyethanol (Sigma) all fish were weighed (nearest g) and measured (nearest mm). Blood samples were taken from the caudal sinus and the gonads were removed and kept in a formol solution (for 1 L solution: 4.5 g NaH_2PO_4 (Merck), 16.4 g Na_2HPO_4 (Merck) in 100 mL 10% formol (Merck) and filled up to 1 L). The blood samples were immediately frozen and kept at −80°C until analysed.

In addition, a laboratory experiment was carried out as a positive and negative control situation for oestrogenic effects. A test population of 50 male roach, bred in a local fish farm, was exposed to a 0.1 mg L^{-1} 17α-ethinylestradiol (EE2) solution as a positive control using semi-static exposure conditions. The negative control population (50 male roach) was kept in semi-static exposure conditions using a 0.01% ethanol solution. After 3, 6, 10 and 15 days, 10 roach from each population were killed and blood samples were taken.

24.2.2 *Gel electrophoresis*

Protein determination was performed following Bradford (1976), using bovine serum albumin as a standard. Blood samples of male roach of the different sites were analysed by one-dimensional electrophoresis, according to Laemmli (1970), on 12% SDS polyacrylamide gels, using a BioRad minisystem. Diluted blood samples were mixed in a 1 : 3 volume ratio with sample buffer (5% w/v 2-mercaptoethanol, 2% SDS, 62.5 mM Tris-HCl, pH 6.8, 25% glycerol, 0.01% bromophenol blue) and heated for 5 min in a boiling water bath. The standard was a SDS-PAGE broad range standard by BioRad (MW 21–200 kDa). Samples were run at 200 V at room temperature until the dye front reached the bottom of the gel. Gels were stained with Coomassie brilliant blue R-250.

24.2.3 *Data analysis*

A new and robust methodology was developed to allow comparative statistical analysis within and across the stained gels (K. Van Campenhout, unpublished). The method allows the ranking of gel lanes according to the density pattern in certain regions of interest, so that meaningful statistical analysis can be applied. Two gel image features were identified, which were called the relative density (RAD) and the contrast ratio (CR). First a region of interest (ROI) was delineated by the 200 and 75 kDa bands, as the molecular weight of cyprinid Vtg is set at 160–170 kDa (Tyler *et al.* 1999). The RAD is the absolute density of the ROI relative to the absolute density of the standard. The absolute densities were derived from 8-bit grey scale gel images provided by an (uncalibrated) flatbed scanner. The CR is the ratio of the total density in the first half of ROI (from approximately 200–100 kDa) to that in the second half (100–75 kDa). The two features, the CR and the RAD, can be combined into a single number using the concept of Fisher's Linear Discriminant (Duda & Hart 1973). This method determines the best straight line separating the two groups – in this case the natural populations of the field study and a positive laboratory control group. The magnitude of distance (D) to the separating line is a measure of how strong the evidence is of belonging to one group. In this case D was used as a semi-quantitative Vtg indicator.

Statistical analysis of the data was performed using the STATISTICA statistical computing package. Normality of data was verified using probability plots, and the homogeneity of variances of the data was checked by the Levene-test. The degree of linear association between the different variables was investigated using Pearson's correlation. Statistical significance was assessed at $P < 0.05$.

24.3 Results

After 6 days of exposure to 0.1 mg L^{-1} EE2, male roach in the laboratory had a significantly higher Vtg content ($P < 0.05$), and a maximum Vtg concentration was reached after 10 days of exposure. The control laboratory population Vtg concentration was insignificant.

Two groups appeared when plotting the natural and the laboratory roach populations in the [RAD, CR]-scatter diagram (Fig. 24.1). The first group consisted of the natural populations, the control laboratory population and the day-3 test population, and was considered as a 'non-vitellogenic' group. The second group consisted of the remaining positive test population (6, 10, 15 days of exposure) of the laboratory, reflecting the significantly higher Vtg blood contents after 6, 10 or 15 days of exposure to 0.1 mg L^{-1} EE2 in comparison with the negative control laboratory group. Based on the two groups a discriminant line was calculated, which served as a reference to calculate the relative Vtg content of the individual blood samples. The amount of Vtg in the blood was expressed as the perpendicular distance towards the discriminant line. Data points below the discriminant line indicated high Vtg levels (D = positive), whereas data points above the separating line refer to low Vtg levels (D = negative). Given that a higher negative distance relative to the discriminant line corresponds with a lower Vtg concentration, a lower mean Vtg content was found in the blood of the roach population of Ghent–Ostend (Fig. 24.1). The Stekene canal roach population contained the

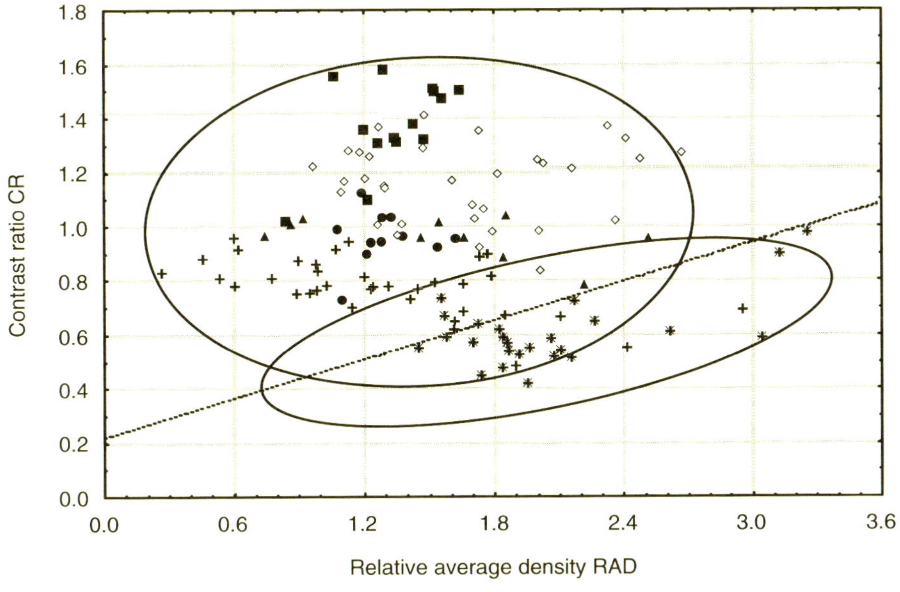

Figure 24.1 Non-vitellogenic and vitellogenic groups formed in the [RAD, CR]-scatter diagram and their optimally separating linear discriminant line. Circles, Bavikhove; squares, Ghent–Ostend; diamonds, Roeselare–Leie; triangles, Stekene; stars, positive control test population; crosses, negative control test and the day-3 test population

highest amount of Vtg in the blood in comparison with other natural populations. The Vtg levels of the natural populations will be lower than those of the strongly exposed (to $0.1\,mg\,L^{-1}$ EE2) laboratory populations, as in the aquatic environment only $ng\,L^{-1}$-levels of EE2 were found (Halling-Sorensen, Nors Nielsen, Lanzky, Ingerslev, Holten Lützhoft & Jorgensen 1998). Nevertheless, during the analysis some individuals showed a higher density of the Vtg band on the electrophoresis gel (Fig. 24.2). These samples were hard to distinguish (even visually) from some of the 6-day exposed laboratory fish, resulting in a small distance to the separating line. The relationships between the different biomarkers: condition factor (CF), gonadosomatic

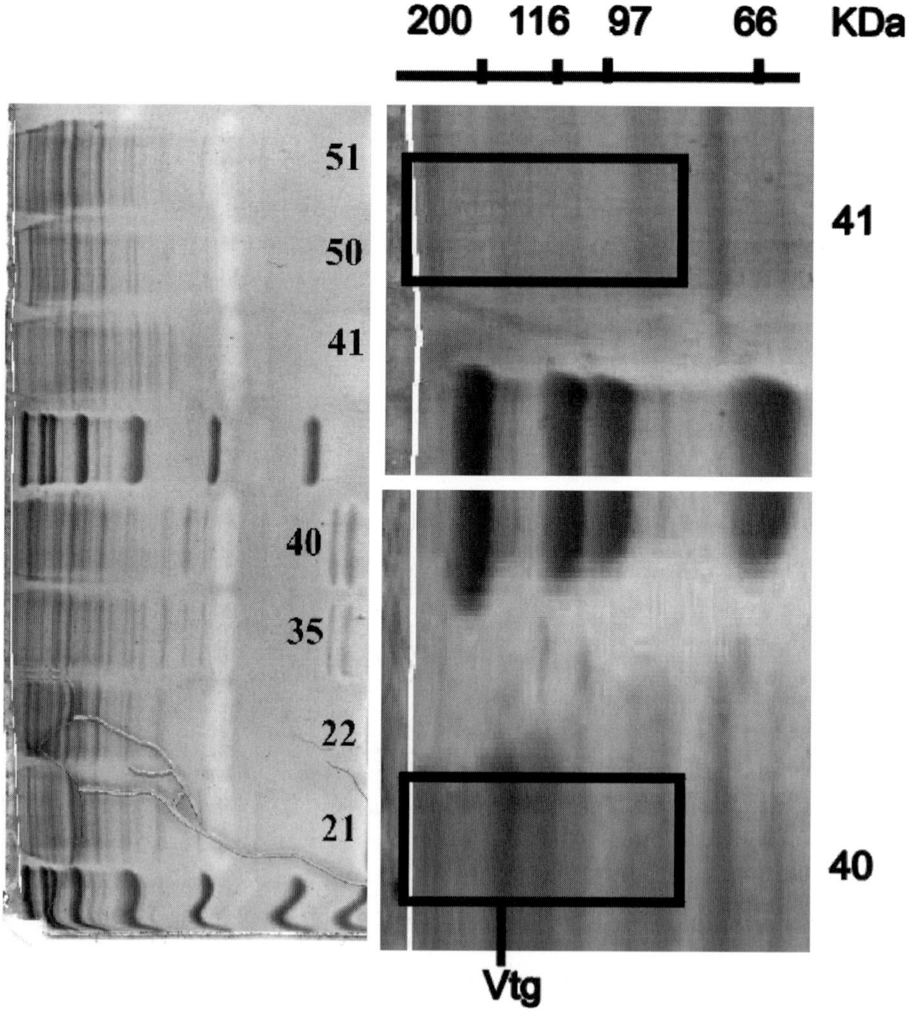

Figure 24.2 Electrophoresis gel showing the protein fractions (including Vtg) in blood from male roach from the Stekene canal. Lane 40, fish with signs of a higher blood Vtg concentration; 41, fish without signs of a higher blood Vtg content

index (GSI) and blood Vtg content (D) for the studied roach populations showed a weak but significant negative correlation ($r = -0.36$, $P < 0.001$) between CF and GSI (Fig. 24.3(a)). A stronger correlation ($r = -0.63$, $P < 0.001$) was found between CF and D (Fig. 24.3(b)); i.e. lower values of D (the higher the Vtg content) are associated with poorer condition of roach. Only a weak significant correlation was found between GSI and D ($r = 0.26$, $P < 0.05$) (Fig. 24.3(c)).

24.4 Discussion

This study provides the first data on oestrogenic effects in resident male roach populations in Flanders. Weak, but significant, correlations between the general condition, testicular weight or blood Vtg content of male roach of the different populations were found. The GSI values of the different natural populations ranged from 1.93 to 2.6, except of the population of Stekene, which had a GSI value of 4.4. Except for the Stekene Canal, the GSI values were within the range of the natural and seasonal GSI fluctuations (1–2.5) reported for male roach caught during the same sampling period in the River Meuse in Belgium (Mattheeuws, Genin, Detollenaere & Micha 1981). The negative relation between the GSI and CF possibly reflects the poor water quality at the sampling sites or other external factors that lead to the loss of total body weight. This change in body weight can be more significant than the loss of testicular weight due to possible endocrine disruption by pseudo-oestrogens. In the latter, the CF value decreases while the GSI remains constant or even rises. Because roach are widely used for restocking programmes, this could also hide possible effects of endocrine disruption, and make it even more difficult to interpret the data. The Stekene canal roach population, however, was different from the other populations because of its higher mean GSI value and its higher mean blood Vtg content, strongly suggesting the presence of endocrine disruptors.

However, in the context of oestrogen-like effects, these results seem to contradict the literature, namely a lower testicular size in combination with a higher blood Vtg content (Jobling *et al.* 1996; Christiansen, Korsgaard & Jespersen 1998). The higher blood Vtg content is probably caused by the presence of oestrogenic chemicals emanating from domestic sources, since the Stekene canal is mainly affected by domestic discharges. The higher testicular weight can be the consequence of several factors. One of them may be the existence in domestic discharges of various types of endocrine disrupting chemicals, which can interact at different levels causing distinct adverse effects. These potentially polluting substances may produce disruption of the reproductive system of the fish that is not mediated through hormonal receptor pathways. Moreover, anti-oestrogenic effects of masculating compounds (e.g. chemicals present in bleached mill effluents, or degradation products from plant sterols) (Bortone & Davis 1994) could interact with the effects of the oestrogenic compounds.

Although some evidence suggests the presence of oestrogenic effects, more work is required before such effects can be confirmed on the canals studied. First, natural variability of the biomarkers, CF, GSI and the blood Vtg content needs examination. In this case fluctuations in these biomarkers due to environmental pollution must be

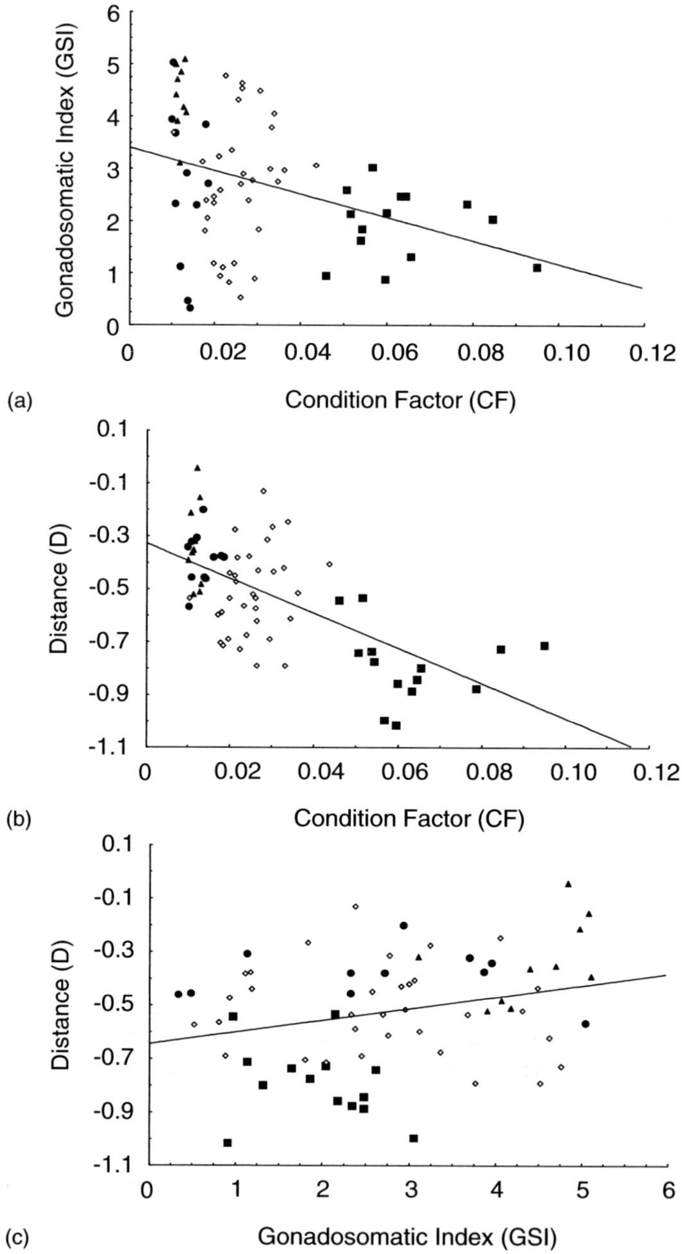

Figure 24.3 (a) Relationship between the GSI and the CF of the roach populations ($r = -0.36$; $n = 68$; $P < 0.001$). (b) Relationship between the distance (*D*), a semi-quantitative Vtg indicator, and the CF of the roach populations ($r = -0.63$; $n = 68$; $P < 0.001$). The more negative the distance the less Vtg is present in the blood. (c) Relationship between the distance (*D*), a semi-quantitative Vtg indicator, and the GSI of the roach populations ($r = 0.26$; $n = 68$; $P < 0.05$). Circles, Bavikhove; squares, Gent–Ostend; diamonds, Roeselare–Leie; triangles, Stekene

distinguished from changes caused by natural (seasonal and ontogenic) variations. Secondly, concentrations of suspected endocrine disruptors should be quantified in the environment and relevant aquatic species. Finally, within the different canals studied the effects of these types of contaminants should be studied at the population level.

From an ecological perspective, the impact of exposure to endocrine disrupting chemicals remains unclear (Arcand-Hoy & Benson 1998). Some studies revealed a reduced fecundity in combination with a lower CF and an impaired cortisol and thyroid hormone level. Other field studies found no effect on fecundity and offspring survival, despite alterations in hormone levels and egg size (Munkittrick, Van der Kraak, McMaster & Portt 1992). Further understanding of the factors influencing the endocrine system in species currently used as wildlife sentinels (e.g. roach) in both laboratory and field studies, are necessary. Finally, the relevance of endocrine disrupting substances for the future conservation of fish species, whether common or endangered, needs assessment.

Acknowledgements

The authors are indebted to Gertie Deneve, Kathy Haustraete and Tim Verslycke for their help in fish sampling.

References

Arcand-Hoy L.D. & Benson W.H. (1998) Fish reproduction: An ecologically relevant indicator of endocrine disruption. *Environmental Toxicology and Chemistry* **17**, 49–57.

Bortone S.A. & Davis W.P. (1994) Fish intersexuality as indicator of environmental stress. *BioScience* **44**, 165–172.

Bradford M. (1976) A rapid and sensitive method for the quantification of microgram quantities of protein utilizing the principle of protein-dye binding. *Analytical Biochemistry* **72**, 248–254.

Christiansen T., Korsgaard B. & Jespersen A. (1998) Effects of nonylphenol and 17β-oestradiol on vitellogenin synthesis, testicular structure and cytology in male eelpout *Zoarces viviparus*. *The Journal of Experimental Biology* **201**, 179–192.

Duda R. & Hart P. (eds) (1973) *Pattern Classification and Science Analysis*. New York: John Wiley & Sons, 482 pp.

Halling-Sorensen B., Nors Nielsen S., Lanzky P.F., Ingerslev F., Holten Lützhoft H.C. & Jorgensen S.E. (1998) Occurrence, fate and effects of pharmaceutical substances in the environment – a review. *Chemosphere* **36**, 357–393.

Jobling S., Sheahan D., Osborne J.A., Matthiessen P. & Sumpter J.P. (1996) Inhibition of testicular growth in rainbow trout (*Oncorhynchus mykiss*) exposed to estrogenic alkylphenolic chemicals. *Environmental Toxicology and Chemistry* **15**, 194–202.

Laemmli U. (1970) Cleavage of structural proteins during assembly of the head of Bacteriophage. *Nature* **277**, 680–685.

Mattheeuws A., Genin M., Detollenaere A. & Micha J.-C. (1981) Etude de la reproduction du gardon (*Rutilus rutilus*) et des effects d'une élévation provoquée de la température en Meuse sur cette reproduction. *Hydrobiologia* **85**, 271–282 (in French).

Munkittrick K., Van Der Kraak G., McMaster M. & Portt C. (1992) Response of hepatic MFO activity and plasma sex steroids to secondary treatment of bleached kraft pulp mill effluent and mill shutdown. *Environmental Toxicology and Chemistry* **11**, 1427–1439.

Rodgers-Gray T., Jobling S., Kelly C., Morris S., Brighty G., Waldock M., Sumpter J. & Tyler C. (2001) Exposure of juvenile roach (*Rutilus rutilus*) to treated sewage effluent induces dose-dependent and persistent disruption in gonadal duct development. *Environmental Science and Technology* **35**, 462–470.

Routledge E., Sheehan D., Desbrow C., Brighty G., Waldock M. & Sumpter J. (1998) Identification of estrogenic chemicals in STW effluent. II. Biological activities. *Environmental Science and Technology* **32**, 1559–1565.

Tyler C., Van Aerle R., Hutchinson T., Maddix S. & Trip H. (1999) An *in-vivo* testing system for endocrine disruptors in fish early life stages using induction of vitellogenin. *Environmental Toxicology and Chemistry* **18**, 337–347.

Tyler C., van der Eerden B., Jobling S., Panter G. & Sumpter J. (1996) Measurement of vitellogenin, a biomarker for exposure to oestrogenic chemicals, in a wide variety of cyprinid fish. *Journal of Comparative Physiology (B)* **166**, 418–426.

Chapter 25
Species introductions to two English lake fish communities of high conservation value: a 10-year study

I.J. WINFIELD,* J.M. FLETCHER and J.B. JAMES
CEH Windermere, Far Sawrey, Ambleside, Cumbria, UK

Abstract

Data gathered between 1991 and 2000 are presented concerning the introductions of roach, *Rutilus rutilus*, ruffe, *Gymnocephalus cernuus*, and dace, *Leuciscus leuciscus*, to the fish communities of the eutrophic Bassenthwaite Lake and mesotrophic Derwent Water in the English Lake District, which are of high conservation value because they contain the last two UK populations of vendace, *Coregonus albula*. Population biology data are presented for each introduced species and for the native perch, *Perca fluviatilis*, populations of both lakes. Possible recent and current species interactions within these two communities are discussed.

Keywords: *Coregonus albula*, *Gymnocephalus cernuus*, *Leuciscus leuciscus*, *Perca fluviatilis*, *Rutilus rutilus*, Bassenthwaite Lake, Derwent Water.

25.1 Introduction

The deliberate or accidental introduction of new fish species is considered to pose one of the most insidious threats to fish conservation around the world (Lowe-McConnell 1990). The fish communities of the Lake District in north-west England are particularly at risk from this threat because they are relatively species-poor when compared with equivalent habitats in the south of the country and elsewhere in Europe (Winfield 1992). Notably, roach, *Rutilus rutilus* (L.), and ruffe, *Gymnocephalus cernuus* (L.), have recently been introduced to a number of water bodies in this area, including Bassenthwaite Lake and Derwent Water which hold the UK's only remaining populations of vendace, *Coregonus albula* (L.) (Winfield, Cragg-Hine, Fletcher & Cubby 1996b). Although these introductions have already been considered in the context of their specific impacts on this nationally rare coregonid (e.g. Winfield, Adams & Fletcher 1996a; Winfield, Fletcher & Cubby 1998), they may also have wider community impacts including interactions with native species such as perch, *Perca fluviatilis* L.

Interactions of roach and ruffe with perch have long been of interest to fundamental and European fisheries ecology, but the introduction of ruffe to North America (Pratt,

*Correspondence: CEH Windermere, Far Sawrey, Ambleside, Cumbria LA22 0LP, UK (email: ijw@ceh.ac.uk).

Blust & Selgeby 1992) with potential impacts on commercially-important fisheries for yellow perch, *Perca flavescans* (Mitchill), has emphasised their applied importance. In their native habitats, competitive interactions between roach and perch (e.g. Persson, 1991), ruffe and perch (e.g. Bergman 1987), and all three species (Bergman & Greenberg 1994) depend on factors including the trophic state of the water body, prey assemblage and temperature. Extensive evidence relating to such interactions has now been amassed from experiments and comparative studies of native habitats, but appropriate time series of data at the community level have not been described for introduced populations.

Bassenthwaite Lake and Derwent Water offer a unique opportunity to collect such data, given that when they were extensively sampled in 1986 and 1987 by Mubamba (1989) the only introduced species recorded was roach in the form of only eight individuals, but by 1991 this species was also found in Derwent Water and in addition a small number of ruffe was found in Bassenthwaite Lake (Winfield *et al.* 1996b). The objectives of the present study were to describe the fish communities and population biology of introduced species and native perch of Bassenthwaite Lake and Derwent Water over the period 1991–2000. Field evidence of potential interactions between native perch and introduced roach and ruffe was a particular focus of interest.

25.2 Materials and methods

25.2.1 *Study sites*

Bassenthwaite Lake (National Grid Reference NY 218 286) and Derwent Water (National Grid Reference NY 264 208) are connected by a 5.5 km length of the River Derwent which flows from the latter into the former. Bassenthwaite Lake (altitude 69 m, surface area 528 ha, maximum and mean depths 19 and 5 m, respectively) has shown a historical increase in trophic state to marked eutrophy in recent decades, although the input of anthropogenic nutrients has very recently declined following an improvement of sewage treatment in 1996. In contrast, Derwent Water (altitude 74 m, surface area 535 ha, maximum and mean depths 22 and 5 m, respectively) has maintained a mesotrophic status. Fishery activities at both lakes are largely limited to non-game, particularly pike, *Esox lucius* L., at Bassenthwaite Lake, and game, particularly brown trout, *Salmo trutta* L., at Derwent Water, recreational angling.

25.2.2 *Survey gill netting and specimen examination*

Sampling, using bottom-set survey gill nets, was undertaken in the autumns of 1997–2000 at both lakes, with an earlier autumn sampling in 1995 at both lakes and in 1996 at Bassenthwaite Lake. On each occasion, one survey gill net was set in the late afternoon at each of five sites at each lake (Fig. 25.1) and lifted the following morning, with all captured fish returned to the laboratory and frozen to await future processing. From 1995 to 1998, the nets used were of multifilament design and each

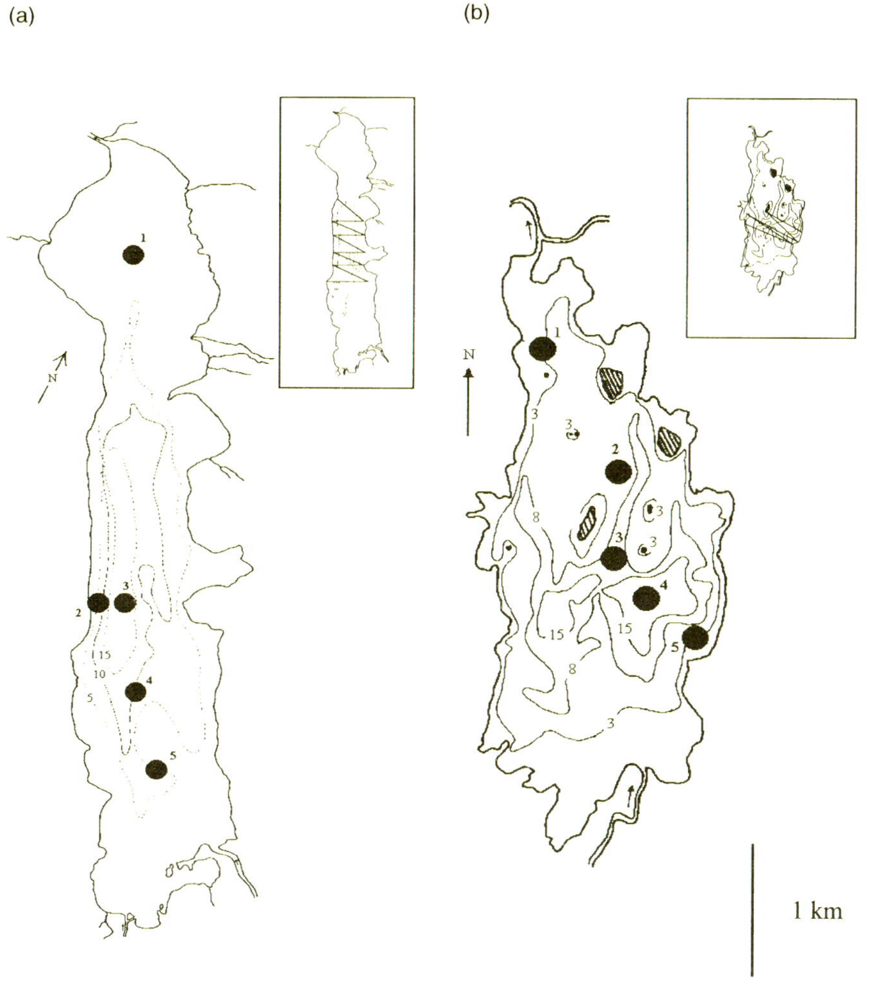

Figure 25.1 Locations of five survey gill net sites (bold numbered closed circles) in (a) Bassenthwaite Lake, and (b) Derwent Water. Echo-sounding transects (continuous lines) are shown on inserts. Depth contours are given in metres. Redrawn with permission from Ramsbottom (1976)

measured 1.5 m deep and 60 m in length with 10 panels of equal length of bar mesh sizes 8, 10, 12, 16, 22, 25, 30, 33, 38 and 43 mm. In 1999 and 2000, monofilament nets were used with the same dimensions and panel number as above, but with mesh sizes of 8, 10, 13, 16, 19, 25, 30, 33, 38 and 45 mm. Sampling dates varied from 13 September to 20 November over the above 6-year period.

All fish were subsequently identified, enumerated, measured (fork length, mm), weighed (total wet, g) and aged by examination of their opercular bones with a nominal birthday of 1 May for all species except vendace. Fish communities were described as composition by numbers, and by the diversity measures of species richness,

Simpson's diversity index and Simpson's equitability index. As von Bertalanffy growth curves could not be fitted to some of the fast growth patterns shown by individual species, a non-linear Gompertz model was fitted to all data sets with at least four age classes present. Such individual data are not yet available for 2000.

In addition, the present study incorporates data from earlier surveys using survey gill nets undertaken from May to September 1991 at two of the above five sites in each lake. Full details for this earlier sampling are given in Winfield *et al.* (1996b).

25.2.3 *Echo sounding*

Echo-sounding surveys were carried out at Bassenthwaite Lake and Derwent Water from 1995 and 1998, respectively, to 2000 using a Simrad EY 200P portable echo sounder with a 200 kHz single beam transducer of beam angle 7°. Data were collected at least 2 h after sunset from a total of ten transects across the deeper areas of Bassenthwaite Lake and five transects across the more limited and less tractable deeper areas of Derwent Water (Fig. 25.1). Surveys were of approximately 1 h duration, and for the area of the lake greater than 10 m in depth gave ratios of coverage (length of surveys : square root of research area) of 4.6 : 1 and 4.0 : 1 for Bassenthwaite Lake and Derwent Water, respectively. Full system, survey and analysis details are given in Winfield, Fletcher and James (2000), although the present study considers only data from September when the fish community was most abundant.

25.3 Results

25.3.1 *Fish communities*

In 1991 the fish community of Bassenthwaite Lake (Fig. 25.2) was dominated by roach (42%), perch (37%) and vendace (16%), with only a relatively few ruffe (2%) and no dace (Fig. 25.2(a)). The latter species was first recorded in 1996, by which time roach had decreased (1%) and ruffe greatly increased (57%). The same situation, with some variation, persisted to 2000 (Fig. 25.2). There were no consistent trends in any of the three diversity indices (Fig. 25.2(b)) and the abundance of the fish community varied between about 560 000 and 1.2 million individuals (Fig. 25.2(c)). In terms of catch per unit effort (CPUE) by numbers, introduced species increased greatly over the study period and have outnumbered native species since 1996 (Fig. 25.2(d)).

In Derwent Water in 1991, the community was dominated by perch (41%), vendace (41%) and brown trout (10%), with only a relatively few roach (3%) and no ruffe or dace (Fig. 25.3(a)). The latter species was first recorded in 1999, by which time roach had greatly increased (68%). There were downward trends in species richness and diversity, but not in equitability (Fig. 25.3(b)), and the abundance of the fish community varied between about 86 000 and 280 000 individuals (Fig. 25.3(c)). As in Bassenthwaite Lake, CPUE of introduced species increased greatly over the study period and was greater than that for native species in 1998 and 1999 (Fig. 25.3(d)).

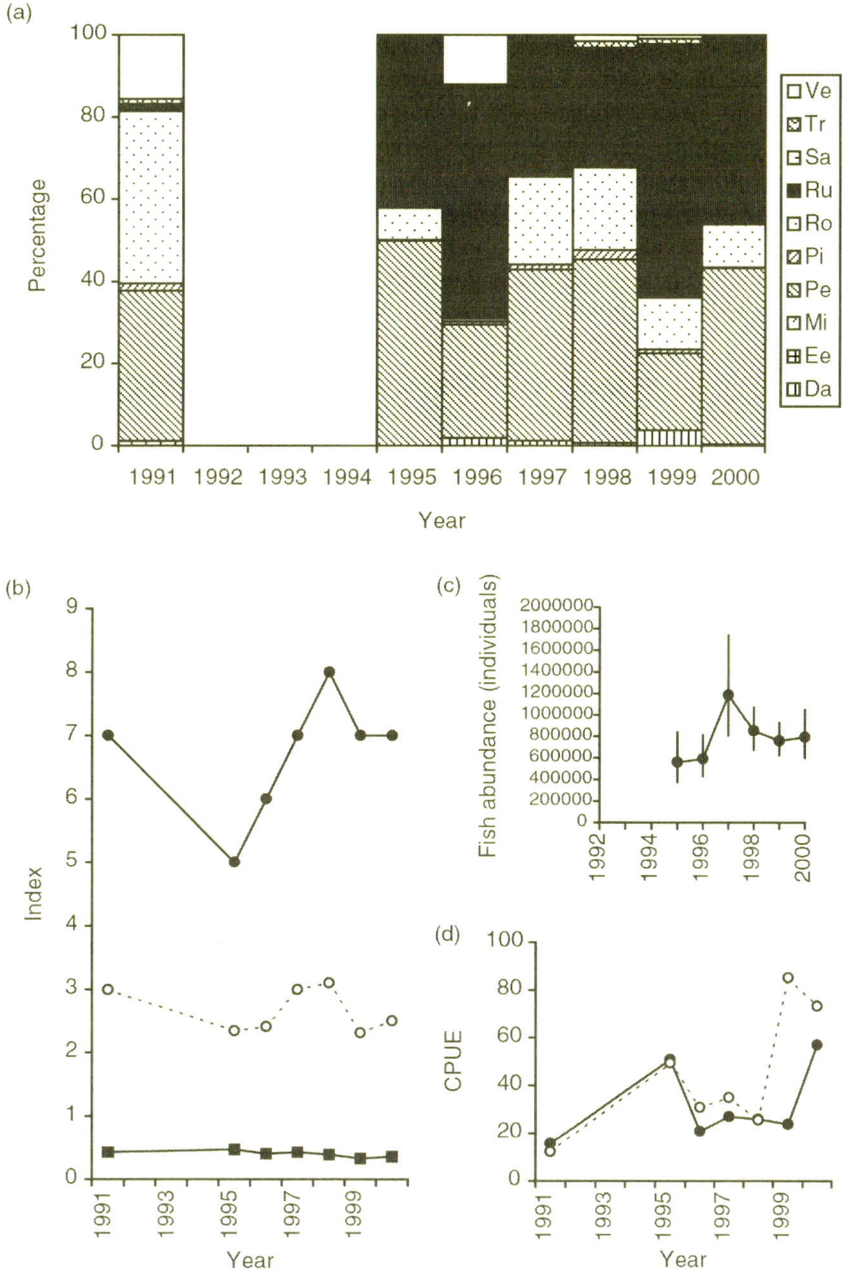

Figure 25.2 Bassenthwaite Lake fish community from the period 1991 to 2000 described as: (a) percentage composition by numbers; (b) species richness (closed circles), diversity index (open circles) and equitability index (closed squares); (c) abundance from echo-sounding as geometric means with 95% confidence limits; and (d) relative abundance from gill netting as CPUE of native (closed circles) and introduced (open circles) species. Abbreviations are as follows: Ve, vendace; Tr, brown trout; Sa, salmon, *Salmo salar* L.; Ru, ruffe; Ro, roach; Pi, pike; Pe, perch; Mi, minnow, *Phoxinus phoxinus* (L.); Ee, eel, *Anguilla anguilla* (L.); Da, dace *Leuciscus leuciscus* (L)

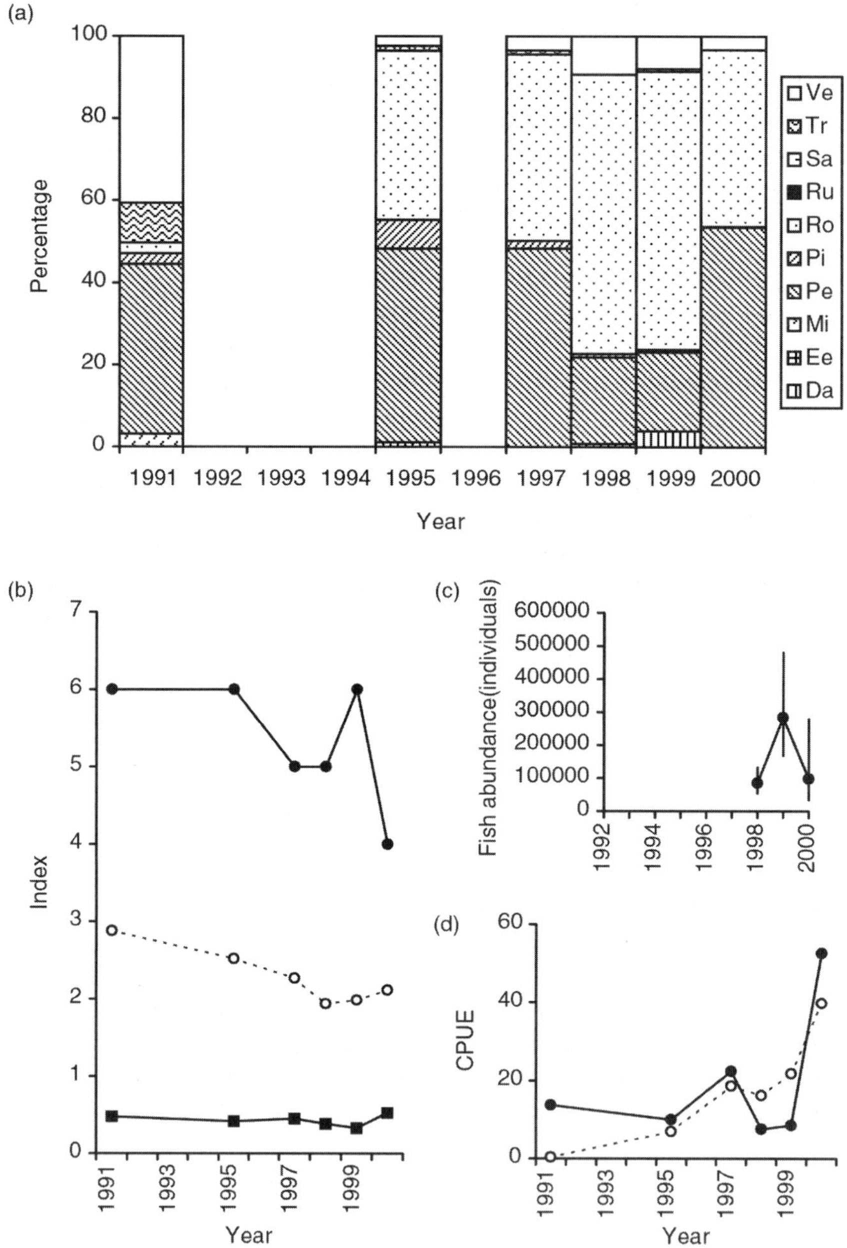

Figure 25.3 Derwent Water fish community from the period 1991 to 2000 described as: (a) percentage composition by numbers; (b) species richness (closed circles), diversity index (open circles) and equitability index (closed squares); (c) abundance from echo-sounding as geometric means with 95% confidence limits; and (d) relative abundance from gill netting as CPUE of native (closed circles) and introduced (open circles) species. Abbreviations are as given in Fig. 25.2

25.3.2 *Population biology*

Growth curves and length frequency distributions for 1991 and 1999 for roach, ruffe, dace (with the exception that initial growth data are from 1996) and perch from Bassenthwaite Lake are given in Figure 25.4. All fits of Gompertz growth models were significant at the 5% level and each explained in excess of 77% of observed variation. Estimated values for L_∞ were 264 and 318 mm for roach and 148 and 462 mm for perch in 1991 and 1999, respectively, while the sole estimate for dace was 440 mm in 1999. Over the study period, the length frequency distributions showed increasingly fewer or no small (young) roach individuals, while in contrast the distribution of perch showed increasing diversity with both small (young) and large (old) individuals well represented. For ruffe, individuals over 140 mm in length were only recorded in 1995.

Corresponding population data for roach, dace (with the exception that no data are available from 1991) and perch from Derwent Water (Fig. 25.5) show that all fits of the Gompertz growth models were significant at the 5% level and each explained in excess of 90% of observed variation. Estimated values for L_∞ for 1999 for roach, dace and perch were 728, 428 and 606 mm, respectively, while the sole estimate for 1991 for perch was 324 mm. Over the study period, the length frequency distributions of the introduced roach rapidly diversified while those of perch were consistently so, reflecting good recruitment and growth.

25.4 Discussion

Within the last 14 years, the fish communities of Bassenthwaite Lake and Derwent Water have become altered since the first records, and subsequent establishments, of roach, ruffe and dace in the former lake in 1986, 1991 and 1996, respectively, and roach and dace in the latter lake in 1991 and 1999, respectively. Although the mode of introduction is unknown in all cases, for reasons discussed by Winfield *et al.* (1996a) the most likely original mechanism is anglers fishing for pike using these species as live-bait. Additional introductions as live-bait may have occurred for roach and dace in Derwent Water, or these species may have migrated into this lake from Bassenthwaite Lake along the connecting River Derwent. Whatever the specific means of introduction, within the last few years both lakes have experienced fish communities numerically dominated by introduced species. While Derwent Water has consequently suffered a decline in overall community diversity, in Bassenthwaite Lake such diversity has been maintained but only because native species have been replaced by introduced species.

The rapid expansion of roach in both lakes was expected given its strong competitive abilities in mesotrophic and eutrophic conditions (Persson 1991), but the reason for its 1990s recruitment failures in Bassenthwaite Lake is less obvious. A negative interaction with the introduced dace population may be hypothesised, but the more recent timing of the latter species' introduction, its still relatively much smaller population size, and the lack of any similar impact on the Derwent Water roach population, argue against this explanation. Similarly, experimental evidence from elsewhere

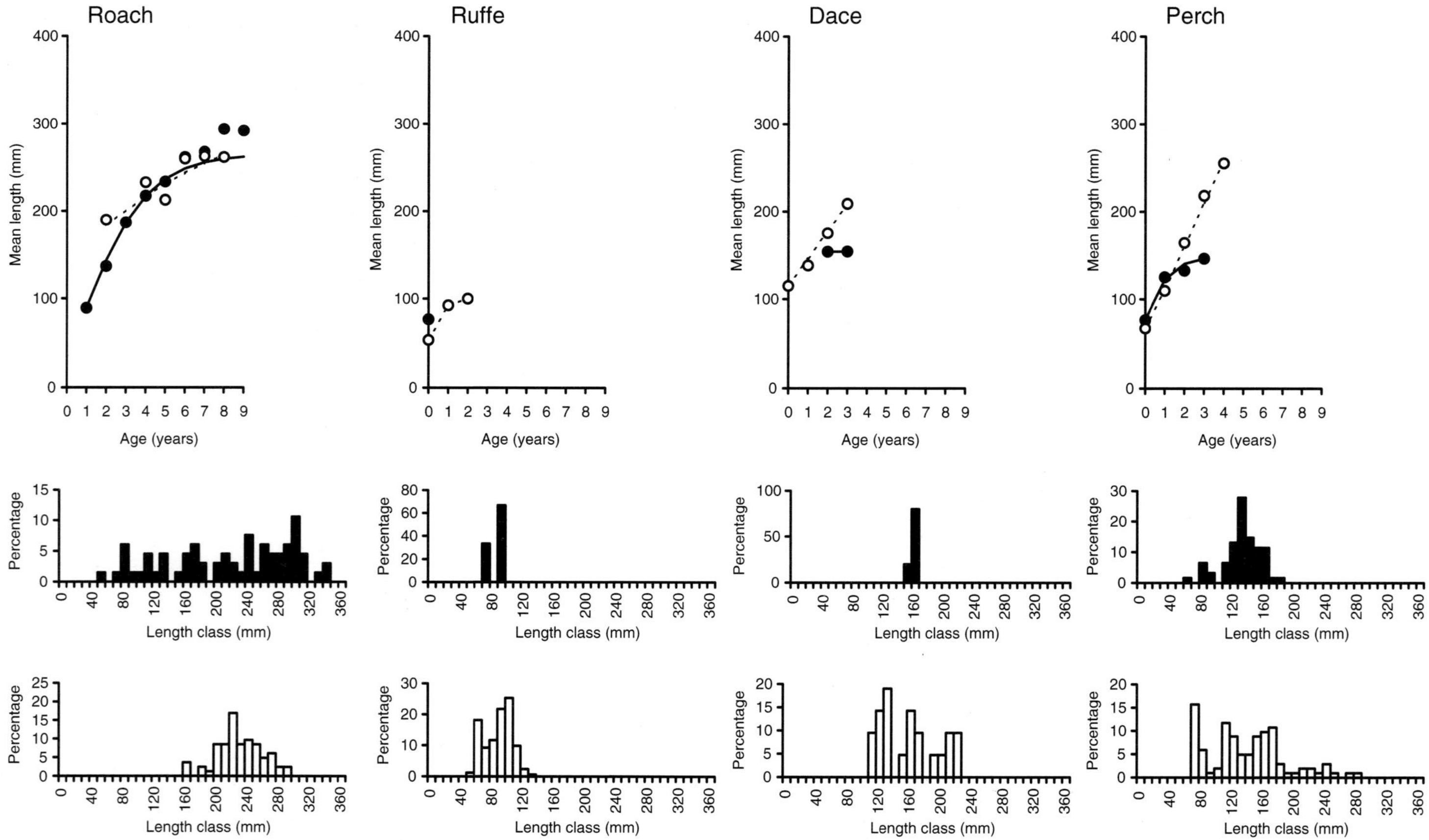

Figure 25.4 Growth curves and length frequency distributions for 1991 (closed circles and columns) and 1999 (open circles and columns) for roach, ruffe, dace (with the exception that initial growth data are from 1996) and perch of Bassenthwaite Lake. Fitted growth curves (continuous and broken lines for earlier and later data, respectively) are Gompertz models (see text for details) for all data sets except for ruffe in 1991 and 1999, and dace in 1996, where points are simply connected

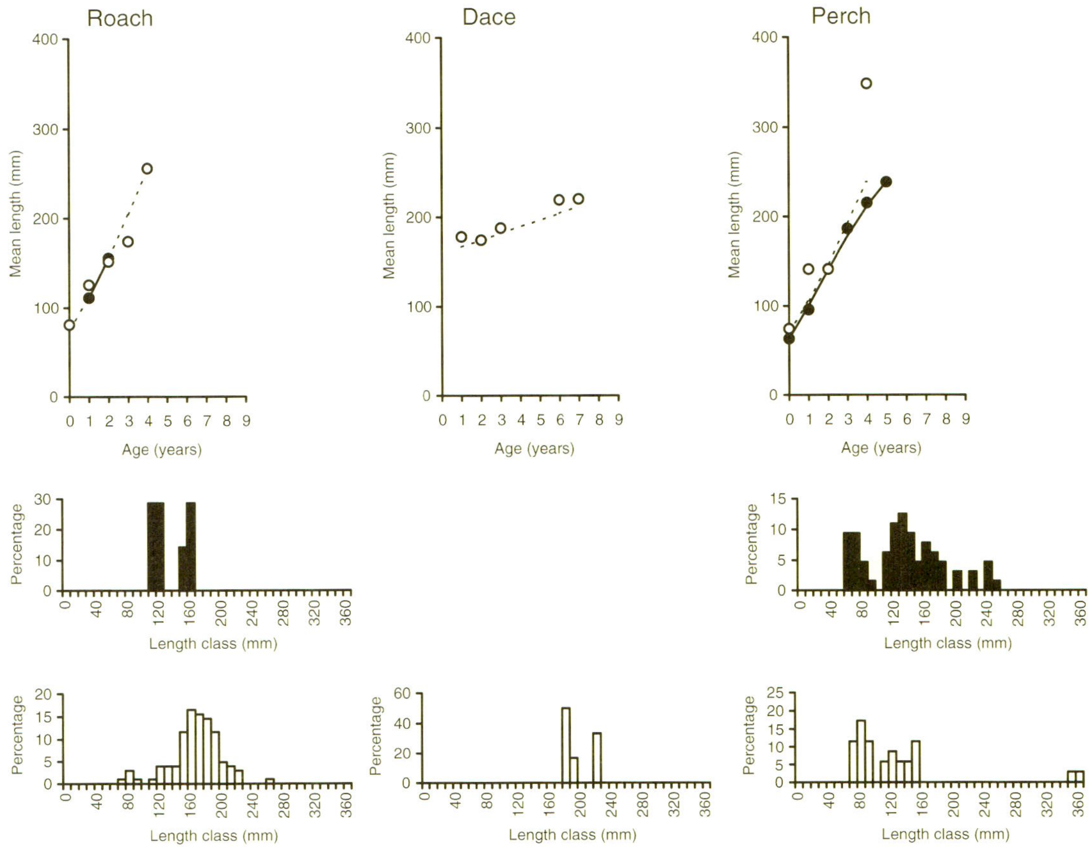

Figure 25.5 Growth curves and length frequency distributions for 1991 (closed circles and columns) and 1999 (open circles and columns) for roach, dace (with the exception that no data are available from 1991) and perch of Derwent Water. Fitted growth curves (continuous and broken lines for earlier and later data, respectively) are Gompertz models (see text for details) for all data sets except for roach in 1991, where points are simply connected

provides no indication of any probable competitive interaction with ruffe to the detriment of roach (Bergman & Greenberg 1994), even though the 1990s demise of roach in Bassenthwaite Lake coincided with the rapid expansion of the ruffe population. A predatory impact of ruffe on roach eggs or larvae is a possible explanation, although this is only speculation. In Derwent Water, the introduced roach population is clearly still expanding and no effects on the perch or other populations have yet been manifested.

As for roach, once introduced, the rapid individual and population growth of ruffe in the eutrophic Bassenthwaite Lake was expected given this species' native habitats. More unexpectedly, a typical individual longevity was only displayed during the mid-1990s, with truncated age and length frequency distributions thereafter. These changes in the late 1990s are real and not a result of sampling artefacts because survey methodology and ageing techniques were consistent with the sampling of an introduced ruffe population in Llyn Tegid, UK, where individuals of up to 7 years old were recorded (Winfield *et al.* 1996a). A possible explanation, discussed below, is a competitive interaction with the native perch population under conditions of a declining roach population and decreasing eutrophy. The future arrival of ruffe in Derwent Water may be anticipated given that this species could enter the lake from Bassenthwaite Lake via the River Derwent, as roach and dace may have already done.

The introduced dace populations of Bassenthwaite Lake and Derwent Water are still at early stages of colonisation, although both are remarkably abundant given their more usual occupancy of flowing water habitats in the UK. While effects on other members of the communities remain unlikely at present, it may be noted for the future that potential competitive interactions with roach (Cowx 1989) and predation on larval coregonids (Shirobokov 1992) have been considered elsewhere.

As the roach and dace introductions to Derwent Water are still in their establishing phases, the consistently good recruitment and growth of the native perch population of this mesotrophic lake was expected. In contrast, the situation in Bassenthwaite Lake, although similar in climate, lake morphology and hence water temperature, has been much more dynamic in terms of lake trophic status and the post-establishing status of the introduced roach and ruffe populations. In the early 1990s, the lake was very eutrophic and its roach population extremely abundant, conditions which typically lead to a depressed perch population (Persson 1991), and the native perch population did indeed show poor growth and short longevity. However, through the 1990s the perch displayed increasing growth rate and longevity, eventually resulting in the common occurrence of individuals in excess of 250 mm in length and a dramatic increase from 1991 to 1999 in L_∞ of 148–462 mm. In the mid-1980s when roach were relatively scarce in Bassenthwaite Lake, a von Bertalanffy growth model fitted to perch data produced a similarly large L_∞ of 364 mm (Mubamba 1989), providing additional support for the competition hypothesis.

In summary, both lake communities have shown dramatic changes as a result of the five species introductions. In the eutrophic Bassenthwaite Lake, it is suggested that the introduction of roach depressed the native perch population, which facilitated the rapid establishment of the subsequently introduced ruffe population. Consistently poor recruitment of roach through the 1990s, which may or may not have been linked to the establishment of the ruffe population, then allowed the perch to recover and in turn impact on the established ruffe population. In the mesotrophic Derwent Water, introduced roach have so far had no impact on the native perch population.

Acknowledgements

We thank all of our colleagues, too numerous to mention individually, for their help with field work. This work was funded by the Natural Environment Research Council, the Environment Agency and North West Water Limited.

References

Bergman E. (1987) Temperature-dependent differences in foraging abilities of two percids, *Perca fluviatilis* and *Gymnocephalus cernuus*. *Environmental Biology of Fishes* **19**, 45–53.

Bergman E. & Greenberg L. (1994) Competition between a planktivore, a benthivore, and a species with ontogenetic diet shifts. *Ecology* **75**, 1233–1245.

Cowx I.G. (1989) Interaction between the roach, *Rutilus rutilus*, and dace, *Leuciscus leuciscus*, populations in a river catchment in south-west England. *Journal of Fish Biology* **35**(Suppl. A), 279–284.

Lowe-McConnell R.H. (1990) Summary address: rare fish, problems, progress and prospects for conservation. *Journal of Fish Biology* **37**(Suppl. A), 263–269.

Mubamba R. (1989) *The ecology of the coregonid fishes in the English Lake District*. Unpublished PhD Thesis. University of Wales, 428 pp.

Persson L. (1991) Interspecific interactions. In I.J. Winfield & J.S. Nelson (eds) *Cyprinid Fishes: Systematics, Biology and Exploitation*. London: Chapman & Hall, pp. 530–551.

Pratt D.M., Blust W.H. & Selgeby J.H. (1992) Ruffe, *Gymnocephalus cernuus*: newly introduced in North America. *Canadian Journal of Fisheries and Aquatic Sciences* **49**, 1616–1618.

Ramsbottom A.E. (1976) Depth charts of the Cumbrian lakes. *Freshwater Biological Association Scientific Publication* No. 33, 39 pp.

Shirobokov I.I. (1992) Estimation of numbers and identification of larval whitefish (Coregonidae) in the diet of some predators. *Journal of Ichthyology* **32**, 77–84.

Winfield I.J. (1992) Threats to the lake fish communities of the U.K. arising from eutrophication and species introductions. *Netherlands Journal of Zoology* **42**, 233–242.

Winfield I.J., Adams C.E. & Fletcher J.M. (1996a) Recent introductions of ruffe (*Gymnocephalus cernuus*) to three United Kingdom lakes containing *Coregonus* species. *Annales Zoologici Fennici* **33**, 459–466.

Winfield I.J., Cragg-Hine D., Fletcher J.M. & Cubby P.R. (1996b) The conservation ecology of *Coregonus albula* and *C. lavaretus* in England and Wales, U.K. In A. Kirchhofer & D. Hefti (eds) *Conservation of Endangered Freshwater Fish in Europe*. Basel: Birkhauser Verlag, pp. 213–223.

Winfield I.J., Fletcher J.M. & Cubby P.R. (1998) The threat to vendace (*Coregonus albula*) eggs from introduced ruffe (*Gymnocephalus cernuus*) in Bassenthwaite Lake, U.K. *Advances in Limnology* **50**, 171–177.

Winfield I.J., Fletcher J.M. & James B.J. (2000) The Urban Waste Water Treatment Directive – Monitoring the vendace populations of Bassenthwaite Lake and Derwentwater, 1999. Unpublished *Report by the Institute of Freshwater Ecology to Environment Agency, North West Region. WI/T11063m7/5*, 23 pp.

Chapter 26
Changes in the Iberian nase, *Chondrostoma willkommii*, population in the Guadalete River (Spain) following impoundment

A. RODRÍGUEZ-RUIZ,* L. ENCINA and C. GRANADO-LORENCIO
Department of Plant Biology and Ecology, Faculty of Biology, University of Seville, Seville, Spain

Abstract

The distribution and population structure of the Iberian nase, *Chondrostoma willkommii*, were analysed before and after construction of a reservoir on the headwaters of Guadalete River (SW, Spain). This stream is characterised by periods of both floods and drought during the year, and the flow regime varies from year to year. Before construction of the reservoir, the fish community comprised only three endemic cyprinids (*Barbus sclateri*, *C. willkommii* and *Leuciscus pyrenaicus*). *C. willkommii* represented 26% of the total specimens caught. This species has a short life span, high growth rate, early age at maturity, and small length at maturity and first breeding. The new reservoir has had a strong effect on the fish community structure. Since 1991, the percentage of *C. willkommii* within the assemblage has fluctuated, and although current values are similar to those before construction of the reservoir, the *C. willkommii* now inhabiting the headwaters, upstream to the dam, are larger; smaller and younger fish were dominant before construction of the reservoir.

Keywords: *Chondrostoma*, Cyprinidae, impoundment, life history strategy.

26.1 Introduction

The fluvial systems of temperate regions are relatively unstable environments for aquatic life, due to the tremendous – and sometimes unpredictable – fluctuations in their hydrology and, thus, in their physico-chemical characteristics. The latter are extreme in Mediterranean rivers, in which torrential flows alternate with periods of high temperature without rainfall, and considerably reduced flow. In addition, many rivers are now impounded, causing irreparable damage to the habitats. One result is the possible proliferation of certain species at the expense of others, although the original ichthyofauna of the river is usually affected (Fernando & Holcik 1991).

This situation is particularly true of the cyprinid, *Chondrostoma willkommii* (Steindachner), which is distributed across the southern Iberian Peninsula in the river basins of the Guadiana, Odiel, Guadalquivir, Guadalete, Guadiaro and Guadalhorce (Doadrio & Elvira 1986; Elvira 1987, 1991, 1997). Unfortunately, little is known about

*Correspondence: Department of Plant Biology and Ecology, Faculty of Biology, University of Seville, Apt 1095, 41080 Seville, Spain (email: dora@cica.es).

its biology: studies have been restricted to its basic life cycle (Herrera & Fernández-Delgado 1994; Rodríguez-Ruiz, Encina & Granado-Lorencio 1998), feeding (Bellido, Hernando, Fernández-Delgado & Herrera 1989; Encina & Granado-Lorencio 1994), production (Rodríguez-Ruiz & Granado-Lorencio 1991) and reproductive migration (Rodríguez-Ruiz & Granado-Lorencio 1992).

This chapter examines the impact of the construction of the Zahara-Gastor reservoir in the headwaters of the River Guadalete on the life history strategy of *C. willkommii*. The chapter compares information on the population ecology of Iberian nase in the headwaters of the River Guadalete before construction of the dam (1991) and again after its complete filling (1996).

26.2 Study area

The River Guadalete, located in the south of the Iberian Peninsula, is 167 km long, and is the axis of an independent hydrographic basin of 3677 km^2 (1020 m asl, 0.611% mean gradient). It has three main tributaries: the rivers Bocaleones and Majaceite on the left bank and Guadalporcún on the right. The river had two reservoirs in its mid-course: at Bornos (constructed in 1961) and Arcos (constructed in 1965). In 1991, a new reservoir was constructed on the headwaters (20 km from the source) at Zahara-El Gastor. Although closed in 1991 (autumn–winter), the reservoir was only half full by the winter of 1995/1996 due to a strong and continuous drought in the region (Rodríguez-Ruiz 1998). Vegetation in this zone is typical Mediterranean scrub, with abundant rushes (*Typha* sp.), oleanders (*Nerium* sp.), and some willows (*Salix* sp.) and poplars (*Populus* sp.).

The Guadalete river basin has a Mediterranean hydrological regime (floods in autumn to early spring, and then a markedly dry summer, with a high irregularity between years). However, its location at the point of the highest rainfall in the Iberian Peninsula (Sierra de Grazalema, average of 1300 mm annually) gives it unique hydrological characteristics within the Mediterranean region. These characteristics include high environmental instability, so that species living in this system have necessarily had to develop adaptive mechanisms for their survival (Granado-Lorencio 1992; Rodríguez-Ruiz *et al.* 1998).

26.3 Materials and methods

Sampling was carried out monthly in the headwaters (8 km from the source) during the years before closure of the dam (1987–1991). After damming, the headwaters were studied between November and June–July from 1994 to 1996 and in 1999. Fish were captured using electric fishing (DC, 220 V, 0.6–1 A and 1200 W), wading upstream with one anode. In all samples, the duration of electric fishing was 30 min (constant effort).

All the fish were identified, counted and measured (nearest mm), and most were returned to the water (99%). A representative sample of each size range was removed monthly for further laboratory examination. Total and standard length (mm), total and eviscerated fresh weight (g) and sex were determined for a total of 354 Iberian nase specimens.

Six to nine scales from the left side of the body between the lateral line and dorsal fin were removed and mounted dry between two slides for determination of age and back-calculated growth. The oral scale radius was measured using an ocular micrometer. Age was also determined from opercular bones. Size groups were established by length–frequency distribution of all specimens caught throughout the study period.

Statistical analyses were based on specimens for which all measurements were available. The data were analysed by two-way ANOVA or ANCOVA (Zar 1984) using the SPSS-X software package (SPSS Inc. 1988). The frequency distribution was studied using Kolmogorov–Smirnov's Z statistic and χ^2.

26.4 Results

The fish assemblage in the Guadalete headwaters from 1987 to 1991 (before construction of the reservoir and upstream of its site) was exclusively three endemic cyprinids: Iberian barbel, *Barbus sclateri* (Günther), Iberian nase, *C. willkommii*, and Iberian chub, *Leuciscus pyrenaicus* (Günther). Between 1994 and 1996 and in 1999 (after construction of the reservoir and upstream of it) the contribution by Iberian barbel and Iberian chub fluctuated, but the Iberian nase population was relatively stable (Table 26.1). The proportion of Iberian nase in the assemblage each year does not differ significantly throughout the study period ($\chi^2 = 8, P = 0.433$), even when comparing pre and post construction of the dam ($\chi^2 = 2.94, P = 0.086$). However, the mean length and weight of Iberian nase specimens captured show considerable variation over the study period (Table 26.2).

26.4.1 *Before closure of the Zahara dam*

The length frequency distribution of Iberian nase in the headwaters before closure of the dam ranged between 43 and 256 mm TL, with a mean of 125.3 mm (±1.49 SE) (Fig. 26.1); 94% of the Iberian nase captured were aged between 1+ and 5+ (from 75 to 185 mm).

The size of the Iberian nase captured varied significantly during the annual cycle ($F_{11,188} = 5.67, P < 0.001$), but not by sex ($F_{1,188} = 3.62, P > 0.05$). Analysis of the somatic weight showed it varied both throughout the study period ($F_{11,187} = 5.62$,

Table 26.1 Changes in the composition (percentages) of fish assemblage in the headwaters before the closing of the dam (1991) and after

	1987	1988	1989	1990	1991	1994	1995	1996	1999
Barbel	48.5	40.3	26.1	10.0	9.0	15.5	4.2	74.9	49.8
Nase	19.0	26.7	28.7	30.0	19.0	9.6	15.7	15.1	28.0
Chub	32.5	33.0	45.2	60.0	72.0	74.9	80.1	10.0	22.2

Table 26.2 Mean weights and lengths of *Chondrostoma willkommii* captured in Guadalete Stream during the study period

	n	Weights (g)	Lengths (mm)
1987	67	21.8 ± 2.8	123 ± 5.4
1988	352	21.2 ± 0.9	127 ± 1.9
1989	99	19.6 ± 2.1	122 ± 3.6
1990	89	16.1 ± 0.6	124 ± 1.6
1991	62	19.2 ± 3.7	122 ± 8.1
1994	85	15.0 ± 0.3	122 ± 0.8
1995	103	15.3 ± 0.9	118 ± 2.5
1996	141	91.4 ± 4.9	202 ± 8.1
1999	137	35.2 ± 2.2	150 ± 3.5

Values are means ± SE; n = number of fish.

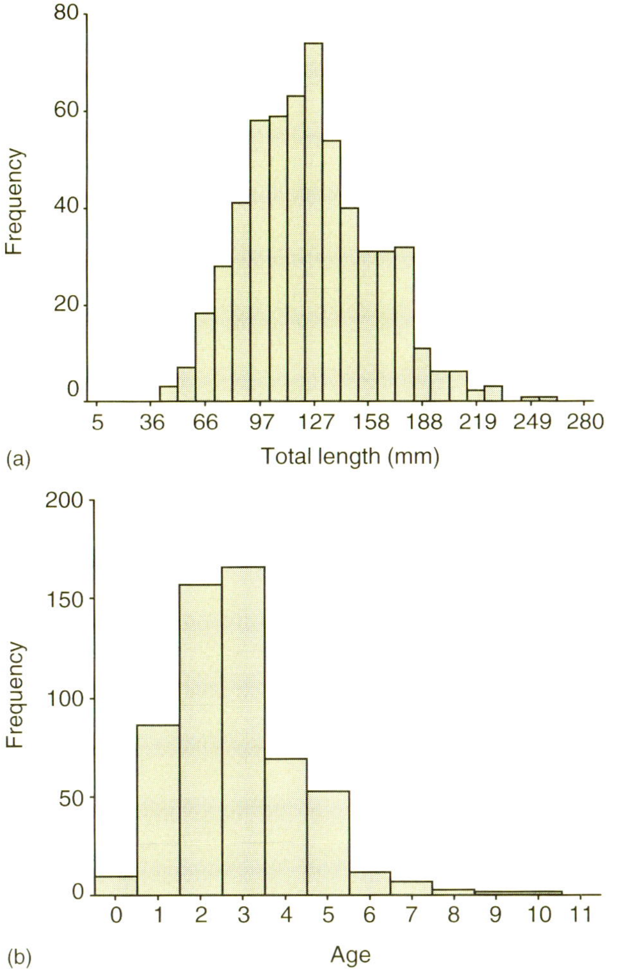

Figure 26.1 Frequency distributions of total length (a) and age (b) for the *C. willkommii* captured in the headwaters before closure of the dam

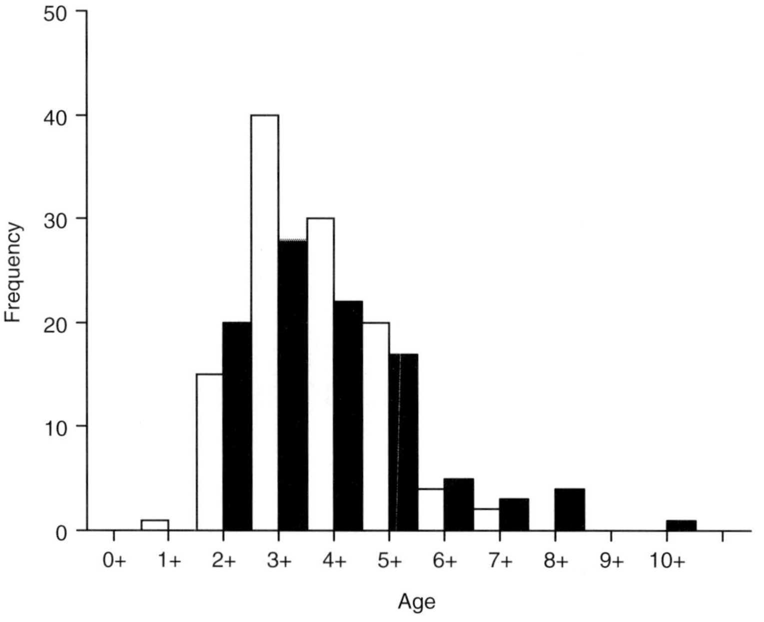

Figure 26.2 The sex distribution throughout the *C. willkommii* life span captured in the headwaters before closure of the dam. Male (white) and female (black)

$P < 0.001$) and by sex ($F_{1,187} = 4.87$, $P < 0.05$), but there was no significant interaction between the factors ($F_{11,187} = 0.90$, $P > 0.05$). As a result, the pattern of each is similar during the study period, with the males always presenting lower values than the females.

The frequency distribution of Iberian nase by sex (Fig. 26.2) showed that males matured earlier (age class 1+) than females, but the latter lived longer (all fish greater than 7+ were females). For age classes in which both males and females were recorded, i.e. from 2+ to 7+, there were no significant differences in sex ratio ($P > 0.05$).

26.4.2 *After closure of the dam*

The frequency distribution of total length for all the Iberian nase captured in the Guadalete headwaters following closure of the dam of the Zahara-El Gastor reservoir is shown in Fig. 26.3. There are two differences from the pre-closure situation. First, the distribution is bimodal and significantly different ($Z = 3.824$, $P < 0.0001$) to the distribution of specimens captured before closure (Fig. 26.1). Secondly, the size range is greater; the length categories most represented were different to those appearing before closure of the dam: 75–185 mm pre-impoundment compared to 100–150 mm, with a secondary peak between 225 and 275 mm after. Before closure, the capture of specimens exceeding 190 mm was exceptional. Moreover, the mean length of the Iberian nase was greater, at 153.39 mm (±1.69 mm, SE), after impoundment than before closure (125.32 ± 1.49 mm, SE) ($F_{1,1033} = 85.984$, $P < 0.0001$).

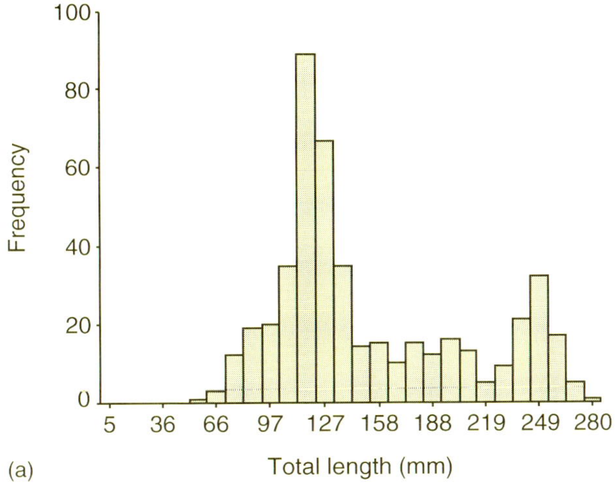

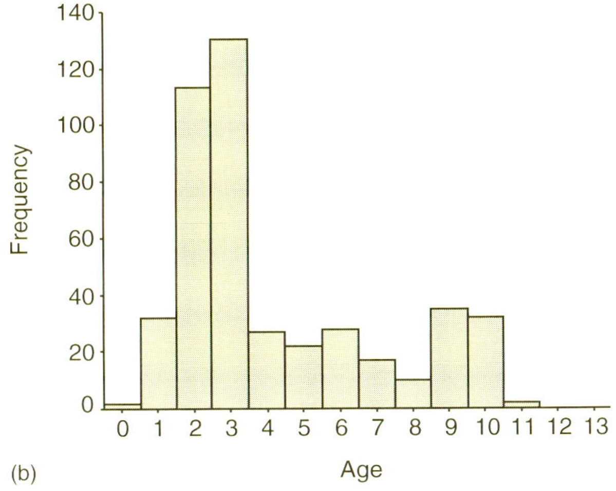

Figure 26.3 Frequency distribution of total length (a) and age (b) for all the *C. willkommii* captured in the River Guadalete headwaters following closure of the dam of the Zahara-El Gastor reservoir

Throughout the study period, from before closure of the dam to after it, the mean length and weight of Iberian nase captured in the headwaters varied significantly, with a maximum in 1996 (Table 26.2), and although the average length and weight declined during 1999, they remained higher than that recorded in previous years.

26.5 Discussion

Because alterations in a river change its conditions, the life history of a population or (in particular) an individual can be modelled by a compendium of biotic and abiotic factors (Przybylski & Borowska 1998). Plasticity in the growth and reproduction of fish is interpreted as an adaptation to local environmental conditions (Mann 1991), with patterns showing great variations between species, populations, and individuals (Roff 1992; Stearns 1992). The Iberian nase population inhabiting the headwaters of the River Guadalete post-impoundment presents a size structure different to that found before closure of the Zahara-El Gastor dam in 1991.

The first event following closure of the Zahara dam in 1991 was a fragmentation of the fish community assemblage (Rodríguez-Ruiz 1998). The Iberian nase is one of the species comprising this assemblage. The population structure in the headwaters (upstream of the dam) varied little until 1996 (with the reservoir filled completely), when differences were detected. The Iberian nase population, localised to small seasonal rivers in the south of the Iberian Peninsula was, before the construction of the reservoir, characterised by a short life span, very rapid growth during the first year of life, early maturity, and possible multiple reproductions with high fecundity (Rodríguez-Ruiz *et al.* 1998). Such characteristics are typical of species inhabiting unstable environments (Bruton 1989; Wootton 1990). However, the great changes in the river, such as the construction of a reservoir just a few kilometres downstream of the studied zone, appear to have altered the life history traits (but not strategies) of Iberian nase inhabiting the headwaters of the River Guadalete.

Before the reservoir was filled, the size of Iberian nase rarely exceeded 190 mm in total length; specimens older than 5+ were also rare. This size structure was consistent with that found in similar southern Iberian rivers (Herrera 1991; Herrera & Fernández-Delgado 1994). The length and weight of the Iberian nase captured from 1996 were larger probably as a result of larger specimens fish moving upstream from the new reservoir. These individuals (originating from the river) had spent several years in the reservoir (a different habitat), and this stable environment could have affected their rate of growth and reproduction (Schlosser 1991). Weatherley and Gill (1987) stated that the absolute size of the body of water affects the maximum length of the fish, so that the greater the space, the greater the growth, bearing in mind that fish growth is considered a plastic process that can change considerably in response to environmental factors (Weatherley 1990).

The first stages of colonisation of the reservoir appears to have favoured the Iberian nase, a pioneer and opportunist species, over the rest of the native ichthyofauna of the River Guadalete. Not only did the population number increase, but also the length of the individuals (Rodríguez-Ruiz 1998). When the large individuals of this species leave the reservoir to breed in the river, they remain there a long time (from February–March to June–July, unpublished data), probably for two reasons. This species exhibits multiple spawning because of its adaptive advantages in Mediterranean-type environments (Rodríguez-Ruiz 1992; Granado-Lorencio, Rodríguez-Ruiz, Encina, Escot, Mellado & Prenda 2000). Also the rich and diverse food resources offered by the river during spring are better (higher energy content) than what is offered by the reservoir (Encina & Granado-Lorencio 1997). The better conditions (better than for other coexisting fish species) enable these individuals to exploit the river's resources (space and food) for a long period. However, when the environmental conditions in the river become adverse (increased temperature and reduced flow), the individuals return to the reservoir, where, although the food resources are poorer, environmental conditions are more favourable (i.e. water quality and quantity). Thus, they probably return to the reservoir when the fluvial environment no longer covers the ecological needs of the species (Rodríguez-Ruiz 1998). The young of the year remain in the river because this age group is able to survive in the stressed environment.

The construction of the reservoir in the headwaters of the River Guadalete has produced changes in the fish population structure. First, the mean size of individuals has increased and second the Iberian nase which inhabit the reservoir exploit the lotic system during part of their annual cycle, when they form part of the fish assemblage of the river.

Acknowledgements

This research was financially supported by Direction General of Research, Science and Technology (DGICYT). We thank Dr Escot for his generous and constant help. Dr I. Cowx and two anonymous reviewers provided comments on an earlier version of this manuscript.

References

Bellido M., Hernando J.A., Fernández-Delgado C. & Herrera M. (1989) Alimentación de la boga del Guadiana (*Chondrostoma polylepis willkommi*, Steind. 1866) en la interfase río-embalse de Sierra Boyera (Córdoba, España). *Doñana, Acta Vertebrata* **16**, 189–201 (in Spanish).

Bruton M.N. (1989) The ecological significance of alternative life-history styles. In M.N. Bruton (ed.) *Alternative Life-History Styles of Animals*. Dordrecht, Netherlands: Kluwer Academic Publications, pp. 503–553.

Doadrio I. & Elvira B. (1986) Nuevas localidades de *Chondrostoma polylepis* Steindachner, 1865 (Ostariophysi: Cyprinidae) en España. *Doñana, Acta Vertebrata* **13**, 163–165 (in Spanish).

Elvira B. (1987) Taxonomy revision of the Genus *Chondrostoma* Agassiz, 1835 (Pisces: Cyprinidae). *Cybium* **11**, 1–14.

Elvira B. (1991) Further studies on the taxonomy of the genus *Chondrostoma* (Osteichthyes: Cyprinidae): species from eastern Europe. *Cybium* **15**, 147–150.

Elvira B. (1997) Taxonomy of the Genus *Chondrostoma*: an updated review. *Folia Zoologica* **46**, 1–14.

Encina L. & Granado-Lorencio C. (1994) Gut evacuation in barbel (*Barbus sclateri* G., 1869) and nase (*Chondrostoma willkommi* S., 1866). *Ecology of Freshwater Fish* **3**, 159–166.

Encina L. & Granado-Lorencio C. (1997) Seasonal variations in condition and energy content of somatic and reproductive tissues of *Chondrostoma popylepis willkommi*. *Folia Zoologica* **4**, 123–133.

Fernando C.H. & Holcik J. (1991) Fish in reservoirs. *International Revue Der Gesamten. Hydrobiologie* **76**, 149–167.

Granado-Lorencio C. (1992) Fish species ecology in Spanish freshwater ecosystems. *Limnetica* **8**, 255–261.

Granado-Lorencio C., Rodríguez-Ruiz A., Encina L., Escot C., Mellado E. & Prenda J. (2000) Ecología del Alto Guadalete. Bases para su conservación. Consejería de Obras Públicas y Transporte. Junta de Andalucía, 98 pp. (in Spanish).

Herrera M. (1991) *Estrategias en los ciclos de vida de una comunidad de ciprínidos en un arroyo de primer orden de la cuenca del Río Guadalquivir*. PhD Thesis, University Córdoba, 99 pp. (in Spanish).

Herrera M. & Fernández-Delgado C. (1994) The age, growth and reproduction of *Chondrostoma polylepis willkommi* in a seasonal stream in the Guadalquivir River basin (southern Spain). *Journal of Fish Biology* **44**, 11–22.

Mann R.H.K. (1991) Growth and production. In I.J. Winfield & J.S. Nelson (eds) *Cyprinid Fishes; Systematics, Biology and Explotation.* London: Chapman & Hall, pp. 456–482.

Przybylski M. & Borowska A. (1998) Reproductive biology of Siberian sculpin, *Cottus poecilopus. Polskie Archiwum Hydrobiologie* **45**, 245–257.

Rodríguez-Ruiz A. (1992) *Relación entre la comunidad íctica y la estructura del hábitat en un río de régimen mediterráneo*. PhD Thesis, University Seville, 400 pp. (in Spanish).

Rodríguez-Ruiz A. (1998) Fish species composition before and after construction of a reservoir on the Guadalete River (SW Spain). *Archiv für Hydrobiologie* **142**, 353–369.

Rodríguez-Ruiz A. & Granado-Lorencio C. (1991) Fish production in a stream with Mediterranean regimen (SW Spain). *Ecology International Bulletin* **19**, 65–78.

Rodríguez-Ruiz A. & Granado-Lorencio C. (1992) Spawning period and migration of three species of cyprinids in a stream with a Mediterranean regimen (SW Spain). *Journal of Fish Biology* **41**, 545–556.

Rodríguez-Ruiz A., Encina L. & Granado-Lorencio C. (1998) Estrategias de vida de las especies ícticas en un río fluctuante en el sur de España: una visión holística. *Boletín de la Sociedad de Biología de Concepción* **69**, 175–189 (in Spanish).

Roff D.A. (1992) *The Evolution of Life-Histories: Theory and Analysis*. New York: Chapman & Hall, 535 pp.

Schlosser I.J. (1991) Stream fish ecology: a landscape perspective. *BioScience* **41**, 704–712.

SPSS Inc. (1998) *SPSS-X User's Guide*. Chicago: SPSS Inc., 1072 pp.

Stearns S.C. (1992) *The Evolution of Life Histories*. Oxford: Oxford University Press, 249 pp.

Weatherley A.H. (1990) Approaches to understanding fish growth. *Transactions of the American Fisheries Society* **119**, 662–672.

Weatherley A.H. & Gill H.S. (1987) *The Biology of Fish Growth.* London: Academic Press, 443 pp.

Wootton R.J. (1990) *Ecology of Teleost Fishes*. London: Chapman & Hall, 404 pp.

Zar J.H. (1984) *Biostatistical Analysis*, second edition. Englewood Cliffs: Prentice-Hall, 718 pp.

Section IV
Conservation genetics

Chapter 27
Conservation of freshwater fishes: integrating evolution and genetics with ecology

C. MORITZ*
Department of Zoology and Entomology, University of Queensland, Brisbane, Australia and Museum of Vertebrate Zoology, University of California, Berkeley, USA

K. McGUIGAN
Department of Zoology and Entomology, University of Queensland, Brisbane, Australia

L. BERNATCHEZ
Department of Zoology and Entomology, University of Queensland, Brisbane, Australia and Departement de Biologie, Pavillon Vachon, Universite Laval, Sainte-Foy, Canada

Abstract

The immediate challenges for conservation of freshwater fish diversity are primarily ecological, but consideration of evolutionary process is important in its own right and can also broaden options available to managers. In the context of conservation, genetic diversity within species can be partitioned into two dimensions, one concerned with the consequences of historical isolation and the other with adaptive evolution of phenotypes. This distinction suggests general principles for defining conservation units within species and managing populations to conserve processes. However, development of strategies for conservation of a specific group or taxon should be based on a clear understanding of evolutionary history. As a broad generalisation, freshwater fish seem capable of rapid and recurrent evolution of specific ecomorphs and are also prone to hybridisation. These concepts are illustrated using three examples, salmonids (*Osmerus* and *Coregonus*) from northern temperate systems and rainbowfish (*Melanotaenia*) from the wet tropics of Australia.

Keywords: adaptation, conservation units, evolutionary processes, freshwater fish, management.

27.1 Introduction

Without doubt, the immediate issues in conservation of freshwater fishes are ecological. Major threatening processes include:

(1) habitat modification and loss, especially disturbance of watershed processes, impoundment or diversion of flows;

*Correspondence: Museum of Vertebrate Zoology, University of California, Berkeley, CA 94720, USA (email: cmoritz@socrates.berkeley.edu).

(2) changes to fish communities and population structure due to release of exotic species or cultured fish, which may contribute to the spread of pathogens or result in increased hybridisation, competition, predation and widespread homogenisation of faunas (Allendorf 1991; Hughes & Noss 1992; Naylor, Goldburg, Mooney, Beveridge, Clay, Folke, Kautsky, Lubcheno, Primavera & Williams 1998; Huxel 1999; Rahel 2000); and
(3) overharvesting leading to local depletion/extirpation or changes in life history (Miller & Williams 1989; Law 2000) (see Cowx, Chapter 20 for review).

It follows that conservation efforts should be focussed on identifying and reducing the impact of such threatening processes. However, the aim should not be to restore just ecological processes, but also to maintain evolutionary processes that underpin population viability and continued diversification (Frankel 1974). Thus, a general goal in conservation is to maintain evolutionary processes responsible for generating biodiversity and the ecological viability of populations and landscapes necessary to achieve this.

In this context, there are three ways in which evolutionary theory and the tools of molecular and/or quantitative genetics should complement ecological approaches to conservation (Moritz 1994a; Avise & Hamrick 1996). The first is to identify key processes that have operated and should be sustained within the taxa concerned. The second is to identify the products of these historical processes, which may or may not correspond with traditionally recognised taxa. The third lies in the use of molecular approaches to augment ecological studies of demography and metapopulation processes (Bernatchez, Choinard & Lu 1999). With this knowledge, we are equipped to develop strategies to achieve our goal that are consistent with both the pressing ecological concerns as well as the broader issue of maintaining evolutionary processes.

In this chapter, two themes are developed. The first is that inclusion of evolutionary thinking can broaden options within a management framework focussed on ecological processes. If the processes that lead to extant products of evolution and underpin current evolutionary dynamics are understood, it may be possible to identify strategies that are more flexible than would be the case if taxa and varieties are considered as discrete and immutable. Second, it is argued that, while conservation goals and principles should be general, the strategy adopted for a given taxon/system should reflect our knowledge of key ecological/evolutionary processes for that group. Thus, there are no management 'rules', just guidelines and tools that need to be adopted judiciously and with explicit justification.

In the following, general principles for maintaining evolutionary processes are identified and then characteristics of freshwater fishes that make their conservation management distinct from that for terrestrial species on which conservation paradigms are typically based are reviewed. Recent work on a genus of Australopapuan rainbowfish, *Melanotaenia*, is then reviewed with particular attention to rare and threatened species from north-east Queensland, using this as a case study to illustrate how specific strategies for conservation can be derived from an understanding of evolutionary history and current processes of adaptation.

27.2 Evolutionary processes and products

Through the joint action of mutation, genetic drift and selection, evolution produces a continuum of diversity, manifest as phenotypic, genetic and phylogenetic diversity. Systematists seek to describe this diversity as discrete categories, with diagnosable and/or reproductively isolated species representing the major currency for biodiversity evaluation and management. Yet debate continues about how to conceptualise and diagnose species, with some authors arguing that the widely adopted biological species concept (Mayr 1963) should be replaced by others focussing on cohesion (Templeton 1989), phenotypic clusters (Mallet 1995), or diagnosable evolutionary lineages (Cracraft 1989). This diversity of views reflects varying perspectives on the processes and diagnosis of a common continuum of divergent speciation (de Queiroz 1998). This is no mere abstraction as the perception of species diversity, and thus the number of taxa to be assessed for conservation and management, can vary widely, typically being much greater when employing the phylogenetic species concept (Cracraft 1997; Kottelat 1997). While a comprehensive description of diversity in the form of named species might appear precautionary and thus, laudable, care is needed to maintain taxonomic rigour and to avoid oversplitting, the latter potentially leading to management strategies that inappropriately isolate historically connected phenotypic variants (Moritz 1999).

Related to the debates about the nature of species and how to diagnose them is the question of how to identify and prioritise evolutionary diversity within species. Some (e.g. Crandall, Bininda-Emonds, Mace & Wayne 2000) argue that attention should be focussed on (presumed) adaptive phenotypic diversity; others (Moritz 1994b) emphasise historically isolated populations, or a combination of the two (Dizon, Lockyer, Perrin, Demaster & Sisson 1992; Vogler & DeSalle 1994; Bernatchez 1995). The debate (reviewed in Nielsen 1995) centres around whether and how to recognise major components of intraspecific diversity, often referred to as 'Evolutionarily Significant Units' (ESUs; Ryder 1986). As interpreted for salmonids (Waples 1991, 1995), and increasingly other taxa, these correspond to discrete population segments that can be listed under the conservation legislation of some countries. One view (Cracraft 1997; Goldstein, DeSalle, Amato & Vogler 2000) holds that application of a more fine-grained approach to identification of species, the phylogenetic species concept, will elevate all diagnosable evolutionary lineages to species and remove the need to recognise ESUs, a category outside of the formal nomenclature. Another view is that the whole approach is flawed in that arbitrary divisions are being created within a continuum of evolutionary divergence (Crandall, Bininda-Emonds, Mace & Wayne 2000). Whatever the outcome of these debates, the need remains to have some way to recognise and manage the evolutionary diversity that resides within formally recognised species.

On the basis of current knowledge of evolutionary processes responsible for generating diversity within species, it would be useful to consider separately the genetic diversity generated through ecological processes leading to adaptive evolution *v.* that resulting from historical isolation *per se* (Fig. 27.1). These two evolutionary processes are not necessarily correlated; substantial phenotypic divergence and even new species can develop in the absence of historical isolation (Endler 1977; Rice & Hostert 1993; Bush 1994; Hendry, Wenburg, Bentzen, Volk & Quinn 2000; Quinn, Unwin & Kinnison 2000a) and, conversely, historical isolation in the absence of divergent selection does not necessarily lead to overt phenotypic divergence (Schneider, Smith, Larison & Moritz 1999; Colborn, Crabtree, Shaklee, Philer & Bowen 2001). Where the two processes are correlated, leading to populations that are both historically isolated and phenotypically divergent, most systematists would recognise discrete species (upper right in Fig. 27.1). The products of historical isolation can be identified readily through analysis of molecular phylogeography (Avise 2000) as sets of populations that form discrete and reciprocally monophyletic lineages in gene trees (Fig. 27.1). Such sets of populations correspond to ESUs under the criteria suggested by Moritz (1994b). By contrast, populations that have diverged through adaptive processes without substantial historical isolation will appear polyphyletic or paraphyletic in gene trees, but should also be distinct for phenotypic traits related to fitness (Fig. 27.1). That said, it should be emphasised that it is hazardous to infer adaptive divergence from phenotypic differences alone as several studies have shown that morphology does not always predict ecology (Douglas & Matthews 1992; Turgeon, Estoup & Bernatchez 1999). A formal demonstration of adaptive values thus requires evidence for fitness differences among heritable phenotypes (Hard 1995; Phillips & Ehlinger 1995). This can be achieved (e.g. Schluter 1994; see also Taylor 1991), although it requires demanding experimental work. Alternatively, recent advances in quantitative molecular genetics may eventually provide more practical surrogates for inferring adaptive differences (Ritland 2000).

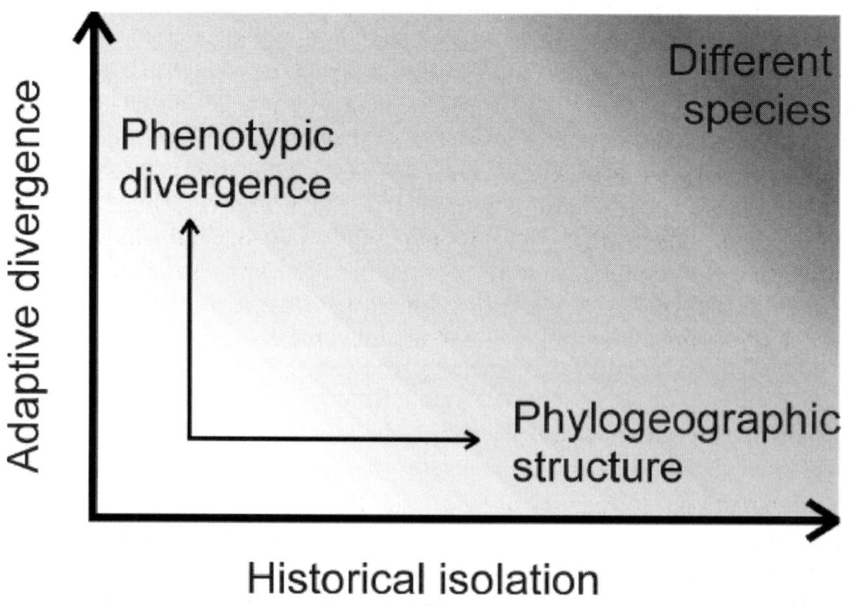

Figure 27.1 Two axes of genetic diversity relevant to conservation (modified from Moritz 1999). See text for explanation

The question remains, how to prioritise and manage populations along two axes of genetic diversity shown in Figure 27.1. The answer to this question probably lies in the extent to which either form of diversity can be recovered following extinction. Whether or not discrete phylogeographic lineages reflect potential adaptive diversity or incipient speciation (Avise, Walker & Johns 1998; *cf.* Hendry, Vamosi, Latham, Heilbuth & Day 2000b), these products of historical isolation are irreplaceable in any sensible timeframe for management. Thus, such sets of populations should be recognised for conservation. By contrast, subject to retention of population viability (Pearse, Lande & Bull 1989; Lande & Shannon 1996), it is possible that adaptive differences among populations can be regenerated through recurrent selection. This capacity to regenerate phenotypes of high fitness, whether through genetic divergence or plasticity, is demonstrated through experimental studies of a variety of organisms (e.g. Malhotra & Thorpe 1991; Reznick, Shaw, Rodd & Shaw 1997; Losos, Warheit & Schoener 1997; Rundle, Nagel, Boughman & Schluter 2000), comparative studies (Taylor 1999) and phylogenetic analysis of recent adaptive radiations (e.g. Givnish & Systma 1997; Losos, Jackman, Larson, de Queiroz & Rodriguez-Schettino 1998).

From the above, a two-stage approach is proposed, wherein historically isolated sets of population within a species are first identified and protected (ESUs *sensu* Moritz 1994b) and then, within these, the potential for adaptive evolution should be maximised. The latter would entail, for example, ensuring that the ecological amplitude of the species is retained in the form of viable populations across heterogeneous environments. Whether or not it is necessary to protect all phenotypically (but not historically) distinct forms will depend on evidence that populations can sustain directional selection to a locally adapted phenotype. If comparative or experimental studies indicate that this is the case, then there is more latitude for maintaining historical patterns of connectivity and phenotypic diversity through carefully planned habitat management, translocations, etc. (Moritz 1999). Bernatchez (1995) proposed that preservation of adaptive potential may also be optimised by ranking phenotypic variants within phylogeographic units according to their level of rarity on the assumption that rarer phenotypes reflect adaptation to environmental selection pressures. In this framework, rarer phenotypes are considered to be less prone to evolve again, should these populations or the environments that harbour them be lost. Consequently, higher priority for conservation should be given to rarer phenotypes. The key issue is to recognise and ensure that phenotypically or genetically distinct 'management units' within an ESU are capable of continued evolution and to maintain the ecological context for this.

27.3 Evolution in freshwater fish

Studies of evolutionary dynamics of freshwater fishes have revealed several properties which make them distinct from birds and mammals, from which many of the paradigms of conservation biology of vertebrates are derived. Genotype–phenotype relationships of fish differ from birds and mammals. Typically, fishes have larger phenotypic variation but lower heritabilities, i.e. they are more strongly affected by environmental fluctuations (Allendorf & Waples 1996). This may also translate to high levels of phenotypic plasticity for eco-morphological traits (reviewed in Robinson & Schluter 1999). Freshwater fishes have unusual capabilities to hybridise (Hubbs 1955; Campton 1987), magnifying problems due to introduced alien species or populations. On the other hand, hybridisation may also be a rich source of genetic variability that may lead to evolutionary novelties (Smith 1992; Dowling & Secor 1997; Svardson 1998).

Freshwater fishes are particularly notable for their propensity for rapid population divergence which may also lead to speciation within a few thousands of generations (Behnke 1972; Schluter 1996; Albertson, Markert, Danley & Kocher 1999; Lu & Bernatchez 1999; Hendry *et al.* 2000a). This may, in turn, translate into the replicated (parallel) evolution of specific adaptive phenotypes (Taylor & Bentzen 1993a; Bernatchez 1995; Bernatchez, Vuorinen, Bodaly & Dodson 1996; Pigeon, Choinard & Bernatchez 1997; Reznick, Shaw, Rodd & Shaw 1997; Douglas, Brunner & Bernatchez 1999; Rüber, Verheyen & Meyer 1999; Taylor 1999), and even recurrent speciation events (Rundle, Nagel, Boughman & Schluter 2000). It is not suggested that these necessarily are general phenomena; nonetheless, the possibility that one or more properties operate in a particular system should be considered.

27.4 Case studies

The potential for rapid and recurrent evolution of specific ecomorphs is well illustrated in recent studies that combine analyses of phylogeny/phylogeography, phenotypic and reproductive discontinuities and ecological traits. Two such examples concern the evolution of freshwater ecotypes of the rainbow smelt (*Osmerus* spp., Osmeridae) and the whitefish *Coregonus* (Salmonidae) species complexes. A third example, rainbowfish of the genus *Melanotaenia* (Melanotaeniidae), further illustrates the use of molecular data, combined with experimental analyses of phenotypic variation, to uncover the history of a regional fauna, species boundaries and the evolutionary processes that continue to operate.

27.4.1 *Rainbow smelt,* Osmerus mordax *Mitchill*

The rainbow smelt, an osmerid species native to watershed tributaries of north-western Atlantic Ocean, exhibits extensive life-history diversity and is composed of anadromous and lacustrine landlocked populations. Lacustrine populations are commonly found as normal and dwarf-sized phenotypes. Populations classified as dwarf or normal ecotypes differ in phenotypic attributes that are characteristic of a limnetic and benthic mode of life, respectively (Taylor & Bentzen 1993a, b). These forms frequently occur sympatrically in lakes and remain reproductively isolated. Consequently, normal and dwarf-sized smelt have been recognised by some as two distinct taxa (e.g. McAllister 1985; Mayden, Burr, Page & Miller 1992), while phenotypic dissimilarity between the two groups has been judged insufficient by others to justify taxonomic distinction (e.g. Robins, Bailey, Bond, Brooker, Lachner, Lea & Scott 1991).

Analyses of mtDNA variation in native smelt populations revealed two major clades which showed a strong geographical pattern of distribution. This phylogeographical pattern can best be explained by an allopatric origin of two monophyletic smelt groups in different glacial refugia, followed by differential recolonisation. These discrete lineages therefore represent major components of the historical axis of diversity (Fig. 27.1), or ESUs (*sensu* Moritz 1994b). A detailed genetic analysis of a contact zone between populations belonging to the distinct evolutionary lineages provided evidence for their strong reproductive isolation (Bernatchez & Martin 1996). Dwarf and normal smelt ecotypes exist within each of these distinct lineages, indicating that they evolved repeatedly and independently subsequent to colonisation of post-glacial lakes (Taylor & Bentzen 1993a; Bernatchez 1997). Additional support for the recurrent evolution of these phenotypes was indicated by detailed analyses within each phylogeographic group (Baby, Bernatchez & Dodson 1991; Taylor & Bentzen 1993a, b). Recurrent evolution of phenotypic variation most likely reflects the predominant role of natural selection related to trophic ecology in promoting population divergence in smelt.

27.4.2 *Whitefish,* Coregonus *species*

This group of fish has a continuous circumpolar distribution in the northern hemisphere, exhibiting pronounced phenotypic variation in morphological and life history traits. Here, the specific Palearctic and Nearctic populations that have been divided into two major nomenclatural species complexes based on continental distribution are referred to; the European whitefish, *Coregonus lavaretus* (L.) and the lake whitefish, *Coregonus clupeaformis* Mitchill. Depending on authorship, these two groups may be either considered as a single polytypic species or split into numerous taxa.

A mtDNA phylogeographic analysis revealed that this species complex is composed of five major evolutionary lineages, corresponding to lineages that survived the Pleistocene glaciations in distinct refugia (Bernatchez & Dodson 1994). As for smelt, these lineages represent major components of the historical axis of diversity (ESUs), and a detailed genetic investigation of a contact zone between populations belonging to the distinct evolutionary lineages provided evidence for their reproductive isolation (Bernatchez & Dodson 1990; Lu, Basley & Bernatchez 2001). Furthermore, experimental studies revealed that embryonic mortality rates were two to five times higher in hybrids between glacial lineages than in intra-lineage crosses, indicating the development of genetic incompatibility (Lu & Bernatchez 1998).

The geographical patterns of whitefish population genetic assemblages provide evidence for the rapid and recurrent evolution of identical phenotypes. Thus, a population tree constructed from molecular data revealed that populations with similar morphology occurred across different evolutionary lineages and sometimes were most closely related to phenotypically dissimilar forms also recognised as distinct taxa (Bernatchez & Dodson 1994). As for smelt, whitefish are characterised by the sympatric occurrence of forms differing in morphology, diets, spawning behaviour and life history traits, a relatively common phenomenon in European lakes (Kottelat 1997). In North America, sympatric forms present two distinct patterns of diversification. In the first, sympatric ecotypes found in four lakes of the Yukon territory (Canada) are distinguished by low (LGR) and high (HGR) counts of gill rakers, this trait being heritable and strongly related to trophic specialisation in coregonids (Lindsey 1981). The LGR ecotype feeds primarily on benthic prey whereas the HGR ecotype consumes mostly pelagic, planktonic food. In the second pattern of diversification, dwarf and normal ecotypes of whitefish are sympatric in several disjunct locations in eastern North America. In contrast to the LGR and HGR forms from the Yukon, which share relatively similar life history traits, dwarf and normal forms differ strikingly in growth, maximum size, age at reproduction and lifespan. As for smelt, a detailed analysis of ecological divergence between dwarf and normal ecotypes provided evidence of a functional link between differential morphology and trophic use (Bernatchez *et al.* 1999). The extent of morphological differentiation (other than size) between ecotypes varied among lakes, and involved traits related to trophic use, the most discriminant one being the number of gill rakers, as for the HGR and LGR forms. Furthermore, the extent of morphological specialisation was correlated with the degree of partitioning between limnetic and benthic niches by dwarf and normal ecotypes, respectively.

Detailed analyses of mtDNA and nuclear DNA (allozymes or microsatellites) provided evidence for the recurrent evolution of these sympatric dwarf and normal ecotypes (Pigeon *et al.* 1997), the HGR and LGR ecotypes from the Yukon (Bernatchez *et al.* 1996), as well as the sympatric European forms (Douglas *et al.* 1999). This strongly suggested that resource-based natural selection was responsible in driving population divergence in whitefish. The possibility that these same ecological processes are also responsible for determining the extent of reproductive isolation between ecotypes was also supported by a negative correlation between the extent of gene flow and that of trophic morphological specialisation between ecotypes in different lakes (Lu & Bernatchez 1999).

The examples of smelt and whitefish highlight three points of relevance for management and conservation:

(1) the use of molecular systematics identified historically isolated groups of populations, not taxonomically recognised, exhibiting distinct geographical distribution;
(2) the evolutionary distinctiveness of these groups was indicated by evidence of reproductive isolation in sympatry; and
(3) replicate evolution or replaceability of phenotypes may be a common phenomenon in these fishes.

As such, these results illustrate that either the criteria of traditional taxonomy, or ecological exchangeability (*sensu* Crandall *et al.* 2000) would fail to identify proper conservation units in an evolutionary based framework. The binary approach presented in Fig. 27.1 provides a better framework for including the evolutionary processes responsible for generating population diversity in both smelt and whitefish. Similar conclusions would also be reached by applying the method proposed by Bernatchez (1995).

27.4.3 *Rainbowfish,* Melanotaenia

Amongst the relatively depauperate freshwater fish fauna of Australia and New Guinea, the colourful rainbowfish of the family Melanotaenidae stand out as an unusually diverse group (Allen & Cross 1982). There are 65 currently recognised species, including 31 described in the past two decades (Allen & Renyaan 1998). Many species are known from very small geographical ranges (Allen 1995). Recent molecular phylogenetic analyses (Zhu, Jamieson, Hugall & Moritz 1994; McGuigan, Zhu, Allen & Moritz 2000), focussed on the most speciose genus (*Melanotaenia*), have identified six major lineages (Fig. 27.2); two basal groups restricted to north-west New Guinea and northern New Guinea, respectively, and four derived clades of which two are shared between southern New Guinea and Australia and two are unique to Australia. These phylogeographic patterns are explicable in the context of the late Miocene–early

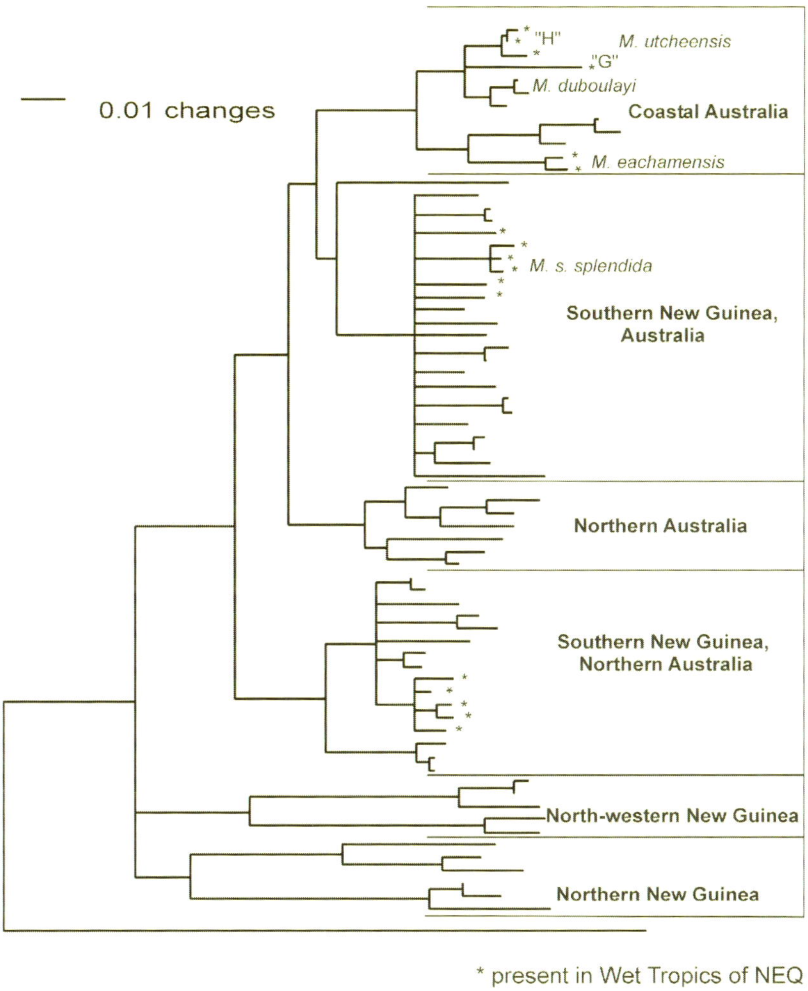

Figure 27.2 Schematic mtDNA phylogeny of rainbowfish *Melanotaenia* (after McGuigan *et al.* 2000) showing the broad geographic distribution of major clades across Australia and New Guinea and the phylogenetic distribution of taxa (*) found in the wet tropics region of north-east Queensland

Pliocene uplift of the central highlands of New Guinea and Pleistocene freshwater connections between southern New Guinea and Australia (McGuigan *et al.* 2000).

This broad phylogeny provides a valuable backdrop to understanding and managing the diverse rainbowfish found in the rivers of the wet tropical rainforests of north-east Queensland. This small region was previously thought to contain four species of *Melanotaenia*; *M. splendida* Peters, *M. trifasciata* Rendahl, *M. maccullochi* Ogilby and *M. eachamensis* Allen & Cross (Herbert & Peeters 1995). Of these, *M. splendida* is a widespread species, ranging from southern New Guinea, through north-east Australia to the ephemeral water bodies of central Australian drainages. Recent molecular studies suggest that this species may have expanded into the streams of the north-east at some stage during the previous interglacial (Hurwood & Hughes 2001). *M. maccullochi* belongs to the same clade as *M. splendida* (Fig. 27.2) and is the only other species of the genus that is found in both New Guinea and Australia, where it has a patchy distribution north and west of the wet tropics (Fig. 27.2). *Melanotaenia trifasciata*, the only Australian representative of a southern New Guinea clade, also has a patchy distribution north and west of the wet tropics.

Of particular interest here is the narrow endemic, *M. eachamensis*. This species was originally described, and only known, from a single crater lake, Lake Eacham, on the Atherton Tableland in north-east Queensland. It disappeared from Lake Eacham following introduction of piscivorous fish to this site and was presumed to be extinct in the wild (but extant in captivity) (Barlow, Hogan & Rodgers 1987). The common species on the Tableland, *M. splendida*, is highly variable in colour and morphology (Allen 1995), and questions were raised as to whether *M. eachamensis* was a divergent form of *M. splendida*, rather than a species in its own right (Crowley & Ivanstoff 1991). Subsequent examination of mtDNA, microsatellites and morphology demonstrated that *M. eachamensis* was distinct from *M. splendida*, being more closely related to a taxon from north-west Australia than to eastern species (Zhu, Degnan & Moritz 1998; Fig. 27.2). Further, these studies revealed several extant populations of *M. eachamensis*, mostly in relatively isolated headwaters or lakes, as well as mixed, and possibly hybridising, populations of *M. eachamensis* and *M. splendida*. A more geographically comprehensive analysis of morphology (Pusey, Bird, Kennard & Arthington 1997) also identified additional populations of *M. eachamensis*, but further phylogenetic analysis of mtDNA (McGuigan, Zhu, Allen & Moritz 2000; Fig. 27.2) revealed that some of these were a fifth, previously unrecognised species (*M. utcheensis* McGuigan, McGuigan 2001) which is morphologically similar to *M. eachamensis* and a member of the same clade, but most closely related to species (*M. duboulayi* Castelnau, *M. fluviatilis* Castelnau) from coastal and inland drainages well to the south (Fig. 27.2).

On the basis of the phylogenetic analysis and subsequent examination of distribution of mtDNA variants (Fig. 27.3), microsatellites and morphology, the rather complex situation with respect to these taxa can now be summarised as follows. *Melanotaenia eachamensis* and *M. utcheensis* are non-sister taxa but members of a geographically discontinuous Australian clade and are presumed to represent 'paleo-endemics' of the region. The former is now represented by a small number of pure populations in relatively isolated upstream sites or crater lakes on the Tableland, and the latter by two

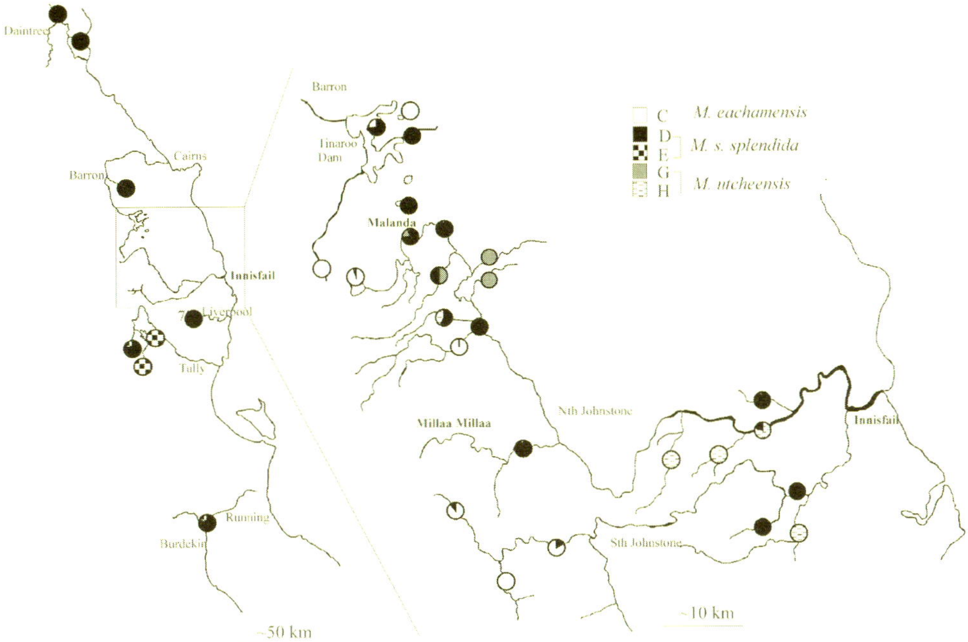

Figure 27.3 Distribution of mtDNA lineages (as diagnosed by RFLPs; Zhu *et al.* 1998; McGuigan, unpublished data) in the wet tropics region of north-east Queensland. Populations fixed for the mtDNA lineages diagnostic of *M. eachamensis* tend to be located in relatively remote upstream areas or isolated crater lakes, and the same is true for some upland populations of *M. utcheensis*

genetically and morphologically discrete lineages (ESUs), one on the Tableland and the other largely confined to the adjacent lowlands (Fig. 27.3). Superimposed on the range of these two species is recent colonisation by *M. splendida*, possibly enhanced by human introduction and manipulation of drainages, which exists in mixed and probably hybridising populations with both *M. eachamensis* and *M. utcheensis* (Fig. 27.3; Zhu *et al.* 1998; K. McGuigan, F. Jones & C. Moritz, unpublished data). Significantly, a thriving population of *M. splendida* has recently been discovered in Lake Eacham, the type locality of *M. eachamensis*, presumably as the result of a recent introduction. The genetic surveys also revealed a pure population of *M. eachamensis* with unusually low heterozygosity (Dirran Ck., Zhu *et al.* 1998) and these fish have been found to have low resistance to parasites, increasing after outcrossing to a more variable population (Gleeson, McCallum & Owens 2000), and poor avoidance of predation by piscivorous fish (Brown & Warburton 1997).

While no longer 'extinct in the wild', *M. eachamensis* remains vulnerable to extinction. Pure populations are known from only four sites in headwaters or isolated lakes (Fig. 27.3); as recent evidence indicates, these populations are vulnerable to introduced predators, pathogens and relatives. The recent introduction of *M. splendida* to Lake Eacham will seriously undermine further efforts to return *M. eachamensis* to its

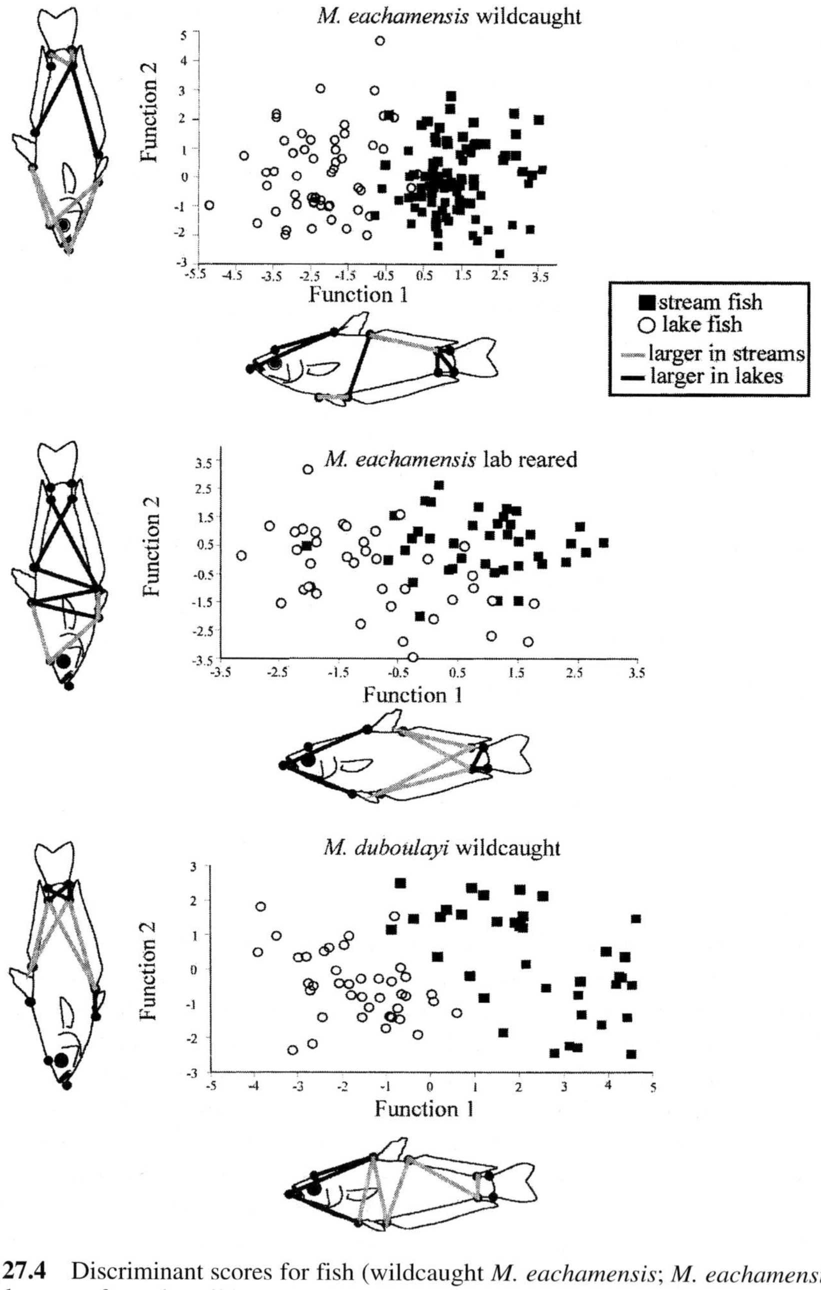

Figure 27.4 Discriminant scores for fish (wildcaught *M. eachamensis*; *M. eachamensis* bred in the laboratory from the wildcaught populations and; wildcaught *M. duboulayi*) on functions 1 and 2 from a discriminant functions analysis with habitat and sex as the discriminator. Fish are identified by habitat and sex (K. McGuigan, unpublished data)

type locality because of the propensity for these two species to hybridise. A priority for management is therefore to avoid any further such introductions to the remaining intact populations of *M. eachamensis* (and *M. utcheensis*). Further, one of the four remaining populations of *M. eachamensis*, Dirran Creek, appears to show some indications of inbreeding depression. This raises the question of whether translocations are warranted to augment variation in the inbred population or to establish additional populations, and if so, what the source of fish for these translocations should be.

Populations of *M. eachamensis* are not divided into sets that are reciprocally monophyletic for mtDNA (Zhu *et al.* 1998; K. McGuigan, unpublished data) and therefore represent a single ESU (*sensu* Moritz 1994b). However, there is evidence of structuring of variation at mtDNA and nuclear microsatellite loci (Zhu *et al.* 1998), and fish show evidence of morphological (Fig. 27.4) and perhaps physiological adaptation to lake *v.* stream environments, suggesting that lake and stream populations represent separate management units. In such circumstances, Moritz (1999) suggested that, if necessary, consideration can be given to translocating between phenotypically distinct management units or, if heterozygosity is low, using admixed populations as sources for reintroductions. This strategy rests on the assumption that any ensuing outbreeding depression or maladaptation can be reversed by natural selection. That the broad pattern of morphological differentiation between lake and stream *M. eachamensis* is repeated in a related species, *M. duboulayi* (Fig. 27.4), suggests that these morphotypes can be regenerated through natural selection.

27.5 Conclusions

The examples reviewed above and others from northern temperate lakes (other salmonid fishes, Gasterosteidae, Centrarchidae and Catostomidae, reviewed in Taylor 1999) but also other biotopes (cichlids from the African Rift Lakes, Stiassny & Meyer 1999; cyprinodonts from Mexican lagoons, Strecker, Meyer, Sturmbauer & Wilkens 1996; percids from Andeans lakes, Ruzzante, Walde, Cussac, Macchi & Alonso 1998) illustrate the potential for rapid and, in several cases, recurrent evolution of adaptive phenotypes in freshwater fish. These studies also stress the importance of integrating genetic, phenotypic and ecological studies to understand the evolutionary processes that drive diversification in freshwater fish and also to ensure that taxonomy reflects natural groupings. Although it was not argued that all fishes have such level of evolutionary flexibility, there is mounting evidence that the potential of freshwater fish for generating and even replicating phenotypic adaptive diversity rapidly in terms of evolutionary time may be more widespread than assumed until recently. Clearly, however, the possibility for such unique evolutionary processes to operate requires the integrated protection or restoration of environmental settings (habitat quality, absence of exotic competitors and predators, etc.) that appears necessary to promote diversification.

From this it is argued that knowledge of evolutionary processes should inform interpretation of current patterns of species and genetic diversity and, thereby, inform population management. It is this knowledge of historical and contemporary evolutionary processes, combined with a consensus on goals and principles of conservation that should underpin a strategy for the conservation of biodiversity. In the case of freshwater fishes, there are pressing ecological issues repeated across continents, effects of changes in land use, development of water resources, introductions of exotic species and cultured fish, etc. The capacity to meet these challenges through effective management of habitats and the species that they sustain is likely to be enhanced by considering evolutionary processes in conservation practice, rather than blindly assuming that existing diversity is fixed in time and space.

References

Albertson R.C., Markert J.A., Danley P.D. & Kocher T.D. (1999) Phylogeny of a rapidly evolving clade: The cichlid fishes of Lake Malawi, East Africa. *Proceedings of the National Academy of Science (USA)* **96**, 5107–5110.

Allen G.R. (1995) *Rainbowfishes in Nature and the Aquarium*. Melle, Germany: Tetra-Verlag, 178 pp.

Allen G.R. & Cross N.J. (1982) *Rainbowfishes of Australia and Papua New Guinea*. Sydney: Angus & Robertson, 141 pp.

Allen G.R. & Renyaan S.J. (1998) Three new species of rainbowfishes (Melanotaeniidae) from Irian Jaya, Indonesia. *Aqua, Journal of Ichthyology and Aquatic Biology* **3**, 69–80.

Allendorf F.W. (1991) Ecological and genetic effects of fish introductions: synthesis and recommendations. *Canadian Journal of Fisheries and Aquatic Sciences* **48**, 178–181.

Allendorf F.W. & Waples R.S. (1996) Conservation and genetics of salmonid fishes. In J.C. Avise & J.L. Hamrick (eds) *Conservation Genetics: Case Histories from Nature*. London: Chapman & Hall, pp. 238–280.

Avise J.C. (2000) *Phylogeography. The History and Formation of Species*. Cambridge, Mass.: Harvard University Press, 447 pp.

Avise J.C. & Hamrick J.L. (1996) *Conservation Genetics: Case Histories from Nature*. London: Chapman & Hall, 504 pp.

Avise J.C., Walker D. & Johns G.C. (1998) Speciation durations and Pleistocene effects on vertebrate phylogeography. *Proceedings of the Royal Society of London B* **265**, 1707–1712.

Baby M.C., Bernatchez L. & Dodson J.J. (1991) Genetic structure and relationships among anadromous and landlocked populations of rainbow smelt (*Osmerus mordax* Mitchill) revealed by mitochondrial DNA analysis. *Journal of Fish Biology* **39**, 61–68.

Barlow C.G., Hogan A.E. & Rodgers L.J. (1987) Implication of translocated fishes in the apparent extinction in the wild of the Lake Eacham rainbow fish, *Melanotaenia eachamensis*. *Australian Journal of Marine and Freshwater Research* **38**, 897–902.

Behnke R.J. (1972) The systematics of salmonid fishes of recently glaciated lakes. *Journal of Fisheries Research Board* **29**, 639–671.

Bernatchez L. (1995) A role for molecular systematics in defining Evolutionarily Significant Units in fishes. In J.L. Nielsen (ed.) *Evolution and the Aquatic Ecosystem: Defining Unique Units in Population Conservation*. Bethesda, Maryland: American Fisheries Society, pp. 114–132.

Bernatchez L. (1997) Mitochondrial DNA confirms the existence of two glacial races of rainbow smelt (*Osmerus mordax*) and their reproductive isolation in the St Lawrence R. estuary (Quebec, Canada). *Molecular Ecology* **7**, 73–83.

Bernatchez L., Choinard A. & Lu G. (1999) Integrating molecular genetics with ecology in studies of adaptive radiation: whitefish, *Coregonus* sp. as a case study. *Biological Journal of the Linnean Society* **68**, 173–194.

Bernatchez L. & Dodson J.J. (1990) Allopatric origin of sympatric populations of lake whitefish (*Coregonus clupeaformis*) as revealed by mitochondrial DNA restriction analysis. *Evolution* **44**, 1263–1271.

Bernatchez L. & Dodson J.J. (1994) Phylogenetic relationships among palearctic and nearctic whitefish (*Coregonus* sp.) populations as revealed by mitochondrial DNA variation. *Canadian Journal of Fisheries and Aquatic Sciences* **51**, 240–251.

Bernatchez L. & Martin S. (1996) Mitochondrial DNA diversity in anadromous rainbow smelt, *Osmerus mordax* Mitchill: a genetic assessment of the member-vagrant hypothesis. *Canadian Journal of Fisheries and Aquatic Sciences* **43**, 424–433.

Bernatchez L., Vuorinen J.A., Bodaly R.A. & Dodson J.J. (1996) Genetic evidence for reproductive isolation and multiple origins of sympatric trophic ecotypes of whitefish (Coregonus). *Evolution* **50**, 624–635.

Brown C. & Warburton K. (1997) Predator recognition and anti-predator responses in the rainbowfish *Melanotaenia eachamensis*. *Behavioural Ecology and Sociobiology* **41**, 61–68.

Bush G.L. (1994) Sympatric speciation in animals: new wine in old bottles. *Trends in Ecology and Evolution* **9**, 285–288.

Campton D.E. (1987) Natural hybridisation and introgression in fishes. In N. Ryman & F. Utter (eds) *Population Genetics in Fisheries Management.* London: Chapman & Hall, pp. 161–192.

Colborn J., Crabtree R.E., Shaklee J.B., Philer E. & Bowen B.W. (2001) The evolutionary enigma of bonefishes (*Albula* spp.): Cryptic species and ancient separations in a globally distributed shorefish. *Evolution* **55**, 807–820.

Cracraft J. (1989) Speciation and its ontology: the empirical consequences of alternative species concepts for understanding patterns and processes of differentiation. In D. Otte & J.A. Endler (eds) *Speciation and its Consequences.* Sunderland, Mass.: Sinauer Associates, pp. 28–59.

Cracraft J. (1997) Species concepts in systematics and conservation biology – an ornithological viewpoint. In M.F. Claridge, H.A. Dawah & M.R. Wilson (eds) *Species: The Units of Biodiversity.* London: Chapman & Hall, pp. 325–339.

Crandall K.A., Bininda-Emonds O.R.P., Mace G.M. & Wayne R.K. (2000) Considering evolutionary processes in conservation biology. *Trends in Ecology and Evolution* **15**, 290–295.

Crowley L.E.L.M. & Ivanstoff W. (1991) Genetic similarity among populations of rainbowfishes (Pisces: Melanotaeniidae) from Atherton Tableland, Northern Queensland. *Ichthyological Exploration of Freshwaters* **2**, 129–137.

de Queiroz K. (1998) The general lineage concept of species, species criteria, and the process of speciation: a conceptual unification and terminological recommendations. In D.J. Howard & S.H. Berlocher (eds) *Endless Forms, Species and Speciation.* Oxford: Oxford University Press, pp. 57–75.

Dizon A.E., Lockyer C., Perrin W.F., Demaster D.P. & Sisson J. (1992) Rethinking the stock concept: a phylogeographic approach. *Conservation Biology* **6**, 24–36.

Douglas M.E. & Matthews W.J. (1992) Does morphology predict ecology? Hypothesis testing within a freshwater fish assemblage. *Oikos* **65**, 213–244.

Douglas M.R., Brunner P.C. & Bernatchez L. (1999) Do assemblages of *Coregonus* (Teleostei: Salmoniformes) in the central alpine region of Europe represent species flocks? *Molecular Ecology* **8**, 589–604.

Dowling T.E. & Secor C.L. (1997) The role of hybridization and introgression in the diversification of animals. *Annual Review of Ecology and Systematics* **28**, 593–619.

Endler J. (1977) Experimental clines and the effects of interaction among different loci. In *Geographic Variation, Speciation and Clines.* Princeton, New Jersey: Princeton University Press, pp. 97–141.

Frankel O.H. (1974) Genetic conservation: our evolutionary responsibility. *Genetics* **78**, 53–65.

Givnish T.J. & Systma K.J. (1997) *Molecular Evolution and Adaptive Radiations.* Cambridge: Cambridge University Press, 621 pp.

Gleeson D.J., McCallum H.I. & Owens I.P.F. (2000) Differences in initial and acquired resistance to *Ichthyophthirius multifiliis* between populations of rainbowfish. *Journal of Fish Biology* **57**, 466–475.

Goldstein P.Z., DeSalle R., Amato G. & Vogler A.P. (2000) Conservation genetics at the species boundary. *Conservation Biology* **14**, 120–131.

Hard J.J. (1995) A quantitative genetic perspective on the conservation of intraspecific diversity. *American Fisheries Society Symposium* **17**, 304–326.

Hendry A.P., Wenburg J.K., Bentzen P., Volk E.C. & Quinn T.P. (2000a) Rapid evolution of reproductive isolation in the wild: evidence from introduced Salmon. *Science* **290**, 516–518.

Hendry A.P., Vamosi, S.M., Latham S.J., Heilbuth, J.C. & Day, T. (2000b) Questioning species realities. *Conservation Genetics* **1**, 67–76.

Herbert B. & Peeters J. (1995) *Freshwater Fishes of Far North Queensland*. Brisbane: Queensland Department of Primary Industries, 74 pp.

Hubbs C.L. (1955) Hybridization between fish species in nature. *Systematic Zoology* **4**, 1–20.

Hughes R.M. & Noss R.F. (1992) Biological diversity and biological integrity: current concerns for lakes and streams. *Fisheries* **17**, 11–19.

Hurwood D.A. & Hughes J.M. (2001) Historical interdrainage dispersal of eastern rainbowfish (*Melantotaenia splendida splendida*) from the Atheron Tableland, northeastern Australia. *Journal of Fish Biology* **58**, 1125–1136.

Huxel G.R. (1999) Rapid displacement of native species by invasive species: effects of hybridization. *Biological Conservation* **89**, 143–152.

Kottelat M. (1997) European freshwater fishes. An heuristic checklist of the freshwater fishes of Europe (exclusive of the former USSR), with an introduction for non-systematists and comments on nomenclature and conservation. *Biologia* **52**, 1–271.

Lande R. & Shannon S. (1996) The role of genetic variability in adaptation and population persistence in a changing environment. *Evolution* **50**, 434–437.

Law R. (2000) Fishing, selection, and phenotypic evolution. *ICES Journal of Marine Science* **57**, 659–668.

Lindsay C.C. (1981) Stocks are chameleons: Plasticity in gill rakers of coregonid fishes. *Canadian Journal of Fisheries and Aquatic Sciences* **38**, 1497–1506.

Losos J.B., Warheit K.I. & Schoener T.W. (1997) Adaptive differentiation following experimental island colonisation in *Anolis* lizards. *Nature* **387**, 70–72.

Losos J.B., Jackman T.R., Larson A., de Queiroz K. & Rodriguez-Schettino L. (1998) Contingency and determinism in replicated adaptive radiations of island lizards. *Science* **279**, 2115–2118.

Lu G. & Bernatchez L. (1998) Experimental evidence for reduced hybrid viability between dwarf and normal ecotypes of lake whitefish (*Coregonus clupeoformis* Mitchill). *Proceedings of the Royal Society of London B* **265**, 1025–1030.

Lu G. & Bernatchez L. (1999) Correlated trophic specialisation and genetic divergence in sympatric lake whitefish ecotypes (*Coregonus clupeaformis*): support for the ecological speciation hypothesis. *Evolution* **53**, 1491–1505.

Lu G., Basley D.J. & Bernatchez L. (2001) Contrasting patterns of mtDNA and microsatellite introgressive hybridization between lineages of whitefish (*Coregonus clupeoformis*); relevance for speciation. *Molecular Ecology* **10**, 965–985.

Malhotra A. & Thorpe R.S. (1991) Experimental detection of rapid evolutionary response in natural lizard populations. *Nature* **353**, 347–348.

Mallet J. (1995) A species definition for the modern synthesis. *Trends in Ecology and Evolution* **10**, 294–299.

Mayden R.L., Burr B.M., Page L.M. & Miller R.R. (1992) The native freshwater fishes of North America. In R.L. Mayden (ed.) *Systematics, Historical Ecology and North American Freshwater Fishes*. Stanford: Stanford University Press, pp. 827–863.

Mayr E. (1963) *Animal Species and Evolution*. Cambridge Mass.: Belknap Press, 797 pp.

McAllister D.A. (1985) A list of the threatened and endangered fishes in Canada. *Syllogeus* **51**.

McGuigan, K. (2001) An addition to the rainbowfish (Melanotaeniidae) fauna of north Queensland. *Memoirs of the Queensland Museum* (in press).

McGuigan K., Zhu D., Allen G.R. & Moritz C. (2000) Phylogenetic relationships and historical biogeography of melanotaeniid fishes in Australia and New Guinea. *Marine Freshwater Research* **51**, 713–723.

Miller R.R. & Williams J.D. (1989) Extinction of north American fishes during the last century. *Fisheries* **14**, 22–38.

Moritz C. (1994a) Applications of mitochondrial DNA analysis on conservation: a critical review. *Molecular Ecology* **3**, 401–411.

Moritz C. (1994b) Defining 'Evolutionary Significant Units' for conservation. *Trends in Ecology and Evolution* **9**, 373–375.

Moritz C. (1999) Conservation units and translocations: Strategies for conserving evolutionary processes. *Hereditas* **130**, 217–228.

Naylor R.L., Goldburg R.J., Mooney H., Beveridge M., Clay J., Folke C., Kautsky N., Lubcheno J., Primavera J. & Williams M. (1998) Nature's subsidies to shrimp and salmon farming. *Science* **282**, 883–884.

Nielsen J.L. (1995) *Evolution and the Aquatic Ecosystem: Defining Unique Units in Population Conservation*. Bethesda, Maryland: American Fisheries Society, 310 pp.

Pearse C.M., Lande R. & Bull J.J. (1989) A model of population growth, dispersal and evolution in a changing environment. *Ecology* **70**, 1657–1664.

Phillips R.B. & Ehlinger T.J. (1995) Evolutionary and ecological considerations in the reestablishment of Great Lakes Coregonid fishes. In J.L. Nielsen (ed.) *Evolution and the Aquatic Ecosystem: Defining Unique Units in Population Conservation*. Bethesda, Maryland: American Fisheries Society, pp. 133–144.

Pigeon D., Choinard A. & Bernatchez L. (1997) Multiple modes of speciation involved in the parallel evolution of sympatric dwarf and normal morphotypes of lake whitefish (*Coregonus clupeoformis*, Salmonidae). *Evolution* **51**, 196–205.

Pusey B.J., Bird J., Kennard M.J. & Arthington A.H. (1997) Distribution of the Lake Eacham Rainbowfish in the Wet Tropics Region, North Queensland. *Australian Journal of Zoology* **45**, 75–84.

Quinn T.P., Unwin M.J. & Kinnison M.T. (2000) Evolution of temporal isolation in the wild: Genetic divergence in timing of migration and breeding by introduced chinook populations. *Evolution* **54**, 1372–1385.

Rahel F.J. (2000) Homogenisation of fish faunas across the United States. *Science* **288**, 854–856.

Reznick D.N., Shaw F.H., Rodd F.H. & Shaw R.G. (1997) Evaluation of the rate of evolution in natural populations of guppies (*Poecilia reticulata*). *Science* **275**, 1934–1937.

Rice W.R. & Hostert E.E. (1993) Perspective: laboratory experiments on speciation: what have we learned in forty years. *Evolution* **47**, 1637–1653.

Ritland K. (2000) Marker-inferred relatedness as a tool for detecting heritability in nature. *Molecular Ecology* **9**, 1195–1204.

Robins C.R., Bailey R.M., Bond C.E., Brooker J.R., Lachner E.A., Lea R.N. & Scott W.B. (1991) Common and scientific names of fishes in the United States and Canada. *American Fisheries Society Special Publication* **20**, 244 pp.

Robinson B.W. & Schluter D. (1999) Natural selection and the evolution of adaptive genetic variation in northern freshwater fishes. In T.A. Mousseau, B. Sinervo & J.A. Endler (eds) *Adaptive Genetic Variation in the Wild*. New York: Oxford University Press, pp. 65–94.

Rüber L., Verheyen E. & Meyer A. (1999) Replicated evolution of trophic specialisations in an endemic cichlid fish lineage from Lake Tanganyika. *Proceedings of the National Academy of Science (USA)* **96**, 10230–10235.

Rundle H.D., Nagel L., Boughman J.W. & Schluter D. (2000) Natural selection and parallel speciation in sticklebacks. *Science* **287**, 306–308.

Ruzzante D.E., Walde S.J., Cussac V.E., Macchi P.J. & Alonso M.F. (1998) Trophic polymorphism, habitat and diet segregation in *Percichthys trucha* (Pisces: Percichthydiae) in the Andes. *Biological Journal of the Linnean Society* **65**, 191–214.

Ryder O.A. (1986) Species conservation and systematics: the dilemma of subspecies. *Trends in Ecology and Evolution* **1**, 9–10.

Schluter D. (1994) Experimental evidence that competition promotes divergence in adaptive radiation. *Science* **266**, 798–801.

Schluter D. (1996) Ecological speciation in postglacial fishes. *Philosophical Transactions of the Royal Society* **351**, 807–814.

Schneider C.J., Smith T.B., Larison B. & Moritz C. (1999) A test of alternative models of diversification in tropical rainforests: ecological gradients vs. refugia. *Proceedings of the National Academy of Science (USA)* **96**, 13869–13873.

Smith G.R. (1992) Introgression in fishes: significance for paleontology, cladistics and evolutionary rates. *Systematic Biology* **41**, 41–57.

Stiassny M.L.J. & Meyer A. (1999) Cichlids of the African Rift Lakes. *Scientific American* **280**, 64–69.

Strecker U., Meyer C.G., Sturmbauer C. & Wilkens H. (1996) Genetic divergence and speciation in an extremely young species flock in Mexico formed by the genus *Cyprinodon* (cyprinodontidae: Teleostei). *Molecular Phylogenetics and Evolution* **6**, 143–149.

Svardson G. (1998) Postglacial dispersal and reticulate evolution of nordic coregonids. *Nordic Journal of Freshwater Research* **74**, 3–32.

Taylor E.B. (1991) A review of local adaptation in Salmonidae, with particular reference to Pacific and Atlantic salmon. *Aquaculture* **98**, 185–207.

Taylor E.B. (1999) Species pairs of northern temperate freshwater fishes: evolution, taxonomy and conservation. *Reviews of Fish Biology and Fisheries* **9**, 299–324.

Taylor E.B. & Bentzen P. (1993a) Evidence for multiple origins and sympatric divergence of trophic ecotypes of smelt (*Osmerus*) in northeastern North America. *Evolution* **47**, 813–832.

Taylor E.B. & Bentzen P. (1993b) Molecular genetic evidence for reproductive isolation between sympatric populations of smelt *Osmerus* in lake Utopia, south-western New Brunswick, Canada. *Molecular Ecology* **2**, 345–357.

Templeton A.R. (1989) The meaning of species and speciation: A genetic perspective. In D. Otte & J.A. Endler (eds) *Speciation and its Consequences.* Sunderland, Mass.: Sinauer Associates, pp. 3–27.

Turgeon J., Estoup A. & Bernatchez L. (1999) Species flock in the north America great lakes: molecular ecology of the Lake Nipigon ciscoes (Teleostei: Coregonidae: *Coregonus*). *Evolution* **53**, 1857–1871.

Vogler A.P. & DeSalle R. (1994) Diagnosing units of conservation management. *Conservation Biology* **8**, 354–363.

Waples R.S. (1991) Pacific salmon, *Oncorhynchus* spp., and the definition of 'species' under the Endangered Species Act. *Marine Fisheries Review* **53**, 11–22.

Waples R.S. (1995) Evolutionarily significant units and the conservation of biological diversity under the Endangered Species Act. In J. Nielsen (ed.) *Evolution and the Aquatic Ecosystem.* Bethesda, Maryland: American Fisheries Society.

Zhu D., Jamieson B.G.M., Hugall A. & Moritz C. (1994) Sequence evolution and phylogenetic signal in control-region and cytochrome b sequences of rainbow fishes (Melanotaeniidae). *Molecular Biology and Evolution* **11**, 672–683.

Zhu D., Degnan S. & Moritz C. (1998) Evolutionary distinctiveness and status of the endangered Lake Eacham rainbowfish (*Melanotaenia eachamensis*). *Conservation Biology* **12**, 80–93.

Chapter 28
Hybridogenous biotypes in spined loaches (genus *Cobitis*) in Germany with implications for conservation

J. BOHLEN,* P. RÁB, V. ŠLECHTOVÁ, M. RÁBOVÁ
Institute of Animal Physiology and Genetics, Academy of Sciences of Czech Republic, Czech Republic

D. RITTERBUSCH
Institute of Inland Fisheries e.V., Jägerhof am Sacrower See, Groß Glienicke, Germany

J. FREYHOF
Leibniz – Institute of Freshwater Ecology and Inland Fisheries, Berlin, Germany

Abstract

A karyotype and allozyme screening of 12 populations of spined loaches (genus *Cobitis*) from eight river systems in Germany tested the hypothesis that *Cobitis taenia* is the only species in these river systems. Three populations contained exclusively *C. taenia*, while in eight populations from the Odra, Elbe, Weser, Issel and Rhine river basins *C. taenia* was found together with hybridogenous biotypes, including genomes of *C. taenia* and *Cobitis elongatoides*. In one population from the Danube basin, hybridogenous biotypes including genomes of *C. elongatoides* and *Cobitis tanaitica* were identified. Altogether, five different combinations of genomes were found in the hybridogenous biotypes, which included diploid, triploid and tetraploid specimens. Diploid hybridogenous biotypes were found for the first time. Depending on locality, up to three different biotypes were associated with *C. taenia*. Several hybridogenous males indicate that gynogenesis is not the only mode of reproduction for the polyploid biotypes. Complexes appear to be a common type of population structure in spined loaches. The widespread occurrence of further hybrid complexes in the whole of Central Europe seems likely. Strong locality-dependent differences in the composition of complexes reflect the independent state of each complex. Consequently, conservation legislation and conservation measures need: (1) proper identification of species and population composition; and (2) a concept to treat complexes as evolutionary units.

Keywords: *Cobitis elongatoides*, *Cobitis taenia*, conservation, hybrids, karyotype, polyploidy.

28.1 Introduction

For a long time, *Cobitis taenia* L. was believed to be the most widespread species of spined loach occurring throughout Europe (Lelek 1987). However, recent studies

*Correspondence: Institute of Animal Physiology and Genetics, Academy of Sciences of Czech Republic, 277 21 Libéchov, Czech Republic (email: bohlen@iapg.cas.cz).

showed that several different species, including *C. taenia*, occur in large parts of Europe (reviewed in Kottelat 1997). In Eastern Europe and eastern Central Europe, these species often coexist with various polyploid hybridogenous biotypes (Vasil'ev & Vasil'eva 1982; Boroń 1992; Ráb & Slavík 1996; Vasil'eva & Vasil'ev 1998). Until now, these biotypes were considered to belong to gynogenetically reproducing unisexual lineages, which have to co-occur with a bisexually reproducing species of *Cobitis* (host-specific sperm parasitism) (Vasil'ev, Vasil'eva & Osinov 1989). Due to the high morphological similarity between diploid and polyploid fishes, the most powerful tool to distinguish species and to identify hybridogenous biotypes is an investigation of the karyotype (Vasil'eva & Vasil'ev 1998), best combined with a study of diagnostic allozymes (Šlechtová, Šlechta, Lusková, Lusk, Halacka & Bohlen 2000).

Spined loaches are recognised as endangered in many European countries (Kotusz 1996) and listed in the appendix of the Bern Convention. Recently, *C. taenia* was included as a species of special concern in the appendix of Freshwater Fisheries Directive of the European Union. This Directive enforces conservation activities in all member states of European Union; therefore an increase of interest in conservation of spined loaches can be expected. But conservation measures are irrelevant unless the distribution of species, regional differences in population structure and the problem of hidden taxonomic diversity are understood. In the case of a hidden diversity within the target species, any conservation measure including transplantation and propagation of specimens risks disturbance of the natural state. Complexes of spined loaches were even considered to have evolutionary significance by allowing reticulate speciation (Vasil'eva, Osinov & Vasil'ev 1989).

In the present study the assumption that *C. taenia* is the only representative of the genus *Cobitis* in a given part of western Central Europe was tested. The study was carried out in Germany, because:

(1) it is located close to known localities with diploid–polyploid complexes in Poland and Czech Republic; and
(2) it covers parts of several major European drainages, allowing predictions to be made for neighbouring countries.

28.2 Material and methods

Sample specimens were obtained by dip net or electric fishing at the following localities (numbers correspond to the numbers in Fig. 28.1, Table 28.2):

(1) backwater of the River Rhine near Wörth, Rhine basin;
(2) River Kleevsche Graaf, Issel basin;
(3) River Lagerhase at Essen (Oldb.), Ems basin;
(4) River Haaren at Oldenburg, Weser basin;
(5) River Ise, Weser basin;
(6) Lake Kleiner Plöner See, Schwentine basin;
(7) River Alte Ilau near Winsen, Elbe basin;
(8) Lake Müggelsee, Elbe basin;

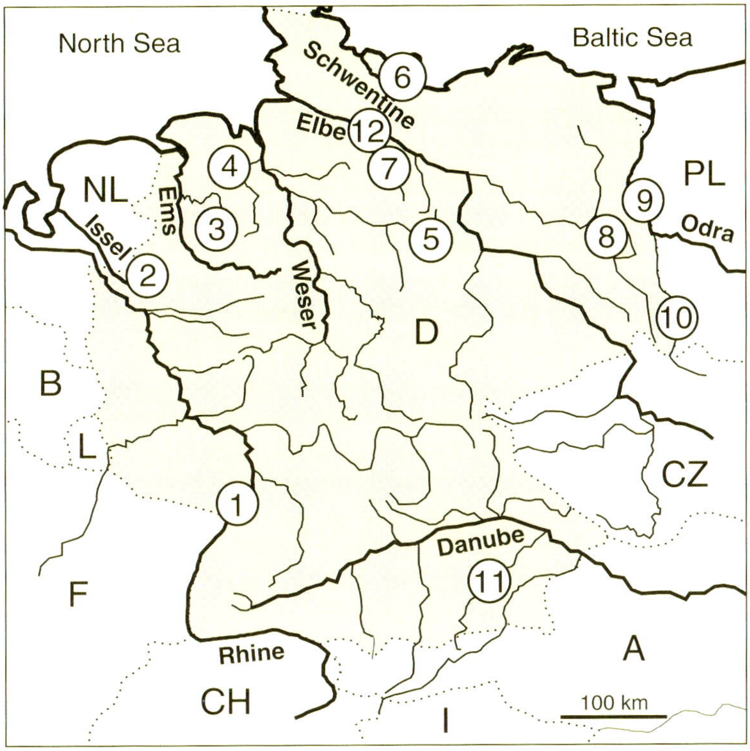

Figure 28.1 Localities of sampling of spined loaches in Germany (shaded area). Dotted lines represent country borders, countries are indicated by their international car code

(9) River Odra at Küstrin, Odra basin;
(10) River Pleißnitz at Hagenwerder, Odra basin;
(11) pond near Erding, Danube basin, aquarium bred specimens (F_1 generation, details about breeding in Bohlen 1999);
(12) River Bille at Hamburg, Elbe basin.

Chromosome preparation followed Ráb and Roth (1989). Six allozyme loci were studied according to Harris and Hopkinson (1976), with modifications described in Šlechtová, Lusková, Šlechta, Lusk and Berrebi (1998): Glucosephosphate isomerase (Enzyme commission number – EC 5.3.1.9, Gpi-A*, liver, Gpi-B*, muscle), Superoxide dismutase (EC 1.15.1.1, Sod*, liver), Aspartate aminotransferase (EC 2.6.1.1, s-Aat*, liver, muscle), Malate dehydrogenase (EC 1.1.1.37, s-Mdh-A*, liver) and Lactate dehydrogenase (EC 1.1.1.27, Ldh-B*, muscle). The distances between allelic bands were expressed as relative position in comparison to the position of the most common allele. Species were identified by comparing their karyotype and allelic composition with those of *C. elongatoides* Băcescu & Maier, *C. taenia* and *C. tanaitica* Băcescu & Maier (Table 28.1), taken from Vasil'ev (1995), Ráb, Rábová, Bohlen and Lusk (2000), Šlechtová *et al.* (2000) and unpublished data. In specimens without successful preparation of

Table 28.1 Diagnostic karyotype characters and alleles for the three species of *Cobitis* mentioned in the present study

	Number of chromosomes				Diagnostic alleles					
Species	2n	m	sm	a	Gpi-A*	Gpi-B*	Sod*	s-Aat*	s-Mdh-A*	Ldh-B*
C. taenia	48	10	20	18	74, 87	100, 300	60	67, 71, 78	60	100, 117
C. elongatoides	50	30	18	2	100, 113	100	100	85, 100	40, 100	100
C. tanaitica	50	10	22	18	87	100	60	71	100	100

m, meta-; sm, submeta-; a, acrocentric chromosomes in diploid chromosome set. Abbreviations of alleles are explained in text.

chromosomes, genetic composition was derived from relative dosage of genes. Usually, at least the rough number of chromosomes could be estimated for the fish (ploidy level), making the relative dosage of genes a reliable tool.

28.3 Results

In ten out of 12 locations diploid pure *C. taenia* were found (Table 28.2). No other forms besides *C. taenia* were found in locations 3, 4 and 6. In eight locations (locations 1, 2, 5, 7, 8, 9, 10 and 12) diploid ($2n = 49$), triploid ($3n = 73$ or 74) and tetraploid ($4n = 97$) hybridogenous biotypes between *C. taenia* and *C. elongatoides* occurred. In all these locations, except location 10, the biotypes occurred together with *C. taenia*. The two triploid biotypes differed in their genome composition (*C.* 1 *taenia*–2 *elongatoides* v. *C.* 2 *taenia*–1 *elongatoides*). Both individuals from location 11 were tetraploid ($4n = 100$) biotypes containing genomes of *C. elongatoides* and *C. tanaitica* (one specimen identified by karyotype, the second by gene dosage). Altogether, 41.5% of all investigated specimens were hybridogenous biotypes. Up to three different biotypes were associated with *C. taenia* in certain populations (localities 7 and 9).

28.4 Discussion

According to the analysis, the situation of *Cobitis* in Germany is as follows. One species, *C. taenia*, is widespread and occurs in all investigated river basins, except the Danube basin. In the Danube basin, *C. elongatoides* or *C. tanaitica* may be expected to occur, since they were the parental species of the biotype found. *C. elongatoides* may also be found in some locations in the upper Elbe basin or upper Odra basin, since this species was already found in these basins in the Czech Republic (Ráb & Slavik 1996) and Poland (Boroń 1999). Kottelat (1997) indicated that another species, *C. spilura* Holandre, may occur in the western parts of Germany, especially in the Rhine

Table 28.2 Composition of karyotypes and diagnostic allozymes of 11 populations of spined loach from Germany

		Karyotypes				Alleles									
River system Sample number	Species or biotype	Chromosome no.	Male	Female	Total	Gpi-A*	Gpi-B*	Sod*	s-Aat*	s-Mdh-A*	Ldh-B*	Male	Female	Juvenile	Total
1 Rhine	*C. taenia*	2n = 48	1	1	2	87	100, 300	60	71	60	117	1	2	0	3
	C. 1 *taenia*–2 *elongatoides*	3n = 74	0	6	6	87, 113	100	60, 100	71, 100	60, 100	100, 117	0	8	0	8
	C. 2 *taenia*–1 *elongatoides*	3n = 73	0	2	2	87, 113	100	60, 100	71, 100	60, 100	100, 117	0	8	0	8
2 Issel	*C. taenia*	2n = 48	8	8	16	87	100	60	71	60	100, 117	8	8	0	16
	C. 2 taenia–1 *elongatoides*	3n = 73	0	11	11	87, 113	100	60, 100	71, 100	60, 100	100, 117	0	11	0	11
3 Ems	*C. taenia*	2n = 48	6	10	16	87	100	60	71	60	100, 117	12	18	0	30
4 Weser	*C. taenia*	2n = 48	2	13	15	87	100	60	71	60	100, 117	13	19	1	33
5 Weser	*C. taenia*	2n = 48	4	3	7	87	100	60	71	60	100	5	3	0	8
	C. 1 *taenia*–1 *elongatoides*	2n = 49	0	12	12	87, 100	100	60, 100	71, 100	60, 100	100	0	12	0	12
	C. 2 taenia–1 *elongatoides*	3n = 73	2	5	7	87, 100	100	60, 100	71, 100	60, 100	100	3	5	0	8
6 Schwentine	*C. taenia*	2n = 48	2	2	4	74, 87	100	60	67, 71	60	100	8	8	8	24
7 Elbe	*C. taenia*	2n = 48	3	5	8	87	100	60	71	60	100, 117	4	6	0	10
	C. 1 *taenia*–1 *elongatoides*	2n = 49	0	3	3	87, 113	100	60, 100	71, 100	60, 100	100	0	5	0	5
	C. 2 *taenia*–1 *elongatoides*	3n = 73	0	2	2	87, 113	100	60, 100	71, 100	60, 100	100	0	2	0	2
	C. 3 *taenia*–1 *elongatoides*	4n = 97	0	1	1	87, 113	100	60, 100	71, 100	?	100	0	1	0	1
8 Elbe	*C. taenia*	2n = 48	0	0	0	87	100	60	71	60	100, 117	3	3	1	7
	C. 2 *taenia*–1 *elongatoides*	3n = 73	4	8	12	87, 100	100	60, 100	71, 100	60, 100	100, 117	6	21	0	27
	C. 1 *taenia*–2 *elongatoides*	3n = 74	0	0	0	87, 100	100	60, 100	71, 100	60, 100	100, 117	0	2	0	2

Table 28.2 continued

		Karyotypes				Alleles									
River system Sample number	Species or biotype	Chromosome no.	Male	Female	Total	Gpi-A*	Gpi-B*	Sod*	s-Aat*	s-Mdh-A*	Ldh-B*	Male	Female	Juvenile	Total
9 Odra	*C. taenia*	2n = 48	1	3	4	87	100	60	71	60	100, 117	3	5	9	17
	C. 1 *taenia*–1 *elongatoides*	2n = 49	0	1	1	87, 100, 113	100	60, 100	71, 100	60, 100	100	0	3	3	6
	C. 2 *taenia*–1 *elongatoides*	3n = 73	0	2	2	87, 100, 113	100	60, 100	71, 100	60, 100	100, 117	0	4	5	9
	C. 3 *taenia*–1 *elongatoides*	4n = 97	1	0	1	87, 100	100	60, 100	71, 100	40, 60,?	100, 117	1	0	1	2
10 Odra	*C.* 1 *taenia*–2 *elongatoides*	3n = 74	0	4	4	87, 100	100	60, 100	71, 100	40, 60, 100	100, 117	0	4	0	4
11 Danube	*C.* 3 *elongatoides*–1 *tanaitica*	4n = 100	0	1	1	87, 100	100	100	71, 85, 100	100	100	0	0	2	2
12 Elbe	*C. taenia*	2n = 48	0	1	1	87, 100	100	60	71	60	100	0	1	0	1
	C. 2 *taenia*–1 *elongatoides*	3n = 74	0	1	1	87, 113	100	60, 100	71, 100	60, 100	100, 117	0	1	0	1

Sample numbers correspond to Fig. 28.1. Question marks indicate cases where the allele position could not be estimated doubtless.

basin. The present data did not support this possibility, but showed *C. taenia* plus triploid biotypes to occur.

Besides diploid bisexual fish, five different biotypes were found in the present samples. Up to four forms with different compositions of genomes occurred together in certain complexes. These complexes exhibited a wide distribution within the investigated area. Therefore, complexes cannot be regarded as rare events of local significance, but have to be considered as the common type of population structure in spined loaches. Complexes from different sampling sites differed considerably in their composition of hybridogenous biotypes, even within one river catchment. Moreover, biotypes with identical karyotype composition (e.g. *C.* 2 *taenia*–1 *elongatoides*) differed between locations in their allele composition. These locality-dependent differences in the composition of complexes reflect the independent state of each complex. Local and/or historical factors probably lead to selection or generation of biotypes in certain complexes. The huge impact of local ecological condition on coexistence and composition of hemiclonal hybrid and sexual hosts was already demonstrated by Hellriegel & Reyer (2000). In the light of freshwater fish conservation, the differences between local populations indicate that no translocations of specimens between populations, even within the same river catchment, should be carried out.

However, hybridogenous biotypes make up a considerable part of spined loach populations. This high abundance of hybrids within the samples demonstrates that they are far from being the 'evolutionary dead-ends' that clonal lineages sometimes were considered to be (Vrijenhoek 1989; Milinski 1995). Until now, little is known about ecological similarity of bisexual and unisexual forms in spined loach, but preliminary studies indicate a high degree of similarity (Ritterbusch & Bohlen 2000). The mechanisms that enable the coexistence of several ecologically similar, but genetically distinct forms in the complexes remain unclear. Thus the impact of hybridogenous forms on the bisexual species cannot be assessed. The consequence for fish conservation should be not to disturb such systems by active measures.

Another paucity of information regards the reproductive system of spined loaches in complexes. The common opinion assumes a past hybridisation to lead towards diploid hybrids which gives rise to unisexual polyploid forms, which reproduce by gynogenesis (Vasil'ev *et al.* 1989). Indeed, diploid hybrids were found that were not previously observed in other studies on complexes of spined loaches. In accordance with the theory of a gynogenetic mode of reproduction, all diploid hybridogenous biotypes and most polyploid biotypes were females. However, ten polyploid specimens were males, which could not have arisen from an exclusively unisexual clonal lineage. Several examples of hybridogenous fish biotypes with a much more advanced system of reproduction than pure gynogenesis have been already described (Kim & Lee 1995, 2000; Alves, Coelho & Collares-Pereira 1998; Zhang & Arai 1998). In these examples, some variability in the reproductive mode of hybridogenous biotypes led to the production of non-clonal offspring, in some cases even to cyclic changes in ploidy and to the genesis of biotypes with yet another mode of reproduction (Kim & Lee 1995, 2000; Alves *et al.* 1998). The presence of males in hybridogenous biotypes suggests that the mode of reproduction in *Cobitis* includes not only a strict gynogenesis, but also some mechanism of 'minimal sex' (Beukeboom & Vrijenhoek 1998). This issue has to be considered in conservation measures for these fishes, especially if they include artificial propagation.

The distribution of *C. taenia* and the hybridogenous biotypes in Germany has to be evaluated with respect to their general distribution in Europe (Fig. 28.2). Pure diploid *C. tanaitica* were, until now, only reported from the north-western Black Sea area (Vasil'eva & Vasil'ev 1998). The current finding of *C. tanaitica* genomes in hybridogenous biotypes from southern Germany shows it to have a much wider distribution than previously believed. *Cobitis tanaitica* may have been one of the parental species of tetraploid specimens from upper Vistula basin in Poland (Boroń & Kotusz 2000) and Danube basin in Czech Republic (Ráb, Rábová, Bohlen & Lusk 2000). *Cobitis elongatoides* was found in the basins of the upper Odra in Poland (Boroń & Kotusz 2000), the upper Elbe in Czech Republic (Ráb *et al.* 2000) and several parts of the Danube (Erös 2000; Povz & Šumer 2000; Ráb *et al.* 2000). Polyploid biotypes containing genomes of *C. taenia* and *C. elongatoides* were reported from Odra and Elbe (Boroń & Kotusz 2000; Ráb *et al.* 2000). The contribution of genomes of *C. elongatoides* to all hybridogenous biotypes in Germany also demonstrates it has a strong impact on the ichthyofauna of

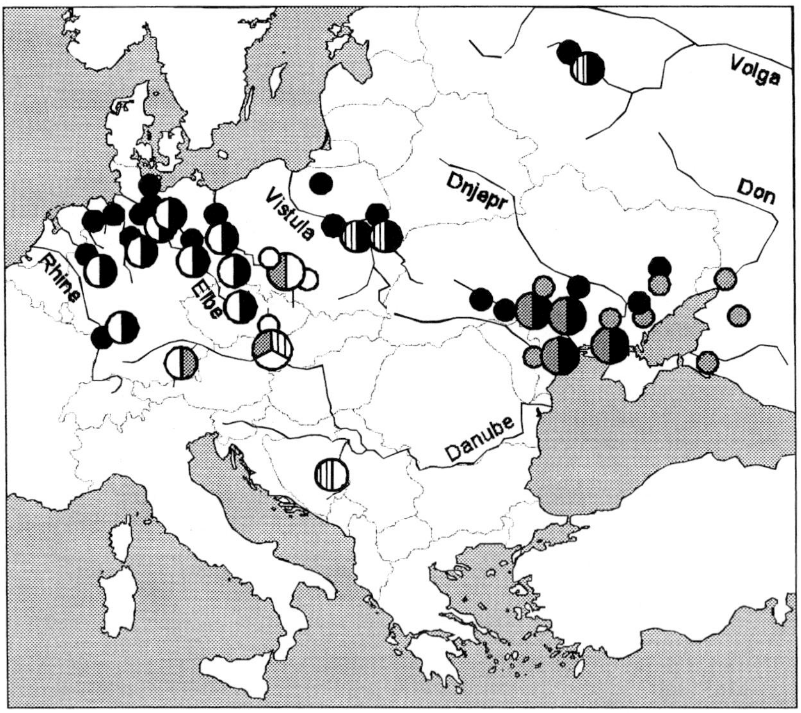

Figure 28.2 Occurrence of *C. taenia*, *C. elongatoides* and *C. tanaitica* and hybrid biotypes in Central and Eastern Europe. Data from Boron 1992, Vasil'ev 1995, Ráb & Slavík 1996, Vasil'eva & Vasil'ev 1998 and present study. Dotted lines represent country borders; the main river systems are indicated. ● *C. taenia*, ◍ *C. tanaitica*, ○ *C. elongatoides*, striped fields indicate unidentified parental species, ◑ hybridogenous biotypes

western Central Europe. Pure *C. taenia* were found in the Ukraine and Russia (Vasil'eva & Vasil'ev 1998) and in Poland (Boroń & Kotusz 2000). The new findings from northern Germany exceed the known distribution area. Irrespective, the western limit of distribution still is not reached in Germany.

Based on this scenario, some predictions may be drawn on the identity of spined loaches in further, not yet investigated, countries. Pure *C. taenia* and hybridogenous biotypes with genomes of *C. taenia* and *C. elongatoides* should be expected along the whole drainage areas of the rivers Rhine and Issel, namely northern Switzerland, the Benelux countries and northern France. *Cobitis taenia* also probably occurs in Scandinavia and the British Isles. *Cobitis elongatoides* and *C. tanaitica*, plus hybridogenous biotypes with genomes of the both species, should be expected in the Danube basin. Consequently, three species make up nearly all spined loach populations in Central Europe. Unfortunately, they were not identified as different species in fish biology and conservation legislation from large parts of its distribution range. Therefore, information about threats as well as considerations in protection are still missing.

Another point to be recognised in conservation of spined loaches in Europe is complexes, which now appear to be frequent. Gyno- or hybridogenic biotypes certainly represent independent evolutionary units, and a discussion has already started whether they shall be protected as species or not (Kraus 1995). Moreover, in the examples of fish biotypes with cyclic change of ploidy, not only the biotypes but the complex as a whole have to be considered as the evolutionary unit. In this context, the complexes in *Cobitis* differ from complexes in some cyprinid fish (e.g. *Carassius* or *Leuciscus*) that they are restricted in their possibility to change the sperm donating species. By that, complexes in *Cobitis* are ecologically more closed systems and can occur only sympatrically with a bisexual species of spined loach. Therefore, the question whether or not hybridogenous bioypes in *Cobitis* should be protected is a little academic. The hybridogenous biotypes will always enjoy the same protection as the associated species, profiting from supporting measures and suffering from threats.

Acknowledgements

We wish to express our thanks to G. Augustin, A. Bischoff, M. Breil, P. Grabe, A. Hartl, C. Lecour, M. Pfeifer and D. Staginnus for providing samples or sampling permissions, to J. Bučková, L. Kulová, J. Pivoňková and V. Šlechtová jun for help with the laboratory work as well as K. Janko, P. Kotlík, A. Nolte, V. Šlechta and two anonymous reviewers for critical discussion of earlier drafts of the manuscript. This study was supported by grant No. 206/00/0668 by GA CR.

References

Alves M.J., Coelho M.M. & Collares-Pereira M.J. (1998) Diversity in the reproductive modes of females of the *Rutilus alburnoides* complex (Teleostei: Cyprinidae): a way to avoid the genetic constraints of uniparentalism. *Molecular Biology and Evolution* **15**, 1233–1242.

Beukeboom L.W. & Vrijenhoek R.C. (1998) Evolutionary genetics and ecology of sperm-dependent parthenogenesis. *Journal of Evolutionary Biology* **11**, 755–782.

Bohlen J. (1999) Reproduction of spined loach, *Cobitis taenia* (Cypriniformes: Cobitidae) under laboratory conditions. *Journal of Applied Ichthyology* **15**, 49–53.

Boron A. (1992) Karyotype study of diploid and triploid *Cobitis taenia* (Pisces: Cobitidae) from Vistula river basin. *Cytobios* **72**, 201–206.

Boron A. (1999) A preliminary report and karyotype of a new *Cobitis* species in the ichthyofauna of Poland. *Cytobios* **98**, 59–64.

Boron A. & Kotusz J. (2000) The preliminary data on diploid–polyploid complexes of the genus *Cobitis* in the Odra River basin, Poland (Pisces: Cobitidae). *Folia Zoologica* **49**(Suppl. 1), 79–84.

Erös T. (2000) Population biology of *Cobitis elongatoides* in a lowland stream of the Middle Danube (Hungary). *Folia Zoologica* **49**(Suppl.1), 151–158.

Harris H. & Hopkinson D.A. (1976) *Handbook of Enzyme Electrophoresis in Human Genetics.* Amsterdam: North Holland Publishers.

Hellriegel B. & Reyer H.U. (2000) Factors influencing the composition of mixed populations of a hemiclonal hybrid and its sexual host. *Journal of Evolutionary Biology* **13**, 906–918.

Kim I.S. & Lee E.H. (1995) Interspecific hybridisation between triploid hybrid fish, *Cobitis sinensis-longicorpus* and two diploid species from Korea. *Korean Journal of Ichthyology* **7**, 71–78.

Kim I.S. & Lee E.H. (2000) Hybridization experiment of diploid-triploid cobitid fishes, *Cobitis sinensis-longicorpus* complex (Pisces: Cobitidae). *Folia Zoologica* **49**(Suppl. 1), 17–22.

Kottelat M. (1997) European Freshwater Fishes. *Biologia* **52**(Suppl. 5), 1–271.

Kotusz J. (1996) Species protection of loaches (Cobitoidea, Cypriniformes) in Poland in relation to their distribution and status in other European countries. *Zoologica Poloniae* **41**(Suppl.), 147–155.

Kraus F. (1995) The conservation of unisexual vertebrate populations. *Conservation Biology* **9**, 956–959.

Lelek A. (1987) *Freshwater Fishes of Europe. Vol. 9 Threatened Fishes of Europe.* Wiesbaden: Aula-Verlag, 343 pp.

Milinski M. (1995) Hybridogenetic frogs on an evolutionary dead-end road. *Trends in Ecology and Evolution* **9**, 62.

Povz M. & Šumer S. (2000) Present status and distribution of the species of the genera *Misgurnus*, *Cobitis* and *Sabanejewia* in Slovenia. *Folia Zoologica* **49**(Suppl. 1), 107–112.

Ráb P. & Roth P. (1989) Cold-blooded vertebrates. In: P. Balicek, J. Forejt & J. Rubeš (eds) *Methods of Chromosome Analysis.* Brno: Czechoslovak Biological Society Publishers, pp. 115–124 (in Czech).

Ráb P. & Slavik O. (1996) Diploid–triploid–tetraploid complex of the spined loach, genus *Cobitis* in the Psovka creek: the first evidence of the new species of *Cobitis* in the ichthyofauna of the Czech Republic. *Acta Universitatis Carolinae-Biologica* **39**, 201–214.

Ráb P., Rábová M., Bohlen J. & Lusk S. (2000) Genetic differentiation of the two hybrid diploid–polyploid complexes of loaches, genus *Cobitis* (Cobitidae) involving *C. taenia*, *C. elongatoides* and *C.* spp. in the Czech Republic: Karyotypes and cytogenetic diversity. *Folia Zoologica* **49**(Suppl. 1), 55–66.

Ritterbusch D. & Bohlen J. (2000) On the ecology of spined loach in Lake Müggelsee. *Folia Zoologica* **49**(Suppl. 1), 187–192.

Šlechtová V., Lusková V., Šlechta V., Lusk S. & Berrebi P. (1998) Genetic variability of common barbel, *Barbus barbus* populations in Czech Republic. *Folia Zoologica* **47**(Suppl. 1), 21–33.

Šlechtová V., Šlechta V., Lusková V., Lusk S., Halacka K. & Bohlen J. (2000) Genetic differentiation of two diploid–polyploid complexes of spined loach, genus *Cobitis* (Cobitidae) involving *C. taenia*, *C. elongatoides* and *C.* spp. in Czech Republic: Allozyme interpopulation and interspecific differences. *Folia Zoologica* **47**(Suppl. 1), 67–78.

Vasil'ev V.P. (1995) (Karyological diversity and taxonomical non-homogenity of *Cobitis 'taenia'* (Pisces: Cobitidae)). *Dokladi Academii NAUK* **342**, 839–842 (in Russian).

Vasil'ev V.P. & Vasil'eva E.D. (1982) (A new diploid-polyploid complex in fish). *Dokladi Academii NAUK* **266**, 250–252 (in Russian).

Vasil'ev V.P., Vasil'eva E.D. & Osinov A.G. (1989) Evolution of a diploid–triploid–tetraploid complex in fishes of the genus *Cobitis* (Pisces: Cobitidae). In R.M. Dawley & J.P. Bogart (eds) *Evolution and Ecology of Unisexual Vertebrates*. Albany, New York: Bulletin of the New York State Museum 466, pp. 153–169.

Vasil'eva E.D. & Vasil'ev V.P. (1998) Sibling species in genus *Cobitis* (Cobitidae). *Cobitis rossomeridionalis* sp. nova. *Journal of Ichthyology* **38**, 580–590.

Vasil'eva E.D., Osinov A.G. & Vasil'ev V.P. (1989) The problem of reticulate speciation in vertebrates: the diploid–triploid–tetraploid complex in the genus *Cobitis* (Cobitidae). I. Diploid species. *Journal of Ichthyology* **29**, 35–47.

Vrijenhoek R.C (1989) Genetic and ecological constrains on the origins and establishment of unisexual vertebrates. In R.M. Dawley & J.P. Bogart (eds) *Evolution and Ecology of Unisexual Vertebrates*. Albany, New York: Bulletin of the New York State Museum 466, pp. 24–31.

Zhang Q. & Arai K. (1998) Distribution and reproductive capacity of natural triploid individuals and occurrence of unreduced eggs as a cause of polyploidization in the loach, *Misgurnus anguillicaudatus*. *Ichthyological Res*earch **46**, 153–161.

Chapter 29
RAPD markers as diagnostic keys for Cyprinidae identification: implications for conservation management

C. CALLEJAS and M.D. OCHANDO*
Departamento de Genética, Facultad de CC. Biológicas, Universidad Complutense, Madrid, Spain

Abstract

Classifications of fish species are generally based on classical morphological and osteological methods. Moreover, accurate identifications are in many cases, such as Cyprinidae species, very difficult because they have considerable infraspecific morphological variability. Nowadays, the development of Random Amplified Polymorphic DNA, based on the polymerase chain reaction (PCR) method (RAPD-PCR), provides a powerful tool in genetic research and particularly in identification problems. In this study RAPD markers were characterised to identify five species of Cyprinidae: *Chondrostoma lemmingii*, *Leuciscus pyrenaicus*, *Barbus bocagei*, *Barbus comizo*, all endemic in the Iberian Peninsula, and the recently introduced *Alburnus alburnus*. Sixty-nine diagnostic markers were detected that accurately identify these species. RAPD analysis revealed one species-specific marker in *B. comizo*, six in *B. bocagei*, 24 in *C. lemmingii*, 19 in *L. pyrenaicus* and 19 in *A. alburnus*. The results demonstrated the usefulness of this technique for identification problems. At present, endemic freshwater fish of the Iberian Peninsula are in decline due to human impacts and their unambiguous identification constitutes the first requisite in any effective conservation programme.

Keywords: diagnostic band, non-destructive technique, RAPD-PCR.

29.1 Introduction

The Iberian fish fauna comprises 64 species grouped in two freshwater families, the Cyprinidae and Cobitidae, and 10 others with a certain tolerance to salt water of marine origin or with a cycle shared between fresh waters and the sea. The family Cyprinidae is important to the Spanish ichthyofauna because it contributes 22 species, and exhibits high endemism (Granado-Lorencio 1992).

Morphological features have traditionally been used to identify cyprinids. However, these features are, in many cases, of limited value for identification purposes because they show considerable variation, mainly those related to growth and biogeographical or ecological changes. Consequently, the individual character ranges frequently overlap and differences among species are small. Likewise, hybridisation often occurs

---

*Correspondence: Departamento de Genética, Facultad de CC. Biológicas, Universidad Complutense, 28040 Madrid, Spain (email: Dochando@bio.ucm.es).

among the cyprinids, and produces specimens with intermediate traits. Consistently, some of the characters traditionally used to distinguish the different species are taxonomically doubtful.

At present, the endemic freshwater fish of the Iberian Peninsula are in decline due to pollution, water abstraction and other human impacts (see Cowx, Chapter 20). This has resulted in the degradation and disappearance of suitable habitats and a decrease in the number of specimens in areas where fishes were formerly plentiful.

Recently, much attention has been focussed on conservation programmes to protect the Iberian freshwater fauna. The conservation of the species requires the discrimination of these taxa (Furman, Grattapaglia, Dvorak & O'Malley 1997), as precise identification of species is a necessary first step to conserve biodiversity (Meffe & Carrol 1997). Therefore, an important issue in the formulation of conservation programmes is the accurate identification of organisms at the species level and at any developmental stage.

Since the early 1990s, several molecular methods have been applied to overcome this problem. Among them is the PCR-based technique of RAPD (Random Amplified Polymorphic DNA) (Williams, Kubelik, Livak, Rafalsky & Tyngey 1990), which is less labour intensive and faster than other DNA procedures, a large number of primers are available or can be designed for analysis, many samples can be analysed simultaneously and the complete genome is screened. In addition, the whole protocol can be carried out without radioactive compounds, only minute quantities of DNA are needed for analysis, a prior knowledge of the organism's genome is not required and, of paramount importance for conservation programmes, non-destructive techniques can be applied for tissue collection.

The aim of this study was to perform a RAPD analysis to test the usefulness of this technique in fishery-management related problems and to generate informative genetic markers for distinguishing different cyprinid species of the Iberian Peninsula and their possible hybrid forms; i.e. roach, *Chondrostoma lemmingii* (Steindachner), barbel, *Barbus comizo* (Steindachner), both threatened of extinction in Central Spain, *B*arbus *bocagei* (Steindachner), chub, *Leuciscus pyrenaicus* (Günther), and the alien bleak, *Alburnus alburnus* (L.) first reported in Spain in 1995 (Elvira 1995).

29.2 Materials and methods

To conduct the RAPD analysis on the species under study, a total of 59 specimens from 10 samples (Table 29.1) were sampled. Since the genus *Barbus* shows the highest specific diversity among all fish genera of the Iberian freshwater fauna and seven out of its eight species are endemic, two samples of *B. bocagei* and three of *B. comizo,* were included in the present study.

Conservation programmes require the collection of tissue samples by employing a non-destructive technique such as scales or fin clipping to avoid the sacrifice of valuable fishes. Hence, total genomic DNA of *C. lemmingii*, *B. comizo*, *L. pyrenaicus* and *A. alburnus* was isolated from scales or fin tissue conserved in ethanol until DNA extraction, adapted from the protocol of Hillis and Moritz (1990).

Table 29.1 Species analysed, sample size (N) and sampling sites

Code	Taxon[a]	N	Sampling site
Aa	*Alburnus alburnus*	6	River Cabriel (Júcar basin). Cofrentes. Valencia
Bb-1	*Barbus bocagei*	2	River Manzanares. (Tajo Basin). El Pardo. Madrid
Bb-2	*Barbus bocagei*	2	River Cea (Duero Basin). Mayorga. Valladolid
Bc-1	*Barbus comizo*	4	River Tajo. Aranjuez. Madrid
Bc-2	*Barbus comizo*	2	Cijara dam River Guadiana. Badajoz
Bc-3	*Barbus comizo*	3	River Guadiana. Puebla de Don Rodrigo. Ciudad Real
Lp	*Leuciscus pyrenaicus*	4	River Tajuña (Tajo basin). Perales de Tajuña. Madrid
Cl-1	*Chondrostoma lemmingii*	5	River Tajuña (Tajo basin). Perales de Tajuña. Madrid
Cl-2	*Chondrostoma lemmingii*	23	River Tajuña (Tajo basin). Perales de Tajuña. Madrid
Of	*Chondrostoma lemmingii*	8	Juvenile specimens, Laboratory crosses

[a]Assigned by morphological identification.

Approximately 40 mg of tissue were incubated in 500-µL of cell lysis buffer (100-mM NaCl, 50 mM Tris–HCl pH 8.0, 10 mM EDTA, plus addition of 15-µL SDS 20% and 15-µL Proteinase K 20 mg mL^{-1}) for 40 min at 50°C. Total DNA was purified with two phenol–chloroform–isoamyl alcohol (25 : 24 : 1) and one chloroform extraction. The DNA was percipitated in isopropyl alcohol, pelleted by centrifugation, washed with 70% ethanol and resuspended in 150 µL of TE buffer (10 mM Tris, 1 mM EDTA, pH 8.0). The DNA concentration was diluted to 5 ng µL^{-1} working solutions for amplification reactions.

To generate RAPD profiles, four decamer primers, from two different kits of Operon Technologies (A-07, C-05, C-11 and C-15), were used. RAPD-PCR amplifications were performed according to procedures of Williams *et al.* (1990), with minor modifications in 12.5-µL-reaction volume containing 12.5 ng of fish DNA, 2.5 pmol of primer, 100-µM of each dNTP, 4 mM $MgCl_2$, 1.25-µL Stoffel buffer 10× and 0.6 units of Stoffel Fragment DNA polymerase (PE Applied Biosystems).

Amplifications were achieved in a thermal cycler M.J. Research PT-100. The reaction mix was preheated at 94°C for 5 min followed by 45 cycles of amplification (1 min at 94°C, 1 min at 36°C and 6 min at 72°C). A final step was made at 72°C for 6 min.

Amplification products were resolved electrophoretically on 2% agarose gels, containing 1-µg-µL^{-1} ethidium bromide, using a TAE buffer system (40 mM Tris, 20 Mm acetic acid and 1 mM EDTA). Gels were visualised with UV light and photographed with a Polaroid camera.

The main drawback of this technique is sensitivity to changes in reaction conditions, which may affect reproducibility of amplification products. Thus, reactions were

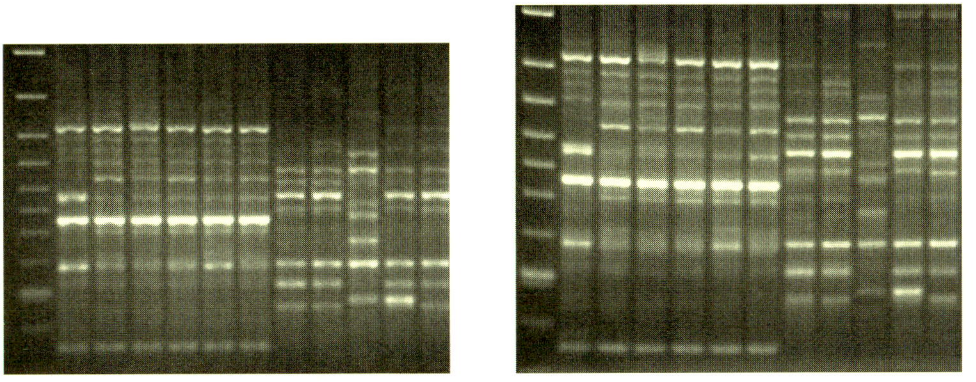

Figure 29.1 Reproducibility of amplification products by RAPD technique with primer C-05. In both gels from two different amplifications and runs: first lane contains a 100 bp ladder molecular weight marker. Lane 2: *B. comizo*. Lanes 3–7: *B. bocagei*. Lanes 8–9 and 11–12: *C. lemmingii*. Lane 10: *L. pyrenaicus*

performed following a strict protocol with standardised conditions and each amplification reaction was repeated at least twice to avoid problems with repeatability. The amplification reactions were considered consistently reproducible (Fig. 29.1).

The sizes of the amplification products were inferred by comparison with a size standard, a 100 bp ladder. All distinct fragments were identified by the primer name followed by its size. For example, the band A-07 870 refers to a band 870 bp long amplified employing primer A-07.

29.3 Results

Amplification of the cyprinid genomic DNA from all 59 specimens with the four primers produced different profiles for each species studied. The bands ranged in size from 300 to 1600 bp. With the primers tested, RAPD analysis revealed a total of 69 DNA diagnostic markers (Table 29.2) for the five species analysed, i.e. bands present in all or in a very high frequency (>95%) of the individuals of one species and absent in the remaining species. The amplification results were routinely repeatable even after the DNA of some samples of *B. comizo* and *B. bocagei* were stored at −20°C for more than 2 years. Representative DNA profiles of the five species, obtained with the four primers are shown in Figure 29.2a–d.

Several anomalies were found in the various samples analysed. The four specimens from the River Tajo, morphologically identified as *B. comizo* (Bc-1, Table 29.1), presented typical RAPD profiles of the species *B. bocagei* (also an endemic species of *Barbus* that inhabits basins of central Spain) when species-specific controls were performed for each set of amplifications. To confirm these results, the number of species-specific controls was increased with individuals from different populations of *B. comizo*

Table 29.2 Sizes (bp) of the diagnostic bands in each cyprinid species, and primers which produced them

Primer	*A. alburnus*	*B. comizo*	*B. bocagei*	*C. lemmingii*	*L. pyrenaicus*
A-07	1081, 870, 590, 520, 510		1120, 355	1300, 1050, 945, 850, 700, 605, 555, 505, 435	1260, 1005, 985, 880, 580, 500, 375
C-05	1350, 1230 1120, 900, 520	852	1157, 711	1065, 940, 865, 805, 510	1095, 1075, 780, 490
C-11	1120, 1030, 850, 760, 580, 570, 470, 400		939, 692	1100, 860, 700, 640, 560, 510, 305	1000, 880, 750, 660, 580
C-15	450			635, 470, 400	1140, 1115, 375

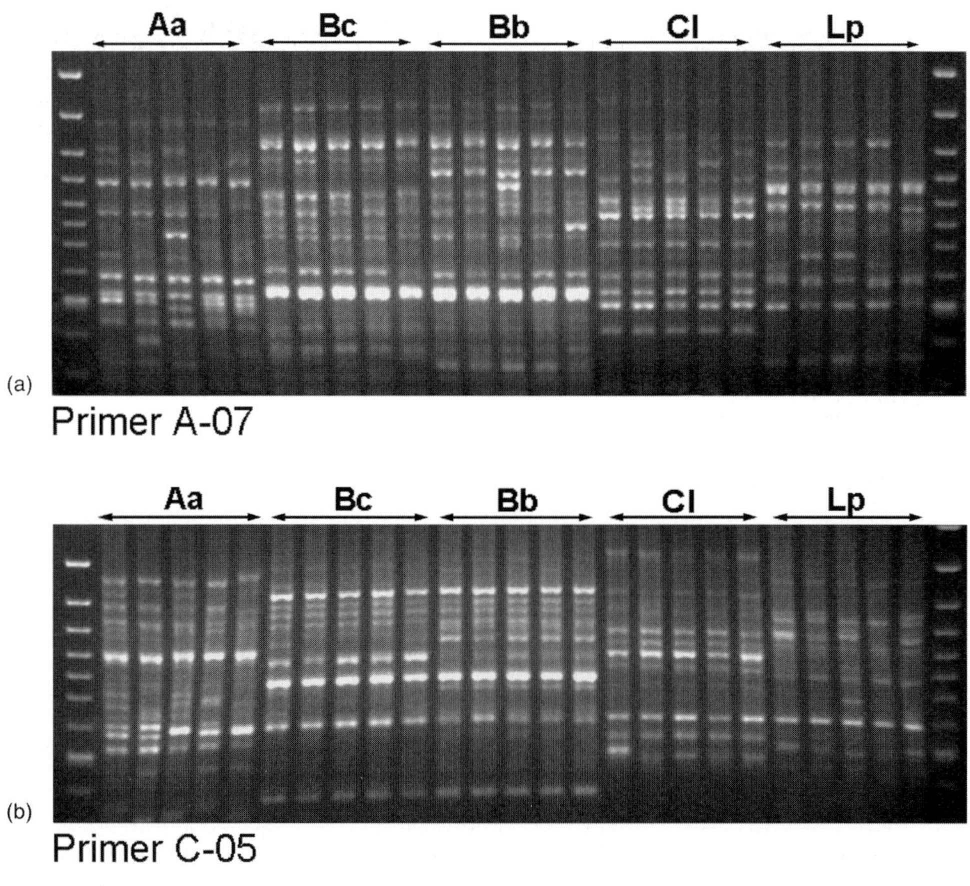

Figure 29.2 RAPD profiles of the five Cyprinidae species analysed with primer (a) A-07; (b) C-05; (c) C-11; (d) C-15. First and last lanes contain a 100 bp ladder molecular weight marker. Aa: *A. alburnus*, Bc: *B. comizo*, Bb: *B. bocagei*, Cl: *C. lemmingii* and Lp: *L. pyrenaicus*

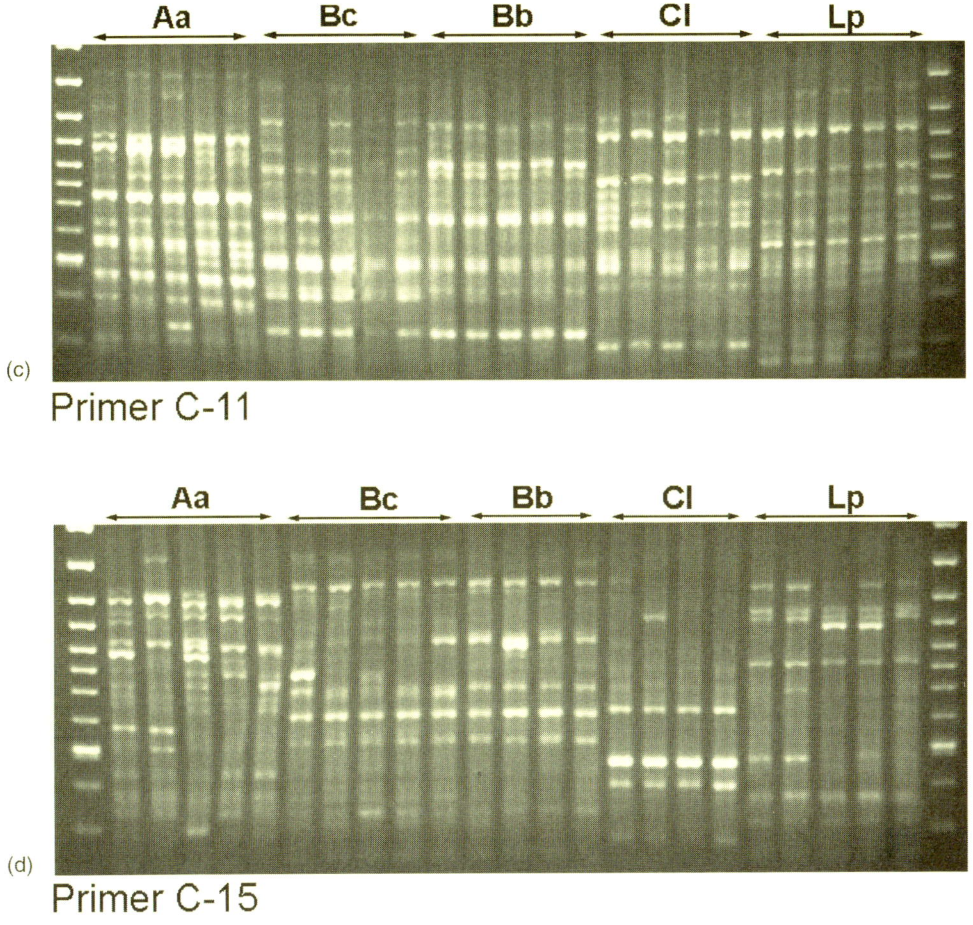

Figure 29.2 continued

and *B. bocagei* whose DNA had been isolated and analysed previously (Callejas & Ochando, 2001). One RAPD marker was detected for *B. comizo* and six for *B. bocagei* that unambiguously differentiated both species, and proved the differentiation observed in the Tajo sample.

RAPD analysis of the five *Chondrostoma* (Cl-1, Table 29.1) and that of *Leuciscus* (Lp, Table 29.1) revealed one of the specimens thought to be *Chondrostoma* exhibited the same RAPD pattern as *Leuciscus* with all the primers tested. The detection of 24 markers in *C. lemmingii* and 19 markers in *L. pyrenaicus* allowed the genetic characterisation of both species.

Conversely, specimens morphological identified *C. lemmingii* (sample Rl-2, Table 29.1) were tested by comparison of their RAPD profiles with those of species-specific controls. From 23 fishes analysed, 16 exhibited the typical RAPD patterns of *C. lemmingii* whereas the seven remaining presented similar RAPD fingerprints to *L. pyrenaicus*.

The first-generation of two different crosses of *C. lemmingii* were also analysed to ensure larvae were not hybrids (sample Of, Table 29.1), with four juvenile individuals of each cross being tested with the four primers. Furthermore, species-specific controls with *C. lemmingii* and *L. pyrenaicus* were achieved for each set of amplifications. Results with all primers tested showed that offspring from these crosses produced the banding patterns typical of pure *C. lemmingii* and absence of the specific RAPD markers of *Leuciscus*. Therefore, the eight juvenile specimens analysed proceeded from crosses between specimens of *C. lemmingii*.

Nineteen diagnostic bands that identified *A. alburnus* were detected with the four primers and unambiguously distinguished the *Alburnus* species from other cyprinids species studied (Table 29.2).

29.4 Discussion

RAPD analysis successfully identified 69 DNA markers, with all primers tested, that accurately differentiated the five species of Cyprinidae surveyed. More precisely, RAPD analysis revealed one diagnostic marker in *B. comizo*, six in *B. bocagei*, 24 in *C. lemmingii*, 19 in *L. pyrenaicus* and 19 in *A. alburnus* (Table 29.2).

As the number of specific markers increases, even when in some cases the number of specimens sampled is reduced, the ability to unambiguously identify the different species as well as to detect hybridisation also increases (Campton 1987). Thus, four specimens of *B. bocagei* erroneously identified as *B. comizo* and eight individuals of *L. pyrenaicus* erroneously identified as *C. lemmingii* were differentiated.

Similarly, RAPD analysis proved to be a useful taxonomic tool for species identification for juvenile life stages. Since the current status of fish taxonomy relies almost exclusively on adult characters, RAPD analysis is particularly useful in cases requiring a prompt identification, such as screenings of larval cultures in conservation programmes. In this study, the progenies of two doubtful crosses, were accurately identified as non-hybrid juveniles.

An important advantage of RAPD technique is that no prior knowledge is required about particular sequences or genes from the fishes of interest. It implies that genetic analysis on new species can be done without a previous database of classical genetics (Haymer 1994). A second advantage of RAPD technique for conservation programmes is the use of non-destructive methods, in this study by fin clipping and scales, to obtain tissue for analysis (Howland & Arnau 1992). Conversely, other techniques employed for identification, such as most allozyme techniques, require the collection of pieces of muscle or viscera for electrophoresis and consequently, the sacrifice of fishes. Furthermore, the RAPD technique is simpler, faster and more cost effective than other techniques of PCR, such as microsatellite or AFLP analysis. Thus, RAPD analysis appears to be a powerful, reliable, dependable, reproducible and relatively easy tool for cyprinid species identification and hybridisation detection (despite the dominance of this kind of markers). Hybridisation is more usual among freshwater than marine fishes, possibly because most freshwater environments are of comparatively recent origin (Mayr 1963; Campton 1987). Consequently, reproductive isolation mechanisms in many freshwater fish species may not have been completely developed (Billington & Hebert 1991), in part due to allopatry that avoids selective pressures against hybrids. Knowledge of hybridisation and introgression are critical for conservation programmes because they are the main mechanisms that compromise the integrity of species as distinct breeding groups (Grant 1981).

Some questions about fish management could be answered by employing RAPD markers. The possible use of diagnostic bands in future SCAR analysis (Sequence-Characterised Amplified Regions, Paran & Michelmore 1993) opens a new avenue in applied fisheries management. On the other hand, the maintenance of genetic diversity over time and genetic differentiation of populations in space are of critical interest (Schierenbeck, Skupski, Lieberman & Lieberman 1997). Consequently, the RAPD technique can be used to quantify the genetic diversity maintained in different populations or broodstocks. Also in this regard, RAPD markers could be used to identify and evaluate correctly the genotypes of broodstock and/or wild fish populations (Billington & Hebert 1991; Williams, Kazianis & Walter 1998). It is now evident that RAPD is an exceptionally useful tool in stock discrimination because sequences amplified by the RAPD technique evolve more quickly than other DNA sequences. RAPD polymorphic sequences are mostly derived from repetitive and non-coding sequences in the genome (Williams *et al.* 1990, Haymer & McInnis 1994) and evolutionary forces acting on this type of DNA appear to have less constraints to change than in coding and regulatory sequences. Therefore, this variation in repetitive sequences of the genome can serve as a basis for stock discrimination.

Fisheries managers have largely overlooked the genetic implications of fish introductions. It is suggested that the detection of genetically-discrete fish stocks and the identification of stock-specific markers is crucial for the conservation of threatened species of cyprinids in the Iberian Peninsula.

In summary, with the RAPD-PCR technique a wide range of diagnostic markers that unambiguous and accurately identified the five Cyprinidae species studied were generated. RAPD analysis has multiple advantages over other DNA techniques and is a powerful and reliable tool for species identification and detection of hybridisation.

Acknowledgements

We are greatly indebted to A. Callejas, L. Campos, A.L. Mora, Dr. García de Jalón, B. Ochando and Dr M. Quirós for their kind help with the fish samples collections.

References

Billington N. & Hebert P.D.N. (1991) Mitochondrial DNA diversity in fishes and its implications for introductions. *Canadian Journal of Fisheries and Aquatic Sciences* **48**, 80–94.

Callejas C. & Ochando M.D. (2001) Molecular identification (RAPD) of the eight species of the genus *Barbus* (Cyprinidae) in the Iberian Peninsula. *Journal of Fish Biology* **57** (in press).

Campton D.E. (1987) Natural hybridization and introgression in fishes. In N. Ryman & F. Utter (eds) *Population Genetics and Fishery Management*. Seattle: University of Washington Press, pp. 161–192.

Elvira B. (1995) Native and exotic freshwater fishes in Spanish river basins. *Freshwater Biology* **33**, 103–108.

Furman B.J., Grattapaglia D., Dvorak W.S. & O'Malley D.M. (1997) Analysis of genetic relationships of Central American and Mexican pines using RAPD markers that distinguish species. *Molecular Ecology* **6**, 321–331.

Granado-Lorencio C. (1992) Fish species ecology in Spanish freshwater ecosystems. *Limnetica* **8**, 255–261.

Grant V. (1981) *Plant Speciation*. New York: Columbia University Press, 435 pp.

Haymer D.S. (1994) Random Amplified Polymorphic DNAs and microsatellites: what are they, and can they tell us anything we don't already know? *Annals of the Entomological Society of America* **87**, 717–721.

Haymer D.S. & McInnis D. (1994) Resolution of populations of the Mediterranean fruit fly at the DNA level using random primers. *Genome* **37**, 244–248.

Hillis D.M. & Moritz C. (eds) (1990) *Molecular Systematics*. Sunderland, MA: Sinauer Associates, 655 pp.

Howland D. & Arnau J. (1992) RAPDs: Random Amplified Polymorphic DNA. In R.J. Berry, T.J. Crawford & G.M. Hewitt (eds) *Genes in Ecology*. Boston: Blackwell Scientific Publications, pp. 466–468.

Mayr E. (1963) *Animal Species and Evolution*. Harvard: Belknap Press, 797 pp.

Meffe G.K. & Carrol C.R. (1997) *Principles of Conservation Biology*. Sunderland, MA: Sinauer Associates, Inc., 673 pp.

Paran I. & Michelmore R.W. (1993) Development of reliable PCR-based markers linked to downy mildew resistance genes in lettuce. *Theoretical and Applied Genetics* **85**, 985–993.

Schierenbeck K.A., Skupski M., Lieberman D. & Lieberman M. (1997) Population structure and genetic diversity in four tropical tree species in Costa Rica. *Molecular Ecology* **6**, 137–144.

Williams D.J., Kazianis S & Walter R.B. (1998) Use of Random Amplified Polymorphic DNA (RAPD) for identification of largemouth bass subspecies and their intergrades. *Transactions of the American Fisheries Society* **127**, 825–832.

Williams J.K.G., Kubelik A.R., Livak K.J., Rafalsky J.A. & Tyngey S.V. (1990) DNA polymorphisms amplified by arbitrary primers are useful as genetic markers. *Nucleic Acids Research* **18**, 6531–6535.

Chapter 30
Contribution of molecular studies to the management of *Anaecypris hispanica*

M.M. COELHO,* P. SALGUEIRO, M.J. ALVES and
M.J. COLLARES-PEREIRA
Centro de Biologia Ambiental/Departamento de Zoologia e Antropologia, Faculdade de Ciências, Lisboa, Portugal

Abstract

Anaecypris hispanica is restricted to the Guadiana River drainage, where its abundance and geographical range have contracted markedly during the last 20 years. Recent ecological data suggest that the species is presently confined to 10 tributaries in Portugal. Mitochondrial DNA (sequence and RFLP analyses of cytochrome *b* and control region) and nuclear DNA (five microsatellites loci) revealed strongly structured populations with substantial genetic variation. Using the mtDNA results, it was possible to define three evolutionarily significant units (ESUs): Odeleite, Foupana and a northern group comprising the Xévora, Caia, Álamo, Degebe, Ardila, Chança and Carreiras. Considering both microsatellite and mtDNA data, it is proposed that populations from the Xévora and Caia, Álamo, Degebe, Ardila, Chança and Carreiras plus Vascão, should be considered as six discrete management units (MUs) within the northern ESU. The more relevant conservation implications arising from the genetic results are: (1) translocations between the three ESUs should be avoided, at least in a short term; (2) active managed restocking actions are advised for populations under catastrophic reductions or extinctions, but the source of donor individuals have to be carefully evaluated; (3) in some rivers of the middle reaches, if the heterozygosity is very low, the use of admixed populations from the same or nearby MUs as broodstock can be an option.

Keywords: *Anaecypris hispanica*, conservation genetics, population fragmentation, mitochondrial DNA, microsatellites.

30.1 Introduction

Anaecypris hispanica (Steindachner) is currently the most threatened non-migratory Iberian fish. It is restricted to the Guadiana River drainage, where it was abundant in the Portuguese section in the 1970s. However, its abundance and geographical range have contracted markedly in the last 20 years in both Spain and Portugal (Collares-Pereira, Cowx, Rodrigues, Rogado & Costa 1999). Recent data suggest that the species is presently confined to 10 tributaries in Portugal: Xévora, Caia, Álamo, Degebe,

*Correspondence: Departamento de Zoologia e Antropologia, Faculdade de Ciências, Campo Grande C2 – Piso 3, 1749-016 Lisboa, Portugal (email: mmcoelho@fc.ul.pt).

Ardila, Chança, Carreiras, Vascão, Foupana and Odeleite (see Fig. 19.1 in Collares-Pereira, Cowx, Rodrigues & Rogado, Chapter 19). In the northern rivers, the population from the River Caia exhibits the highest densities, while in the southern rivers all populations from the rivers Vascão, Odeleite and Foupana attain comparatively high densities. The tributaries in the middle reaches the Guadiana in Portugal exhibit very low densities, with the exception of the River Chança.

When stock decrease is coupled with fragmentation, conservation actions are particularly complex due to the possible reduction of the gene pool (Carvalho & Cross 1998). The relevance of genetic information to species conservation planning has long been recognised (e.g. Lande & Barrowclough 1987; Simberloff 1988), therefore conservation of genetic diversity is an essential component of many species management programmes (Templeton 1994; O'Connel & Wright 1997).

To obtain a more accurate perspective of the population structure of *A. hispanica* and to define conservation units (Moritz 1994), two genetic approaches, respectively based on mtDNA and nuclear markers analyses, were used in the 10 sub-basins where *A. hispanica* still persists. The overall implications of the results from these studies for the conservation of the species are presented.

30.2 Materials and methods

In the genetic studies, mtDNA (sequence and RFLP analyses of cytochrome *b* and control region) and nuclear DNA (five microsatellite loci) were analysed from specimens caught in the 10 tributaries of Guadiana River Basin in Portugal where the species exists: Xévora, Caia, Álamo, Degebe, Ardila, Chança, Vascão, Carreiras, Foupana and Odeleite. The River Álamo was not analysed in the mtDNA study and the Xévora and Carreiras in the microsatellites analysis.

DNA was obtained from small fin-clips from the pelvic fins, which were fixed in absolute ethanol. The methods followed in mtDNA and microsatellites studies are described in Alves, Coelho, Collares-Pereira and Coelho (2001) and Salgueiro, Carvalho, Collares-Pereira and Coelho (in press), respectively.

30.3 Results

The genetic studies revealed strongly structured populations with substantial genetic variation.

The mitochondrial DNA analysis showed geographic structuring, with migration only occurring between some geographically close rivers (Xévora and Caia, Degebe and Ardila, Carreiras and Vascão). The populations from extreme southern rivers (Foupana and Odeleite) are monophyletic entities, suggesting that they have spent an appreciable amount of time in isolation. Sequence data and RFLP data revealed, in general, high within-river haplotype diversity for *A. hispanica*. This indicates that the contraction in the abundance of the species in the last two decades is not severe enough to cause the extinction of haplotypes and consequent reduction of haplotype diversity.

The only exception was the population from the River Carreiras, which exhibited low haplotype diversity estimates.

In the study with microsatellites, significant differences for pairwise comparisons between all populations were observed. High allelic diversity was also found within each population, with the population of Caia showing the highest number of alleles over all loci. In addition, almost all sampled populations contained unique alleles, which reinforces the idea that effective dispersal between populations is very restricted and that they have been genetically isolated from each other for some time.

The population from the River Álamo should be given special consideration, because it exhibits a considerable high level of heterozygosity but a reduced average number of alleles. These values suggest a possible recent bottleneck effect (Luikart & Cornuet 1998) that is consistent with a population presently isolated in the upper reaches of a highly polluted area downstream to Reguengos de Monsaraz municipality.

These genetic results indicate that *A. hispanica* is not a genetically depauperate species, which is probably related to a larger historically effective population size. In general, the observed population genetic variability showed a positive relationship with the observed densities. The exception is the Vascão population, one of the largest extant populations, which presented a comparatively lower genetic diversity.

30.4 Discussion

The success of any management actions will be generally enhanced by knowledge about the genetic population structure. Evolutionary significant units (ESUs) are geographically discrete sets of populations that have been identified by genetic studies as having evolved separately for a substantial period of time (Moritz 1994; Moritz, McGuigan & Bernatchez, Chapter 27). This is the case of the three ESUs defined by the mitochondrial DNA study: i.e. the Odeleite, Foupana and a northern group constituted by Xévora, Caia, Álamo, Degebe, Ardila, Chança, Vascão and Carreiras. Microsatellite results pointed out a higher differentiation between all studied populations.

By combining the findings from the microsatellites (Salgueiro *et al.*, in press) and the mtDNA data (Alves *et al.* 2001), it can be proposed that populations from the Xévora and Caia, Álamo, Degebe, Ardila, Chança and Carreiras plus Vascão, constitute six discrete management units (MUs) within the northern ESU (Fig. 30.1), and that conservation efforts should be directed towards preserving the genetic integrity of each unit.

It may be prudent in the short term to avoid translocations between the three ESUs, following Moritz (1994). Another practical implication of the present results is that if populations suffer catastrophic reductions or extinctions, natural repopulation through immigration will be unlikely. If so, active managed restocking actions might be necessary to re-establish populations, as those of the low density populations identified by Collares-Pereira *et al.* (Chapter 19). Given the distinctiveness of individual populations, the source of donor individuals for such actions would have to be carefully evaluated if maintaining the current genetic structure is deemed an important conservation goal. Therefore restocking actions should avoid protocols that could reduce variation

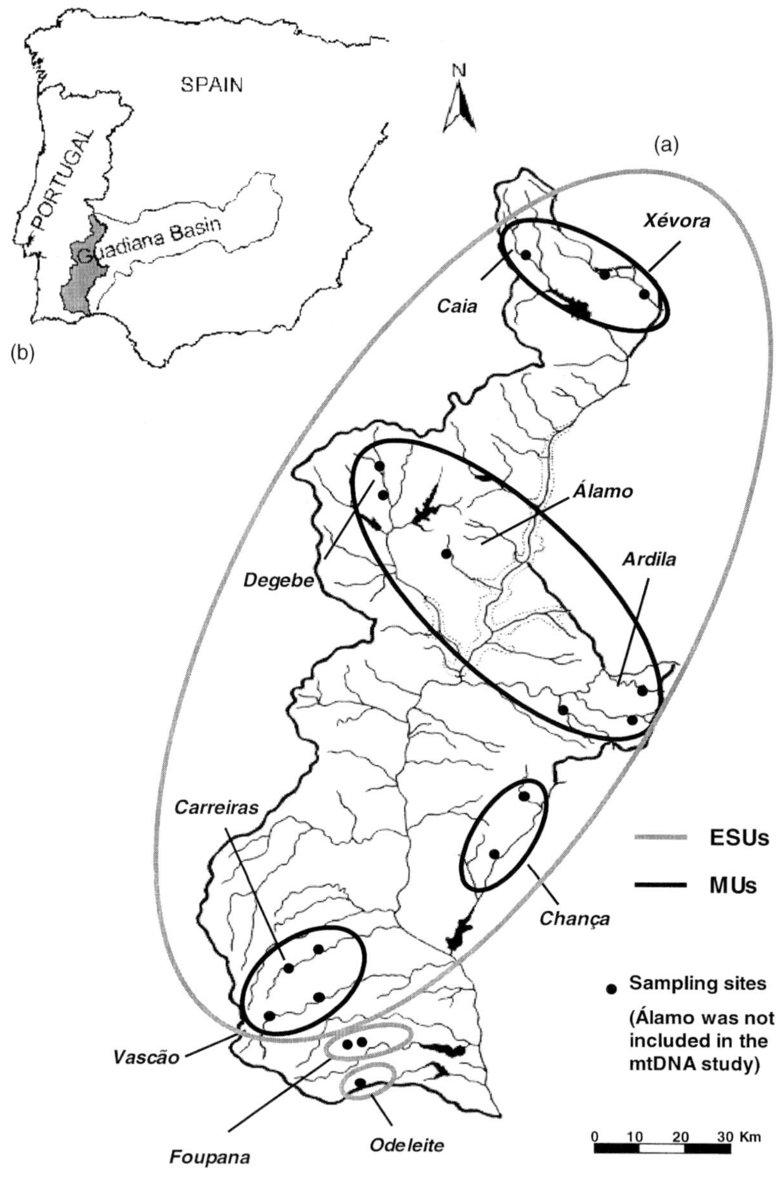

Figure 30.1 (a) Distribution area of *A. hispanica* in the Iberian Peninsula; (b) map of the Guadiana River basin showing the sampling site locations, and the established ESUs and MUs according to Salgueiro *et al.* (in press)

within and among populations. Nevertheless, if heterozygosity is of low level, as observed in the rivers of the middle reaches (with the exception of the River Chança) the use of admixed populations as sources for stocking can also be an option (Moritz 1999) for the rehabilitation of *A. hispanica* populations, but preferably from the same or nearby MUs. One example in the northern populations group is the River Álamo, presently considered the most isolated and threatened population. This river could be stocked with specimens from the River Caia, although both were considered distinct MUs. A different situation can be considered in the south, where it is desirable to design a restocking plan for the River Carreiras, the population with a lower genetic variability, with individuals from the Vascão broodstock as it is the population with the highest variability, belonging to the same MU.

However, Collares-Pereira *et al.* (Chapter 19) indicated that the priority for the recovery of *A. hispanica* is not just specifically aimed at the species, but more at the conservation and rehabilitation of natural habitats. As pointed out by Meffe and Carroll (1994), 'if the habitat is not available, and materials and energy do not flow through the ecosystem, the maintenance of genetic diversity will ultimately degrade to an exercise in *ex situ* care; without suitable ecosystems and dynamic ecological processes, genetic diversity in isolation may be nearly worthless'. In relation to the populations of *A. hispanica*, this is particularly important because if the threats to the habitats persist (e.g. alterations to flow regime and loss of longitudinal connectivity through water resource development schemes, overabstraction of water in summer periods when the species is in refuge pools, pollution and the spread of exotic species) the genetic variability of the species may be drastically affected.

In summary, after removing the present habitat bottlenecks, there is a need to make genetic management an integral part of an *A. hispanica* conservation plan. The necessary genetic indicators are now available but they require careful interpretation when being used to derive management advice.

Acknowledgements

This study was carried out under the project 'A conservation strategy for *Anaecypris hispanica*', EU/LIFE contract B4-3200/97/280.

References

Alves J., Coelho H., Collares-Pereira M.J. & Coelho M.M. (2001) Mitochondrial DNA variation in the highly endangered cyprinid fish *Anaecypris hispanica*: importance for conservation. *Heredity* **87** (in press).

Carvalho G.R. & Cross T.F. (1998) Enhancing fish production through introductions and stocking: genetic perspectives. In I.G. Cowx (ed.) *Stocking and Introduction of Fish*. Oxford: Fishing News Books, Blackwell Science, pp. 437–449.

Collares-Pereira M.J., Cowx I.G., Rodrigues J.A., Rogado L. & da Costa L.M. (1999) The status of *Anaecypris hispanica* in Portugal: problems of conserving a highly endangered Iberian species. *Biological Conservation* **88**, 207–212.

Lande R. & Barrowclough G.F. (1987) Effective population size, genetic variation, and their use in population management. In M.E. Soulé (ed.) *Viable Populations for Conservation*. Cambridge: Cambridge University Press, pp. 87–123.

Luikart G. & Cornuet J.M. (1998) Empirical evaluation of a test for identifying recently bottlenecked populations from allele frequency data. *Conservation Biology* **12**, 228–237.

Meffe G. & Carroll R. (1994) Genetics: conservation of diversity within species. In G. Meffe & R. Carroll (eds) *Principles of Conservation Biology*. Sunderland: Sinauer Associates, pp. 143–178.

Moritz C. (1994) Defining 'Evolutionary Significant Units' for conservation. *Trends of Ecology and Evolution* **9**, 373–375.

Moritz C. (1999) Conservation units and translocations: strategies for conserving evolutionary processes. *Hereditas* **130**, 217–228.

O'Connel M. & Wright J.M. (1997) Microsatellite DNA in fishes. *Reviews in Fish Biology and Fisheries* **7**, 331–363.

Salgueiro P., Carvalho G., Collares-Pereira M.J. & Coelho M.M. Microsatellite analysis of genetic population structure of the endangered cyprinid *Anaecypris hispanica* in Portugal: implications for conservation. *Biological Conservation* (in press).

Simberloff D. (1988) The contribution of population and community biology to conservation science. *Annual Review of Ecology and Systematics* **19**, 437–511.

Templeton A.R. (1994) Biodiversity at molecular genetic level: experiences from disparate macroorganisms. In D.L. Hantsworth (ed.) *Biodiversity, Measurement and Estimation*. London: Chapman & Hall, pp. 59–64.

Chapter 31
Effects of fishery management on populations of brown trout, *Salmo trutta*, in central Spain

A. ALMODÓVAR,* G.G. NICOLA and J. SUÁREZ
Agricultural Research Institute of Madrid, Madrid, Spain

Abstract

Brown trout, *Salmo trutta*, in central Spain is currently threatened by overfishing, habitat destruction and stocking with allochthonous stocks. The effects of angling exploitation were assessed in five rivers between 1992 and 1998 by comparing population and catch statistics within exploited and unexploited areas. The main effects of angling were a decrease in density, biomass and production in the exploited stocks. Likewise, the number of legal-sized trout was <10% of the total populations. Sport fishing parameters showed that fishing pressure was high in the majority of these rivers, resulting in an alteration of population maintenance. A depletion in the breeding stock and egg production was also observed in the exploited sections. Moreover, the stocking of allochthonous stocks caused a significant decrease of the original intraspecific variability. Two-thirds of populations analysed showed genetic introgression and gene flow between native and allochthonous strains. A summary of conservation guidelines are proposed in relation to particular threats.

Keywords: angling impact, conservation, genetic introgression, management, *Salmo trutta*.

31.1 Introduction

Brown trout, *Salmo trutta* L., is one of the most genetically sub-structured vertebrate species, and several genetic studies suggest that a large part of the evolutionary diversity corresponds to southern European countries bordering the Mediterranean (Allendorf & Leary 1988; Ferguson 1989). In Spain, this species is an important economic and social resource, which provides valuable sport fisheries. Brown trout is considered as 'vulnerable', in the Spanish and Portuguese Red Books of Vertebrates. Likewise, this species is considered of 'Special Concern' in some regions of Spain. However, the brown trout is not threatened at the international conservation level. On the other hand, the high genetic diversity and fragmentation of brown trout populations in central Spain requires management at the population level. In central Spain, brown trout is threatened by habitat destruction, water pollution and introduction of exotic species (Elvira 1996). Flow modification is one of the most widespread human disturbances in the headwaters of Spanish rivers, and an important cause of the decline of

*Correspondence: Agricultural Research Institute of Madrid (IMIA), El Encín, PO Box 127, E-28800 Alcalá de Henares, Madrid, Spain (email: ana.almodovar@imia.comadrid.es).

central populations (Almodóvar & Nicola 1999). Additionally, the Spanish populations are currently overexploited (Braña, Nicieza & Toledo 1992; Reyes-Gavilán, Garrido, Nicieza, Toledo & Braña 1995; Almodóvar & Nicola 1998) and their genetic diversity is threatened by stocking with non-native trout from central and northern Europe (García-Marín, Jorde, Ryman, Utter & Pla 1991; Morán, Pendás, García-Vázquez, Izquierdo & Lobón-Cerviá 1995; García-Marín & Pla 1996; Cagigas, Vázquez, Blanco & Sánchez 1999; Machordom, García-Marín, Sanz, Almodóvar & Pla 1999; Machordom, Suárez, Almodóvar & Bautista 2000). In this study, the current impacts on brown trout populations in central Spain are assessed. The primary objective was to evaluate the effects of sport fishing on population structure and production. Additionally the genetic impact of stocking with allochthonous stocks on local populations was determined. Finally, some proposals for conservation and management are given.

31.2 Materials and methods

31.2.1 *Sport fishing and trout populations*

This study was carried out on five rivers, which are tributaries to the River Tagus (Hoz Seca, Cabrillas, Gallo, Dulce and Jarama) (Fig. 31.1). These are non-polluted streams and brown trout is the dominant species throughout the study area. The crystalline River Jarama originates at elevations close to 1200 m above sea level. The rest of the rivers flow through limestone catchments at elevations from 900 to 1400 m above sea

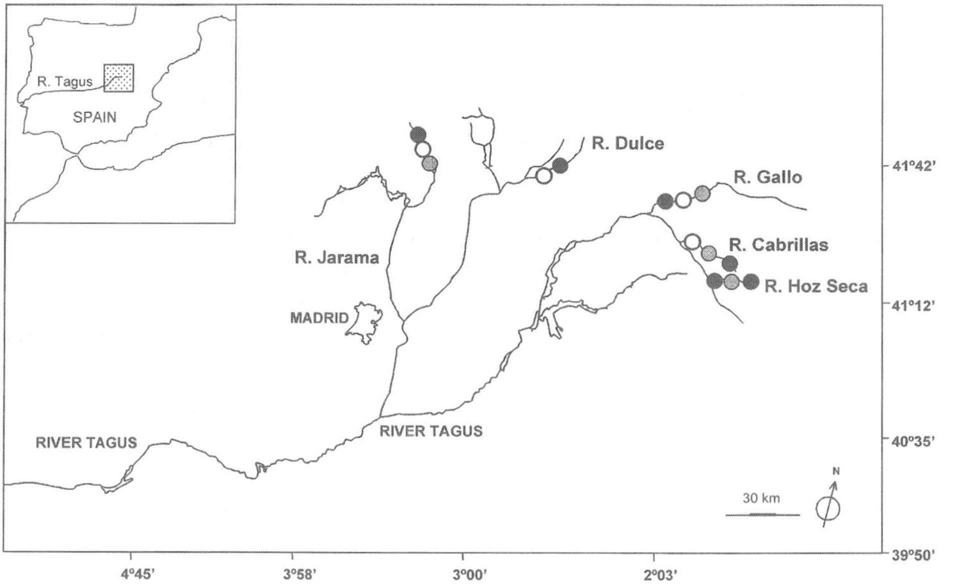

Figure 31.1 Map of the study area showing sampling sites in the studied rivers (black, preserved, catch and release; grey, restricted regulation; white, open regulation)

level. Two or three sampling sites were selected in each river with enough distance between them to avoid any migration, to be quite homogeneous with respect to habitat structure and to cover different angling regulations.

The sampling sites included exploited and unexploited sections, with four different angling regulations concerning daily bag limit, number of anglers per day and type of gears:

- preserved sections: unexploited, fishing activities are forbidden;
- catch and release sections: unexploited, no limitation in number of anglers per day, fly-fishing only;
- restricted regulation sections: exploited, daily bag limit of five trout, four to six anglers per day, only artificial lures permitted;
- open regulation sections: exploited, daily bag limit of six trout, no limitation in number of anglers per day, only artificial lures permitted.

During the 1996 and 1997 fishing seasons (March–August) a creel-survey was conducted to determine the use of the fishery and to evaluate angler impacts. A total of 415 anglers were surveyed. Fishing pressure was estimated by the number of anglers per hectare per year and the number of hours fished per hectare per year. Harvest rate was calculated as the mean number of trout kept per hectare per year. Exploitation rate was determined considering the biomass of legal-sized trout and the total harvest during the fishing season. A Kruskal–Wallis analysis of variance was applied to compare these parameters among rivers. Multiple comparisons were carried out using Dunn's procedure (Zar 1999).

Fish were sampled every third month from 1992 to 1998 at 14 localities by electric fishing. Trout were anaesthetised with tricaine methane-sulphonate (MS-222 SANDOZ) and their fork length (nearest mm) and weight (nearest g) were measured. Scales were taken for age determination. Fish density (trout ha^{-1}) was estimated by applying the removal method (Zippin 1956). Biomass ($kg\,ha^{-1}$) was calculated following Mahon, Balon and Noakes (1979). Production ($kg\,ha^{-1}\,year^{-1}$) was estimated using the increment summation method from Newman and Martin (1983). Measures of variance were computed according to Newman and Martin (1983). Fecundity and density values were used in conjunction with percentages of sexually mature trout (Nicola & Almodovar, in press) to compute the egg output of each cohort and the density of the breeding stock during each year (Crisp 1994). One-way ANOVA was used to compare mean values between unexploited and exploited sections. Assumptions concerning data distributions were tested using a Shapiro–Wilk test and homogeneity of variances using a Levene test. The significance level for all statistical tests was set at $\alpha = 0.05$.

31.2.2 *Genetic introgression*

Estimates of genetic introgression in five Spanish rivers were determined to assess the effect of stocking on native populations. In addition, published data from six populations from central Spain (Machordom *et al.* 1999) are presented. Introgression of allochthonous genes was estimated using the diagnostic *LDH-5** locus. The *LDH-5*90* allele at this locus is routinely used in southern European regions to identify the presence of commercial strains of hatchery-reared fish (García-Marín *et al.* 1991). Hardy–Weinberg (H–W) equilibrium was tested and an exact-test was employed to analyse heterozygote deficiency (Rousset & Raymond 1995) using GENEPOP v. 3.2a.

31.3 Results

The mean number of anglers (fishing pressure measured as number of fishing hours) did not differ significantly ($P > 0.05$) between the rivers studied (Table 31.1). However, despite the harvest rate being similar among the different rivers, the exploitation rate was significantly different ($\chi^2 = 22.30$, $P < 0.001$).

The age structure was significantly different between harvested trout and the legal-sized stock in nearly all the studied rivers (Fig. 31.2), with harvest being size selective towards on the larger specimens. Age 3+ brown trout made up 45.9 ± 24.6% of the harvest, but 56.5 ± 16.2% of the populations. Age 4+ accounted for 40.0 ± 17.0% of the harvest and a similar proportion in the populations (38.5 ± 13.6). Ages 5+ onwards accounted for the remaining annual harvest. The 2+ trout were not harvested, probably due to the late recruitment of this age class into the legal-sized stock during the fishing season. The length of the harvested trout was also significantly different among rivers, but a subsequent comparison of means revealed that the differences were due to the lower values found in rivers Dulce and Jarama, which were significantly different from the rest (Dunn test, $P < 0.05$). The scarcity of large trout in the River Dulce suggests that angling may remove large numbers of fish before they have the opportunity to grow to a large size, because the growth in this river is the highest within the study area (unpublished data). Overexploitation of the largest trout had a direct consequence on the reproduction in these rivers. The breeding stock showed significant differences ($F_{1,43} = 7.23$, $P < 0.05$) between exploited (401.5 trout ha^{-1}) and unexploited sections (745.9 trout ha^{-1}). Likewise, mean egg production was significantly lower ($F_{1,43} = 5.23$, $P < 0.05$) in exploited areas (148 518 eggs) compared with unexploited ones (276 677 eggs). This negative effect of angling on reproduction was

Table 31.1 Estimated means and ranges of sport fishery statistics at the five rivers studied

River	No. of anglers (ha^{-1}year^{-1})	Fishing hours (ha^{-1}year^{-1})	Harvest (ha^{-1}year^{-1})	Harvest length (cm)	Exploitation rate (%)
Hoz Seca	7.7 (4.6–23.0)	19.0 (1.2–36.9)	4.6 (0–27.6)	26.7 (24.0–35.0)	19.7
Cabrillas	24.4 (7.1–63.5)	86.6 (3.0–241.5)	12.8 (0–42.3)	26.5 (22.0–42.0)	11.0
Gallo	26.0 (9.3–101.8)	84.2 (2.3–419.4)	18.8 (0–101.8)	26.7 (22.0–46.0)	19.0
Dulce	19.6 (7.6–114.6)	50.8 (7.6–120.3)	9.5 (0–84.0)	23.9 (22.0–32.0)	8.3
Jarama	20.1 (9.8–88.3)	77.6 (5.7–496.6)	13.4 (0–68.7)	25.4 (22.0–40.5)	34.3
	$H_{4,162} = 17.22$**	$H_{4,157} = 8.87$ n.s.	$H_{4,162} = 6.75$ n.s.	$H_{4,265} = 13.37$**	

The results of the Kruskal–Wallis test are given (** = $P < 0.01$, n.s. = non-significant).

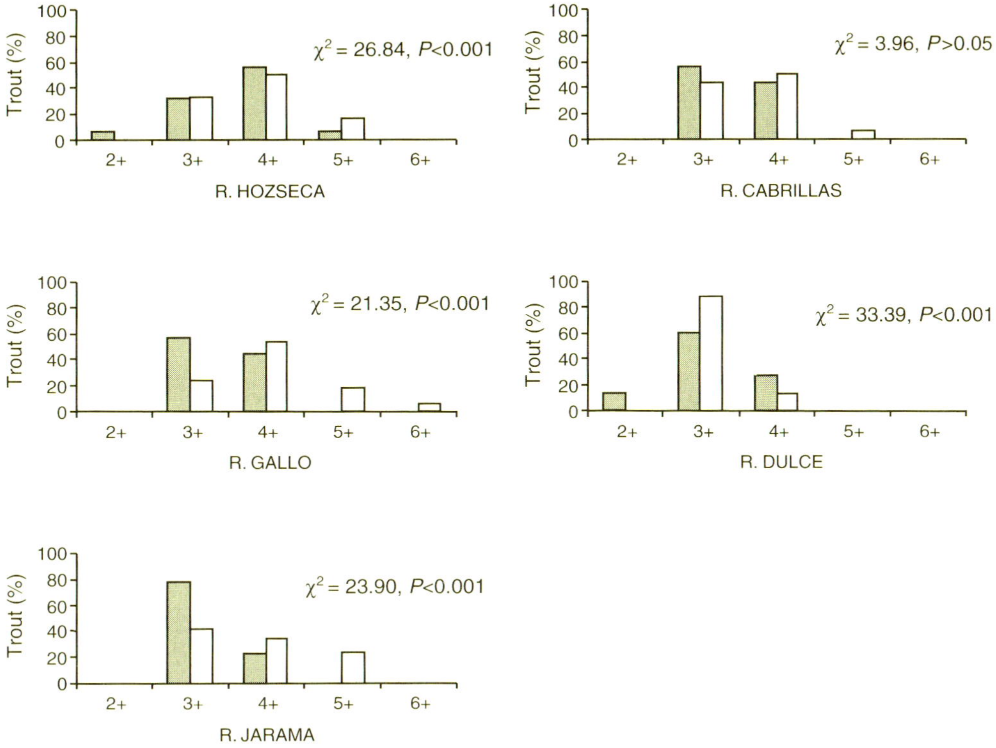

Figure 31.2 Per cent angler harvest by age group (creel-surveys, white) and age distribution of legal-sized trout (sampling, grey) in the five rivers studied. The results of the chi-square test are given

also evident when comparing the change in the breeding stock and egg production during the study period. A marked reduction of these characteristics was observed in the exploited areas throughout the 5-year study.

Angling pressure and overexploitation contributed to a serious decline of brown trout abundance in the exploited sections (Table 31.2). In addition, production was considerably lower in the exploited areas. Only the turnover ratio did not show any consistent differences. The density and biomass of legal-sized trout in all the study rivers were low, averaging about 10% of the total population in each river, with the lowest values in exploited areas.

Stocking with allochthonous strains has disrupted the original genetic integrity of wild trout populations in central Spain. The analysis of the *LDH-5** locus in the studied populations revealed four different electromorphs. Three of them (*LDH-5*85*, *LDH-5*100* and *LDH-5*110*) indicate native trout, while *LDH-5*90* was fixed in all stocked trout. Only five populations did not exhibit genetic contamination, while 10 had some frequency of the non-native allele *LDH-5*90* (Table 31.3). Most of the introgressed populations did not show significant deviations from H–W equilibrium. Only the River Cega (Navafría) displayed a significant heterozygote deficiency.

Table 31.2 Mean (± s.e.) population statistics for brown trout in unexploited and exploited sections in the studied rivers

	Unexploited	Exploited		Probability
Density (trout ha^{-1})	2099.0 ± 188.5	1234.4 ± 85.9	$F_{1,162} = 17.42$	<0.001
Biomass (kg ha^{-1})	104.3 ± 8.8	61.0 ± 5.4	$F_{1,162} = 17.61$	<0.001
Production (kg ha^{-1} year^{-1})	96.0 ± 16.5	62.5 ± 7.6	$F_{1,48} = 3.92$	<0.05
Turnover ratio (year^{-1})	1.2 ± 0.1	1.1 ± 0.1	$F_{1,48} = 0.09$	>0.05
Density legal-sized trout (trout ha^{-1})	236.0 ± 25.3	128.4 ± 17.1	$F_{1,138} = 12.45$	<0.001
Biomass legal-sized trout (trout ha^{-1})	41.9 ± 4.6	20.9 ± 3.1	$F_{1,138} = 14.39$	<0.001

The results of the one-way ANOVA tests are given.

Table 31.3 Introgression frequencies for the diagnostic *LDH-5** locus from the studied populations

Sampling sites	*n*	Basin	*LDH-5*90*	H–W
Cega (Los Chorros)	19	Duero	0.000	–
Cega (Navafría)	19	Duero	0.053	$P < 0.05$
Hoz Seca (Peralejos)	20	Tagus	0.000	–
Hoz Seca (Orea)	20	Tagus	0.050	n.s.
Cabrillas (Checa)	20	Tagus	0.000	–
Cabrillas (Taravilla)	22	Tagus	0.045	n.s.
Gallo	20	Tagus	0.075	n.s.
Jarama (Hayedo)	20	Tagus	0.000	–
Jarama (Hiruela)	24	Tagus	0.000	–
[a]Eresma (Casa de Peñón)	25	Duero	0.020	n.s.
[a]Eresma (Boca del Asno)	20	Duero	0.025	n.s.
[a]Aguisejo	19	Duero	0.211	n.s.
[a]Dulce	20	Tagus	0.100	n.s.
[a]Bornova	17	Tagus	0.059	n.s.
[a]Guadiela	19	Tagus	0.237	n.s.

Goodness-of-fit of the observed introgression frequency values to the H–W equilibrium are given (n.s.=non-significant). [a]Data from Machordom *et al.* (1999).

31.4 Discussion

The fishing pressure observed in the different rivers studied was comparable to that found in other brown trout populations (Wiley & Dufek 1980; Avery 1990; Hunt 1991, review). The harvest and exploitation rates were analogous to those described by Wiley and Dufek (1980), Avery and Hunt (1981) and Avery (1990), who found mean values between 18% and 37%. Avery (1990) suggested that the annual angler exploitation of trout should not exceed 40% of the legal-sized fish if the population is to remain healthy. Therefore, the high rate found in the River Jarama suggests an overexploitation of this population, which was more evident when analysing the present breeding stock.

The decrease in the breeding stock and total egg production observed over the study period denotes that angling harvest is depleting the mature stock in the studied rivers. If this trend continues, the number of parent cohorts would not be sufficient to ensure full utilisation of the available spawning and juvenile habitat.

Angling exploitation was found to be responsible for the lower values in abundance parameters observed in the exploited sections. This finding agrees with Anderson and Nehring (1984) who found that fish abundance in exploited areas was comparatively lower than the unexploited ones in a Colorado river. In contrast, Braña *et al.* (1992) observed no apparent changes in abundance between fished and unfished sections in northern Spanish rivers. Angling effects seem to show different patterns among areas and thus fishing regulations should be adapted to the diverse ecological conditions of the populations.

The genetic introgression by allochthonous genes of hatchery origin occurred in 67% of the analysed populations. Most of the introgressed populations were at H–W equilibrium, which suggests that gene flow between non-native and wild stocks exists. The lack of *LDH-5*90/100* heterozygotes observed in the River Cega could imply a very low survival of hatchery fish (Morán, Pendás, García-Vázquez & Izquierdo 1991; Martínez, Arias, Castro & Sánchez 1993) or strong assortive mating between hatchery and wild stocks in this population.

31.4.1 *Conservation and management proposals*

Overexploitation and stocking with allochthonous trout have become so great in some rivers that urgent conservation measures are needed. Guidelines for the conservation and management of central Spain brown trout populations are needed and these should include the following elements:

(1) The implementation of more restrictive harvest regulations, like catch and release or slot limits (harvest of fish between two specified lengths, Favro, Kuo & McDonald 1980; Jensen 1981), are recommended. Catch and release seems to be effective in maintaining the abundance of large trout within the fishery (Anderson & Nehring 1984; Hunt 1991; Almodóvar & Nicola 1998). Slot limits also reduce the harvest of larger trout (Power & Power 1996), which would contribute to higher fecundity and presumably increases in the number and quality of recruits.

(2) The use of creel-surveys should be encouraged, since they proved to be an essential tool to evaluate fishing pressure and harvest. Although this type of information is invaluable for effective management of populations, its application in Spanish rivers is still rare.

(3) A research-management cooperation is needed to achieve a more comprehensive approach to the conservation of trout populations. Harvest regulations should be focused on biological data gained from long-term monitoring.

(4) Management actions should not imply losing or disrupting genetic diversity, as such effects are irreversible. Therefore, stocking with allochthonous strains should be avoided. When populations must be stocked it is best to use material from the population itself.
(5) Conservation and management efforts should be centred on the genetic differences among populations within the species, as proposed by Machordom *et al.* (2000) for some Spanish rivers.

Acknowledgements

This study was supported by the Spanish INIA under Project SC-95/005 and a post-graduate fellowship (to J.S.). Dr B. Elvira provided intellectual stimulus, thoughtful advice, and helpful review of the manuscript. We are especially indebted to Dr A. Machordom for her contribution to genetic analyses. We are most grateful to the regional Environmental Agencies who have permitted populations studies to be conducted over many years.

References

Allendorf F.W. & Leary R.F. (1988) Conservation and distribution of genetic variation in a polytypic species, the cutthroat trout. *Conservation Biology* **2**, 170–184.

Almodóvar A. & Nicola G.G. (1998) Assessment of a brown trout *Salmo trutta* population in the River Gallo (central Spain): angling effects and management implications. *Italian Journal of Zoology* **65**, 539–543.

Almodóvar A. & Nicola G.G. (1999) Short-term effects of a small hydropower station upon brown trout (*Salmo trutta* L.) and macrobenthos in the Hoz Seca river (Tagus River, Spain). *Regulated Rivers, Research and Management* **15**, 477–484.

Anderson R.M. & Nehring R.B. (1984) Effects of a catch and release regulation on a wild trout population in Colorado and its acceptance by anglers. *North American Journal of Fisheries Management* **4**, 257–265.

Avery E.L. (1990) The White River trout population and sport fishery: an exploratory study, 1984–1986. *Wisconsin Department of Natural Resources, Research Report* 150, 25 pp.

Avery E.L. & Hunt R.L. (1981) Population dynamics of wild brown trout and associated sport fisheries in four central Wisconsin streams. *Wisconsin Department of Natural Resources, Technical Bulletin* **121**, 26 pp.

Braña F., Nicieza A.G. & Toledo M.M. (1992) Effects of angling on population structure of brown trout, *Salmo trutta* L., in mountain streams of Northern Spain. *Hydrobiologia* **237**, 61–66.

Cagigas M.E., Vázquez E., Blanco G. & Sánchez J.A. (1999) Genetic effects of introduced hatchery stocks on indigenous brown trout (*Salmo trutta* L.) populations in Spain. *Ecology of Freshwater Fish* **8**, 141–150.

Crisp D.T. (1994) Reproductive investment of female brown trout, *Salmo trutta* L., in a stream and reservoir system in northern England. *Journal of Fish Biology* **44**, 343–349.

Elvira B. (1996) Endangered freshwater fish of Spain. In A. Kirchhofer & D. Hefti (eds) *Conservation of Endangered Freshwater Fish in Europe*. Basel: Switzerland, Birkhäuser Verlag, pp. 55–61.

Favro L.D., Kuo P.K. & McDonald J.F. (1980) Effects of unconventional size limits on the growth rate of trout. *Canadian Journal of Fisheries and Aquatic Sciences* **37**, 873–876.

Ferguson A. (1989) Genetic differences among trout, *Salmo trutta*, stocks and their importance for the conservation and management of the species. *Freshwater Biology* **21**, 35–46.

García-Marín J.L., Jorde P.E., Ryman N., Utter F. & Pla C. (1991) Management implications of genetic differentiation between native and hatchery populations of brown trout (*Salmo trutta*) in Spain. *Aquaculture* **95**, 235–249.

García-Marín J.L. & Pla C. (1996) Origins and relationships of native populations of *Salmo trutta* (brown trout) in Spain. *Heredity* **77**, 313–323.

Hunt R.L. (1991) Evaluation of a catch and release fishery for brown trout regulated by an unprotected slot length. *Wisconsin Department of Natural Resources, Technical Bulletin* **173**, 39 pp.

Jensen A.L. (1981) Optimum size limits for trout fisheries. *Canadian Journal of Fisheries and Aquatic Sciences* **38**, 657–661.

Machordom A., García-Marín J.L., Sanz N., Almodóvar A. & Pla C. (1999) Allozyme diversity in brown trout (*Salmo trutta*) from Central Spain: genetic consequences of restocking. *Freshwater Biology* **41**, 707–717.

Machordom A., Suárez J., Almodóvar A. & Bautista J.M. (2000) Genetic differentiation and phylogenetic relationships among Spanish brown trout (*Salmo trutta*) populations. *Molecular Ecology* **9**, 1325–1338.

Mahon R., Balon E.K.G. & Noakes D.L.G. (1979) Distribution, community structure and production of fishes in the upper Speed River, Ontario: a preimpoundment study. *Environmental Biology of Fishes* **5**, 343–360.

Martínez P., Arias J., Castro J. & Sánchez L. (1993) Differential stocking incidence in brown trout (*Salmo trutta*) populations from Northwestern Spain. *Aquaculture* **114**, 203–216.

Morán P., Pendás A.M., García-Vázquez E. & Izquierdo J.I. (1991) Failure of stocking policy, of hatchery reared brown trout, *Salmo trutta* L., in Asturias, Spain, detected using *LDH-5** as a genetic marker. *Journal of Fish Biology* **39**, 117–122.

Morán P., Pendás A.M., García-Vázquez E., Izquierdo J.I. & Lobón-Cerviá J. (1995) Estimates of gene flow among neighbouring populations of brown trout. *Journal of Fish Biology* **46**, 593–602.

Newman R.M. & Martin F.B. (1983) Estimation of fish production rate and associated variances. *Canadian Journal of Fisheries and Aquatic Sciences* **40**, 1729–1736.

Nicola G.G. & Almodovar A. (in press) Reproductive traits of stream-dwelling brown trout *Salmo trutta* in contrasting neighbouring rivers of Central Spain. *Freshwater Biology*.

Power M. & Power G. (1996) Comparing minimum-size and slot limits for brook trout management. *North American Journal of Fisheries Management* **16**, 49–62.

Reyes-Gavilán F.G., Garrido R., Nicieza A.G., Toledo M.M. & Braña F. (1995) Variability in growth, density and age structure of brown trout populations under contrasting environmental and managerial conditions. In D.M. Harper & A.J.D. Ferguson (eds) *The Ecological Basis for River Management*. Chichester: John Wiley & Sons, pp. 389–406.

Rousset F. & Raymond M. (1995) Testing heterozygote excess and deficiency. *Genetics* **140**, 1413–1419.

Wiley R.W. & Dufek D.J. (1980) Standing crop of trout in the Fontenelle tailwater of the Green River. *Transactions of the American Fisheries Society* **109**, 168–175.

Zar J.H. (1999). *Biostatistical Analysis*, fourth edition. London: Prentice Hall, 663 pp.

Zippin C. (1956) An evaluation of the removal method of estimating animal population. *Biometrics* **12**, 163–189.

Chapter 32
Proposals for the reassessment of threatened status of freshwater fishes in France

H. PERSAT*
UMR CNRS 5023 Ecologie des hydrosystèmes fluviaux, Université Lyon 1, Lyon, France

P. KEITH
Institut d'Ecologie et de Gestion de la Biodiversité, Muséum National d'Histoire Naturelle, Paris, France

Abstract

French territory often represents the limits in the distribution range of many European fishes. Here, they appear as distinct populations segments (DPSs) that might have reached the status of evolutionarily significant units (ESUs) or even species. Unfortunately, because of lack of data about their actual distribution range and status, the risk of extinction is probably underestimated. Therefore, further investigations are urgently needed to detect all the possible endangered species, ESUs and DPSs. In addition, French policy only considers the species level, and thus infra-specific units will not be conserved by existing rules. Also territorial borders do not necessarily match the needs of freshwater management, and new rules should be established at the European Community level, applied at the river drainage basins level, and follow the principle of subsidiarity in the decision-making process. Pending these major improvements, inter-basin translocations should be prohibited to protect French and European species and infra-species biodiversity.

Keywords: biodiversity, conservation, France, freshwater fish, policy.

32.1 Introduction

Bioconservation is promoted from either an economic point of view (protection of a valuable resource), a geographic point of view (protection of the species richness in a given area), or a biological point of view (protection of a given endangered species). The first perspective does not specifically worry about habitat or life diversity: it may even be interested in reducing it to favour the selected stock. It is promoted by collectors, farmers and managers within the area they are exploiting or controlling. The second perspective does not specifically worry about the biological distinctiveness of the local species: it may even be interested in artificially increasing the local richness by supplementing the area with alien species or populations. It is promoted by landowners or land managers within the district they are controlling. The third perspective does not specifically worry about the local richness in any place: it may even be interested

*Correspondence: UMR CNRS 5023 Ecologie des hydrosystèmes fluviaux, Bat. 403, Université Lyon 1, 69622 Villeurbanne Cedex, France (email: persat@biomserv.univ-lyon1.fr).

in reducing it to improve the survival of the selected species at a single site. It is promoted by narrow-minded conservationists influenced by land managers claiming the freedom to alter biodiversity everywhere for any immediate profits, as it is always possible to keep the genetic diversity (in case it might be profitable in the future) in a few conservation sites (nature reserves, zoos, cryo-banks, or even, in the near future, data banks [full genome sequences] only).

Thus, each one of these conservationist perspectives may be detrimental to the others, which might happen when the main objectives of conservation policies, that are often trying to pool the approaches, are not clearly defined. Whatever the aims, applied conservation measures always require definition of the objective on two dimensions, a biological one and a spatial one, i.e. which life form is to be protected in which area? Determining the biological units to protect is the task of biologists who are not supposed to be tied to administrative borders, whereas the management of these units is the task of land managers (landowners, administrations or politics), the action range of whom is strictly limited to their jurisdictional territory. International conservation policy, as expressed in the IUCN rules (1990, 1995), are based mainly at the taxonomic level of species: those species having very limited extension or very small populations in the world must be protected. In this way, they should gain protection whatever the country. By contrast, at the country level, the status of the species will depend on their local abundance or extension, whatever their world abundance or extension. Thus the protection status of one species in one country will strongly depend on the size and geographic position of that country. Consequently, many southern species tend to be protected in northern countries and vice versa (and, in general, a given species has more chance of being protected in a small, less diverse, rather than in a large, more diverse, country). The case of the bald eagle, *Haliaeetus leucocephalis* (L.), in the USA highlighted this paradox: in Alaska, the species is so common that it does not require any protection, but in the more southern states it does (USFWS 1995). This leads to the question of protection at the infra-specific level. In interpreting the expression 'distinct population segment' used in the US Endangered Species Act (USHR 1973), two contrasted opinions were developed, some considered the biological units and others the geographical units to be the management units. The first proposed the concept of evolutionarily significant unit (ESU), that, however, underwent a significant drift from a initially rather holistic point of view 'populations that actually represent significant adaptative variation based on concordance between sets of data derived by different techniques' (Ryder 1986), to a more genetically focused point of view 'reproductively separated populations with unique or different adaptation' (Waples 1991, 1995, 1999), ending in Moritz's (1994) definition: 'ESUs should be reciprocally monophyletic for mtDNA alleles and show significant divergence of allele frequencies at nuclear loci'. Crandall, Bininda-Edmonds, Mace and Wayne (2000) highlighted how this shift towards the prevalence of genetics can induce false assumptions, especially when based on molecular techniques promoting the use of neutral genetic data, which are by definition evolutionarily not significant. To restore the ESU concept in a more effective way, the latter proposed a framework based on both genetic and ecological exchangeability criteria during historical and recent times to calibrate the population distinctiveness. Being mainly based on reproductive isolation, the ESU narrows the 'population' level on one side, but with genetic bases in addition that might raise it to the species level on the other (Mayden & Wood 1995).

This is considered to be too restrictive by those promoting the notion of distinct population segment (DPS), as initially conceived in the US, ESA (Pennock & Dimmick 1997; 2000). DPSs are populations that are markedly separated from other populations of the same taxon as a consequence of physical, physiological, ecological or behavioural factors that are significant to the species to which they belong. The ESU concept does not allow for protection of a locally endangered population within a globally non-endangered ESU. Thus the ESU concept cannot be used for differential management according to different situations, essentially different places inhabited by separate populations, but sometimes also to different times such as migration or spawning times. By contrast, the DPS concept is much more flexible and can be applied to a peculiar subset of a given species, ESU, or even population (for example, one life-stage in one particular place). Thorough scientific knowledge is not required to consider a population segment as distinct: any criterion can be used, provided it evolved from natural processes. Thus, the DPS notion is strongly related to a spatial dimension without minimising the biological dimension. It makes the connection between biological and spatial units at a quite applied level.

By contrast, with the species-based conservation approach, and because species conservation cannot be envisaged in the long-term outwith of the protection of the relevant habitats, an alternative strategy is to discriminate different areas according to their biological contents: i.e. eco-regions (Dinerstein, Olson, Graham, Webster, Primm, Bookbinder & Ledec 1995). An eco-region is defined by its biological distinctiveness, a combination of species richness (quantitative biological parameter), species distinctiveness (qualitative biological parameters), community distinctiveness (structural biological parameter), habitat types (qualitative physical parameter), rarity (quantitative physical parameter) and evolutionary process (dynamic parameter) (Abell, Olson, Dinerstein, Hurley, Diggs, Euchbaum, Walters, Wettengel, Allnut, Loucks & Hedao 2000). Each identified eco-region would then require specific management to keep its natural characteristics. However, eco-region conservation can be considered in the same restrictive point of view as species conservation. Like a species, an eco-region is unique, and like a species, that can theoretically be conserved because of a single representative population, the main features of a given eco-region might be protected through the conservation of a single representative sub-region, such as national parks or reserves. If that is possible for some terrestrial eco-regions, riverine eco-regions raise many problems. Indeed, they require first, the protection of the entire terrestrial area that determines their water regime, mainly throughout the drainage basin, and second, the conservation of the river continuum (Vannote, Minshall, Cummins, Sedell & Cushing 1980) that is a major physical attribute that aquatic life, and especially fish, is adapted to. Therefore, freshwater eco-regions often cannot be split into pieces because they are made up of a single major drainage basin, the integrity of which is obligatory for the conservation of its environmental distinctiveness. River drainages constitute the main spatial management units in freshwater conservation, within which the possible sub-units related to the ESUs or DPSs, with their peculiar habitats being more or less dependent on the whole hydrosystem, have to be considered.

Finally, for freshwater fishes, the spatial and biological dimensions of biodiversity are so intimately associated that a unilateral approach is necessarily condemned to failure. The acuity of these problems can be illustrated by the status of the French freshwater fish fauna.

32.2 Biogeographical status of the French fish fauna

The French freshwater fish fauna comprises many species with a distribution widespread across continental Europe. However, the large watersheds from the Rhine (north-east) to the Garonne (south-west) rivers represent the limit of the range of many mid-European species. France is also the northern limit for a few southern species (*Chondrostoma toxostoma* (Vallot), *Barbus meridionalis* (Risso), *Salaria fluviatilis* (Asso). Only one native species has its distribution extending in all directions beyond French territory, the brown trout *Salmo trutta* L., and only one species is endemic to France, the apron, *Zingel asper* (L.) (see Mari, Labonne & Gaudin, Chapter 15), but this represents the western limit of the Danubian genus *Zingel*.

If *Z. asper* has evolved at the western edge of the distribution range of the genus *Zingel*, there is a high probability that other species may have evolved at the edge of their distribution range in France, if they have existed for a long time. This is particularly relevant where discrete populations have been isolated from populations in large drainage basins, thus restricting gene flows. Therefore, the time of colonisation of each species in each basin is important from a conservation perspective because that may provide clues to identify potential ESUs (Persat & Keith 1997).

Morphological studies suggest the presence in France of different stocks for species such as dace, *Leuciscus leuciscus* (L.), (Blanchard 1866; Spillmann 1961), toxostome, *C. toxostoma* (Blanchard 1866; Spillmann 1961), blageon *Telestes souffia* (Risso) (Cépède 1905; d'Aubenton, Daget & Spillmann 1971), but have not been investigated further until the recent development of genetic methods (Chappaz, Gilles, Miquelis, Cavalli & Martin 1998; Gilles, Chappaz, Cavalli, Lortscher & Faure 1998). The later have made it possible to investigate population diversity in a more accurate way, and despite the rather limited number of species examined, illustrate the number of DPSs and ESUs present, and thus potentially threatened, in France.

32.3 Case studies

32.3.1 *Genetic evidence of co-existence of different ESUs within the French extension area of some freshwater fish species*

Brown trout, Salmo trutta *(L.)*

With a widespread distribution outside French territory, this species was not considered, until recently, to have developed lineages in France that can often be observed at the periphery of the distribution range of many species. As probably the most widespread species in France, the brown trout appears far from threatened, but it has nevertheless been ranked among the species to be protected in the French 1988 decree (FME 1988). This status was not based on biological, but on socio-economic, reasons: the aim was rather to protect a resource than a species. Indeed, brown trout has been heavily stocked everywhere in the last century, regardless of the possible existence of different populations according to the river basin. The existence of a major difference between the Atlantic and Mediterranean stocks has only recently been confirmed by genetic investigations (Guyomard 1989; Bernatchez, Guyomard & Bonhomme 1992; Poteaux, Bonhomme & Berrebi 1998). Unfortunately, the stocks reared in fish farms are of Atlantic origin only (Krieg & Guyomard 1985), and stocking has induced a high level of introgression within many Mediterranean populations. Thus, there were two different brown trout ESUs in France, and one is threatened because of stocking. Conservation laws appear inappropriate to protect such biological diversity. Further investigations have now indentified a third ESU in the central mountains of Corsica (Berrebi 1995), which is among the most endangered in Europe, and will require special attention.

Grayling, Thymallus thymallus *(L.)*

France marks the south-west limit for this species, and local variations would be expected, since, by contrast with brown tout, most of its populations are isolated from each other (Persat, Pattee & Roux 1977). Allozymic investigations on a large number of populations discriminated three main groupings corresponding to the three main drainages in which the species is native in France, i.e. the Rhine, Rhone and Loire (Table 32.1) (Persat & Eppe 2000). Unfortunately stocking has resulted in introgression in many populations, and in some cases the populations have been replaced by stocked fish, mainly from Rhine, but sometimes also from more distant origins (Persat & Eppe 1997). Thus, three ESUs exist which must be managed separately, and morphological divergence can even be observed within some of them (Surre, Persat & Gaillard 1986; Persat 1996). If the species is classified as vulnerable in France, the Loire and Rhone ESUs should be considered endangered, and many of their local populations, that may be considered as distinct population segments according to criteria other than allozymes, should be classified as critical.

32.3.2 *Genetic evidence of co-existence of different species within the French extension area of some common freshwater fish*

Sculpin, Cottus gobio *(L.)*

The sculpin is perhaps one of the most widespread fish in France after the brown trout. Little attention was paid to it until Bacescu and Bacescu-Mester (1964) described a new species *Cottus petiti* from a small river in South France based on morphological features. Recent genetic investigations to validate this description provided a much more complicate pattern (Eppe & Persat 1999; Eppe, Persat, Beaudou & Berrebi 1999). Instead of a single isolated species, a large set of highly divergent populations that constitute several distinct species and ESUs based on interspecific genetic

Table 32.1 Observed allozymic variability within the French populations of the European grayling *Thymallus thymallus*, and speculation on the basic genotypes involved in it (Persat & Eppe, unpublished compilation)

Loci		Ck-1			Ldh-1		Pgdh-2		Gpi-3		Pgm-1		Pgm-2	
Alleles	N	100	107	97	0	100	100	112	100	105	75	100	100	115
Observed genotypes														
Rhône drainage basin														
Doubs	10	20	0	0	0	20	20	0	20	0	0	20	20	0
Dessoubre	10	9	11	0	1	19	15	5	20	0	0	20	20	0
Cusancin	10	20	0	0	0	20	20	0	20	0	0	20	20	0
Loue 99	10	20	0	0	0	20	20	0	20	0	0	20	20	0
Loue 86	19	38	0	0	0	38	38	0	38	0	0	38	38	0
Upper Ain[a]	40	54	26	0	26	54	55	25	75	5	17	63	71	9
Bienne	20	40	0	0	0	40	39	1	38	2	0	40	40	0
Pt Ain/Gévrieux	60	114	6	0	11	109	111	9	119	1	6	114	113	7
Miribel	22	43	1	0	3	41	43	1	44	0	0	44	44	0
Seran	10	20	0	0	2	18	17	3	20	0	0	20	20	0
Pont de Lucey[a]	41	47	34	1	32	50	53	29	79	3	19	63	76	6
Furans	29	53	5	0	15	43	39	19	58	0	5	53	58	0
Chautagne→95[a]	46	11	81	0	70	22	50	42	92	0	47	45	72	20
Chautagne 96→[a]	47	2	92	0	82	12	46	48	93	1	47	47	56	38
Versoix[a]	18	8	28	0	25	11	10	26	30	6	9	27	16	20
Rhine-Maas drainage basin														
French Rhine	10	0	20	0	16	4	17	3	16	4	10	10	16	4
Doller[a]	29	58	0	0	0	58	57	1	58	0	0	58	58	0
Moselle	27	1	53	0	51	3	1	53	54	0	49	5	48	6
Swiss Rhine	30	0	60	0	55	5	55	5	60	0	24	36	52	8
Ourthe (Belgium)	29	0	58	0	58	0	21	37	58	0	28	30	58	0
Loire drainage basin														
Huisne[a]	10	0	20	0	20	0	1	19	20	0	9	11	13	7
Loire	18	0	0	36	0	36	36	0	36	0	36	0	36	0
Allagnon	29	0	0	58	0	58	58	0	58	0	22	36	48	10
Hatchery stocks														
Augerolles	93	97	87	2	48	138	37	149	149	37	20	166	93	93
Aug. Allier	21	6	15	21	13	29	16	26	34	8	7	35	31	11
Finland	28	15	41	0	37	19	0	56	53	3	0	56	55	1
Austria	32	57	7	0	0	64	0	64	1	63	0	64	0	64
Bavaria	22	33	11	0	1	43	2	42	18	26	4	40	1	43
Extrapolated basic genotypes involved in the biodiversity of the above populations														
Rhine type		0	2	0	2	0	1	1	2	0	1	1	2	0
			Mosel fixation? →				0	2			2	0		
Rhône type		2	0	0	0	2	2	0	2	0	0	2	2	0
Loire type		0	0	2	0	2	2	0	2	0	2	0	2	0
Danube type		2	0	0	0	2	2	0	0	2	0	2	0	2
								Variant? →	1	1				
Finland type		1	1	0	1	1	0	2	2	0	0	2	2	0
Augerolles type		1	1	0	1	1	1	1	1	1	1	1	1	1
(hatchery stock)	With trend →				0	2	0	2	2	0	0	2		

[a]Heavily or entirely stocked populations.

distance criteria. Moreover, examination of the distribution of these fish highlighted the quite limited presence of many in south France. Thus, there are a large number of isolated populations at the ESU level that are either endangered or critical. This observation lead to the conclusion that several others might already be extinct. Consequently, each of those remaining in south France should be fully protected, whereas many northern French ESUs probably do not require special attention.

Loaches, genus Cobitis

Loaches (genus *Cobitis*) was among the most noted species group in the regional fish faunas of France during the 19th century. Nowadays, its distribution appears much more limited: it is almost extinct in the majority of the Rhone and Loire drainages, and often rare elsewhere. Consequently, the *Cobitis* is classified as vulnerable. Nevertheless, the dispersed distribution of the remaining populations raises the question of the actual homogeneity of these stocks. Recent investigations in Central Europe pointed out the marked diversity within this genus (Bohlen, Ráb, Šletchová, Rábová Ditterbusch & Freyhof, Chapter 28). If this scenario is applicable to France, then the co-existence of different species is highly likely. A visual examination of those fish living in the River Durance, South France, (J. Bohlen, personal communication) suggests they look more closely related to the Italian *Cobitis bilineata* (Canestrini) than to the nominal *Cobitis taenia* (L.). Therefore, rather than one vulnerable taxon, France might be inhabited by several endangered or critical species. An overall survey of the biological diversity within this group is necessary for effective conservation of the French patrimony.

32.3.3 *Species in distinct population segments without genetic evidence of ESU value*

Many French freshwater fish species are distributed in discrete population segments. This might be due to recent partitioning, because of river development, e.g. the apron in the Rhone river system, or invasion of competitors like for the toxostome, which has been lost from the main watercourses of the same system because of colonisation of the nase, *Chondrostoma nasus* (L.), during the 19th century. Consequently, no genetic divergence should be expected, but some species range partitioning can be much more ancient, for example species depending on peculiar habitats such as mountain streams such as the meridional barbel, *Barbus meridionalis* (Risso).

Barbus meridionalis only lives in the rivers entering the Mediterranean Sea, from the Spanish Catalunia to the French Côte d'Azur, i.e. in a series of discrete drainages (Tsigenopoulos & Berrebi 2000). In the largest hydrosystem of the area, the Rhone River basin, it only occurs in small southern tributaries. This distribution is based on peculiar ecological requirements (with a preference for streams of first orders) reinforced by competition from the common barbel *Barbus barbus* (L.) (Persat & Berrebi 1990). Partitioning is thus rather ancient in this species, but allozyme studies did not identify significant divergences among populations: all appear quite uniform and almost fixed, except possibly in the south-western aspect of its dsitribution range where some variability was noticeable (Berrebi, Lamy, Cattaneao-Berrebi & Renno 1988). Thus, the species, considered as a whole, is only classified as LR/nt (low risk, near threatened) but, despite numerous populations, many of them, taken independently, should be considered endangered or critical, with some peripheral ones being close to or already extinct.

This example highlights the dilemma between biological and regional conservation: how many local extinctions can be tolerated before drastic conservation measures are taken for the remaining populations; or, alternatively, which populations are to be protected or otherwise, and why?

32.3.4 *Global overview*

Based on the case studies, a large number of freshwater fish species in France should not be considered as homogeneous unit stocks. The genetics of the large native cyprinid *Leuciscus cephalus* (L.), (Durand, Bouvet & Persat 1999) suggest they colonised French waters in recent times (interglacial Riss-Würm period, or post Würm time). Consequently, different species may exhibit the same characteristics. Thus according to what might be hypothesised from previous studies (Persat & Keith 1997; Keith 1998), a putative framework can be drawn for what might be expected in terms of population structure, number of discrete populations segments and risk levels, for the different species in France (Table 32.2).

Many species might be made up of two or three ESUs, with several to numerous DPSs in their peripheral distribution. Those species having potentially survived the last glaciations are more liable to have split into different units. Moreover, the smaller-sized species are more disposed to that trend than large ones. Therefore, there are strong probabilities that species such as the stone loach, *Barbatula barbatula* (L.), the gudgeon, *Gobio gobio* (L.), or the minnow, *Phoxinus phoxinus* (L.) are also polytypic in France. Indeed, many species inhabiting the Loire and Garonne river basins, the most distant from the Rhine system, the main source of the French fish fauna, might differ somewhat from oriental stocks.

Whatever their biological distinctiveness, in many places many fishes only appear in isolated populations. All populations sufficiently isolated to not be enhanced by natural dispersion are potentially vulnerable, endangered or critical according to either their size or the size of the area they inhabit. Keith (1998) used the partition of the French territory according to the main hydrographic units proposed by Persat and Keith (1997) to highlight the number of extinctions and the extension of the risks for the 'official' nominal species (Table 32.3). Combining Tables 32.2 and 32.3 makes it possible to evaluate the real conservation status of freshwater fishes in France.

Compared to the size of the problem, the French legislation is correspondingly weak. First, the biological dimension of biodiversity is only considered at the species level, and second, only one species is theoretically fully protected, the sturgeon *Acipenser sturio* (L.) since 1982 (FME 1982), with another, the apron, being so at the

Table 32.2 Possible number of ESUs and DPSs in the French native freshwater fish species with their probable status

Species (not exhaustive list)	ESU Number	ESU VU	ESU EN	ESU CR	ESU Ex	DPS Number	DPS VU %	DPS EN %	DPS CR %
Lampetra planeri	1–>n?					>100	20	30	30
Salmo trutta	3		1	1		>100	20	10	10
Salvelinus alpinus	2(3)	1	(1)		1[a]				
Coregonus 'lavaretus'	3	1			2[b]				
Thymallus thymallus	3	1	2			±30	33	33	33
Esox lucius	≥1?	1							
Alburnus alburnus	≥1?								
Leuciscus leuciscus	1 (–>3?)	1				>100	50	25	25
Leuciscus cephalus	1 (–>2?)								
Telestes souffia	3	1	1		1?[c]	>50	25	10	10
Phoxinus phoxinus	1 (–>n?)					>100	25	10	5
Chondrostoma nasus	1	1							
Chondrostoma toxostoma	2?	1	1			±50	50	30	20
Alburnoides bipunctatus	≥1?								
Blicca bjoerkna	≥1?								
Abramis brama	≥1?								
Rhodeus amarus	≥1?								
Rutilus rutilus	≥1?								
Scardinius erythrophthalmus	≥1?								
Gobio gobio	1 (–>n?)					>100	20	20	10
Barbus barbus	1 (–>2?)								
Barbus meridionalis	1(2)	1				±50	50	30	20
Tinca tinca	≥1?								
Barbatula barbatula	1–>n?					>100	20	20	10
Cobitis taenia	1–2–>n?					>100	20	30	50
Misgurnus fossilis	1 (–>2?)					±20	20	30	50
Lota lota	1 (–>3?)					±20	33	33	33
Perca fluviatilis	1 (>1?)								
Gymnocephalus cernuus	1								
Cottus gobio	3–>n?					>100	20	20	10
Zingel asper	1			1		±20	33	33	33
Salaria fluviatilis	1 (–>2?)	(1?)				±20	33	33	33

[a]Paladru lake; [b]Leman lake; [c]Annecy lake.

habitat level through the Habitats Directive (EC 1992). Some protections may also be obtained with respect to the reproductive periods and habitats, or fishing rules for a series a species listed in other French legislation (FME 1988) (Fiers, Gauvrit, Gavazzi, Haffner & Maurin 1997), but this is rather limited and ineffective. Finally, few reserves, and national or regional natural parks have been created for freshwater fish conservation (Keith 2000).

Table 32.3 Status of endangered freshwater fish in France (Corsica excluded) per reference basin and national status (Keith 1998)

Endangered species in France	Rhine-Meuse	Artois-Picardie	Seine	Brittany	Loire	Vendée-Charente	Adour-Garonne	Rhône	Languedoc-Roussillon	Côte d'Azur	National status
Sea lamprey	–	EX	EN	VU	VU	–/I	VU	VU	VU	–	VU
River lamprey	CR	EN	VU	VU	EN	VU	VU	EN	CR	CR	VU
Sturgeon	EX	EX	EX	–	EX	EX	CR	EX	–	–	CR
Allis shad	EX	–	CR	EN	VU	EX	VU	EX	EX	–	VU
Twaite shad	EX	–	CR	EN	VU	VU	VU	VU	VU	–/I	VU
Sea trout	–/I	VU	VU	VU	VU	VU	VU	–/I	–/I	–/I	VU
Salmon	CR	CR	CR	VU	CR	CR	EN	–	–	–	EN
Arctic charr	VU	–	–	–	–/I	–	–/I	NM	–/I	–/I	LR/nt
Grayling	VU	–/I	–/I	–	VU	–	–/I	VU	–	–/I	VU
Lavaret	LR/nt	–	–	–	NM			LR/nt	–	–	LR/nt
Houting	EX	–	–	–	–	–	–	–	–	–	EX
Pike	VU	VU	NM	NM	NM	VU	–/I	NM	–/I	–/I	VU
Meridional barbel	–	–	–	–	–	–	VU	NM	NM	NM	LR/nt
Toxostome	–/I	–	–/I	–	–/I	–	NM	NM	NM	VU	LR/nt
Soufie	EN	–	–/I	–	–	–	–	NM	NM	NM	VU
Spirlin	NM	NM	NM	EN	NM	NM	NM	NM	NM	NM	NM
Bitterling	VU	–/I	VU	–	–/I	–	–/I	VU	–/I	–	VU
Spined loach	VU	–/I	VU	–/I	EN	–/I	VU	CR	–/I	–	VU
Weatherfish	EN	–/I	–	–	–	–	–	CR	–	–	EN
Eel	VU	VU	VU	VU	VU	VU	VU	VU	VU	VU	VU
Burbot	NM	–/I	NM	–	VU	–	–	VU	EN	–	VU
South europe toothcarp	–	–	–	–	–	–	–	–	–	EX	EX
Spanish toothcarp	–	–	–	–	–	–	–	–	EX	–	EX
Valencia toothcarp	–	–	–	–	–	–	–	–	EX	–	EX
Apron	–	–	–	–	–	–	–	CR	–	–	CR
Freshwater blenny	–	–	–	–	–	–	VU	NM	VU	VU	VU
Lez bullhead	–	–	–	–	–	–	–	–	CR	–	CR
Total endangered	15	7	10	8	11	8	11	13	10	5	26

*Ex: extinct; CR: critically endangered; EN: endangered; VU: vulnerable; LR: low risk (UICN 1995); –/I: introduced species (without UICN status); NM: non endangered species.

32.4 Conclusions

This overview of the freshwater fish status in France leads to the following statements:

(1) the overall biological biodiversity of the French freshwater fish fauna is underestimated;
(2) infra-species biodiversity is not considered by French legislation;
(3) spatial biodiversity cannot be taken into account by policies based on state borders.

These statements can probably be extended to many other European countries, but are more relevant in France because of the rather common aspect (many species found elsewhere in the region) of its freshwater fish fauna and the highly centralised organisation that drives policy in the state. Currently, French legislation only recognises the species level. Consequently, any fish belonging to the so-called species can be imported in France from any parts in Europe. This issue is exacerbated by free trade movements within the European Community, and with other countries through 'globalisation' of markets. Thus, for example, gudgeon (*Gobio* sp.) are freely imported from south east Europe, because of its low price there, and rudd (*Scardinius* sp.) are imported from Italy because the local form (sub-species or species?) is much more resistant than the French conspecific. None of these imports are controlled, either at the European Community borders or at the French border. Irrespective of the introgression of natural populations, these movements have also increased the possibility of introductions of aliens, such as *Pseudorasbora parva* (Temminck & Schlegel) and *Pachychilon pictum* (Heckel & Kner). The number of introductions are thus likely to increase rapidly in the near future, implying stronger risks of extinctions within the native French fauna, especially when introduced fishes are close relatives of native species (for example the case of the *Chondrostoma* species mentioned above).

By ignoring the natural geographical borders within its own territory, French legislation also allows uncontrolled, inappropriate translocations between French river basins, with similar negative effects to those found with international translocations/introductions.

To protect fish biodiversity in France, like in other countries, a new policy should be adopted based on the following main principles.

(1) Precautionary principle: translocations between basins should not be authorised until the biological similarity between local and introduced stocks has been proven. This principle prohibits stocking in most rivers and connected water bodies.
(2) Subsidiarity principle: conservation should be managed on an appropriate geographical scale. This needs to match physical units with biological units. Consequently, a single rule per species or per site cannot exist, but rules must exist of each species at each site. Since national borders are irrelevant in this aim: a general regulation should be established at the level of the European Community, and rules should be set at the functional level of the taxa-habitat management unit under consideration. In that way, drainages should be considered as the major management units, and the maintenance of connectivity within these units should be protected.

(3) Regional rather than global conservation principle: conservation should not focus on keeping alive one or a few sets of each taxa in one or a few sites (ESU strategy), but as much native taxa as possible in as many sites as possible (DPS strategy). This principle denies the right for any authority to destroy openly one population or one site by arguing that other populations or sites exist elsewhere: any destructive project must be quantified and debated in advance before being carried out.
(4) Accurate knowledge of the overall biodiversity within the French and European fish faunas is essential for appropriate management. The promotion of extensive research into biodiversity within the rare as well as the assumed common species is urgently required, to identify all possible ESUs present. This requires initiating genetic studies to characterise populations, to enhance the morphometric studies that were prematurely interrupted at the beginning of the 20th century, with the misconception that fish biodiversity census work was then completed for France.

References

Abell R.A., Olson D.M., Dinerstein E., Hurley P.T., Diggs J.T., Euchbaum W., Walters S., Wettengel W., Allnutt T., Loucks C.J. & Hedao P. (eds) (2000) *Freshwater Ecoregions of North America. A Conservation Assessment*. Washington: WWF USA, Island Press, 319 pp.

Bacescu M. & Bacescu-Mester L.E. (1964) *Cottus petiti* sp. n., un chabot nouveau récolté en France. Considérations zoogéographiques et données comparatives sur d'autres *Cottus* d'Europe. *Vie et Milieu* Suppl. **17**, 431–446 (in French).

Bernatchez L.K., Guyomard R. & Bonhomme F. (1992) DNA sequence variation of the mitochondrial control region among geographically and morphologically remote European brown trout, *Salmo trutta* populations. *Molecular Ecology* **1**, 161–173.

Berrebi P. (1995) Etude génétique des truites de Corse. Rapport Final de contrat Parc Naturel Régional de Corse, novembre 1995, 36 pp. + annexes (in French).

Berrebi P., Lamy G., Cattaneo-Berrebi G. & Renno J.F. (1988) Variabilité génétique de *Barbus meridionalis* Risso (Cyprinidae): une espèce tétraploïde quasi monomorphe. *Bulletin Français de Pêche et de Pisciculture* **310**, 77–84 (in French).

Blanchard E. (1866) *Les Poissons des Eaux Douces de la France*. Paris: Baillière, 656 pp. (in French).

Cépède C. (1905) Le blageon et la suiffe bouchesse (*Squalius agassizi* Heck.) dans les lacs et les cours d'eau de la région delphino-savoisienne. *Annales de l'Université de Grenoble* **17**, 301–306 (in French).

Chappaz R., Gilles A., Miquelis A., Cavalli L. & Martin J.F. (1998) Différenciation génétique et hybridation chez le cyprin *Leuciscus leuciscus*. Comptes Rendus de l'Académie des Sciences, Series 3, *Sciences de la Vie* **321**, 933–940 (in French).

Crandall K.A., Bininda-Edmonds O.R.P., Mace G.M. & Wayne R.K. (2000) Considering evolutionary processes in conservation biology. *Trends in Ecology and Evolution* **15**, 290–295.

d'Aubenton F., Daget J. & Spillmann J. (1971) Classification numérique des blageons *Leuciscus (Telestes) soufia* (Pisces, Cyprinidae). *Bullletin du Muséum National d'Histoire Naturelle*, Paris **42**, 839–848 (in French).

Dinerstein E., Olson D.M., Graham D.J., Webster A.L., Primm S.A., Bookbinder M.P. & Ledec G. (1995) *A Conservation Assessment of Terrestrial Ecoregions of Latin America and the Caribbean*. Washington: The World Bank, 148 pp.

Durand J.-D., Bouvet Y. & Persat H. (1999) Phylogeography and postglacial dispersion of the chub (*Leuciscus cephalus*) in Europe. *Molecular Ecology* **8**, 989–997.

Eppe R. & Persat H. (1999) Etude biogéographique et évolutive du complexe taxonomique des Cottidés ouest-européens apparentés à l'espèce *Cottus gobio*: implications dans le domaine de la conservation de la biodiversité. *Technical Report to the French Minister of Environment*, Université Lyon 1, 24 pp. (in French).

Eppe R., Persat H., Beaudou D. & Berrebi P. (1999) Genetic variability in sculpin (genus *Cottus*) from southern France, with reference to the taxonomic status of an endemic species, *C. petiti*. *Heredity* **83**, 533–540.

European Commission (1992) Directive 'Habitats-Faune-Flore' n° 92/43/CEE du Conseil du 21 May 1992 concernant la conservation des habitats naturels ainsi que de la faune et de la flore sauvages. *Journal Officiel de la Communauté Européenne*, 22/07/92.

Fiers V., Gauvrit B., Gavazzi E., Haffner P. & Maurin H. (1997) *Statut de la faune de la France metropolitaine – statuts de protection, degrés de menace, statuts biologiques*. Paris: Muséum National d'Histoire Naturelle, 225 pp.

FME (French Minister of the Environment) (1982) Arrêté du 25 janvier 1982 relatif à la protection de l'espèce *Acipenser sturio*. *Journal Officiel de la République Française*, 14/02/82 (in French).

FME (French Minister of the Environment) (1988) Arrêté du 8 décembre 1988 fixant la liste des espèces de poissons protégés sur l'ensemble du territoire national. *Journal Officiel de la République Française*, 22/12/88 (in French).

Gilles A., Chappaz R., Cavalli L, Lörtscher M. & Faure E. (1998) Genetic differentiation and introgression between putative subspecies of *Leuciscus soufia* (Telesotei: Cyprinidae) of the region of the Mediterranean Alps. *Canadian Journal of Fisheries and Aquatic Sciences* **55**, 2341–2354.

Guyomard R. (1989) Diversité génétique de la truite commune. *Bulletin Français de la Pêche et de la Pisciculture* **314**, 118–135 (in French).

IUCN (1990) *IUCN Red List of Threatened Animals*. Gland: IUCN, 228 pp.

IUCN (1995) *Catégories de l'UICN pour les listes rouges*. Gland: UICN, 22 pp.

Keith P. (1998) *Evolution des peuplements ichtyologiques de France et stratégies de conservation*. PhD thesis, University Rennes 1, 235 pp. (in French).

Keith P. (2000) The part played by protected areas in the conservation of threatened French freshwater fish. *Biological Conservation* **92**, 265–273 (quoted page 14).

Krieg F. & Guyomard R. (1985) Population genetics of French brown trout (*Salmo trutta* L.): large geographical differentiation of wild populations and high similarity of domesticated stocks. *Génétique Sélection et Evolution* **17**, 225–242.

Mayden R.L. & Wood R.M. (1995) Systematics, species concepts, and the evolutionrily significant unit in conservation biology. In L. Nielsen (ed.) *Evolution and the Aquatic Ecosystem: Defining Unique Units in Population Conservation*. Bethesda, Maryland: American Fishery Society Symposium **7**, 58–113.

Moritz C. (1994) Defining 'evolutionarily significant units' for conservation. *Trends in Ecology and Evolution* **9**, 373–375.

Pennock D.S. & Dimmick W.W. (1997) Critique of the evolutionarily significant unit as a definition for 'distinct population segments' under the U.S. Endangered Species Act. *Conservation Biology* **11**, 611–619.

Pennock D.S. & Dimmick W.W. (2000) Distinct population segments and congressional intent: reply to Waples. *Conservation Biology* **14**, 567–569.

Persat H. (1996) Threatened populations and conservation of the European grayling, *Thymallus thymallus* (L., 1758). In A. Kirchhofer & D. Hefti (eds) *Conservation of Endangered Freshwater Fishes in Europe*. Basel: Birkhäuser Verlag, pp. 233–247.

Persat H. & Berrebi P. (1990) Relative ages of present populations of *Barbus barbus* and *Barbus meridionalis* (Cyprinidae) in southern France: preliminary considerations. *Aquatic Living Resources* **3**, 253–263.

Persat H. & Eppe R. (1997) Allevinage, pollution et cloisonnement de l'espace fluvial dans les structures génétiques des populations de poisson: l'ombre commun, *Thymallus thymallus*, dans le Rhône au niveau de la Savoie. *Bulletin Français de la Pêche et de la Pisciculture* **344/345**, 287–299 (in French).

Persat H. & Eppe R. (2000) Inventaire du patrimoine génétique des populations françaises d'Ombre commun en vue de leur conservation. *Technical Report to the French Ministry of Environment*, Université Lyon 1, 14 pp. (in French).

Persat H. & Keith P. (1997) La répartition géographique des poissons d'eau douce en France: qui est autochtone et qui ne l'est pas? *Bulletin Français de la Pêche et de la Pisciculture* **344/345**, 15–32 (in French).

Persat H., Pattee E. & Roux A.L. (1978) Origine et caractéristiques de la distribution de l'Ombre commun, *Thymallus thymallus* (L., 1758) en Europe et en France. *Verhandlungen International Vereiningung für Limnologie* **20**, 2117–2121 (in French).

Poteaux C., Bonhomme F. & Berrebi P. (1998) Differences between nuclear and mitochondrial introgressions of brown trout populations from a restocked main river and its unrestocked tributary. *Biological Journal of the Linnean Society* **63**, 379–392.

Ryder O.A. (1986) Species conservation and systematics: the dilemma of subspecies. *Trends in Ecology and Evolution* **1**, 9–10.

Spillmann J.L. (1961) *Faune de France. 65. Poissons d'eau douce*. Paris: Lechevalier, 303 pp. (in French).

Surre C., Persat H. & Gaillard J.-M. (1986) A biometric study of three populations of the European grayling, *Thymallus thymallus* (L.) from the French Jura mountains. *Canadian Journal of Zoology* **64**, 2430–2438.

Tsigenopoulos C. & Berrebi P. (2000) Mitochondrial phylogeny of North Mediterranean freshwater barbs (genus *Barbus*: Cyprinidae) inferred from cytochrome b sequences: biogeographic and systematic implications. *Molecular Phylogenetics and Evolution* **14**, 165–179.

USFWS (United States Fish and Wildlife Service) (1995) Final rule to reclassify the Bald Eagle populations as endangered and threatened in all of the lower 48 states. *Federal Register* **60**, 36000–36010.

USHR (United States House of Representatives) (1973) Report 412, 93rd Congress, 1st session. In Endangered Species Act of 1973. Publication L. 93–205, 81 Stat. 884, Decembre 28.

Vannote R.L., Minshall G.W., Cummins K.W., Sedell J.R. & Cushing C.E. (1980) The river continuum concept. *Canadian Journal of Fisheries and Aquatic Sciences* **37**, 130–137.

Waples R.S. (1991) Pacific salmon, *Oncorhynchus* spp., and the definition of 'species' under the Endangered Species Act. *Marine Fisheries Review* **53**, 11–22.

Waples R.S. (1995) Evolutionarily Significant Units and the conservation of biological diversity under the Endangered Species Act. In L. Nielsen (ed.) *Evolution and the Aquatic Ecosystems: Defining Unique Units in Population Conservation*. Bethesda, Maryland: AFS Symposium 17, pp. 8–27.

Waples R.S. (1999) Evolutionarily Significant Units, Distinct Population Segments, and the Endangered Species Act: reply to Pennock and Dimmick. *Conservation Biology* **12**, 718–721.

Section V
Management aspects of fish conservation

Chapter 33
Connecting science to management and policy in freshwater fish conservation

G.K. MEFFE*
Department of Wildlife Ecology and Conservation, University of Florida, Gainesville, Florida, USA

Abstract

Effective conservation consists of three sides of a triangle: science, management, and policy, all working within a larger social, political and economic framework. Science, management and policy each have their own interests and modes of operation, and they respond to different motivations, constituencies and reward systems. However, each must be engaged and must work effectively with the others as a cooperative unit if conservation is to succeed. Conservation often fails because these three components do not work together well, and even work at cross purposes. Science is only one aspect of the overall process of conservation, and is often a small component or absent, at least partly because scientists fail to understand the larger decision-making arena. Understanding the so-called 'success triangle' will help in that regard: successful interactions consist of substance, process, and relationships. Scientists tend to focus on substance while ignoring process and relationships, a recipe for marginalisation or failure. A conceptual model of effective conservation being developed in the United States illustrates the role of scientific knowledge relative to institutions and socio-economic interests in conservation. Learning how to understand and interact with players from these other spheres of influence is critical to ensure that conservation is based on good science and engaged pragmatically in a world of diverse interests and goals. More effective conservation requires scientists to learn new skills to interact in a complex, disorderly and confusing arena, and not simply rely on producing scientific information in the vain hope that it will be used by managers and policy makers.

Keywords: ecosystem, management, policy, science, success triangle.

33.1 Introduction

Because the theme being addressed in these proceedings is '*Freshwater Fish Conservation: Options for the Future,*' this presentation focuses on how to better exercise those options to make real changes in real places. The main concern deals with how much of the scientific information contained in these pages (and in other such proceedings) will actually be put into practice. The crucial question is, how much scientific information will be incorporated into policy to help determine management actions? The prediction is, very little, and this needs to be changed. The main goal in

*Correspondence: Department of Wildlife Ecology and Conservation, University of Florida, Box 110430, Gainesville, Florida 32611, USA (email: meffe@gnv.ifas.ufl.edu).

this essay is to connect what we do as scientific researchers to the world of practical conservation, and ask how we can be more effective in making real conservation advances.

It is the scientists' responsibility to ensure that a transition occurs; a better job must be done in moving scientific understanding into the policy and management arenas, or all the science will amount to little more than accumulated information in journals, temporarily impressing colleagues, but mostly collecting dust on shelves. We need to change the paradigm of the role of the academic scientist in the world at large. It is no longer enough to simply do the science and put it in a journal; it must be carried further, into effective policy and management. Most scientists do not know how to do that, but they must learn. This is what it will be addressed here.

I begin with some simple observations of historical, global trends. Nobody would disagree with the observation that scientific knowledge and understanding of the natural world increased enormously in the 20th century, probably at an ever-accelerating rate. Much more was known about how the natural world works, its patterns and processes, at the end of the 20th century than at the beginning; this could be measured as papers published, species described, habitats explored, high-level reports written, number of ecological scientists practising, and so forth. One might predict *from the scientific information alone* that the condition of the world's natural resources also would have improved in the century. After all, better scientific information should lead to better management and policy regarding renewable resources. However, most major metrics about the natural world are in decline, as measured by species loss, loss of primary forest, soil loss, desertification, coral reef declines, spread of exotic species, and so on. This raises a fundamental enigma: as scientific understanding has increased, the condition of the world's resources has declined, in direct opposition to what might be predicted if human activities and global trends were a function only of scientific knowledge. Obviously, science alone is not enough.

Therein lies the problem: humanity's actions and the resultant global trends are a function of complex biophysical, socio-economic, and institutional interrelationships that go well beyond scientific information. Obviously, we are not being as successful in transferring our collective scientific understanding into these human activities (policy and management) as we need to be, and thus are not affecting global trends to the degree necessary. Why are we ineffective? Partly because the process is not well understood. Practical conservation success consists of much more than scientific or substantive knowledge; the failure is not in the science but in its effective application in a very complex world. To understand this better, I will attempt to connect several lines of thought into a coherent vision of how science, policy, and management need to join in a synergistic fashion to address the greatest threats humanity faces.

33.2 Effective conservation

There are three components of effective conservation that form a triangle (Fig. 33.1(a)): scientific research, informed policy (legislation), and management (implementation of legislative mandates). These are applied within a larger social, economic and political

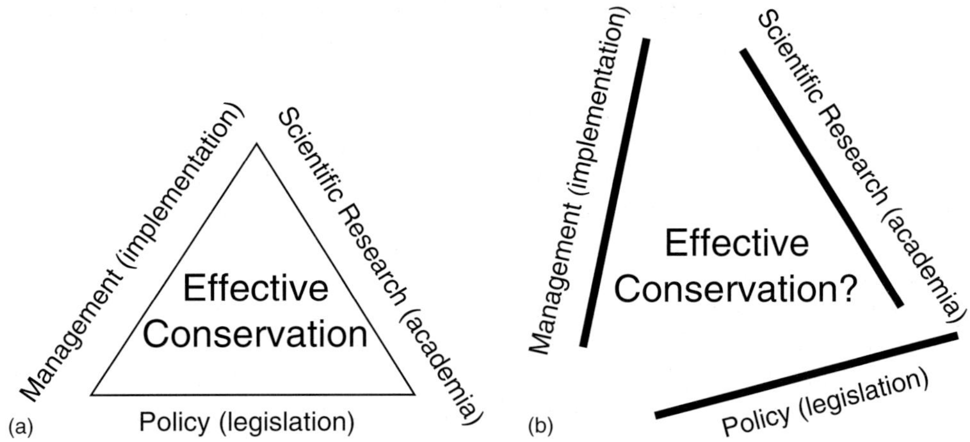

Figure 33.1 (a) Effective conservation consists of scientific research (largely, but not entirely conducted in academia), policy development conducted by legislators, and implementation of policy by natural resource managers. There should be a tight linkage and interdependence among these three. (b) The actual situation in many places is a disconnect between the three components, leading to ineffective conservation actions

framework that has enormous influences, but which is beyond the bounds of this discussion.

Effective conservation is not equivalent to research, which only contributes the guideposts or instructions for good activities, but not the actions; informed policy and management must come into play. Ideally, policy is based on good science and then management carries out the mandates of policy. These three components should lean on and support and rely upon each other to result in effective applied conservation.

Are there good connections among these three? At least in the USA, historically, the answer is a resounding no. There are major disconnections between scientific knowledge and what policy directs management to do (Fig. 33.1(b)). Perhaps this is improving in recent years but there remains a long way to go. There continue to be fundamental disconnections between the ecological information produced by the scientific process, information used to make environmental policy and data needed by natural resource managers. It is not much of a triangle, not very strong, and not effective.

My central argument is that science is only part of a larger process of environmental decision making and often only a small part or not used at all. Scientists would like to think that science drives effective conservation. But in reality, science may have little or no input in decision making. As a group, scientists generally are ineffective in influencing public policy and management; there is a need to learn to play the game or be left out.

How can the three components be pulled together to make a more effective triangle, thus leading to more effective conservation? We need to understand what motivates people and how successful interactions are conducted. This involves a different triangle, called the success triangle.

33.3 The success triangle

Success in any substantive activity or interchange (and certainly including effective conservation) consists of three components (Fig. 33.2).

(1) *Substance* includes the objective and factual data of most interest to scientists and engineers. Scientists usually prefer that decisions be based solely on factual content, and are most comfortable in this arena. But most people in the world are not scientists, so this does not readily happen.
(2) *Process* refers to the explicit and formal steps used to reach decisions through proper adherence to laws and including all the proper parties. This is the realm of administrators, lawyers, legislators and advocacy groups, all of whom typically have more power and influence than scientists.
(3) *Relationships* are about human behaviour, trust building, and formal and informal networks in communities of interest or communities of place. It is often of great influence in directing the process and is the speciality of politicians, journalists, the clergy and community leaders.

Scientists tend to focus only on substance, leaving the other two components to others and to chance. For example, they publish papers and assume that the information is available to policy makers or managers. In reality, it is not; they do not read scientific journals nor would they generally understand them if they did. Consequently, to policy makers and managers the information scientists so carefully provide does not exist unless a special effort is made to move it through policy circles. This is one reason why egregious errors and ecologically foolish decisions in policy and management are made and one wonders how they can occur. It is because the scientific perspective is only one actor in the play, and sometimes a very minor actor. Real progress will never be made in affecting environmental policy and management until the success triangle

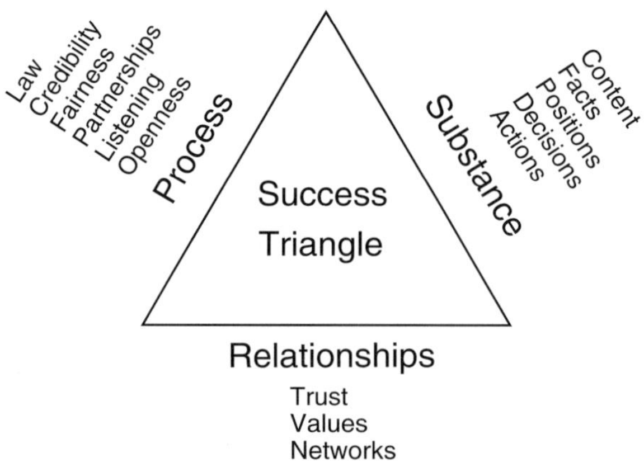

Figure 33.2 The success triangle consists of substantive knowledge, effective and legitimate process, and relationships based on trust and common values all working together toward a common vision

is understood and scientists become more effective in the process and relationship parts of the triangle. Scientists must, therefore, engage at these other levels for science to affect policy and management.

The critical point is to understand that the entire triangle is necessary for successful interactions and effective conservation. Scientists have to learn these other skills – process and relationships – if they want to be effective. They cannot simply provide substance and assume it will be integrated by someone else into effective conservation; it probably will not be. It must be recognised that a much larger game is being played out, and that the game will be played with or without substance.

In sum, the three worlds of science, policy and resource management need to be more strongly interconnected and closely integrated. Scientists need to make concerted efforts to learn these other skills and become more active players in ensuring that good science is a real part of policy and management.

A direct comparison of these three participants – scientists, resource managers and policy makers – will help to understand them better (Table 33.1). The interests, modes of operation, motivations, constituencies and reward systems of these three types of participants in effective conservation are very different. What motivates a scientist, for example (discovery, professional fame) is quite foreign to a resource manager or policy maker. Likewise, the modes of operation or reward systems of the policy maker are largely irrelevant to the scientist. Unless, and until, scientists begin to appreciate and understand why others do the things and think the way they do, little progress will be made in working with these other fields of interest. The scientists' way of life and

Table 33.1 Comparison of the words of the scientist, resource manager, and policy maker

	Scientist	Resource manager	Policy maker
Interests	Data, facts, deduction	Meet performance standards; make tangible changes on the land	Laws, demonstrable actions, popularity
Modes of operation	Reason, objectivity, controlled experimentation	Regulation, persuasion, compromise	Personal interactions and persuasion; collection and integration of information from diverse sources
Motivations	Truth, discovery, professional fame and respect	Taking positive actions; meeting bureaucratic goals	Re-election; public satisfaction
Constituencies	Other scientists, students, universities, public	Bureaucracy, public, legislators	Public; other policy makers
Rewards	Publications, professional recognition, career advancement, grants	Public and personal satisfaction, career advancement, compliance	Re-election; satisfaction with new directions; public legacy

way of doing business cannot be simply translated over to these other worlds and be expected to be effective; it will not work.

How can changes to the *status quo* be made? The answer is simple in concept, although far from easy in practice: scientists need to understand these other worlds and work on their terms. They need to learn the worlds of the managers and policy makers, learn their motivations, modes of operation and reward systems, to appreciate and understand their professional lives. This involves spending time in policy and management, focusing on the process and relationship parts of the success triangle, while bringing our science to the table. It involves working (for example) with the media, talking with political leaders, offering their services, being willing to translate science in simple and understandable terms to the public, and working directly with managers on an equal footing. It is not easy (because it is so foreign to many scientists) and it can be frustrating. But the learning curve is steep and the rewards can be great.

One reason that scientists have been inhibited in pursuing such activities on a large scale is that the reward systems in academia generally do not recognise these efforts as valuable and as contributing toward scholarly achievement. The major metrics of achievement remain publications, grants and overhead brought into universities. Consequently, if scientists wished to contribute to sound environmental decision making they have had to ignore the reward system and perhaps sacrifice some career advances to pursue what they consider to be important activities. Many have done so, but the reward systems must be changed so that more scientists can be not only tolerated, but encouraged to and rewarded for, entering the complex arena that constitutes human decision making.

Scientists run into another potential problem when they become involved in policy- and management-related issues: retaining their scientific objectivity *v.* falling into advocacy. Where is the line drawn? How do we not cross from offering objective science to becoming advocates and thus losing professional credibility? The answer is rather simple: scientists need to stay within the bounds of their scientific understanding and advocate the use of the best science in problem solving rather than advocating particular positions. Scientists should all be strong advocates for employing the best science in a more effective way both in policy development and management of natural resources, and in seeing that the science is used properly. There is nothing wrong or compromising about ensuring that the best science is used by society to make decisions.

33.4 Effective community-based conservation

The several points discussed thus far may be brought together in a simple model of effective, community-based conservation (also called ecosystem management) being used in the United States. This is a conceptual model to help think about how to do effective, community-based work in a complex modern society. It will be used to further discuss and try to understand the critical importance of non-scientific aspects of conservation. This helps to define the playing field for effective conservation.

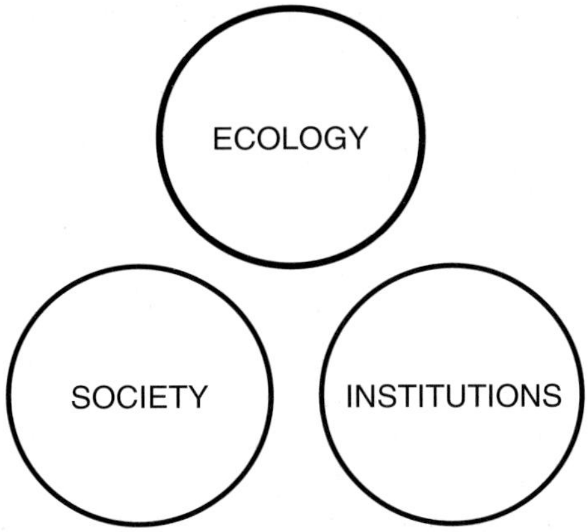

Figure 33.3 The three contexts of effective conservation. Each context has something important to contribute to good conservation actions and none can be ignored

The model consists of three very different perspectives, or contexts, from which to approach effective conservation (Fig. 33.3).

(1) The ecological context represents our physical basis for existence. It includes objective scientific knowledge of the ecosystem – its composition, structure, and function – as understood through ecological principles, models, and theories.
(2) The socio-economic context encompasses all the needs, wants, desires and actions of the human system. It is everything that humans do and want and take from the physical world.
(3) The institutional context represents the agencies empowered to regulate society through punishment, rewards and co-operation. It includes their missions, laws and policies of those cultures.

Within each context people have different viewpoints, limited possession of facts, and different capacities to pursue or prohibit actions that affect ecosystems. The single best decision or course of action as seen from any single context may prove unworkable, politically impossible or ecologically foolish. Effective conservation cannot be accomplished by thinking within any single circle; all three circles – their opportunities and limitations – must be understood to pursue effective conservation. For example, if one only focuses on the ecological circle – protecting the integrity of an ecosystem through heavy regulation, for instance – and ignores the socio-economics of the area, there is likely to be a backlash against that protected area, and a strong push to change the laws (through the institutional circle) to open up the resource. This has been repeatedly seen in the USA. Or, if we concentrate solely on economic development to the exclusion of aesthetic or biodiversity concerns we eventually find that the ecological support services degrade so severely that society rejects the activities and we see backlashes from environmentally-oriented thinkers. The point is, when dwelling only within any a single context, even the ecological context, 'win–lose'

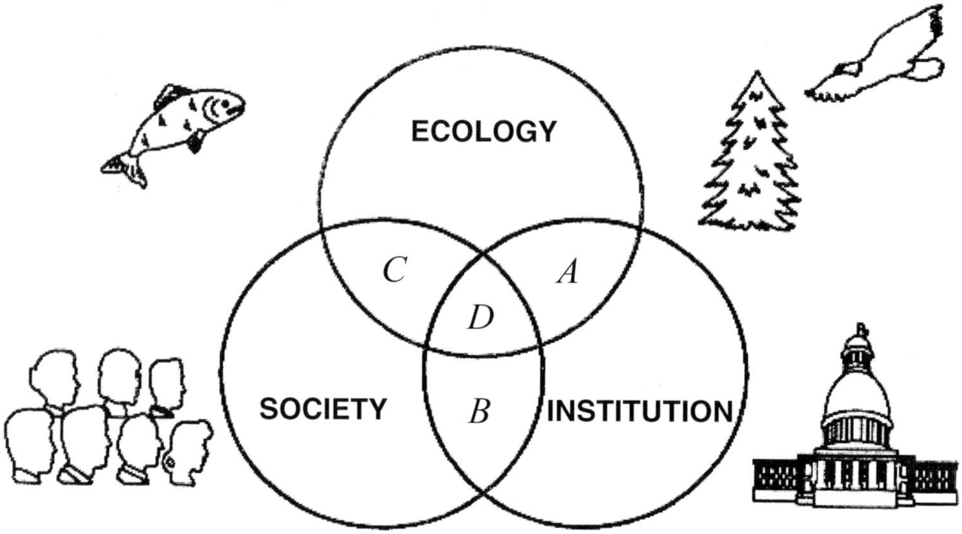

Figure 33.4 The three contexts of effective conservation when partially overlapped. The resultant zones of overlap result in interesting and important conservation opportunities. Zone *D* is the 'target' and results in 'win–win–win' situations

situations, court battles and social unrest are set up, and this strains the ability of the institutions to deal with all of it. Blending of these three contexts, knowledge bases, and realities represents the real challenge in conservation.

When these three contexts are partially driven toward each other, good and interesting things begin to happen. In particular, four areas of contextual overlap result (Fig. 33.4). These provide important insights to explain the roles of influence, regulations and partnerships as tools for effective conservation.

A. Zone of regulatory or management authority – At the interface of institutions and ecosystems, society has granted its government agencies the specific legal authority to make decisions that may be counter to the goals or short-term interests of some members of society. Such laws allow agencies, for example, to stop or modify land use and development proposals, prohibit the discharge of pollutants or contaminants, or directly manage natural resources on public lands. Agencies are required to exercise their authority through formal processes and they retain that authority until it is redefined by a court or changed through legislation. This zone typically occurs on public lands, although regulations certainly extend to the private sector as well.

B. Zone of societal obligations – This is the interface of institutions and society, separate from direct influences on ecosystems. This is of less direct interest to scientists, but demonstrates that natural resource management and environmental protection agencies also must serve the broader interests of society by implementing civil service laws, ensuring equal opportunity in employment or accessibility for the disabled, or protecting cultural resources.

C. Zone of influence – This is the interface of society and the environment apart from direct institutional influences; mostly, this occurs on private lands not subject to direct institutional control. This zone represents an area of decision making where interpersonal relationships and informal processes, rather than institutional control and regulation, are important. It is an area of opportunity that is often used by grass-roots organisers or politically connected developers and provides opportunities not available through regulatory authority. This approach relies on informal processes among interested stakeholders and is accomplished largely through interpersonal relationships and development of trust – the relationship part of the success triangle.

D. Zone of win–win–win partnerships – This is the main focal area for effective conservation, the place where the three contexts are best served. It represents the region where the needs, demands and constraints of each of the three contexts are satisfied, with no permanent damage done to any. It means decisions are reached that are ecologically sensible, are socio-economically satisfying and institutionally feasible. It usually involves strong partnerships among the three contexts, and usually works best when the major stakeholders agree on common goals and directions. It uses regulatory control by institutions as well as voluntary agreements among people to reach goals.

This general model is being pursued throughout many parts of the USA as the best means of avoiding expensive lawsuits and extended court battles, and is especially effective at the local community level when people with different value systems come together to find mutual interests and understanding. It may provide one of the great hopes for the next century as a method of problem solving in a complex environment.

33.5 Synthesis

To summarise, these several ideas will be superimposed for an overall synthesis. First, effective conservation was defined as consisting of research, policy, and management. The success triangle, consisting of substance, process, and relationships was then added. Finally, a three-circle model of community-based conservation, consisting of ecological, socio-economic, and institutional contexts was brought in. If superimposed (Fig. 33.5), they all point in the same direction toward an unavoidable conclusion: to make real and lasting differences in the conservation of fishes (or any taxa), to have more effective conservation and reverse the negative environmental trends of 20th century, then scientists will need to learn new skills and approaches. They cannot cling to old, comfortable models of academic isolation and live in the ivory tower. The scientific information available – as demonstrated in this collection of chapters and many others, in books and journals, and the classrooms – needs to experience a life far beyond the halls of academia. It needs to become resident in the halls of legislative bodies and management agencies. It needs to be a central part of society in every form. And it is up to scientists to make that happen.

Where better to begin this process than with fishes and watersheds? Everyone wants clean water, beautiful lakes and streams, and recreational opportunities, including

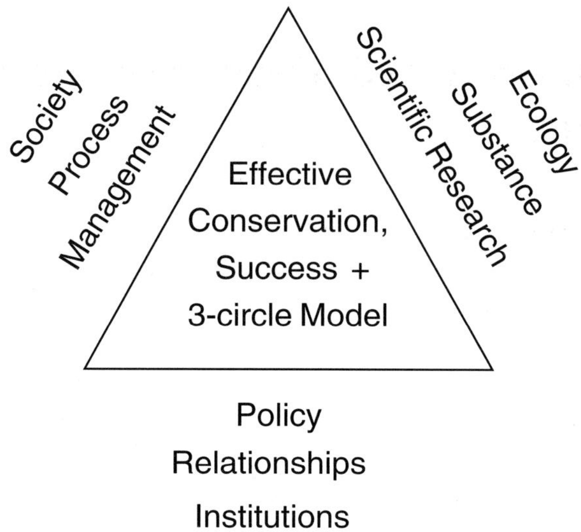

Figure 33.5 The conceptual models of effective conservation, the success triangle, and the three-circle contextual model can be overlapped with good concordance among the individual components

fishing. Our fate as a species is tied to that of fishes and other aquatic species, and this should be an easier sales job than for many taxa. Freshwater fish conservation may be the proving grounds for the rest of the world and a true test of effectiveness in reversing a centuries-long global decline in the natural world. The challenge is to move beyond the traditional academic circles, to learn new skills, and to begin to take science to a higher level – that of truly serving humanity and the species we are trying to save.

Acknowledgements

Many of these ideas were developed in concert with Richard Knight, Larry Nielsen and Dennis Schenborn, colleagues in a US Fish and Wildlife Service (National Conservation Training Center) training course on ecosystem management that we have co-instructed for 6 years. I especially recognise and thank Dennis Schenborn for development of the three-circle contextual model presented. I also thank the organisers of the symposium for their invitation to speak and for their travel support.

Chapter 34
The role of protected areas in freshwater fish conservation

A.J. CRIVELLI*
Station Biologique de la Tour du Valat, Le Sambuc, Arles, France

Abstract

The number of protected areas established specifically for fish conservation in continental waters of Europe is much smaller than in the marine environment. There has been little effort directed towards evaluating impacts of protected status in fresh waters; however, two studies that surveyed the status afforded by nature reserves on freshwater fishes in Europe concluded that protected areas are often of little value for fish conservation. The effect of reserves for fish conservation, on fish diversity and on the dynamics of fish communities are reviewed. It is recognised that the fundamental landscape unit for all freshwater fish conservation efforts should be the catchment area. In Europe, the creation of protected areas for fish conservation must be complemented by an integrated catchment management approach. Such an integrated approach should be applied first to catchments that still maintain rich native communities and relatively intact watersheds. Since conservation and rehabilitation of aquatic systems are fund limited, priority catchment areas should be identified.

Keywords: conservation strategies, ecosystem management, freshwater fishes, protected areas.

34.1 Introduction

Environmentalists working on the management of wetlands and aquatic systems have often focused on birds and plants. Fish and macro-invertebrates have been relatively neglected (Ryman, Utter & Laikre 1995; Leidy & Moyle 1998), although they probably are the most conspicuous components of the biodiversity of aquatic ecosystems (Moyle & Leidy 1992). It is unfortunate that our knowledge of freshwater fish and their conservation lags far behind that of other taxa (Maitland 1995; Richter, Braun, Mendelson & Master 1997; Leidy & Moyle 1998; Cambray & Bianco 1998).

Eschmeyer (1998) recognised about 23 250 valid fish species in the world, and new species are discovered or recognised at a rate of approximately 200 per year (Eschmeyer 1998). About 40% of these species are freshwater fishes (Nelson 1994).

*Correspondence: Station Biologique de la Tour du Valat, Le Sambuc, 13200 Arles, France (email: a.crivelli@tour-du-valat.com).

The status of these freshwater fishes is alarming: last century a third of known species were estimated to have become extinct or are in danger of extinction in the world, and about 3–5% are included in the IUCN list (1996) of endangered animals (Pollard, Ingram, Harris & Reynolds 1990; Moyle & Leidy 1992; Osse & Hollingworth 1992; Bruton 1995; Crivelli & Maitland 1995; Crivelli 1996; Kirchhofer & Hefti 1996; Cambray & Bianco 1998; Leidy & Moyle 1998; McDowall 1999; Ricciardi & Rasmussen 1999). Kottelat (1997), using a phylogenetic concept, recognised 358 species of freshwater fishes in Europe (excluding the former USSR), of which 78 (22%) are included in the IUCN list (1996). It should be noted that the IUCN (1996) recognised that fish were not comprehensively assessed and that it is therefore impossible to have an overall idea of their status.

Game fishes were the subject of the early management of conservation efforts by biologists (Pearse 1988; Maitland & Campbell 1992; Burger 2000). It was only in the early 1970s that the first conservation efforts directly targeted at native non-game species began in the United States (see Minckley & Deacon 1991; Moyle 1995) and a little later in Europe (1980s and 1990s depending on the country, Maitland 1995).

In contrast to other vertebrates for which the emphasis has been on the creation of protected areas, the first actions in terms of freshwater fish conservation mainly consisted of enacting, amending and strengthening national and international legislation on the species and their habitats and more rarely creating areas with various degrees of protection status. Even when aquatic habitats containing endangered fish species were given protected status, established fishing rights and fish stocking frequently continued so that fish were effectively excluded from any protection (e.g. Lake Mikri Prespa in Greece: Crivelli, Catsadorakis, Malakou & Rosecchi 1997).

The choice, design, number and size of protected areas in terrestrial habitats have led to the development of several model approaches (e.g. 'single large or several small (SLOSS) reserves' corridors) of much debate (Diamond 1975; Pickett & Thompson 1978; Simberloff & Abele 1982; Kirkpatrick 1983; Soule & Simberloff 1986; Burkey 1989; Bedward, Pressey & Keith 1992; Simberloff, Farr, Cox & Mehlman 1992; Barrett & Barrett 1997; Pressey & Logan 1998; Schwartz 1999; Emneborg & Götmark 2000; Margules & Pressey 2000). Marine and inland aquatic environments differ greatly from terrestrial ecosystems (Allison, Lubchenco & Carr 1998; Moss 2000), and knowledge of the requirements for establishing effective protected areas of freshwater habitats is poorer than that for terrestrial habitats. Few authors have attempted to develop the guiding principles for the creation of protected areas in marine and inland aquatic habitats (Moyle & Sato 1991; Rowley 1994; Bohnsack & Ault 1996; Allison *et al.* 1998; Roberts 1998; 2000; Carr 2000).

In this chapter, an assessment is first made of how many of the existing protected areas containing species of freshwater fish, and especially species in need of conservation, were created in Europe. The causes of fish decline and extinction are then reviewed, together with the types of protected areas in Europe, and their effectiveness for fish conservation. Finally alternatives, such as an integrated watershed management approach, which could be used in the future to better conserve freshwater fish are discussed.

34.2 Are protected areas an efficient way of protecting freshwater fish?

34.2.1 *The causes of fish decline and extinction*

Anthropogenic disturbance seems to underlie the decline and extinction of many fish species (Beverton 1992; Lande 1998; Cowx, Chapter 20). Perturbations include:

(1) habitat destruction (by water abstraction, building of dams, industrial pollution, channelisation, eutrophication, acidification) and fragmentation (Jennings & Kaiser 1998; Allibone 1999; Côte, Vinyoles, Reynolds, Doadrio & Perdices 1999; Gehrke, Astles & Harris 1999; Smith, Tilman & Nekola 1999; Strevens 1999);
(2) exploitation (Welcomme 1992; Fujita, Foran & Zevos 1998) and
(3) the introduction and the translocation of species (Minckley 1995; Carey, Moyle, Rejmanek & Vermeij 1996; Claudi & Leach 2000; Mack, Simberloff, Lonsdale, Evans, Clout & Bazzaz 2000; Myers, Simberloff, Kuris & Carey 2000).

More recently, global warming has been advocated as a factor of concern for the future of some fishes, although the direction and magnitude of change is likely to vary with species and context.

In the future, conservation action may be most effective if directed towards mitigating the effects of habitat destruction and resource exploitation. Effects of global warming are largely unknown at this stage, and therefore efforts might be best directed at predicting the rate and magnitude of climatic change. As far as introduced species are concerned, it must be recognised that conservationists and scientists have failed completely in their struggle to reduce introductions, which have never been as numerous as now, and whose eradication is still beset with problems (Claudi & Leach 2000; Mack *et al.* 2000; Myers *et al.* 2000). Identifying and limiting future introductions of fish species and taking effective measures to prevent their establishment is a enormous challenge for conservation. For example, the recent European Union 'Shengen' agreement will facilitate transportations of fish, legally or illegally, in Europe and therefore increase fish introductions and translocations.

34.2.2 *Reserves for freshwater fish?*

The declared aim of creating protected areas is to minimise human pressures on the habitats to maintain or restore natural balance and dynamics. In this way short- and long-term natural processes can function and maintain a breeding stock of certain species that will ensure the survival of the species within the protected area and potentially facilitate colonisation of unprotected surrounding areas.

Although many national red books have been published in the past 10–20 years listing endangered species of freshwater fish of varying status, none of these have made any mention of reserves created for these fishes or their occurrence in protected areas established for other vertebrates. There are only two studies in Europe, one in France (Keith 2000) and the other in the UK (Lyle & Maitland 1992) that have analysed the representation of fish in the system of protected areas in these two countries.

There are many types of protected areas in the France and UK with widely varying protection status (national parks, regional parks, nature reserves, etc.). Among existing protected areas in France and the UK, only nature reserves were analysed in these studies, with freshwater aquatic habitats (marshes, lakes, rivers, etc.) occurring in 60% and 75% of them, respectively (Table 34.1). Forty-two per cent and 60% of nature reserves in France and the UK harbour fish, respectively. Only two endangered species out of six in France and three out of 10 in the UK occurred in at least one of these reserves. Virtually none of the reserves in these two countries were specifically created for fish.

In many of the countries of Europe, there are protected areas, other than national parks and nature reserves, that have been created especially for fish, and which can be called 'fishing reserves'. These are often small areas whose main aim is to create areas protected from degradation such as gravel extraction, and where fish populations can complete their life cycle, especially during the breeding season. Sometimes these fishing reserves can also be of use for the protection of species whose numbers have seriously declined. For example, in France, of 468 Biotope Protection Orders issued up to June 1997 (Keith 2000), 46 were created specifically to protect freshwater fish. These 46 areas account for about 1200 km of rivers and streams and concern 17 fish species out of the 71 recorded in France. Among these 17 species, three are endangered species (=critically endangered and threatened) out of the six recorded in France, namely *Acipenser sturio* L., *Salmo salar* L. and *Zingel asper* L. (Keith 1998). Three protected areas for *Valencia hispanica* (Val.) were created in Spain in the Valencia region in the 1980s (Planelles & Reyna 1996). In Poland six 'ichthyological reserves' covering more than 2173 ha, have existed since World War II. These covered 2% of the total area of the 937 nature reserves in Poland at the end of 1987 (Witkowski 1992). These reserves are especially intended for species with a high economic value (mainly salmonids) and protect the spawning habitats of these species. There are probably other examples in Europe, and this list of examples has no pretence of being exhaustive.

Table 34.1 Status of fish in nature reserves (NRs) in France (Keith 2000) and in Great Britain (Lyle & Maitland 1992)

	Number of NRs	Number of NR with fresh waters present	Number of NR having fish present	Number of NR with threatened[a] fish species	Number of threatened species in NR according to total number of species threatened
France	122	84 (69%)	51 (42%)	33 (27%)	2/6
UK	235	176 (75%)	141 (60%)	?	3/10

[a]Threatened, critically endangered and threatened species.

34.2.3 *The effects of reserves*

For conservation. Keith (2000) and Lyle & Maitland (1992) concluded that the existing network of reserves was ineffective for freshwater fish conservation, and especially for endangered fish species in their respective countries, because few endangered species occurred in these protected sites, and none of these protected areas were specifically created for fish. Only a fraction of the home range, usually representing only a single type of habitat, of the fish species was protected, and outside of the protected areas the fish continued to face factors responsible for their decline. Furthermore, factors responsible for decline, such as water pollution or the introduction of exotic species, have often spread to the protected area whatever the limits of the protected area or the regulations in force. A recent study (Aparicio, Vargas, Olmo & de Sostoa 2000) conducted in Catalonia (Spain) showed that only 11.3% of the watercourses in the region had a protection status and that only four of 13 sites with a high value for fish conservation were included in protected areas. They concluded that the current system of nature reserves within their study area does not offer protection to threatened fish species or communities, because the protected areas were almost all in the headwaters and not in the lowlands where sites with highest priorities for native fish species were located.

In the United States, Williams (1991) attempted to assess the effectiveness of protected areas in aquatic habitats situated in the western States, irrespective of whether they were created for fish conservation. He found evidence to suggest that national parks and wilderness areas had a long-term beneficial action for protection of native fish species, and especially for species living in springs, marshes and headwaters. In contrast, their effectiveness was less in larger aquatic systems, which were often invaded by introduced species. Finally he claimed that protected areas for fish reaching large sizes are ineffective. He concluded: 'Protecting the increasing numbers of imperiled fishes will be difficult, and our goal should be to ensure protection of fishes and aquatic communities before there is any need to invoke the ESA (Endangered Species Act)'.

On fish diversity and on the dynamics of fish communities. Since fishing is so intensive in marine systems and is the main cause of the decline of many fish species, the creation of protected areas for spawning, for the protection of juveniles, or for attempting to rebuild breeding stocks is a management method that is widely used in these systems (see Jennings & Kaiser 1998; Caddy 2000). For example, there are 68 marine protected areas (MPAs) in the Mediterranean Sea, including 47 in Spain, France and Italy (Boudouresque 2000). In the absence of any scientific studies evaluating the effect of protected areas on freshwater fish diversity and on the dynamics of freshwater fish communities, it is necessary to examine what has been done in the marine environment where there are several available quantitative studies, usually on a single species or more rarely on communities (Goni, Polunin & Planes 2000).

Most assessments have been made on exploited species, and many studies have concluded that effectiveness of MPAs is debatable (Kramer & Chapman 1999; Caddy 2000; Carr 2000; Dayton, Sala, Tegner & Thrush 2000; Garcia Charton, Williams, Perez Ruzafa, Milazzo, Chemello, Marcos, Kitsos, Koukouras & Riggio 2000; Mosquera, Côté, Jennings & Reynolds 2000).

Reserves can vary in effectiveness, depending on the species or functional guild. For example, sedentary fish are more favoured than migratory ones, and species with a small home range will be favoured compared to those with large home ranges. In many cases it has been shown that protected areas have a larger number of species than unprotected areas (Jennings, Marshall & Polunin 1996). However, this is not always the case, as it depends on the types of habitat present in the study sites (Garcia Charton *et al.* 2000), the families of fish that are taken into account in the analysis, and other factors. One of the major objectives of setting up marine reserves is to allow resident fish to produce a large number of young so that areas subjected to fishing can be recolonised. The density of fish is usually equal to and/or greater in protected areas compared to unprotected ones, and rarely lower (Sanchez Lizaso, Goni, Renones, Garcia Charton, Galzin, Bayle, Sanchez Jerez, Perez Ruzafa & Ramos 2000). This is especially true for heavily exploited species, large specimens of which often occur at higher densities within reserves. Nevertheless, there is no consensus on whether protection within reserves has any effect on abundance of fish outside reserves (Rowley 1994; Russ & Alcala 1996; Kramer & Chapman 1999; Hasting & Botsford 1999; Caddy 2000; Planes, Galzin, Garcia Rubies, Goni, Harmelin, Le Direach, Lenfant & Quetglas 2000). The average size of fish species is often higher in protected areas (Roberts & Polunin 1991; Rowley 1994; Roberts 1995; Ticco 1995; Letourneur 1996; Sasal, Faliex & Morand 1996; Mangel 1998; Jouvenel & Pollard 2001). The most successful marine reserves for fish species have been where the target species were subjected to extremely heavy commercial or sporting fishing pressure before reserve establishment.

There are no similar studies on the effects of freshwater reserves comparable to those on marine reserves. Its seems probable that the conclusions derived from studies in the marine environment could apply to inland waters, and especially lakes, but probably much less to rivers.

The challenge ahead. In the few protected areas created for freshwater fishes, the effects of protection measures usually vary from one protected area to another, and from one species to another and are difficult to determine. Most protected areas were established without the necessary studies (almost always there was no knowledge of the initial state of the fish communities before the reserve status and no monitoring after the establishment of the reserve), which could have helped in assessment of the impact and effectiveness of the protection status (Mosquera *et al.* 2000). To evaluate the effects of a protected area, a number of replicates (=sites evaluated) are needed both in the protected area and outside of it, sampled several times a year both before and after the establishment of the reserve. Statistical issues (e.g. pseudo-replication) must be considered carefully in designing such an impact study (Garcia Charton *et al.* 2000). Finally, many reserves have no firmly established aims and very few have management plans, hindering evaluation of their effectiveness.

Many people think that reserves are necessary, or even essential, but by themselves are insufficient for habitat conservation. It is generally agreed that parallel policies taken in areas outside individual reserves or networks of reserves are also required to ensure the conservation of aquatic habitats and their resources (Moyle & Leidy 1992; Allison *et al.* 1998; Caddy 2000; Margules & Pressey 2000).

34.3 What are the prospects for freshwater fish conservation?

It is evident that the protection of large or small areas for freshwater fish will not resolve the problem of their conservation, even though such measures could help towards this aim. Faced with continuing habitat degradation, severe overfishing, an ever-increasing number of introduced species, and evidence of global warming (Bethoux, Gentili & Tailliez 1998), the emphasis should not be on short-term objectives but on the long term – a much more difficult challenge to accomplish, since it requires the continuous involvement of many players and funding bodies.

Freshwater fish conservation strategies in future years should go beyond the measures currently applied that are based on the species and on the concept of protected areas (exclusion of man) and of corridors. With freshwater aquatic ecosystems, we have the opportunity to easily identify the limits of a dynamic ecosystem unit that integrates all the ecological functions, processes and interactions, which alone can guarantee the long-term conservation: i.e. the catchment (also referred to by some authors as basin, drainage or watershed; Fig. 34.1). Each such catchment should be regarded as a separate, potentially manageable ecosystem unit, relatively isolated from adjacent catchments. Each will have characteristics peculiar to itself (Reynolds 1993). The

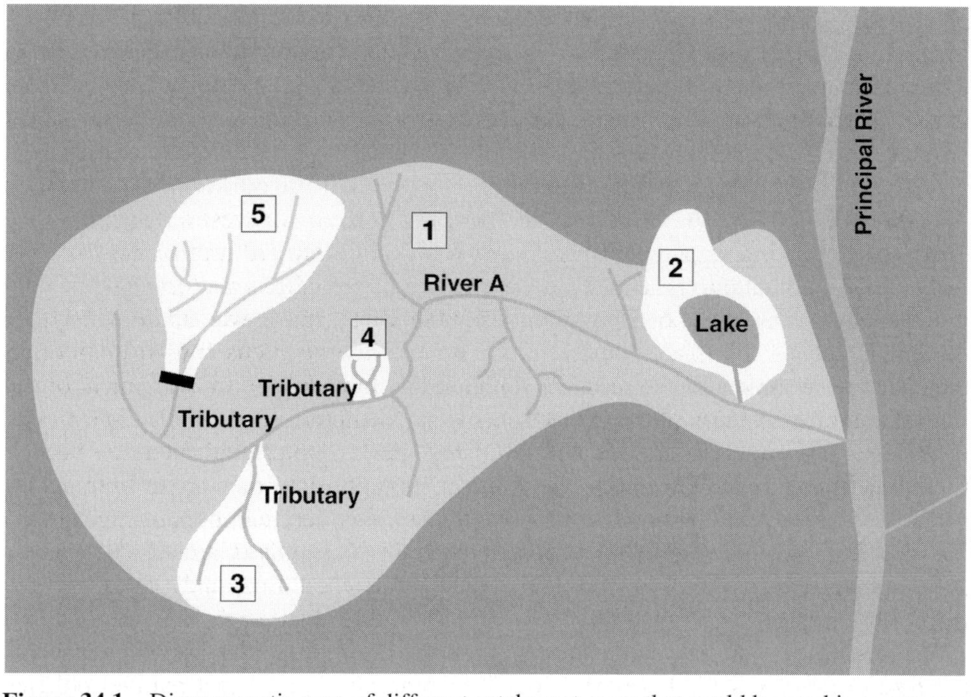

Figure 34.1 Diagrammatic map of different catchment areas that could be used in ecosystem management: on a large scale catchment area (principal river); on a medium scale, catchment of River A (1); on a lake watershed scale (2); on a tributary catchment scale (3 and 4); and on an impounded tributary catchment scale (5)

concept of integrated management must therefore be applied at the ecosystem level (Schofield 1992; Franklin 1993; Grumbine 1994; Christensen, Bartuska, Brown, Carpenter, D'Antonio, Francis, Franklin, MacMahon, Noss, Parsons, Peterson, Turner & Woodmansee 1996; Dombeck, Williams & Wood 1997; Simberloff 1998; Moss 2000; Naiman & Turner 2000; Poiani, Richter, Anderson & Richter 2000) through management plans that will guarantee the compatibility of human activities with the preservation of biodiversity and ecosystem functioning. Protected areas could be included within these catchments, but they would not be the basis of our freshwater fish conservation strategy. As for the size of the catchment, it can be very variable and will depend mainly on the characteristics of the fish communities in question, the mobility of the species in the community, and the heterogeneity of the environment.

Human developments occurring in the catchment often have direct and indirect effects on fish communities leading to changes in assemblage structure and the extinction of species by altering catchment functioning (Ward & Stanford 1989; Carignan & Steedman 2000). Increasing human populations in the catchment are the driving force for environmental degradation, therefore the catchment approach cannot exclude human activities, values and priorities, but must include them in such a way that catchment functioning is maintained (Bickford & Tisa 1992). This approach also has the great advantage of falling in line with the new European Union directive entitled 'The Water Framework Directive' (European Commission 1999), of which one of the obligations is to establish river basin management plans. Many programmes that have used this dynamic ecosystem approach have attempted, often with much subsequent debate, to identify and describe the most important attributes to be taken into account in ecosystem-based management (Christensen *et al.* 1996; Lackey 1998, 1999a; Poiani *et al.* 2000). When a river flowing into the sea contains anadromous or catadromous species, it is obvious that the catchment approach will not resolve the management problems of these species during the marine stage of their life cycle (McDowall 1999; Miller 2000).

An ecosystem management or ecosystem-based management approach is certainly more difficult to put into practice than the creation of protected areas, but it is the only one that can conserve our fish communities in the long term. It also has the advantage of protecting not only fish, but also other aquatic organisms such as invertebrates, plants, riverine forests and amphibians living in the catchment. Such an approach does not preclude protected area approaches being undertaken within the catchment as a complementary measure. In the cases of small, close or semi-closed water bodies, awarding them protective status may be the most appropriate measure to be taken for effective conservation. Our philosophy should be prevention rather than cure, because mitigation cannot be expected to solve all our problems (Meyer 1992). Our current general conservation goal for ecosystem management is therefore to focus on less-degraded catchments that still function correctly and which often have fish communities consisting mainly of native species, so as to ensure their long-term integrity and dynamics of the catchments and their fish communities (Sheldon 1988; Moyle & Sato 1991; Williams 1991; Muhar, Schwartz, Schmutz & Jungwirth 2000). Later we may be able to concentrate on rehabilitating more degraded catchments. As funding for the conservation and rehabilitation of aquatic systems is limited, priority catchment areas should therefore be identified within larger geographic regions to fulfil this general conservation goal in accordance with societal values of inhabitants.

The critical phase of this approach is to consider the geographical scale that may vary in size from a region within a country to a biogeographical zone. For example, the Mediterranean Basin, which is one of the 25 biodiversity hotspots for conservation priorities listed by Myers, Mittermeier, Mittermeier, Da Fonseca and Kent (2000) and harbours a large number of endemic freshwater fish species (Crivelli & Maitland 1995; Crivelli 1996). To establish a priority list of catchments within this region, methods for measuring the biodiversity of fish and the relative integrity of the various catchments must be available. Many methods and other indices have been proposed for individual sites in the past 20 years in the United States (Simon 1999; Karr & Chu 2000), and then in Europe (Hughes & Oberdorff 1999; Boon 2000) and more recently for catchments (Lyons, Navarro-Perez, Cochran, Santana & Guzman-Arroyo 1995; Nehlsen 1997; Moyle & Randall 1998; Moyle & Marchetti 1999). The most well known of these is Index of Biotic Integrity (IBI) for watercourses that focuses on fish communities, but other methods have been used for other wetlands and terrestrial systems (Frissell & Bayles 1996; Kiester, Scott, Csuti, Noss, Butterfield, Sahr & White 1996; Harig & Bain 1998; Muhar *et al.* 2000; Verdonschot 2000).

Such indices could be applied in the Mediterranean region, with modification to some of the special features of the region, such as the shortage of water (Gasith & Resh 1999; Postel 2000), the type of fish assemblage and available knowledge.

Once this exercise has been accomplished, we could choose the catchment to be the subject of a freshwater fish conservation project. Although the method described above may enable identification of catchments that best meet the general conservation goal, the final choice still depends on the presence of a dependable and motivated local partner and the willingness of local, regional and possibly national government to support the project (Dombeck *et al.* 1997).

Once a catchment has been selected, the stage of preparing an integrated management plan can start (Hooper & Margerum 2000). This should be managed by a steering committee including, as far as possible, representatives of all the major stakeholders. During the preparation stage, existing data on various biological and socio-economic aspects of the catchment need to be collated. These include: human population trends, water quality, discharges and extractions, a survey of the main economic activities and their relative importance in terms of employment, income and impacts on aquatic ecosystems and the determination of the structure and abundance of the populations of the main fish species in the catchment at time zero, plus preliminary studies on specific problems on which information is scarce. A very important point in this preparatory stage, is a major work of intense public awareness among people living in the catchment, among all the socio-economic sectors and government representatives, so that they are fully involved in the preparation and production of the future management plan and not feel that the plan is imposed upon them (Rhoads, Wilson, Urban & Herricks 1999; Griffin 1999; Hooper & Margerum 2000; Korfmacher 2001). The challenges and the aims of the project must also be clearly stated and transparent. This stage may take several years, and is fundamental for the subsequent success of the project. It would provide the opportunity for all those involved in the management of the catchment to meet, talk and exchange ideas. It would also allow studies to be undertaken to clarify some difficult points or to collect missing data that appear important for making decisions on the production of the management plan or for defining objectives.

Once the preparation stage has been completed, arbitration for fulfilling the management plan objectives can begin. Whenever possible a 'multiple water resources uses' strategy should be applied, but sometimes one use must be given priority over others, or a given activity must be reduced or prohibited from part of the catchment, if it is too damaging for the functioning of the ecosystem. All these decisions will be made not by scientists but by all the community living in the catchment in question, and will reflect the values of that society (Lackey 1999b, c). These societal values can change with time and may differ between catchments, countries, or between peoples of different cultures. The role of scientists will be to demonstrate and explain the consequences of the various options chosen on how the catchment functions and on its integrity, in terms of the announced objectives. Once the management plan has been produced it should be officially ratified for a defined period, usually 3–5 years, after which it should be revised and renewed. For this revision to take place effectively, it is very important that actions described in the management plan should be accompanied by quantitative targets and that a monitoring programme should be installed that would provide an objective evaluation. Evaluations are often overlooked or poorly planned, and this jeopardises assessment and validity of the programme.

34.4 Conclusions

The creation of protected areas for fish conservation must be complemented by an integrated catchment management approach if the structure and the function of fish communities are to be conserved. This integrated approach should be applied first to catchments that still maintain rich native communities and relatively intact watersheds. Human activities must be integrated into the processes of producing the management plan that will be developed cooperatively by all major stakeholders. Such an approach means that biologists will take a consultative role in the decision-making process. Lande (1998) suggested that conservation problems could almost never be solved by one discipline on its own and that conservation biologists from various branches should be intelligent and humble enough to admit that their speciality may not be the most important. They must be willing to work with other specialist biologists and also with social scientists, economists, political activists and environmental lawyers. This is often the only effective means of conserving and restoring biological diversity against powerful economic and political interests.

Acknowledgements

I thank Peter Maitland for information on the subject, G. Poizat for his comments on the manuscript, and two anonymous reviewers for their helpful comments.

References

Allibone R.M. (1999) Impoundment and introductions: their impacts on native fish of the upper Waipori River, New Zealand. *Journal of the Royal Society of New Zealand* **29**, 291–299.

Allison G.W., Lubchenco J. & Carr M.H. (1998) Marine reserves are necessary but not sufficient for marine conservation. *Ecological Applications* **8**(Suppl.), 79–92.

Aparicio E., Vargas M.J., Olmo J.M. & de Sostoa A. (2000) Decline of native freshwater fishes in a Mediterranean watershed on the Iberian Peninsula: a quantitative assessment. *Environmental Biology of Fishes* **59**, 11–19.

Barrett N.E. & Barrett J.P. (1997) Reserve design and the new conservation theory. In S.T.A. Pickett, R.S. Ostfeld, M. Shachak & G.E. Likens (eds) *The Ecological Basis of Conservation. Heterogeneity, Ecosystems, and Biodiversity*. New York: International Thomson Publishing, pp. 236–251.

Bedward M., Pressey R.L. & Keith D.A. (1992) A new approach for selecting fully representative reserve networks: addressing efficiency, reserve design and land suitability with an iterative analysis. *Biological Conservation* **62**, 115–125.

Bethoux J.-P., Gentili B. & Tailliez D. (1998) Warming and freshwater budget change in the Mediterranean since the 1940s, their possible relation to the greenhouse effect. *Geophysical Research Letters* **25**, 1023–1026.

Beverton R.J.H. (1992) Fish resources: threats and protection. *Netherlands Journal of Zoology* **42**, 139–175.

Bickford W.E. & Tisa M.S. (1992) Flawless fisheries through watershed management. *American Fisheries Society Symposium* **13**, 95–103.

Bohnsack J.A. & Ault J.S. (1996) Management strategies to conserve marine biodiversity. *Oceanography* **9**, 73–82.

Boon P.J. (2000) The development of integrated methods for assessing river conservation value. *Hydrobiologia* **422/423**, 413–428.

Boudouresque C.F. (ed.) (2000) Impact de l'homme et conservation du milieu marin en Méditerranée. Marseille: GIS Posidonie Publications, 243pp. (in French).

Bruton M.N. (1995) Have fishes had their chips? The dilemma of threatened fishes. *Environmental Biology of Fishes* **43**, 1–27.

Burger C.V. (2000) The needs of salmon and Steelhead in balancing their conservation and use. In E.E. Knudsen, C.R. Steward, D.D. MacDonald, J.E. Williams & D.W. Reiser (eds) *Sustainable Fisheries Management: Pacific Salmon*. Boca Raton: CRC Press, pp. 15–29.

Burkey T.V. (1989) Extinction in nature reserves: the effect of fragmentation and the importance of migration between reserve fragments. *Oikos* **55**, 75–81.

Caddy J.F. (2000) A fisheries management perspective on marine protected areas in the Mediterranean. *Environmental Conservation* **27**, 98–103.

Cambray J.A. & Bianco P.G. (1998) Freshwater fish in crisis, a Blue Planet perspective. *Italian Journal of Zoology* **65**, 345–356.

Carey J.R., Moyle P., Rejmanek M. & Vermeij G. (eds) (1996) Invasion biology. *Biological Conservation* **78**, 1–214.

Carignan R. & Steedman R.J. (eds) (2000) Impacts of major watershed perturbations on aquatic ecosystems. *Canadian Journal of Fisheries and Aquatic Sciences* **57**(Suppl. 2), 1–174.

Carr M.H. (2000) Marine protected areas: challenges and opportunities for understanding and conserving coastal marine ecosystems. *Environmental Conservation* **27**, 106–109.

Christensen N.L., Bartuska A.M., Brown J.H., Carpenter S., D'Antonio C., Francis R., Franklin J.F., MacMahon J.A., Noss R.F., Parsons D.J., Peterson C.H., Turner M.G. & Woodmansee R.G. (1996) The report of the Ecological Society of America Committee on the scientific basis for ecosystem management. *Ecological Applications* **6**, 665–691.

Claudi R. & Leach J.H. (eds) (2000) *Non-indigenous Freshwater Organisms. Vectors, Biology and Impacts.* Boca Raton: Lewis Publishers, 464 pp.

Côte I.M., Vinyoles D., Reynolds J.D., Doadrio I. & Perdices A. (1999) Potential impact of gravel extraction on Spanish populations of river blennies *Salaria fluvialtilis* (Pisces: Blenniidae). *Biological Conservation* **87**, 359–367.

Crivelli A.J. (1996) *The Freshwater Fish Endemic to the Northern Mediterranean Region. An Action Plan for Their Conservation.* Arles: Tour du Valat Publication, 172 pp.

Crivelli A.J. & Maitland P.S. (eds) (1995) Endemic freshwater fishes of the northern Mediterranean region. *Biological Conservation* **72**, 1–337.

Crivelli A.J., Catsadorakis G., Malakou M. & Rosecchi E. (1997) Fish and fisheries in the Prespa lakes. *Hydrobiologia* **351**, 107–125.

Dayton P.K., Sala E., Tegner M.J. & Thrush S. (2000) Marine reserves: parks, baselines, and fishery enhancement. *Bulletin of Marine Sciences* **66**, 617–634.

Diamond J. (1975) The island dilemma: lessons of modern biogeographic studies for the design of natural reserves. *Biological Conservation* **7**, 129–146.

Dombeck M.P., Williams J.E. & Wood C.A. (1997) Watershed restoration: social and scientific challenges for fish biologists. *Fisheries* **22**(5), 26–27.

Emneborg H. & Götmark F. (2000) The role of threat and initiative from actors for establishment of nature reserves in southern Sweden 1926–1996. *Biodiversity and Conservation* **9**, 727–738.

Eschmeyer W.N. (ed.) (1998) *Catalogue of Fishes,* Vol. 1. San Francisco: California Academy of Sciences, 958 pp.

European Commission (1999) Commission Proposal for a Council Directive Establishing a Framework for European Community Water Policy – 6404/99.

Franklin J.F. (1993) Preserving biodiversity: species, ecosystems, or landscapes? *Ecological Applications* **3**, 202–205.

Frissell C.A. & Bayles D. (1996) Ecosystem management and the conservation of aquatic biodiversity and ecological integrity. *Water Resources Bulletin* **32**, 229–240.

Fujita R.M., Foran T. & Zevos I. (1998) Innovative approaches for fostering conservation in marine fisheries. *Ecological Applications* **8**(Suppl.), 139–150.

Garcia Charton J.A., Williams I.D., Perez Ruzafa A., Milazzo M., Chemello R., Marcos C., Kitsos M.S., Koukouras A. & Riggio S. (2000) Evaluating the ecological effects of Mediterranean marine protected areas: habitat, scale and the natural variability of ecosystems. *Environmental Conservation* **27**, 159–178.

Gasith A. & Resh V.H. (1999) Streams in Mediterranean climate regions: abiotic influences and biotic responses to predictable seasonal events. *Review of Ecology and Systematics* **30**, 51–81.

Gehrke P.C., Astles K.L. & Harris J.H. (1999) Within-catchment effects of flow alteration on fish assemblages in the Hawkesbury-Nepean River system, Australia. *Regulated Rivers: Research and Management* **15**, 181–198.

Goni R., Polunin N.V.C. & Planes S. (2000) The Mediterranean: marine protected areas and the recovery of a large marine ecosystem. *Environmental Conservation* **27**, 95–97.

Griffin C.B. (1999) Watershed councils: an emerging form of public participation in natural resource management. *Journal of the American Water Resources Association* **35**, 505–518.

Grumbine R.E. (1994) What is ecosystem management? *Conservation Biology* **8**, 27–38.

Harig A.L. & Bain M.B. (1998) Defining and restoring biological integrity in wilderness lakes. *Ecological Applications* **8**, 71–87.

Hasting A. & Botsford L.W. (1999) Equivalence in yield from marine reserves and traditional fisheries management. *Science* **284**, 1537–1538.

Hooper B.P. & Margerum R.D. (2000) Integrated watershed management for river conservation: perspectives from experiences in Australia and the United States. In P.J. Boon, B.R. Davies & G.E. Petts (eds) *Global Perspectives on River Conservation: Science, Policy and Practice.* New York: John Wiley & Sons Ltd, pp. 509–517.

Hughes R.M. & Oberdorff T. (1999) Application of IBI concepts and metrics to waters outside the United States and Canada. In T.P. Simon (ed.) *Assessing the Sustainability and Biological Integrity of Water Resources Using Fish Communities*. Boca Raton: CRC Press, pp. 79–93.

International Union for Conservation of Nature and Natural Resources (IUCN) (1996) *IUCN Red List of Threatened Animals*. Gland, Switzerland. IUCN, 368 pp.

Jennings S. & Kaiser M.J. (1998) The effects of fishing on marine ecosystems. *Advances in Marine Biology* **34**, 201–352.

Jennings S., Marshall S.S. & Polunin N.V.C. (1996) Seychelles' marine protected areas: comparative structure and status of reef fish communities. *Biological Conservation* **75**, 201–209.

Jouvenel J.-Y. & Pollard D.A. (2001) Some effects of marine reserve protection on the population structure of two spearfishing target-fish species, *Dicentrarchus labrax* (Moronidae) and *Sparus aurata* (Sparidae), in shallow inshore waters, along a rocky coast in the northwestern Mediterranean Sea. *Aquatic Conservation: Marine and Freshwater Ecosystems* **11**, 1–9.

Karr J.R. & Chu E.W. (2000) Sustaining living rivers. *Hydrobiologia* **422/423**, 1–14.

Keith P. (1998) *Evolution des peuplements ichtyologiques de France et strategies de conservation*. PhD Dissertation, University of Rennes, France, 239 pp. (in French).

Keith P. (2000) The part played by protected areas in the conservation of threatened French freshwater fish. *Biological Conservation* **92**, 265–273.

Kiester A.R., Scott J.M., Csuti B., Noss R.F., Butterfield B., Sahr K. & White D. (1996) Conservation prioritization using GAP data. Conservation Biology **10**, 1332–1342.

Kirchhofer A. & Hefti D. (eds) (1996) *Conservation of Endangered Fish in Europe*. Bâle: Birkhauser, 341 pp.

Kirkpatrick J.B. (1983) An iterative method for establishing priorities for the selection of nature reserves: an example from Tasmania. *Biological Conservation* **25**, 127–134.

Korfmacher K.S. (2001) The politics of participation in watershed modeling. *Environmental Management* **27**, 161–176.

Kottelat M. (1997) European freshwater fishes. *Biologia, Bratislava* **52**, 1–271.

Kramer D.L. & Chapman M.R. (1999) Implications of fish home range size and relocation for marine reserve. *Environmental Biology of Fishes* **55**, 65–79.

Lackey R.T. (1998) Seven pillars of ecosystem management. *Landscape and Urban Planning* **40**, 21–30.

Lackey R.T. (1999a) Radically contested assertations in ecosystem management. *Journal of Sustainable Forestry* **9**, 21–34.

Lackey R.T. (1999b) The savy salmon technocrat: life's little rules. *Environmental Practice* **1**, 156–161.

Lackey R.T. (1999c) Salmon policy: science, society, restoration, and reality. *Renewable Resources Journal* **17**, 6–16.

Lande R. (1998) Anthropogenic, ecological and genetic factors in extinction and conservation. *Researches Population Ecology* **40**, 259–269.

Leidy R.A. & Moyle P.B. (1998) Conservation status of the world's fish fauna. In P.L. Fiedler & P.M. Kareiva (eds) *Conservation Biology for the Coming Decade*. London: Chapman & Hall, pp. 187–227.

Letourneur Y. (1996) Réponses des peuplements et populations de poissons aux réserves marines: le cas de l'île de Mayotte, Ocean Indien occidental. *Ecoscience* **3**, 442–450 (in French).

Lyle A.A. & Maitland P.S. (1992) Conservation of freshwater fish in the British Isles: the status of fish in National Nature Reserves. *Aquatic Conservation: Marine and Freshwater Ecosystems* **2,** 19–34.

Lyons J., Navarro-Perez S., Cochran P.A., Santana E., & Guzman-Arroyo M. (1995) Index of biotic integrity based in fish assemblages for the conservation of streams and rivers in West-Central Mexico. *Conservation Biology* **9**, 569–584.

Mack R.M., Simberloff D., Lonsdale W.M., Evans H., Clout M. & Bazzaz F.A. (2000) Biotic invasions: causes, epidemiology, global consequences, and control. *Ecological Applications* **10**, 689–710.

Maitland P.S. (1995) The conservation of freshwater fish: past and present experience. *Biological Conservation* **72**, 259–270.

Maitland P.S. & Campbell R.N. (1992) *Freshwater Fishes*. London: Harper Collins Publishers, 368 pp.

Mangel M. (1998). No-take areas for sustainability of harvested species and a conservation invariant for marine reserves. *Ecology Letters* **1**, 87–90.

Margules C.R. & Pressey R.L. (2000) Systematic conservation planning. *Nature* **405**, 243–253.

McDowall R.M. (1999) Different kinds of diadromy: different kinds of conservation problems. *ICES Journal of Marine Science* **56**, 410–413.

Meyer P.A. (1992) The argument for prevention. Prevention, not mitigation. *American Fisheries Society Symposium* **13**, 147–154.

Miller K.A. (2000) Pacific salmon fisheries: climate, information and adaptation in a conflict-ridden context. *Climatic Change* **45**, 37–61.

Minckley W.L. (1995) Translocations as a tool for conserving imperiled fishes: experiences in western United States. *Biological Conservation* **72**, 297–309.

Minckley W.L. & Deacon J.E. (eds) (1991) *Battle against Extinction. Native Fish Management in the American West*. Tucson: The University of Arizona Press, 517 pp.

Mosquera I., Côté I.M., Jennings S. & Reynolds J.D. (2000) Conservation benefit of marine reserves for fish populations. *Animal Conservation* **4**, 321–332.

Moss B. (2000) Biodiversity in freshwaters – an issue of species preservation or system functioning? *Environmental Conservation* **27**, 1–4.

Moyle P.B. (1995) Conservation of native freshwater fishes in the Mediterranean-type climate of California, USA: a review. *Biological Conservation* **72**, 271–279.

Moyle P.B. & Leidy R.A. (1992) Loss of biodiversity in aquatic ecosystems: evidence from the fish faunas. In P.L. Fielder & S. Jain (eds) *Conservation Biology: The Theory and Practice of Nature Conservation, Preservation and Management*. London: Chapman & Hall, pp. 127–169.

Moyle P.B. & Marchetti M.P. (1999) Applications of indices of biotic integrity to California streams and watersheds. In T.P. Simon (ed.) *Assessing the Sustainability and Biological Integrity of Water Resources Using Fish Communities*. Boca Raton: CRC Press, pp. 367–380.

Moyle P.B. & Randall P.J. (1998) Evaluating the biotic integrity of watersheds in the Sierra Nevada, California. *Conservation Biology* **12**, 1318–1326.

Moyle P.B. & Sato G.M. (1991) On the design of preserves to protect native fishes. In W.L. Minckley & J.E. Deacon (eds) *Battle against Extinction. Native Fish Management in the American West*. Tucson: The University of Arizona Press, pp.155–169.

Muhar S., Schwarz M., Schmutz S. & Jungwirth M. (2000) Identification of rivers with high and good habitat quality: methodological approach and applications in Austria. *Hydrobiologia* **422/423**, 343–358.

Myers J.H., Simberloff D., Kuris A.M. & Carey J.R. (2000) Eradication revisited: dealing with exotic species. *Trends in Ecology and Evolution* **15**, 316–320.

Myers N., Mittermeier R.A., Mittermeier C.G., Da Fonseca G.A.B. & Kent J. (2000) Biodiversity hotspots for conservation priorities. *Nature* **403**, 853–858.

Naiman R.J. & Turner M.G. (2000) A future perspective on North America's freshwater ecosystems. *Ecological Applications* **10**, 958–970.

Nehlsen W. (1997) Prioritizing watersheds in Oregon for salmon restoration. *Restoration Ecology* **5**, 25–33.

Nelson J.S. (1994) *Fishes of the World,* third edition. New York: John Wiley & Sons, 600 pp.

Osse J.W.M. & Hollingworth C.E. (eds) (1992) The threatened world of fish. *Netherlands Journal of Zoology* **42**, 130–524.

Pearse P.H. (1988) *Rising to the Challenge. A New Policy for Canada's Freshwater Fisheries.* Canadian Wildlife Federation, 180 pp.

Pickett S.T.A. & Thompson J. (1978) Patch dynamics and the design of nature reserves. *Biological Conservation* **13**, 27–37.

Planelles M. & Reyna S. (1996) Conservation of Samaruc, *Valencia hispanica* (Valenciennes, 1846), (Pisces: Cyprinoddontidae), an endemic and endangered species, in the community of Valencia (east Spain). In A. Kirchhofer & D. Hefti (eds) *Conservation of Endangered Fish in Europe*. Bâle: Birkhauser, pp. 329–335.

Planes S., Galzin R., Garcia Rubies A., Goni R., Harmelin J.-G., Le Direach L., Lenfant P. & Quetglas A. (2000) Effects of marine protected areas on recruitment processes with special reference to Mediterranean littoral ecosystems. *Environmental Conservation* **27**, 126–143.

Poiani K.A., Richter B.D., Anderson M.G. & Richter H.E. (2000) Biodiversity conservation at multiple scales: functional sites, landscapes, and networks. *Bioscience* **50**, 133–146.

Pollard D., Ingram B., Harris J. & Reynolds L. (1990) Threatened fishes in Australia – an overview. *Journal of Fish Biology* **37**, 67–78.

Postel S.L. (2000) Entering an era of water scarcity: the challenges ahead. *Ecological Applications* **10**, 941–948.

Pressey R.L. & Logan V.S. (1998) Size of selection units for future reserves and its influence on actual vs targeted representation of features: a case study in western New South Wales. *Biological Conservation* **85**, 305–319.

Reynolds C.S. (1993) The ecosystems approach to water management. The main features of the ecosystems concept. *Journal of Aquatic Ecosystem Health* **2**, 3–8.

Rhoads B.L., Wilson D., Urban M. & Herricks E.E. (1999) Interaction between scientists and nonscientists in community-based watershed management: emergence of the concept of stream naturalization. *Environmental Management* **24**, 297–308.

Ricciardi A. & Rasmussen J.B. (1999) Extinction rates of North American freshwater fauna. *Conservation Biology* **13**, 1220–1222.

Richter B.D., Braun D.P., Mendelson M.A. & Master L.L. (1997) Threats to imperilled freshwater fauna. *Conservation Biology* **11**, 1081–1093.

Roberts C.M. (1995) Rapid build-up of fish biomass in a Caribbean marine reserve. *Conservation Biology* **9**, 815–826.

Roberts C.M. (1998) Sources, sinks, and the design of marine reserve networks. *Fisheries* **23**(7), 16–19.

Roberts C.M. (2000) Selecting marine reserve locations: optimality versus opportunism. *Bulletin of Marine Science* **66**, 581–592.

Roberts C.M. & Polunin N.V.C. (1991) Are marine reserves effective in management of reef fisheries? *Reviews in Fish Biology and Fisheries* **1**, 65–91.

Rowley R.J. (1994) Marine reserves in fisheries management. *Aquatic Conservation: Marine and Freshwater Ecosystems* **4**, 233–254.

Russ G.R. & Alcala A.C. (1996) Do marine reserves export adult fish biomass? Evidence from Apo Island, central Philippines. *Marine Ecology Progress Series* **132**, 1–9.

Ryman N., Utter F. & Laikre L. (1995) Protection of aquatic biodiversity. In D.P Philipp, J.M.Epifanio, J.E. Mardsen & J.E. Claussen (eds) *Protection of Aquatic Biodiversity, Proceedings of the World Fisheries Congress, Theme 3.* New Delhi: Oxford & IBH Publishing Co. Pvt. Ltd, pp 11–35.

Sanchez Lizaso J.L., Goni R., Renones O., Garcia Charton J.A., Galzin R., Bayle J.T., Sanchez Jerez P., Perez Ruzafa A. & Ramos A.A. (2000) Density dependence in marine protected populations: a review. *Environmental Conservation* **27**, 144–158.

Sasal P., Faliex E. & Morand S. (1996) Population structure of *Gobius bucchichii* in a Mediterranean marine reserve and in an unprotected area. *Journal of Fish Biology* **49**, 352–356.

Schofield C.L. (1992) The watershed as an experimental unit in fisheries research. *American Fisheries Society Symposium* **13**, 69–79.

Schwartz M.W. (1999) Choosing the appropriate scale of reserves for conservation. *Annual Review of Ecology and Systematics* **30**, 83–108.

Sheldon A.L. (1988) Conservation of stream fishes: pattern of diversity, rarity and risk. *Conservation Biology* **2**, 149–156.

Simberloff D. (1998) Flagships, umbrellas, and keystones: is single-species management passé in the landscape era? *Biological Conservation* **83**, 247–257.
Simberloff D. & Abele L.G. (1982) Refuge design and island biogeographic theory: effects of fragmentation. *American Naturalist* **120**, 41–50.
Simberloff D., Farr J.A., Cox J. & Mehlman D.W. (1992) Movement corridors: conservation bargains or poor investments? *Conservation Biology* **6**, 493–504.
Simon T.P. (ed.)(1999) *Assessing the Sustainability and Biological Integrity of Water Resources Using Fish Communities*. Boca Raton: CRC Press, 671 pp.
Smith V.H., Tilman G.D. & Nekola J.C. (1999) Eutrophication: impacts of excess nutrient inputs on freshwater, marine and terrestrial ecosystems. *Environmental Pollution* **100**, 179–196.
Soule M.E. & Simberloff D. (1986) What do genetics and ecology tell us about the design of nature reserves? *Biological Conservation* **35** 19–40.
Strevens A.P. (1999) Impact of groundwater abstraction on the trout fishery of the River Piddle, Dorset; and an approach to their alleviation. *Hydrological Processes* **13**, 487–496.
Ticco P.C. (1995) The use of marine protected areas to preserve and enhance marine biological diversity: a case study approach. *Coastal Management* **23**, 309–314.
Verdonschot P.F.M. (2000) Integrated ecological assessment methods as a basis for sustainable catchment management. *Hydrobiologia* **422/423**, 389–412.
Ward J.V. & Stanford J.A. (1989) Riverine ecosystems: the influence of man on catchment dynamics and fish ecology. *Canadian Special Publication of Fisheries and Aquatic Sciences* **106**, 56–64.
Welcomme R.L. (1992) The conservation and environmental management of fisheries in inland and coastal waters. *Netherlands Journal of Zoology* **42**, 176–189.
Williams J.E. (1991) Preserves and refuges for native western fishes: history and management. In W.L Minckley & J.E. Deacon (eds) *Battle against Extinction. Native Fish Management in the American West*. Tucson: The University of Arizona Press, pp.171–189.
Witkowski A. (1992) Threats and protection of freshwater fishes in Poland. *Netherlands Journal of Zoology* **42**, 243–259.

Chapter 35
The role of legislation, institutions and policy making in fish conservation in Switzerland: past, present and future challenges

A. KIRCHHOFER*
Institute of Zoology, University of Berne, Bern, Switzerland

Abstract

The history of fishing in Switzerland reveals a change in interest in fish which can be related to conservation status. Before the 19th century nearly all fish species were exploited for food, and regulations covered economic aspects of fishing, targeting the conservation of self-reproducing stocks. Religious institutions and nobility, fishery corporations and township authorities set the rules. Later, interest in fish focused on a few angling species, alien species were introduced and hatchery rearing for stocking was promoted. New and revised legislation on water protection and fisheries in the past 30 years included aspects of species and aquatic habitat conservation. Conservation of all fish species and the restoration of habitats is now well founded in a legal framework and aided by public financial support. Habitat restoration, risk assessment for most endangered populations, identification of substances toxic to fish and the implications of climate change will be the main challenges for fish conservation in the future. The key factor for the success of fish conservation will be public awareness about fish as indicators of environmental quality.

Keywords: fish conservation, habitat restoration, history, law, Switzerland.

35.1 Introduction

Freshwater fish spend their lives hidden under the glittering water surface, most are not colourful, and they do not sing when spawning or cry when harmed. Therefore many people simply do not realise how many fish species (still) live in our lakes and rivers. Fish will rarely gain the status of a 'charismatic megafauna' and this makes the conservation of the many endangered species a long, difficult and laborious task. Therefore, institutions defending the interest of fishes and a legal background forcing society to restore what has been destroyed in the past, play a crucial role in fish conservation. The following case-study examines the situation in Switzerland concerning the history of regulations related to fish and fisheries and the driving forces behind them, and highlights possible mechanisms for successful fish conservation.

*Correspondence: WFN–Water Fish Nature, Murtenstrasse 52, CH-3052 Gümmenen, Switzerland (email: a.kirchhofer@bluewin.ch).

35.2 Fishery regulation in the past – first step towards fish conservation

Conservation of fish species is closely related with human culture, religion and economy. The role of fish as a valuable food was most important for the way people regarded fish and their species associations, as well as their biology and habitats. When the catholic religion was a decisive force in medieval society, and monasteries were numerous in middle Europe, fish was an important food source, because no meat but only fish (and beaver) was permitted to be eaten during fasting periods between 100 and 150 days, depending on the strictness of the monastic order (Von Liebenau 1897). With this economic importance, knowledge about fish, fisheries and aquaculture was highly developed.

From the Middle Ages until the 19th century, almost all river and lake fish species served as a natural resource for exploitation. The use of more than just a few 'valuable' species is documented, e.g. in the prayers at table in the monastery of St Gallen, written around 1000, 17 freshwater species were listed and the clergymen had to pray 'God save all the thousands of little fish' ('*Millia coctorum benedic dec pisciculorum*' cited in Von Liebenau 1897). Nobility also considered fish very highly and many tenants had to pay their tribute to princes, knights or land owners in the form of fish. The high esteem of almost all fish species living in fresh water is also illustrated in the fish book by Gessner (1670), where each species description includes a chapter about the use, the advantages or disadvantages of the meat, and the best season for alimentary purposes. In other books, species' descriptions are accompanied by recipes for their use in the kitchen (Mangolt 1557 cited in Ribi 1942). The importance of fish in the meals of clergy and higher society resulted in the furthering of aquaculture. First ponds were initially built just to keep fish alive until they were used. Later the culture of carp, *Cyprinus carpio* L., and other species became more important, and in certain regions a network of artificial ponds covered the landscape (Amacher 1996). However, fish was not very important as food for ordinary people in continental regions compared to coastal regions. Fish was only served to improve the common meals and the proportion of fish as a protein source in food was very low (Amacher 1996). After the religious reformation in the 16th century with the cancellation of fasting, the importance of fish as food was reduced to a valuable ingredient and as a supplement for special occasions.

Besides religion, the legal situation – the ownership of watercourses and lakes – was also an important factor in the regulation of fisheries. In Switzerland the Roman tradition was applicable, such that the right to fish in bigger rivers and lakes belonged to the authorities. Only small brooks and ponds and the related fishing rights were private. This allowed the institutions to regulate fishing and sustain the natural resources. However, from these early times, no records about regulations to limit catch seasonally, spatially or by legal sizes are known.

In the 14th century fishermen in Switzerland united to form fishery-corporations – the first in 1336 in Zürich (Von Liebenau 1897) – and began to regulate their business, together with the townships, the owners of fishing rights. The first regulations were mainly concerned with territorial restrictions for fishing to avoid conflicts between fishermen or described the proper use of fishing gear. In the first written fishery regulations for Lake Zürich from 1386, certain fishing methods and gear were forbidden to protect young fish from being caught (Amacher 1996). In the 15th and 16th centuries periodic meetings of fishermen from all over the country with representatives of the townships – the so-called 'Fischermeyen' – gave a regulatory background to the professionals. The protocol of the 'Fischermeyen' compiled in 1510 in Fribourg, documents fishing gear, legal sizes, close seasons (almost identical with today's regulations!) and banned fishing on spawning places for many species (Von Liebenau 1897). One single species was explicitly allowed to be caught during spawning – the nase, *Chondrostoma nasus* (L.). This might be related with the enormous abundance of this species (Gessner 1670), but strongly contrasts with the actual situation throughout Europe. Besides the fishery practises, there was a ban on migration barriers and commercial regulations were defined, with prices for each species, market places and suppliers. A comparison of prices reveals that burbot, *Lota lota* (L.), tench, *Tinca tinca* (L.), bream, *Abramis brama* (L.), or barbel, *Barbus barbus* (L.), were as valuable as perch, *Perca fluviatilis* L., and sold at prices only slightly cheaper than grayling, *Thymallus thymallus* (L.), trout, *Salmo trutta* L. or coregonids (Aeschbacher 1923). Even sculpins, *Cottus gobio* L., were sold in the market. The comparison with recent markets shows that many of these species are not being sold today and preferences by the consumer – as shown in the market prices – have changed considerably (Table 35.1). Comparable regulations were released for Lake Constance by the Abbey of St. Gall in 1534 after a long-term decrease in catches (Wulpillier 1896).

Although these regulations focused principally at maintaining self-reproducing stocks to secure human use of the natural resource, this can also be interpreted as the first legal framework for fish conservation. However, the background for this legislation was not conservation, but a standardisation of law and the assertion of territorial rights (Amacher 1996). Nevertheless the fisheries, the fishermen's corporations and the township authorities made important contributions to species conservation.

35.3 The decline of fish and fishery

During the following centuries, fishing, the related professions and attention to sustainability somewhat lost their importance. During times of war in Europe in the 17th and 18th centuries, fishing was performed by everybody without regard to the biology of the species or sustainability. After the French Revolution, which also emanated into Switzerland, fishing was no longer regarded as the right of nobility, but as a right of all. This resulted in overfishing and many stocks declined, e.g. in Lake Sempach, where the catch of coregonids strongly decreased after 1730 (Fig. 35.1).

Catches further declined in most rivers and lakes during the first half of the 19th century, partially due to fishing without any regulations, partially due to increasing pollution by the growing human population and the development of industry, partially because of river channelling where a lot of fluvial habitats and tributaries were destroyed (Vischer 1989). With the declining attraction of fishing, knowledge of the

Table 35.1 Relative market prices (1 = price of the most expensive species) of fish species in different periods and regions

	1510 Lake Biel[a]	1692 Monastery of St Gall[b]	1902 Lake Zürich[c]	1928 Lake Constance[d]	2000 Lake Biel[e]
Anguilla anguilla (L.)	1	–	–	0.47	–
Salmo trutta (L.)	1	–	1	1	0.7
Thymallus thymallus (L.)	1	–	–	0.6	–
Coregonus sp.	0.9	0.64	0.35	0.52	0.56
Cyprinus carpio (L.)	0.9	0.82	–	0.3	–
Lota lota (L.)	0.8	1	0.5	0.37	0.38
Barbus barbus (L.)	0.8	–	0.15	0.36	–
Perca fluviatilis (L.)	0.8	0.55	–	0.3	1
Esox lucius (L.)	0.8	0.64	0.5	0.5	0.67
Abramis brama (L.)	0.8	–	0.2	0.36	0.36
Leuciscus cephalus (L.)	0.8	–	0.15	0.1	–
Leuciscus leuciscus (L.)	0.8	–	–	0.1	–
Rutilus rutilus (L.)	0.8	–	0.15	0.1	0.53
Tinca tinca (L.)	0.8	–	0.15	0.34	–
Chondrostoma nasus (L.)	0.6	–	–	–	–
Cottus gobio (L.)	0.4	–	–	–	–
Salvelinus alpinus (L.)	–	1	–	0.91	0.83
Sander lucioperca (L.)	–	–	0.8	0.56	0.82
Alburnus alburnus (L.)	–	–	0.5	–	–
Silurus glanis (L.)	–	–	–	0.27	–

Data sources: [a]Von Liebenau 1897, [b]Wulpillier 1896, [c]Anonymous 1902, [d]Surbeck 1930, [e]personal communication of fishermen.

biology of the species also partially disappeared. An international agreement on the fishery in the River Rhine from 1868 even contained a ban for catches of spawning eel, *Anguilla anguilla* (L.) (Fehlmann 1952). During this time, institutions and legislation played no decisive role in the management of fish stocks, and thus in species conservation.

Following the invention of artificial reproduction of trout in 1725 in Germany, fish hatcheries were built all over the country (Fehlmann 1952), and artificial propagation led to slightly increasing catches again. However, that only a few species (the angler's 'valuable' species) could be furthered with this intervention was forgotten.

35.4 Fish-related legislation in the past 150 years

Following the poor status of the fisheries in Switzerland in the mid-19th century, the first federal fishery law was created in 1875. However, the content was almost identical with the agreement about the River Rhine fishery (see above). In 1883 the Fishery Association of Switzerland was founded, and it was this private organisation of anglers and commercial fishermen who furthered the knowledge about fish and their habitats

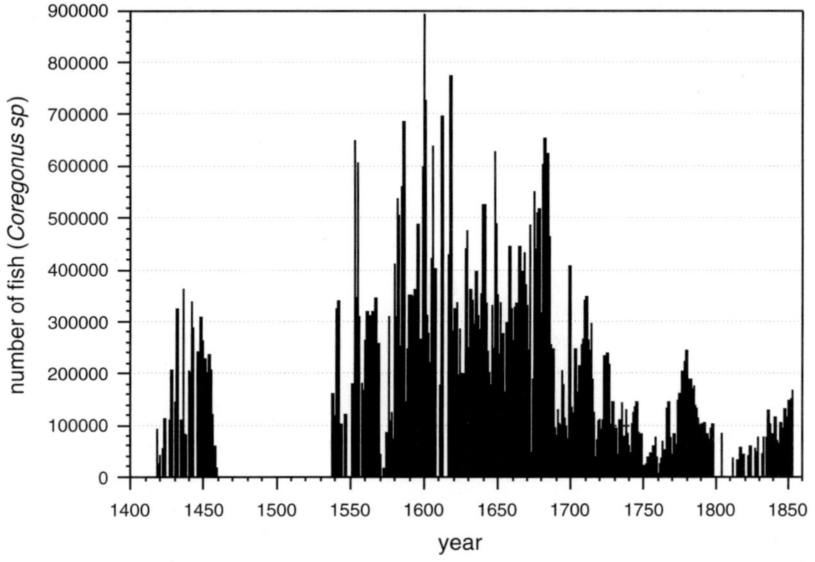

Figure 35.1 The rise and decline of coregonid fishing in Lake Sempach (14.5 km^2) as revealed by the accounting-books of the governor of Sempach/Sursee (Von Liebenau 1897)

with a journal, periodic public exhibitions and political influence of its leading members. This led to the revision of the federal fishery law in 1889. In this legal framework fishing was strictly regulated, with enumeration of forbidden gear, temporal and spatial restrictions, legal sizes for eight species, a general close season at lake shores from 15 April to 31 May and a differentiated closed season for a few species. Later this law was enhanced with articles on civil engineering in rivers, fish migration at hydropower stations, pollution with waste waters from industries and artificial fish propagation (Mathey-Doret 1952).

The changing interest from 'fishing for survival' to 'fishing as a sport' focused much more on angling species. Fishery regulations were made up primarily for a few species of economic interest and many other species were designated as 'unwanted'. The support of alien fish species, which were strongly propagated in Europe and Switzerland since the second half of the 19th century, was especially mentioned and no concern about ecological consequences of these introductions on resident species was foreseen in this fishery law (Cowx 1997).

Increasing land use by agriculture, public infrastructure, settlements and increased use of rivers for hydroelectric purposes led to further developments of legislation concerning fish conservation. The main concern was the worsening of water quality, as many lakes became eutrophic and the populations of salmonids and coregonids were threatened by decreasing oxygen due to algae-blooms (Roth & Geiger 1972). In polluted rivers stocking with hatchery-reared salmonids was greatly increased and measures to avoid fish kills were evaluated. Additionally, research on the treatment of waste water and the effects of harmful substances was intensified (Jaag 1952; Wuhrmann &

Woker 1952). Again the main driving force at the political level was the Fishery Association of Switzerland, which pushed Parliament to formulate a legislation on water protection, finally set into force in 1955 (Slominski 1983). With this new law, another milestone was reached in fish conservation, but again mainly species valuable for fishing purposes were considered (Table 35.2).

The 1970s brought several important changes in the legal basis for fish and their habitats. After nearly 20 years of lobbying, and after a positive plebiscite, an article was finally added to the Constitution of Switzerland in 1975, engaging the authorities to fix minimum flow quantities in rivers used for electricity production. The federal laws on water protection and on fishing were revised in 1971 and 1973, taking into account the latest developments in research on eutrophication and waste-water treatment. Concerning fish conservation, these legal adjustments had two sides. The favourable one was that according to the law on water protection, water quality had to be improved and fish habitats conserved, the unfavourable one was the introduction of the term 'valuable species' in the revised law on fishing (in force since 1976) and thus separating the fish fauna into two categories. The elimination of unwanted species was financially supported to reduce the increasing cyprinid populations in the eutrophic lowland lakes. With this action, not only the catch of roach, bream and bleak was intensified, but also tench, rudd, carp and European catfish, *Silurus glanis* (L.), populations were unwantedly reduced. To fight against prolific growth of algae and aquatic plants, several grass carp species from Asia were introduced to selected rivers or channels. In one river system even Danube salmon, *Hucho hucho* (L.), was introduced to

Table 35.2 Milestones in legislation and activities related to fish conservation in Switzerland

Year	Legal action
1875	First fishery law (regulation of fishery in the River Rhine)
1889	First revision of fishery law (partially 'banning' migration barriers)
1955	Law on water protection with water quality standards
1971	Conservation of fish habitats included in fishery law
1974	Minimum discharge in rivers used for hydroelectricity added to constitution
1983	Law on environmental protection (including environmental impact assessment for new or renewed hydro-power stations)
1985–1989	First nation-wide distribution study
1990	Red List of fish and cyclostomes in Switzerland (revised 1994)
1991	Atlas of fish distribution in Switzerland
1991	Conservation of natural fish diversity and their habitats included in fishery law (with integrated red list, in force since 1994)
1992	Ordinance on protection and restoration of floodplains
1994	Conservation of habitat structures mentioned in laws on water quality and on river engineering
1998–1999	Ordinances on water protection with ecological goals and on river engineering with spatial requirements of rivers
1995–2001	Regional distribution studies
2001	Revision of Atlas and Red List
2001	Implementation of grayling rivers of national importance

reduce nase populations. After an evaluation of preliminary results, these experiments were not continued (Müller 1989).

Due to rising concern about environmental aspects and increasing ecological knowledge in the 1980s, policy-makers became aware of the disappearance of many species and their habitats. This resulted in a new law on environmental protection (1983), including environmental impact assessment for hydropower stations, revisions of the law on water protection with precise minimum flows and the obligation to improve ecologically-stressed waters (1991, in force since 1992), and a law on river engineering, including explicitly the conservation of aquatic biodiversity as well as the restoration of rivers (1991, in force since 1993). Later a new ordinance on the protection of floodplains was set into force in 1992, enhancing the restoration of natural dynamics of rivers, and the new legislation on fishing (1991, in force since 1994) mentions all species as equally valuable, restricts stocking with alien species and accentuates the need for conservation and restoration of natural fish diversity and local populations.

Today Switzerland has a powerful legal framework for fish conservation and many different laws are concerned with species diversity or habitat quality in the aquatic environment (Fig. 35.2). The focus for fish conservation is now set on habitat restoration and many projects are under way or already realised, especially in relation with other aspects of nature conservation like the restoration of floodplains. However, many constraints oppose the fulfilment of the postulated goals such as conservation of all

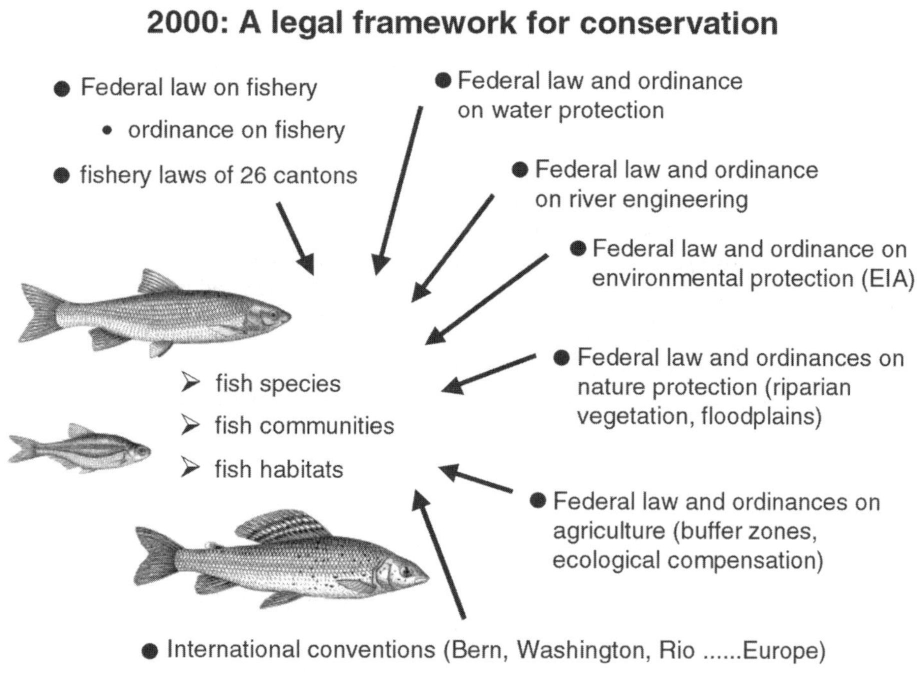

Figure 35.2 A framework of laws on an international, national and regional level helps to conserve the fish fauna and their habitats in Switzerland today

species and local populations, natural species associations and their habitats. Not only financial aspects, but much more the space needed for effective river restoration in the intensively used landscape, restrict many projects.

35.5 Science for fish conservation

A nationwide distribution study (Pedroli, Zaugg & Kirchhofer 1991) further increased public awareness of fish as indicators of environmental quality and resulted in the Red List of the Swiss fish fauna (Kirchhofer, Pedroli & Zaugg 1994; Kirchhofer 1997). To update and refine the knowledge about species distribution and abundance, regional distribution studies are under way and will be compiled to a new fish atlas for Switzerland in 2001/02. A lack of knowledge about the biology of the most threatened species showed up during these studies. To elaborate conservation plans for priority species, research on their biology and habitat requirements was intensified. These studies resulted in some popular monographs including ecological profiles, present distribution, state of populations, threats and suggestions for conservation. To date, the following species have been covered: *Chondrostoma* species (Maier, Zeh, Ortlepp, Zbinden & Hefti 1995), lampreys (Kirchhofer 1996), *Leuciscus souffia* Risso (Schwarz 1998), *Alburnoides bipunctatus* (Bloch) (Breitenstein & Kirchhofer 1999), *Zingel asper* (L.) (Zaugg, Pedroli, Zaugg & Calvino 1999). These monographs should function as guidelines to institutions concerned with fish conservation (fishery administrations, angler societies, river engineers). The federal fishery administration fulfils an important task in supporting these studies, editing and distributing the publications among the concerned institutions, and encouraging local actors with financial support for rehabilitation projects.

Besides habitat rehabilitation, captive breeding of the most endangered species has yet to be considered to conserve the genetic basis of local populations, e.g. for *Z. asper* or *C. nasus* which seem to be at the edge of extinction in Swiss waters. As monitoring of nase reproduction (Hefti, Chapter 38) and intensified studies on several populations have shown (Huber & Kirchhofer 2001), the formerly stated decline is still persisting and nase may disappear from several more rivers in the next few years (A. Kirchhofer & M. Breitenstein, unpublished data). As these declines can be regarded as local events, depending on river structure and discharge situation, the initiative of regional fishery administrations is critical to take action to conserve this species and to formulate specific conservation plans (Fig. 35.3).

35.6 Challenges for the future

Besides data on fish distribution, other databases such as river morphology or spatial requirements of running waters are operational and will cover the whole country in a few years (BUWAL 1998). A further countrywide database being prepared concerns water quality estimations using aquatic invertebrates, and this will help to complete the picture about our waters. One important task for the future will be the integration of

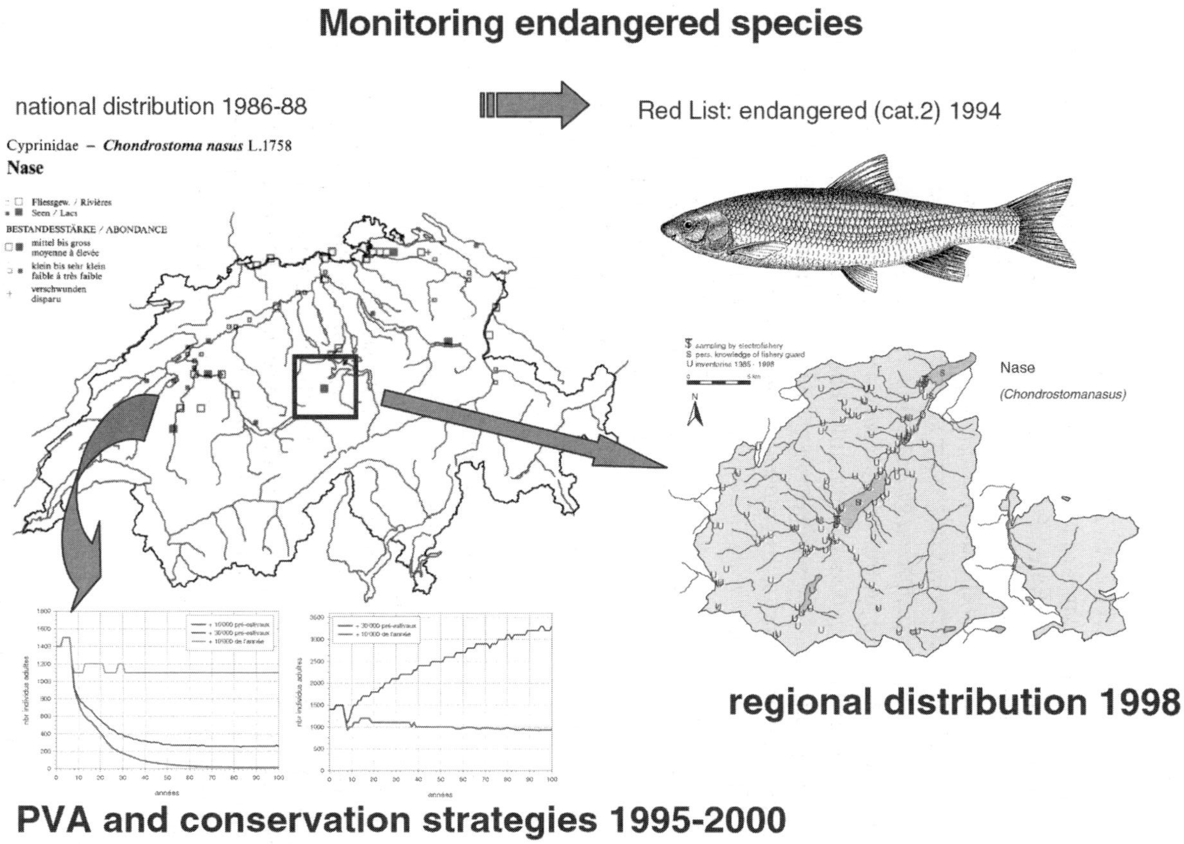

Figure 35.3 The development of nase (*C. nasus*) populations in Switzerland between 1985 and 2000 illustrates the need for species-specific conservation plans for priority species: national distribution as revealed in Pedroli *et al.* (1991) leading to a vulnerability assessment and status 'endangered' in the Red List (Kirchhofer, Pedroli & Zaugg 1994; Kirchhofer 1997); refined distribution studies on a regional scale show further decrease in 1998 (Kirchhofer & Breitenstein, unpublished), and Population Vulnerability Assessment (PVA) for selected sites in 2000 show the need for urgent restocking in order to conserve the population on a short term (A. Kirchhofer & M. Breitenstein, unpublished data)

these databases on fish distribution, habitat requirements, discharge, water quality, river morphology and land use, and the analyses with geographical information systems to highlight ecological deficits of rivers and to integrate these into landscape and urban planning. This will enable politicians and regional administrations to designate priorities for rehabilitation action.

The recent decline of trout catches in many rivers (Friedl 1999) has alarmed fishery associations and authorities. Quite an extensive programme of research in the field of fish health and effects of chemical substances on fish populations has been initiated and the first results show that in many rivers parasites and diseases (e.g. PKD – proliferative kidney disease) seem to be most damaging to trout stocks (Wahli, Meier, Segner & Burkhardt-Holm 1998; Escher, Wahli, Büttner, Meier & Burkhardt-Holm 1999). Further aspects related to climate change, e.g. the impact of flood frequency and intensity on trout reproduction or the change of species associations with increasing temperature in rivers (Jakob 2000), will have to be included in these studies.

Angling societies, which were the driving force for several legal revisions, now tend partially to favour again reinforced stockings with alien species to compensate for the decreasing trout catches. The control of such potential stocking activities will be another important task to conserve the integrity of local populations.

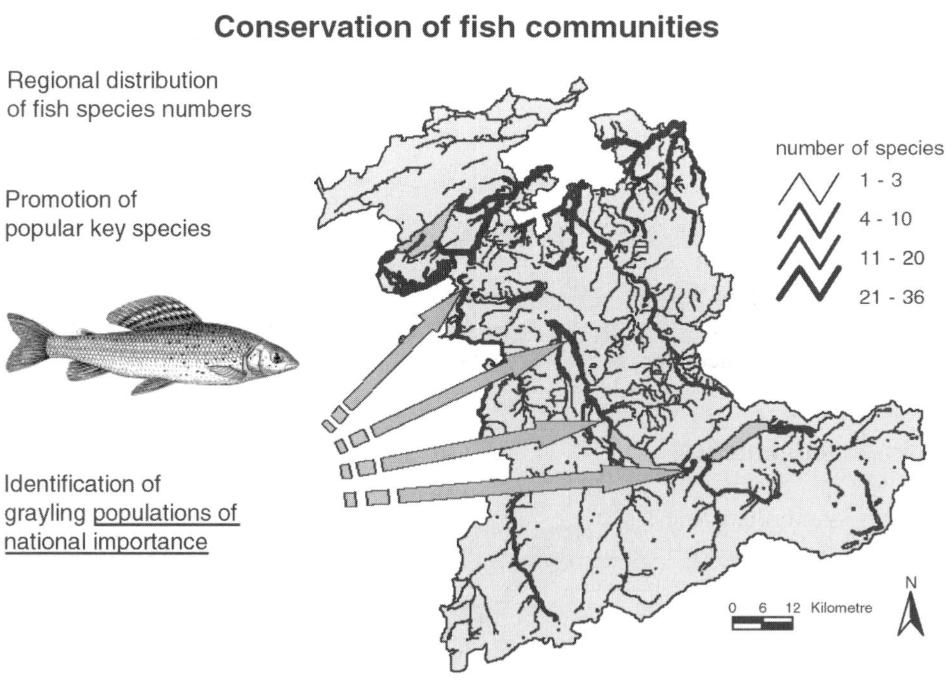

Figure 35.4 Three steps to the conservation of fish communities from mapping species richness in the rivers of Canton Berne (Kirchhofer & Breitenstein 2000) to the identification of grayling populations of national importance in these river sections (Kirchhofer, Breitenstein & Guthruf, unpublished data)

With the focus on the conservation of single species, other views often did not get the attention deserved, e.g. the conservation of whole fish communities (Maitland 1985) or the conservation of dynamic processes. The identification of grayling populations of national importance is a first attempt to use a popular (charismatic!) key species for the conservation of whole fish communities (Fig. 35.4). Furthermore, the ordinance for the conservation of floodplains (1992) obliges the cantons to restore the dynamic river systems as far as possible, which again will help the most threatened riverine species associations.

35.7 Conclusions

Despite all the available information, the most important action for fish conservation remains the restoration of connectivity and rehabilitation of aquatic habitats destroyed over the past 200 years. The biggest challenge to fulfil this task will be financial support. Different approaches have been tried in Switzerland, e.g. through subsidy as a normal public task, or as special restoration funds fed with contributions from electricity companies, or as part of the ordinary budget of the public administration. In many cases, however, private institutions (nature conservation, fishery associations) have to contribute to the realisation of rehabilitation projects. New perspectives arise with so-called 'green power' – a certification scheme for ecologically-sound production of hydroelectricity which minimises impacts on the aquatic system. Market studies have shown that the electricity consumer is ready to pay more, if production is qualified with an eco-label and if this extra price is invested in local rehabilitation schemes (Markard & Truffer 2000), which again is helpful to fish conservation.

As Cambray and Bianco (1998) proposed, a toolkit for fish conservation is urgently needed. Among the many tools they suggested, the support from interest groups, environmental education and the need to put aquatic habitats and fish as indicators of their quality on the political agenda are very important to implement the good legal framework now existing in Switzerland. But the most important and decisive factor for the success of fish conservation in the future will be the attraction of public awareness to the problems fish encounter in a disturbed habitat.

Acknowledgements

I would like to thank the organisers for their invitation to present this contribution at the symposium, the financial support and the warm welcome in Portugal. Further, I thank M. Breitenstein for her critical discussions and two anonymous reviewers for helpful comments and linguistic improvements of the manuscript.

References

Aeschbacher P. (1923) *Geschichte der Fischerei im Bielersee und dessen Nebenflüssen.* Bern: Buchdruckerei G. Grunau, 119 pp. (in German).

Amacher U. (1996) Zürcher Fischerei im Spätmittelalter. *Mitteilungen der Antiquarischen Gesellschaft Zürich* **63,** 1–392 (in German).

Anonymus (1902) Marktpreise der Fische am Zürichsee. *Schweizerische Fischerei Zeitung* **10**, 1–2 (in German).

Breitenstein M. & Kirchhofer A. (1999) Biologie, Gefährdung und Schutz des Schneiders (*Alburnoides bipunctatus*) in der Schweiz. *Mitteilungen zur Fischerei* **62**, 1–46 (in German and French).

BUWAL (ed.) (1998) Methoden zur Untersuchung und Beurteilung der Fliessgewässer: Modul-Stufen-Konzept. *Mitteilungen zum Gewässerschutz* **26**, 1–41 (in German and French).

Cambray J.A. & Bianco P.G. (1998) Freshwater fish in crisis, a Blue Planet Perspective. *Italien Journal of Zoology* **65**(Suppl.), 345–356.

Cowx I.G. (1997) L'introduction d'espèces étrangères de poissons dans les eaux douces européennes: succès économiques ou désastre écologique? *Bulletin Français de la Pêche et de la Pisciculture* **344/345**, 57–78 (in French).

Escher M., Wahli T., Büttner S., Meier W. & Burkhardt-Holm P. (1999) The effect of sewage plant effluent on brown trout (*Salmo trutta fario*). *Aquatic Sciences* **61**, 93–110.

Fehlmann W. (1952) Die Geschichte der Fischerei. In G. Schmid (ed.) *Fisch und Fischerei.* Winterthur: Verlag Gottfried Schmid, pp. 401–415 (in German).

Friedl C. (1999) Fischfangrückgang in schweizerischen Fliessgewässern. *Mitteilungen zur Fischerei* **63**, 1–34 (in German and French).

Gessner C. (1670) *Historia animalium T. 4/5. Vollkommenes Fisch-Buch darstellend eine Abbildung aller grosser und kleiner Fische und seltzamer Meerwunder sammt einer umbstaendlichen Beschreibung.* Unveränderter Nachdruck der Ausgabe Frankfurt am Main, Serlin 1670–1995 Hannover: Schlütersche Verlagsanstalt und Druckerei, 199 pp. (in German).

Huber M. & Kirchhofer A. (2001) Reproductive success of nase (*Chondrostoma nasus* L.) and its influence on population dynamics. *Archiv für Hydrobiologie* **135**, 307–330.

Jaag O. (1952) Der derzeitige Zustand der schweizerischen Gewässer. In G. Schmid (ed.) *Fisch und Fischerei.* Winterthur: Verlag Gottfried Schmid, pp. 343–354 (in German).

Jakob A. (2000) Temperaturentwicklung in Fliessgewässern. *Mitteilungen zur Fischerei* **66**, 29–40 (in German).

Kirchhofer A. (1996) Biologie, Gefährdung und Schutz der Neunaugen in der Schweiz. *Mitteilungen zur Fischerei* **56**, 1–51 (in German and French).

Kirchhofer A. (1997) The assessment of fish vulnerability with distribution data in Switzerland. *Biological Conservation* **80**, 1–8.

Kirchhofer A. & Breitenstein M. (2000) Fische und Krebse des Kantons Bern. Bern: Fischereiinspektorat des Kantons Bern, 112 pp. (in German and French).

Kirchhofer A., Pedroli J.-C. & Zaugg B. (1994) Rote Liste der Fische und Rundmäuler der Schweiz. In P. Duelli (ed.) *Rote Listen der gefährdeten Tiere der Schweiz.* Bern: BUWAL-Reihe Rote Listen, EDMZ, pp. 35–37 (in German and French).

Maier K.-J., Zeh M., Ortlepp J., Zbinden S. & Hefti D. (1995) Verbreitung und Fortpflanzung der in der Schweiz vorkommenden Chondrostoma-Arten: Nase (*C. nasus*), Sofie (*C. toxostoma*) Savetta (*C. soetta*). *Mitteilungen zur Fischerei* **53**, 1–62 (in German and French).

Maitland P.S. (1985) Criteria for the selection of important sites for freshwater fish in the British Isles. *Biological Conservation* **31**, 335–353.

Markard J. & Truffer B. (eds) (2000) Umweltmanagement und Ökolabeling für die Wasserkraft – Markt, Zertifizierung und Praxiserfahrungen. *EAWAG – Ökostrom Publikationen Band* **4**, 1–99 (in German).

Mathey-Doret A. (1952) Fischereigesetzgebung. In G. Schmid (ed.) *Fisch und Fischerei.* Winterthur: Verlag Gottfried Schmid, pp. 325–340 (in German).

Müller R. (1989) Fischereiwissenschaftliche Untersuchungen über die Auswirkungen des Einsatzes ostasiatischer pflanzenfressender Fischarten in verschiedene Gewässer der Schweiz, Schlussbericht. *Schweizerische Fischereiwissenschaft* **6**, 3–6 (in German).

Pedroli J.C., Zaugg B. & Kirchhofer A. (1991) Verbreitungsatlas der Fische und Rundmäuler der Schweiz. *Documenta Faunistica Helvetiae* **11**, 1–206 (in German and French).

Ribi A. (1942) *Die Fischbenennungen des Unterseegebietes, mit dem erstmaligen Abdruck des Fischbüchleins von Gregor Mangolt nach den Original-Handschriften 1557.* Rüschlikon: Baublatt AG, 158 pp. (in German).

Roth H. & Geiger W. (1972) Brienzersee, Thunersee and Bielersee: effects of exploitation and eutrophication on the salmonid communities. *Journal of the Fisheries Research Board of Canada* **29**, 755–764.

Schwarz M. (1998) Biologie, Gefährdung und Schutz des Strömers (*Leuciscus souffia*) in der Schweiz. *Mitteilungen zur Fischerei* **59**, 1–60 (in German and French).

Slominski P. (1983) *100 Jahre Schweizerischer Fischerei Verband.* Bern: Buri Verlag, 158 pp. (in German).

Surbeck P. (1930) Die Schweizerische Bodenseefischerei im Jahre 1929. *Schweizerische Fischerei Zeitung* **38**, 49–55 (in German).

Vischer D. (1989) Impact of 18th and 19th century river training works: Three case studies from Switzerland. In G.E. Petts, H. Möller & A.L. Roux (eds) *Historical Change of Large Alluvial Rivers: Western Europe.* Chichester: John Wiley & Sons, pp. 19–40.

Von Liebenau T. (1897) *Geschichte der Fischerei in der Schweiz.* Bern: Eidgenössisches Departement des Innern, 207 pp. (in German).

Wahli T., Meier W., Segner H. & Burkhardt-Holm P. (1998) Immunohistochemical detection of vitellogenin in male brown trout for Swiss rivers. *Histochemical Journal* **30**, 753–758.

Wuhrmann K. & Woker H. (1952) Abwässer – eine Gefahr für die Fischerei. In G. Schmid (ed.) *Fisch und Fischerei.* Winterthur: Verlag Gottfried Schmid, pp. 355–379 (in German).

Wulpillier P. (1896) St.Gallisches Fischereiwesen vom Mittelalter bis Ende des 18.Jahrhundert. *Schweizerische Fischerei Zeitung* **4**, 343–350 (in German).

Zaugg C., Pedroli J.-C., Zaugg B. & Calvino C. (1999) Schutzkonzept des Aprons (*Zingel asper*): Bestandesaufnahme im Doubs. *Mitteilungen zur Fischerei* **64**, 1–44 (in German and French).

Chapter 36
The role of aquaculture in fish conservation: a case study of *Barbus barbus* in Belgium

P. PONCIN*
Department of Ethology and Animal Psychology, University of Liège, Liège, Belgium

J.C. PHILIPPART
Laboratory of Fish Demography and Aquaculture, University of Liège, Tihange, Belgium

Abstract

Breeding and rearing methods commonly used in aquaculture are important when considering the artificial regeneration of endangered fish populations. Initially baseline data on the population biology and ecology of the endangered species are collected before developing studies in captivity. Collection of sperm and ova from wild spawners, or induced breeding of wild spawners under controlled condition, were commonly used to produce fry in hatchery system. The latter has to be properly reared in captivity conditions (density, temperature, photoperiod and health) to produce mature fish which can constitute a breeding stock (and sometimes represents the first step of domestication). In some fish, particularly cyprinids, growth can be accelerated in captivity (especially in heated water) and both sexes reach maturity at an earlier age and often at a smaller size in captivity than in the wild. Controlled conditions also allow manipulation of the reproductive periods as well as reproductive output characteristics, allowing selection of fish (sometimes on genetic criteria) and choice of the time of year for releasing fish into the wild. Finally, the importance of restoration stocking in the wider context of fisheries management in rivers is emphasised in relation to long-term rehabilitation actions.

Keywords: artificial rearing, cyprinids, reproduction, restocking.

36.1 Introduction

With fish, as with other groups of animals, captive breeding and its logical extension – reintroduction to the wild – can contribute to the conservation of endangered species or populations (see the review by Philippart 1995). This approach is pertinent in species which suffer from severe *ad hoc* pollutions, water diversion, overfishing or habitat (particularly spawning sites) degradation (see Cowx, Chapter 20) as reported, for example, in various areas in Europe (Kirchhofer & Hefti 1996), France (Keith, Allardi & Moutou 1992) and Belgium (Philippart & Vranken 1983).

Captive breeding and stocking of hatchery-reared fish have produced positive results with fishes of all sizes and from a wide variety of habitats in all the regions of

*Correspondence: Department of Ethology and Animal Psychology, 22 quai Van Beneden, B-4020 Liège, Belgium (email: P.Poncin@ulg.ac.be).

the world (Table 36.1) (Philippart 1995). Acipenseridae, Salmonidae and Cyprinidae are the species groups of importance in Europe. In Belgium, for example, barbel, *Barbus barbus* (L.), four other rheophilous cyprinids (nase, *Chondrostoma nasus* (L.), chub, *Leuciscus cephalus* (L.), dace, *Leuciscus leuciscus* (L.) and river bleak, *Alburnoïdes bipunctatus* (L.)) and some small-size cyprinids such as the gudgeon, *Gobio gobio* (L.), and the minnow, *Phoxinus phoxinus* (L.), have been studied. This chapter will focus on Belgian activities to illustrate the range of problems that fish conservationist and managers face with respect to the role of aquaculture and stocking as conservation tools. The chapter describes what is considered an effective strategy to investigate whether artificial propagation and rearing of fishes is feasible for fish species under threat from human activities. The strategy is broken down into three phases and illustrated with reference to studies on barbel in the Belgium River Meuse.

Phase 1 (1971–1981) comprised basic research on population dynamics and production of barbel in the relatively undisturbed barbel zone of the River Ourthe (main tributary of the River Meuse) and a study of the geographical distribution, habitat and population status (normal, in regression, extinct, vulnerable) in the Belgian fresh waters. Phase 2 (1982–1989) studied controlled reproduction in captivity and the mass rearing of fry, fingerlings and adults for stocking including post-stocking studies of the enhanced populations to determine the efficacy of stocking operations. Phase 3 (1988–ongoing) is an in-depth investigation of the ecology of wild largely undisturbed or stressed populations, focusing on key components such as the physical habitat and micro-habitat utilisation and requirements, the migratory and reproductive behaviour (using radio-tracking techniques and surveys of fish passes) and ecology of eggs, larvae and juveniles of the year (biology of recruitment).

36.2 Basic studies on the ecology and status of populations

Barbus barbus is a long-lived (maximum 25 years), gregarious, rheophilic and benthic cyprinid which exhibits nocturnal habits and mainly feeds on aquatic invertebrates. Natural spawning occurs in May–June when temperature reaches 14–18°C. It is a lithophilous spawner which buries eggs in gravelly bottoms in fast flowing waters. Detailed investigation carried out between 1971 and 1975 in the lower course of the River Ourthe (width: 28 m; average annual discharge: 23 $m^3 s^{-1}$; mean slope: 1.6‰; fish community comprising 24 indigenous species and a total biomass of about 300 kg ha^{-1}) showed females mature at a large size (minimum 30 cm fork length (FL) and 6-year old; most often 35 cm FL and 8-year old), thus there are very few females in the population (only 500 mature females were estimated among a total population of 26 000 fish >15 cm in a 27-km (28 ha) stretch of river). Individual fecundity was extremely low (8000 ova in a 40 cm FL female) compared with other rheophilic cyprinids (chub, dace, nase). These demographic traits make the barbel vulnerable to angling mortality (legal size is 30 cm total length (TL) or 27.5 cm FL), which represent most mature females (Philippart 1987).

Studies between 1979 and 1981 in southern Belgium showed barbel to be extinct in the River Scheldt basin, and a strong regression of the stocks in the navigable canalised River Meuse and in several of its sub-basins. Abundant populations were recorded

Table 36.1 List of European fishes currently bred in captivity (source: Philippart 1995)

Species	Reference
Acipenseridae	
A. güldenstaedti	Barannikova (1987)
A. naccarii	Giovannini *et al.* (1991)
A. nudiventris	Barannikova (1987)
A. ruthenus	Barannikova (1987)
A. sturio	Rochard, Castelnaud & Lepage (1990)
A. huso	Barannikova (1987)
Cyprinidae	
Alburnoides bipunctatus[a]	Philippart (1992)
Aspius aspius	Kaukoranta & Pennamen (1990)
Barbus barbus[a]	Philippart (1982); Philippart *et al.*(1989)
Barbus meridionalis[a]	Philippart & Berrebi (1990)
Chondrostoma nasus[a]	Poncin & Castelli (1990); Philippart (1992)
Cyprinus carpio[a]	Billard & Marcel (1986)
Gobio gobio[a]	Kestemont (1988); Kestemont & Mélard (1990)
Leuciscus cephalus[a]	Poncin, Mélard & Philippart (1989)
Leuciscus leuciscus[a]	Philippart (1992)
Leuciscus idus[a]	Kestemont, Mélard, Poncin, De Backer, Micha & Philippart (1991)
Phoxinus phoxinus[a]	Stalmans & Kestemont (1991)
Rutilus rutilus[a]	Worthington, MacFarlane & Easton (1982)
Rutilus pigus	Povz & Ocvirk (1990)
Scardinius erythrophthalmus[a]	Klein Breteler (1979)
Tinca tinca[a]	Von Lukowicz, Tamas & Horvath (1986)
Vimba vimba	Huisman & Hogedoorn (1979)
Salmonidae	
Hucho hucho	Jungwirth (1978); Holcik (1990)
Salmo trutta[a]	Many breeding stations in Europe
Salmo salar	Many breeding stations in Europe
Salmo marmoratus	Budhina & Ocvirk (1990)
Salvelinus alpinus	Maitland & Lyle (1990)
Stenodus leucichthys	Lelek (1987)
Coregonidae	
Coregonus spp.	Maitland & Lyle (1990)
Thymallidae	
Thymallus thymallus	Carmie & Jonard (1988)
Percidae	
Perca fluviatilis[a]	Mélard & Philippart (1984)
Stizostedion lucioperca[a]	Huet (1970)
Siluridae	
Silurus glanis	Epler & Bienarz (1989)
Esocidae	
Esox lucius[a]	Chauveheid & Billard (1983)

[a]Species held in captivity and/or bred at the Tihange Fish Breeding Centre, Belgium.

only in the least affected rivers Ourthe and Semois. Reasons for the decline include water pollution (barbel requires high oxygen level and needs clean shallow gravel beds to spawn efficiently) and habitat modification, mainly channelisation and the subsequent destruction of the spawning grounds, but also river damming and regulation or alteration of flow regime and erosion-sedimentation processes.

36.3 Captive breeding programme

36.3.1 *Rearing facilities used*

Barbel rearing trials were conducted in the experimental Fish Research Station of the Study Centre for Waste Heat Utilisation at Tihange (Huy), on the River Meuse. The facilities comprised square fibre-glass tanks, circular aluminium tanks and flow-through earth ponds (150–200 m^3) simulating a river with a gravel bed. These tanks and ponds were supplied with cooling water from the Tihange nuclear power plants (three plants on the site), water at ambient temperature pumped from the River Meuse or a mixing of both to obtain an optimal temperature of 18–24°C all year round. Chemical quality of the River Meuse water was good and the natural temperature sometimes reached 24°C, but with an oxygen concentration rarely falling below 6 $mg\,L^{-1}$.

36.3.2 *Constitution of brood stocks in captivity*

To establish a brood stock in captivity, wild barbels were electric fished on a spawning ground in the River Ourthe in May 1982. About 8000 ova were stripped from three females in the field and inseminated with sperm (dry method). The fertilised eggs were brought for incubation and hatching in a small hatchery. Larvae were first reared in aquaria, then fry were transferred to the Tihange Fish Research Centre and placed in a tank supplied with cool (heated) water (20–24°C).

Under optimal temperature and feeding conditions, young barbel exhibited an accelerated growth rate and females first attained sexual maturity in captivity as early as January 1984, at 22–25 months and 23–34 cm FL (*cf.* in the wild: minimum 6 years and 30 cm FL). These brood stock fishes were used for artificial rearing (stripping of females which ovulated spontaneously without application of hormonal treatment or hypophysation), and represented the first production of barbel reared entirely in captivity (for details see Poncin 1991). This captive-bred barbel stock was placed under conditions of fast growth to allow it to spawn 2 years later, in the spring of 1986. Brood stock fish were tagged with conventional external numbered tags and later with PIT tags.

36.3.3 *Controlled reproduction and production of eggs*

Stocks of barbel managed for egg production were maintained in 1.5–4 m^3 tanks in a small semi-closed recirculating system at 20°C and >70% dissolved oxygen saturation.

Under these thermal conditions, and a natural photoperiod, males and females mature early in the year (January–March), compared with the natural spawning time (May–June) and spawning in captivity stops in July. Females ovulate spontaneously, allowing ova to be extracted by stripping and fertilisation using the dry method. Spontaneous expulsion of eggs in tanks was never observed because of the lack of a gravelly substratum. Under such rearing conditions, female barbel exhibit repeated spawning and can be stripped up to a maximum of 15 times at 10–20 day intervals during the reproductive season. These females exhibit successive losses and gains of weight as a result of their repeated spawnings. Outside the reproductive periods, rapid somatic growth occurs. During the reproductive periods, oocytes at all stages of development are found in the ovaries.

Captive-bred barbel were used to determine the effect of environmental (photoperiod, temperature and congenerics) and hormonal factors on reproduction. When submitted to an annual photoperiod contracted into 6 months (2 cycles per year), females exhibited two reproductive periods (January–May and September–November) (Fig. 36.1). This results from inhibition of a decreasing photoperiod on the sexual maturity of both sexes. Under natural or constant photoperiods, spawning comes to an end spontaneously (exhaustion of the females). In wild barbel, it is likely that the photoperiod change by the end of June prevents fish from continuing to spawn in August when water temperature is still favourable (15°C). Reproduction in August would not allow sufficient time for the larvae to grow to a minimum size-weight to survive overwinter. These photoperiod treatments were used to manage barbel egg production all year round (Poncin & Castelli 1990).

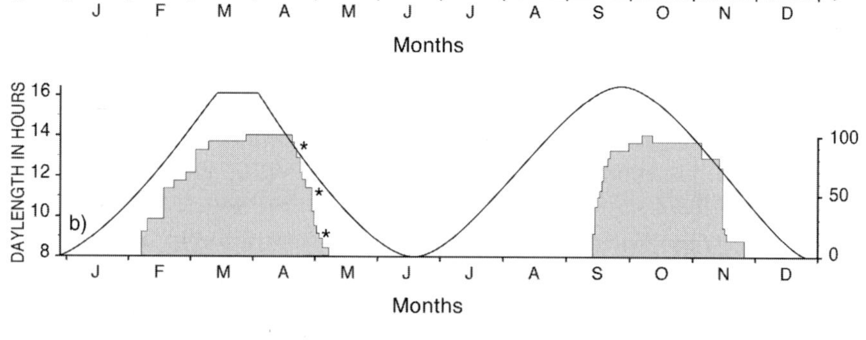

Figure 36.1 Daily percentage of spawning female and weekly percentage of spawning male barbel in two experimental groups submitted to photoperiodic treatments. Group A, simulated natural photoperiod (as for Belgium: 51°N, 5°E); group B, annual natural cycle contracted to 6 months. *Significant results ($P < 0.05$) χ^2 test. From Poncin (1989)

36.3.4 *Mass production of juveniles and adults*

Fertilised eggs are incubated in Zug jars at 17–20°C. Hatching success may be as high as 98% with ova stripped from captive females, but between 67% and 84% with eggs extracted from wild females in the field. The size at which larvae move to exogenous feeding is reached by in 9–14 days (198–253 degree-days) after fertilisation and 5–8 days (113–150 degree-days) after hatching. During that stage of yolk sac reabsorption, mortality is less than 1% when ova are of good initial quality. Larvae have a size ranging from 8 to 12 mm depending on female body length. Newly-hatched larvae are reared in ponds or tanks where they feed on natural preys (rotifers) and artificial (trout) feed. The 100–700 mg fry are grown on in various facilities at optimum temperatures (20–24°C) and dissolved oxygen levels, and are fed to satiation with trout pellets. These fish grow at a very fast rate, attaining a size of 15 cm (40 g) within 10 months, four times faster than under ambient temperature and feeding regimes in rivers (Fig. 36.2). Further information on the intensive culture of barbel in heated water for restocking are provided by Philippart, Mélard and Poncin (1989).

36.3.5 *Role of aquarium studies*

The spawning behaviour of cultured barbels was observed for the first time in aquaria by Poncin (1994). A true spawning (spontaneous and synchronised expelling of sexual products) only happens when spawning substratum (gravel) is available. Behavioural

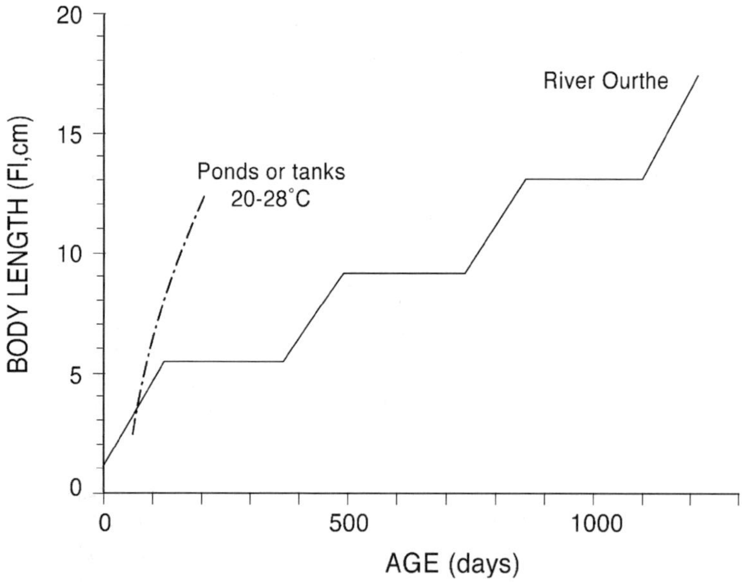

Figure 36.2 Growth pattern of *B. barbus* under a natural (R. Ourthe) and an optimal artificial (Tihange Fish Research Centre) thermal regime

studies in aquaria may produce valuable information for management because spawning behaviour is often difficult to observe in the wild.

In small fish species, studies in aquaria are much easier than in tanks or direct observation in the wild. Aquarium studies can be the first step in establishing a breeding stock, particularly in species in which artificial fertilisation is not easy. This method was successfully used to reproduce in captivity the small meridional barbel, *Barbus meridionalis* (Philippart & Berrebi 1990), and to hybridise it with *B. barbus* (Poncin, Jeandarme & Berrebi 1994). The same approach was used to induce spawning (using carp pituitaries extracts injection) in gudgeon (Kestemont & Mélard 1990; Poncin, Jeandarme, Rinchard & Kestemont 1997), plus tropical and ornamental fish species (Poncin, Libois & Philippart 1988).

36.4 Efficiency of restoration stockings

Restocking Belgian rivers with hatchery-reared barbel started on a very small scale in October 1983 and by the end of 1989, about 85 000 cultured individuals had been released into 14 rivers and streams over Wallonia (Philippart 1990). The three main sizes of barbel used for restocking were 5–15 cm (1-year old), 15–30 cm (2-year old) and >30 cm (3–4-year old). Most stocking operations take place in May–June when river temperature are >10°C and exhibit a rising-trend, so that the fish start growing at a time of the year when production of invertebrates is maximum. When stocked in the autumn, reared barbel have to tolerate adverse environmental conditions (decreasing temperature, low population densities of invertebrate preys) and are less likely to survive over winter. The number of barbel to be released in a particular situation is fixed according to the area of suitable habitats (riffles and runs) available.

Most barbel stockings performed in Belgium from 1983 to 1989 were submitted to a scientific evaluation of their necessity and success or failure. Monitoring was carried out using electric fishing (small streams) and/or tag returns of fish caught by anglers. When the recipient stream is of good chemical quality (high-medium oxygen content levels), the growth rate of hatchery-reared barbel of different sizes reverts to that found in the wild, the majority stay near the release point and survival rate is high irrespective of the origin of the stock (from wild or 100% captive-bred spawners) and despite the intensive culture conditions.

Experiments carried out in the Belgian River Meuse basin in the 1980s demonstrated that restocking rivers and streams with hatchery-reared barbel could be a short-term compensation measure to overcome recruitment bottlenecks in degraded areas. However, there was concern that stock enhancement (especially as it involved intensively-reared fish using heated water) would disturb the gene pool of the native stocks and consequently the adaptation of the small relic populations by the process of selection and micro-evolution (see Philippart 1995). Consequently, most future research and management efforts aimed at long-term and sustainable preservation of barbels resource should concentrate on improving the population ecology and genetics of the species. Some data on the genetics of wild and cultivated barbel is available (P. Berrebi, unpublished data; Castelli, Philippart, Vassart & Georges 1990) but this is inconclusive. Most research effort has been directed towards the barbel eco-behavioural requirements for physical habitat features that can influence the success of recruitment of young and the carrying capacity of streams.

To this end, in the late 1980s a long-term investigation on the ecology of wild, almost undisturbed, or stressed barbel populations, focusing on key features such as the reproductive and migratory behaviour (study of barbel mobility and migration using biotelemetry techniques, inventory of physical obstacles to migration, efficiency of ancient fishing ways and design of new ones) (Baras & Cherry 1990; Baras, Lambert & Philippart 1994; Philippart & Baras 1996), physical habitat and micro-habitat utilisation and requirements and the ecology of eggs, larvae and juveniles of the year (biology of recruitment) (Baras, Nindaba & Philippart 1995), was initiated. This should provide a better knowledge of all these aspects of barbel population ecology and a solid basis for a better conservation and sustainable management of the species. It will also give information on how the species/populations will respond to environmental changes relating to water temperature (global warming), river discharge and eutrophication.

36.5 Conclusions

This research on captive breeding of barbel has demonstrated the interactions between several factors influencing cyprinid fish reproduction. Environmental factors (e.g. photoperiod and temperature) act through intrinsic mechanisms (photosensitivity, hormones) to control reproduction, and consequently the production of eggs and recruitment of young which may be used for stocking (Fig. 36.3). The large scale production of barbel for restoration stocking in Belgian rivers is now possible.

Rearing and stocking techniques and methodologies developed for barbel may also be applied to other running water cyprinids and fish species, taking into account the reproductive pecularities of these species needed for induced spawning (hypophysation, egg size, optimum temperature for development).

Despite its considerable potential and the existing role that artificial propagation of fish has played for a long time in maintaining inland fisheries must never be considered as an effective means for the long-term sustainability of most fish populations. In particular, as reported by Philippart (1995), artificial rearing poses risks in conserving the genetic integrity resources. It must be used only when all other possibilities for conserving a species and its natural habitat have been explored.

Acknowledgements

The work on the barbel was financially supported by grants from IRSIA, the National Foundation for Scientific Research, Electrabel private company, the Fond Piscicole de Liège and the Ministry of Environment of the Région Wallonne. Many thanks to I. Cowx for improving the English version of this manuscript.

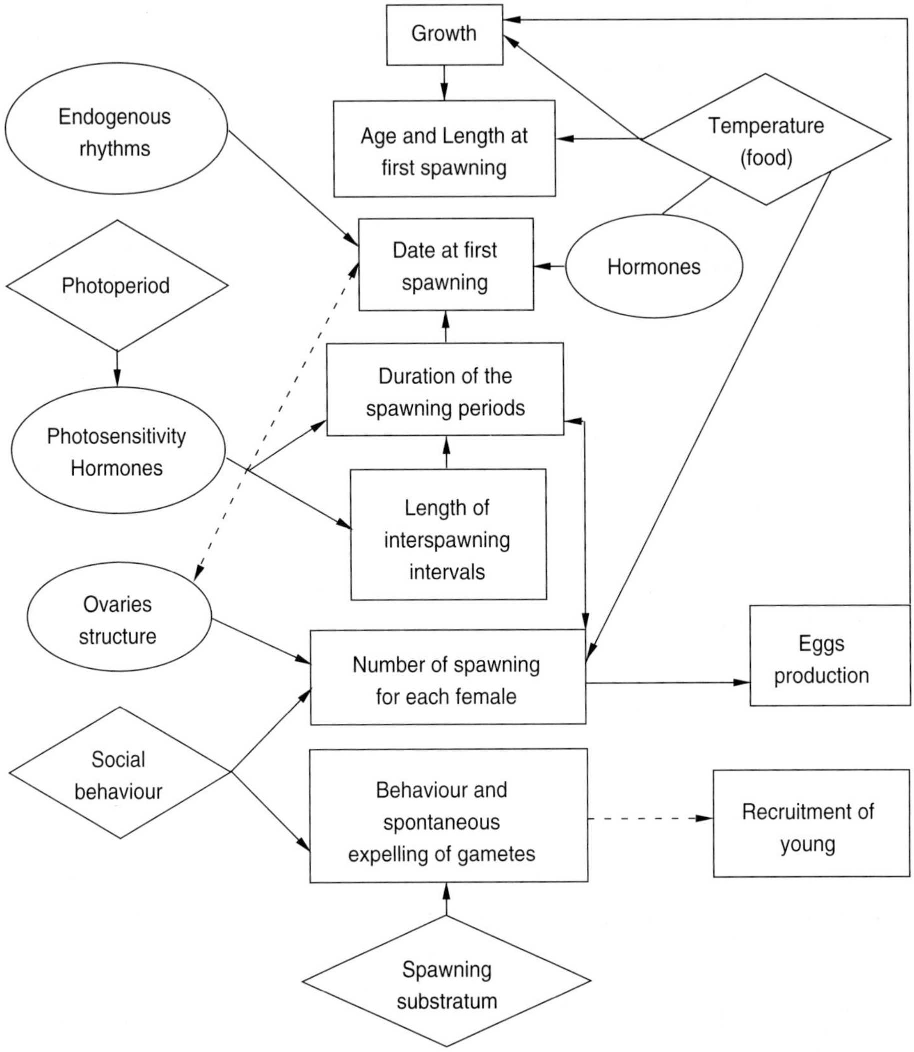

Figure 36.3 Summary of the influence of environmental factors (◇) on reproductive feature (□) of barbel *B. barbus*. Endogenous factors are also indicated (○). From Poncin (1991)

References

Barannikova I.A. (1987) Review of sturgeon farming in the Soviet Union. *Journal of Ichthyology* (C), 62–71.

Baras E. & Cherry B. (1990) Seasonal activities of female barbel *Barbus barbus* (L.) in the River Ourthe (Southern Belgium), as revealed by radio tracking. *Aquatic Living Resources* **3**, 283–294.

Baras E., Lambert H. & Philippart J.C. (1994) A comprehensive assessment of the failure of *Barbus barbus* (L.) migrations through a fish pass in the canalized River Meuse (Belgium). *Aquatic Living Resources* **7**, 181–189.

Baras E., Nindaba J. & Philippart J.C. (1995) Microhabitats utilisés par les alevins 0+ d'une communauté de cyprins rhéophiles: analyse de l' évolution de la communauté et des densités de population. *Bulletin Français de la Pêche et de la Pisciculture* **2**, **3** & **4**, 241–247 (in French).

Billard R. & Marcel J. (1986) *Aquaculture of cyprinids – l'aquaculture des Cyprinidae.* Collection Hydrobiologie et Aquaculture, INRA, Paris, France, 502 pp.

Budhina N. & Ocvirk A. (1990) Breeding and restocking of salmonid fishes in Slovenia. *Journal of Fish Biology* **37**(Suppl. A), 239–40.

Carmie H. & Jonard (1988) Starling of grayling (*Thymallus thymallus*) larvae and production of estival juveniles exlusively using dry food. *Bulletin Français de la Pêche et de la Pisciculture* **296**, 2–16 (in French).

Castelli M., Philippart J.C., Vassart G. & Georges M. (1990) DNA fingerprinting in fish: a new generation of genetic markers. *American Fisheries Society Symposium* **7**, 514–520.

Chauveheid A. & Billard R. (1983) Incubation et éclosion des œufs de brochet et résorption vitelline des larves. In R. Billard (ed.) *Le Brochet: Gestion dans le Milieu Naturel et Élevage*. Paris: INRA Publ., pp. 163–176 (in French).

Epler P. & Bienarz K. (1989) Gonad maturation and hormonal stimulation of spawning in wels (*Silurus glanis* L.). *Polskie Archiv für Hydrobiology* **36**, 417–429.

Giovannini G., Colombo L., Bronzi P. & Arlata G. (1991) Growth of hatchery produced juveniles of Italian sturgeon, *Acipenser naccarii* Bonaparte, reared intensively in freshwater. In G. Willot (ed.) *Proceedings of the First International Symposium on Sturgeon*. Paris; CEMAGREF, pp. 401–404.

Holcik J. (1990) Conservation of the huchen, *Hucho hucho* (L.) (Salmonidae) with special reference to Slovakian rivers. *Journal of Fish Biology* **37**(Suppl. A), 113–121.

Huet M. (1970) *Traité de Pisciculture*. Ch. de Wyngaert, Bruxelles, 718 pp.

Huisman E. & Hogendoorn H. (1979) EIFAC workshop on mass rearing of fry and fingerlings of freshwater fishes. *EIFAC Technical Pap*er **35**(Suppl. 1), 202 pp.

Jungwirth M. (1978) Some notes on the farming and conservation of the Danube salmon *Hucho hucho*. *Environmental Biology of Fishes* **2**, 231–234.

Kaukoranta M. & Pennamen J.T. (1990) Propagation and management of the asp, *Aspius aspius* (L.) in Finland. In W.L.T. Van Desen, B. Steinmetz & R.H. Hughes (eds) *Management of Freshwater Fisheries*. Wageningen: Pudoc, pp. 67–73.

Keith P., Allardi J. & Moutou B. (1992) *Livre Rouge des Espèces Menacées de Poissons d'eau Douce de France et Bilan des Introductions*. Coll. Patrimoines Naturels, vol. 10, SFF-MNHN, CSP, CEMAGREF, Ministère de l' Environement, Paris, 111 pp. (in French).

Kestemont P. (1988) Etude des potentialités d'elevage intensif du *goujon Gobio gobio* L. Comparaison de plusieurs types d'aliments. *Bulletin Français de la Pêche et de la Pisciculture* **297**, 48–54 (in French).

Kestemont P. & Mélard Ch. (1990) About the culture of a small cyprinid, the gudgeon *Gobio gobio* L. *European Aquaculture Quarterly Newsletter* **56**, 39–41.

Kestemont P., Mélard Ch., Poncin P., De Backer L., Micha J.C. & Philippart J.C. (1991) Controlled and intensive larval rearing of freshwater cyprinids (ide, gudgeon, minnow and goldfish) at a pilot scale. In N. De Pauw & J. Joyce (eds) *Aquaculture and Environment*. European Aquaculture Society Special Publications **14**, 169–170.

Kirchhofer A. & Hefti D. (eds) (1996) *Conservation of Endangered Freshwater Fish in Europe.* Switzerland: Birkhaüser Verlag, 360 pp.

Klein Breteler J.G.P. (1979) First experiments concerning controlled reproduction and rearing of fry and fingerling of rudd, *Scardinius erythrophthalmus* L. In E. Huisman & H. Hogendoorn (eds) *EIFAC workshop on Mass Rearing of Fry and Fingerlings of Freshwater Fishes,* Rome: FAO, pp. 184–188.

Lelek A. (1987) Threatened fishes of Europe. In *The Freshwater Fishes of Europe*, Vol. 9. Wiesbadden: AULA-Verlag, 343 pp.

Maitland P.S. & Lyle A.A. (1990) Practical conservation of British fishes: current action on six declining species. *Journal of Fish Biology* **37**(Suppl. A), 255–256.

Mélard Ch. & Philippart J.C. (1984) Essai d'élevage semi-intensif en bassins d'alevins de perche fluviatile *Perca fluviatilis*, obtenus par reproduction artificielles. *Cahiers d'Ethologie Appliquée* **41**(1), 59–66 (in French).

Philippart J.C. (1982) Mise au point de l'alevinage contrôlé du barbeau *Barbus barbus* (L.) en Belgique. Perspectives pour le rempoissonnement des rivières. *Cahiers d'Ethologie appliquée* **2**, 173–202 (in French).

Philippart J.C. (1987) Démographie, conservation et restauration du barbeau fluviatile, *Barbus barbus* (Linné) (Teleostei: Cyprinidae) dans la Meuse et ses affluents. Quinze années de recherches. *Annales de la Société royale zoologique de Belgique* **117**, 57–69 (in French).

Philippart J.C. (1990) La reconstitution d'une population de barbeau fluviatile dans la Méhaigne au moyen de poissons produits en pisciculture. *Travaux de la Conservation de la Nature* **15/2**, 759–770 (in French).

Philippart J.C. (1992) Mise au point de méthodes d'élevage du hotu *Chondrostoma nasus*, du chevaine *Leuciscus cephalus*, de la vandoise *Leuciscus leuciscus* et de l'ablette spirlin *Alburnoides bipunctatus* pour le repeuplement et la restauration des rivières. *Rapport de Recherche aux Fonds Piscicoles*, Université de Liège, 24 pp. (in French).

Philippart J.C. (1995) Is captive breeding an effective solution for the preservation of endemic species? *Biological Conservation* **72**, 281–295.

Philippart J.C. & Baras E. (1996) Comparison of tagging and tracking studies to estimate mobility patterns and home range in *Barbus barbus*. In E. Baras & J.C. Philippart (eds) *Underwater Biotelemetry, Proceedings of the First Conference and Workshop on Fish Telemetry in Europe*. Liège: University of Liège, Belgium, pp. 3–12.

Philippart J.C. & Berrebi P. (1990) Experimental hybridization of *Barbus barbus* and *Barbus meridionalis*: physiological, morphological, and genetic aspects. *Aquatic Living Resources* **3**, 325–362.

Philippart J.C., Mélard Ch. & Poncin P. (1989) Intensive culture of the common barbel *Barbus barbus* (L.) for restocking. In N. De Pauw, E. Jaspers, H. Ackefors & N. Wilkins (eds) *Aquaculture – a Biotechnology in Progress*. Bredene: Belgium, European Aquaculture Society, pp. 483–491.

Philippart J.C. & Vranken M. (1983) Atlas des poissons de Wallonie. Distribution, écologie, éthologie, pêche, conservation. *Cahiers d'Ethologie Appliquée* **3**, 395 pp. (in French).

Poncin P. (1989) Effects of different photoperiods on the reproduction of the barbel, *Barbus barbus* (L.) reared at constant temperature. *Journal of Fish Biology* **35**, 395–400.

Poncin P. (1991) Environmental and endocrine control of barbel (*Barbus barbus* L.) reproduction. In A.P. Scott, J.P. Sumpter, D.E. Kime & M.S. Rolfe (eds) *Reproductive Physiology of Fish*. Sheffield: Published by Fish Symposium 91, pp. 148–150.

Poncin P. (1994) Bilan de 4 années d'étude du comportement reproducteur et de l'hybridation chez *Barbus barbus* et *Barbus meridionalis*. *Bulletin Français de la Pêche et de la Pisciculture* **364**, 169–176 (in French).

Poncin P. & Castelli M. (1990) Manipulation photopériodique des saisons de reproduction chez le barbeau (*Barbus barbus*). Bilan de deux années d'application des techniques. *Cahiers d' Ethologie appliquée* **10**(3–4), 447–450 (in French).

Poncin P., Jeandarme J. & Berrebi P. (1994) Behavioural study of hybridization between *Barbus barbus* and *Barbus meridionalis*. *Journal of Fish Biology* **45**, 447–451.

Poncin P., Jeandarme J., Rinchard J. & Kestemont P. (1997) Le comportement de reproduction du goujon, *Gobio gobio*, en aquarium. Premiers résultats. *Bulletin Français de la Pêche et de la Pisciculture* **346**, 547–555 (in French).

Poncin P., Libois M. & Philippart J.C. (1988) Reproduction artificielle réussie de *Labeo bicolor* et de *Labeo frenatus*. *Revue Française d'Aquariologie* **2**, 39–42 (in French).

Poncin P., Mélard Ch. & Philippart J.C. (1989) Control of the reproduction of the chub, *Leuciscus cephalus* (L.) in captivity. In N. De Pauw, E. Jaspers, H. Ackefors & N. Wilkins (eds) *Aquaculture – a Biotechnology in Progress.* Bredene, Belgium: European Aquaculture Society, pp. 567–571.

Povz M. & Ocvirk A. (1990) Breeding and restocking of Danubian roach, *Rutilus pigus virgo* (Heckel). *Journal of Fish Biology* **37**(Suppl. A), 245–246.

Rochard E., Castelnaud G. & Lepage M. (1990) Sturgeons (Pisces: Acipenseridae), threats and prospects. *Journal of Fish Biology* **37**(Suppl. A), 123–132.

Stalmans J.M. & Kestemont P. (1991) Production de juvéniles de vairons européens (*Phoxinus phoxinus*) à partir de larves obtenues in conditions contrôlées. *Bulletin Français de la Pêche et de la Pisciculture* **320**, 29–37 (in French).

Von Lukowicz, Tamas G. & Horvath I. (1986) Aquaculture of Tench. In R. Billard & J. Marcel (eds) *Aquaculture of Cyprinids*. Paris: INRA, pp. 357–367.

Worthington A.D., MacFarlane N.A.A. & Easton K.W. (1982) Controlled reproduction in the roach *Rutilus rutilus*. In H.J.J. Goos & C.J.J. Richter (eds) *Proceedings of the International Symposium on Reproductive Physiology of Fish.* Wageningen: Pudoc, pp. 220–223.

Chapter 37
The role of scientists in creating public awareness for the conservation of fish species: African and American case studies

J.A. CAMBRAY*
Albany Museum, Grahamstown, South Africa

E.P. PISTER
Desert Fishes Council, Bishop, USA

Abstract

Habitat degradation, alien fish introductions and activism from scientists in preventing further loss of freshwater fish biodiversity are discussed, especially in relation to two endangered species in South Africa and California. The Eastern Cape rocky, *Sandelia bainsii*, endemic to several rivers in the Eastern Cape, South Africa is on the verge of extinction due to habitat degradation and the introduction of alien fish species. The golden trout, *Oncorhynchus aguabonita*, endemic to two small streams in the upper reaches of the Kern River in California is seriously threatened by predation from alien brown trout, *Salmo trutta*, hybridisation with introduced rainbow trout, *Oncorhynchus mykiss*, and by habitat destruction. A paradigm shift to pro-active involvement in environmental education and conservation is required to conserve these fish species.

Keywords: conservation, habitat degradation, public involvement, species introductions.

37.1 Introduction

In the IUCN Red List of Threatened Animals (Groombridge 1993) it was noted that wherever surveys are made, fishes are found to be threatened, usually by introduced non-native fish species or habitat modifications. Cambray (2000a) confirmed this trend by analysing the papers submitted to the Threatened Fishes of the World Series in the journal *Environmental Biology of Fishes*. This chapter uses African and American case studies to consider the role of scientists in reversing these trends. With 30% of freshwater fish species facing extinction within 20–30 years (Cambray & Bianco 1998), turning the tide of extinctions requires scientific involvement at the popular level to create public awareness (e.g. Pister 1999; Cambray 2000b). 'It is not enough that the scientific community has information that will affect the future of our environment. The baton has to be passed from the scientists to the public and then be translated into political and legislative action' (Redford 1991). This chapter examines the importance of increasing public awareness about the relevance of habitat degradation

*Correspondence: Albany Museum, Grahamstown, South Africa (email: J.Cambray@ru.ac.za).

and species introductions in the conservation of freshwater fishes, with particular reference to case studies in South Africa and California.

37.2 Habitat degradation

37.2.1 *The importance of habitat conservation*

Cambray & Bianco (1998), amongst others, noted that 'Conserving freshwater fish is challenging because aquatic ecosystems are poorly understood, relatively unmanageable, shared by multiple users and highly vulnerable to human interventions.' Aquatic systems appear to deteriorate faster than terrestrial systems and are therefore more vulnerable (Stiassny 1998).

Most recent extinctions are associated with human activities (Smith, May, Pellew, Johnston & Walter 1993; Magin, Johnson, Groombridge, Jenkins & Smith 1994), and this has been confirmed for freshwater fishes (Harrison & Stiassny 1999). Habitat modification was considered responsible for the extinction of 71% of the fish species in the world (Harrison & Stiassny 1999) and 73% of the 40 taxa of North American extinct fishes (Miller, Williams & Williams 1989). This shows that habitat management is a critical action in the conservation of fish species. 'Conservation efforts must be derived from an evolutionary perspective that appreciates the need for continued adaptive change in all populations' (Meffe 1990).

Two guidelines were proposed for rare, native fish conservation activities in the United States. Firstly, natural resource agencies should manage the entire ecosystems, which is a commitment to long-term conservation management, rather than the recovery of individual species. Secondly, long-term monitoring programmes should be established to provide baseline data for future assessment of any changes in fish populations and aquatic habitats (Williams, Johnson, Hendrickson, Contreras-Balderas, Williams, Navarro-Mendoza, McAllister & Deacon 1989).

Aquatic scientists must encourage the long-term protection of entire river corridors with regard to their hydrological patterns, quantity and quality of water, floodplain, forests, rare species and natural communities. This requires raising awareness in all communities whose socio-economic requirements are inseparable from the quality of the environment.

37.2.2 *South African case study –* **Sandelia bainsii** *Castelnau habitat*

The river corridor approach is being used in combination with a Red Flag species, *S. bainsii*, to demonstrate the deterioration of the Kowie River catchment (Cambray 2000b). Invasive terrestrial alien vegetation has desiccated the system; alien aquatic vegetation (fed by nutrient loading) has smothered refuge pools; excessive removal of natural riverine bush for pineapple plantations has led to further desiccation of the catchment; farmers allowed temporary grazing rights have damaged riverine areas during drought periods; and excessive water abstraction for cash crops has changed the river. The deterioration of the habitat has led to the dramatic decrease in *S. bainsii* over the past 20 years.

Positive developments have seen the South African Government Working-for-Water Project clear alien plant invaders, the 1998 Water Act takes cognisance of the ecological functioning of rivers, and the invasive water plant has been controlled through biological control.

37.2.3 *Californian case study –* **Oncorhynchus aguabonita** *(Jordan) habitat*

A similar situation is found with the golden trout habitat of the Kern River Plateau, California (E. Pister, unpublished data). This is public land, the Golden Trout Wilderness, administered by the US Forest Service. Since 1860 vast numbers of cattle and sheep were allowed to graze on these high meadows. Damage to the watershed was enormous and still persists. Riparian growth, primarily willow (*Salix* spp.), was destroyed throughout the drainage, water temperatures increased through insolation, spawning gravels became silted up, and other perturbations occurred, creating habitat conditions much more suitable to brown trout than to native goldens.

This public land was leased to private grazing interests, and these conditions persisted until the 1960s, when government and academic scientists coordinated rehabilitative management focussed around: reduced cattle numbers, removal of sheep from the high meadows, rehabilitation of eroded areas, and fencing the stream to preclude cattle and allow willows and riparian grasses to reinstate shade and reduce siltation and pollution. This demonstrated to the grazing permittees the value of sound management actions.

37.3 Alien fish introductions

37.3.1 *Legacy of hatcheries and alien sport fishes*

In many countries there has been a progressive decrease in the percentage of native *v.* alien taxa (Cambray & Bianco 1998) and competition and/or predation from these alien species accounts for 51% of the species extinctions in the world (Harrison & Stiassny 1999). In Africa, problems caused by invasive aquatic animals are well documented (e.g. Welcomme 1984; CIFA 1985; Bruton & van As 1986; MacDonald, Kruger & Ferrar 1986; de Moor & Bruton 1988; Moureau, Arrignon & Jubb 1988; Cambray & Bianco 1998). Hatcheries, built to produce and distribute alien sport fishes, contribute to this loss of native fish species.

37.3.2 **Sandelia bainsii** *and alien fish species*

Hey (1977) referring to the South African Cape Province stated '... of these (i.e. alien introductions), the trout were undoubtedly the best and most successful, although it is unfortunate they preyed on the small indigenous fishes and eliminated them from many rivers. It is regrettable that a number of endemic species of particular scientific interest have consequently become rare, but in those days one did not think of such matters.' Hey (1977) also referred to Pirie Trout Hatchery, now funded by the Eastern Cape Government at the upper end of the Buffalo River, which still produces trout in spite of conservation policy changes. The hatchery is situated close to streams which contain two of the most endangered species in the Eastern Cape Province, *S. bainsii* and *Barbus trevelyani* Günther. Hatchery staff, over the past 50 years, have done little to study these two endangered native fish species but instead have stocked the rivers with alien sport fish. It is essential these hatcheries do not release any more aliens into such sensitive areas.

Sandelia bainsii are also threatened by the sharptooth catfish, *Clarias gariepinus* (Burchell), which was translocated by an interbasin transfer connecting the Orange and Fish rivers (Cambray & Jubb 1977). *Clarias gariepinus* are now moved from one Eastern Cape river to the next by anglers and prey on *S. bainsii*. *Micropterus salmoides* (Lacépède) and *Micropterus dolomieu* (Lacépède) as well as *Tilapia sparrmanii* Smith are also impacting on the few remaining populations of *S. bainsii*.

37.3.3 Oncorhynchus aguabonita *and alien fish species*

The underlying problem involving California's native golden trout began during the early 1900s, when alien fishes were introduced widely throughout the western United States with no consideration of the biological consequences. Both brown and rainbow trout were stocked in the lower reaches of the South Fork Kern River of California's southern Sierra Nevada, downstream from what were thought to be impassable natural barrier falls. Both species made their way past these barriers, likely in buckets carried by uninformed anglers more interested in angling for big fish than in maintaining aquatic biodiversity. When the magnitude of the problem was recognised during the 1960s, brown trout had nearly eradicated the smaller and more vulnerable golden trout, outnumbering them in some locations by a ratio of nearly 100 : 1.

Adding to the problem was a similar invasion of rainbow trout, which hybridise readily with the closely related golden trout, creating the insidious problem of hybridised offspring virtually impossible to detect without sophisticated genetic analysis.

A massive campaign (including extensive barrier construction and chemical eradication of all fishes, then restocking with pure golden trout) was initiated in 1965 and is still underway on more than 160 km of stream. Significant progress has been made in resolving the problem, but annual monitoring will be necessary to detect any new invaders.

37.3.4 *Role of anglers*

In South Africa there are endeavours to wean trout and bass anglers to indigenous fish species which have angling potential. Motivational elements which a trout angler seeks can be found when angling for indigenous species, i.e. solitude, companionship, escapism, nature appreciation, outdoor skills, the wish to outwit a large fish and to get exercise, and this needs emphasising through extension procedures. These extension processes must involve the angling public, perhaps best accomplished through media advertising, organised angling clubs and, ultimately, peer pressure.

In California the major support for the management and retention of native fish fauna is provided by organised groups. The North American Native Fishes Association, the Aguabonita Fly Fishers and others fly fish for indigenous fishes, primarily golden trout – and show disdain for the pursuit of any alien fish. Unfortunately *S. bainsii* in Cape Province is not known as an angling species and so the task of awareness of its conservation value is all the more difficult.

Resistance to any change in the angling *status quo* may be expected to be widespread and strong, but changes are required in the interests of the basic resource and the aspirations of future generations. These goals are likely to be accomplished in generations rather than years unless scientists actively bring the problems to the attention of the public.

37.4 The required activism

37.4.1 *Scientists*

When the two case studies in the two countries are compared, there are similarities and differences (Table 37.1), but human impacts remain a common theme. How these are dealt with may be construed as politics.

Sandelia bainsii is slipping into extinction whilst the golden trout has received a measure of protection due to good government financial backing and legislation to prevent their extinction. Consequently, the need is not merely more scientific information but sharing it with all sections of the community, including politicians.

Scientists are only part of this holistic perspective and their most important role is to provide the factual information on which value-decisions can be made. This information should then be used to encourage the general public to take action to conserve biodiversity within the social and political framework available (e.g. Endangered Species Acts, Environmental Impact Assessments).

Scientists must define the problems clearly and suggest solutions. Developers can plan wisely under the watchful eye of the electorate, provided they are aware of the issues at stake. Consequently, training of scientists must include elements of social and economic sciences to provide the necessary skills to tackle the problem of information dissemination.

37.4.2 *Challenging the comfort zone of scientists*

The kind of activism in popular articles and at controversial meetings requires a bold conservation stand and a clear understanding of the issues involved. Pister (1999)

Table 37.1 Comparative table of the similarities and dissimilarities of management aspects of conserving *S. bainsii* in South Africa and *O. aguabonita* in the USA

	South Africa	USA
Species	*S. bainsii*, Eastern Cape rocky	*O. aguabonita*, golden trout
Government		
Type of government	New democracy (since 1994)	Established democracy
Rio convention	Signatories	Signatories
Responsibility	Provincial government	State and federal
Endangered Species Act	Specific act required Species status is highlighted in Red data list	Positive and strong act
Government action	Very little	Expensive eradication of alien fish campaign
Factors limiting intervention	Lack of funds Lack of trained manpower Lack of good policy	Lack of funds Bureaucratic inertia
Public		
Publicity potential	Named after a local chief, Chief Sandile, largest African anabantid Moving alien fish into *S. bainsii* habitat, e.g. *C. gariepinus*	Good profile, California's state fish – angling
Anglers – negative	*S. bainsii* not an angling species	Move alien fish above barriers and into *O. aguabonita* habitat
Anglers – positive	Notified scientists of new distribution record	Form clubs for angling for this species, members against alien trout
Public awareness	Not receptive to aquatic environmental issues	Receptive to environmental concerns
Farmers/ranchers	Mostly negative as the species does not generate income	Farmers which graze animals in watershed have reduced stock numbers to protect rivers
Scientists	Actively studying the species, genetics, life history Need long-term monitoring study	Studies on genetics, life history Constant monitoring of populations
General		
Financial resources available	Minimal resources to manage endangered species and their habitats	Resources are available to manage endangered species and their habitats
Alien fish threats	Trout, bass, catfish	Brown and rainbow trout
The future scenarios	Not good, too few resources, too many threats and growing human population as well as spread of alien fish	Reasonably good if constantly monitored
Recommendations	Strong endangered species act National government support Good provincial government backing Good government policy and action Extend Working-for-Water project to include ichthyofauna conservation Long-term monitoring Continued public awareness campaign	Enforcement of Endangered Species Act Continued monitoring Continued public awareness campaign Allocation of more funds

argued that some scientists will shy away from getting involved with conservation issues as this might 'compromise both objectivity and reputation'. However, a biologist should '... pass on to future generations a biota essentially undiminished from that inherited from my predecessors' (Pister 1985).

37.4.3 *The powerful alliance – public understanding supported by scientific findings*

'We live in a time characterised by intense self-centeredness. When society begins to place its short-term self-interest ahead of the integrity of its own environment, it approaches the dangers of going the wrong way on a freeway during rush hour. The entire situation assumes a long-term significance comparable to nuclear war' (Erman & Pister 1989). To conserve some of the genetic diversity of aquatic organisms, Pullin (1990) suggested that there should be increases in funding for training, field research, database development and awareness literature.

If scientists communicate their concerns about disappearing freshwater fishes to the public then public understanding, backed by research findings, can form a powerful alliance. Holcík (1990) stated that 'In general no protective measure, except education, was effective ...' when he was discussing huchen, *Hucho hucho* (L.), conservation.

Well established philosophical frameworks or paradigms driving a system are always difficult to overcome and change (Kuhn 1970). There is a need to change people's attitude to fish. Pullin (1990) suggested that if we answer the question 'Why bother?' with the answer 'Because it is the right thing to do' this will not work with most people. The answer should be, 'because, if we fail, it will cost in diminished quality of life ... '. Creating public awareness for freshwater fish is not easy, thus it should be a duty of the scientific community to devote time to public awareness and changing the public's attitude. Cambray (1993, 2000b) and Cambray and Bianco (1998) showed that various aquarist groups are willing to assist conservation efforts. Aquarists themselves have suggested they take a wider perspective and understand what is happening in nature (Huntley 1996). Organisations such as the Anabantoid Association of Great Britain and the International Labyrinth Society are already giving support to the *S. bainsii* awareness project (Cambray 2000b). The World Wide Web is a powerful tool to achieve this objective, see the WWF-SA web page on *S. bainsii* (http://www.ru.ac.za/affiliates/am/wwf/wwf.htm) and the Desert Fishes Council page (http://www.desertfishes.org/) for examples.

37.5 Conclusions

Differences between the conservation efforts to conserve *S. bainsii* in South Africa and *O. aguabonita* in the USA lie mainly in the government approaches to endangered freshwater fish species although allocation of sufficient funds is a common problem (Table 37.1). South Africa has more complex social patterns, which are more difficult for a scientist to grapple with. The USA has a more sophisticated political system with a long history of environmental activism. In South Africa the conservation issues are

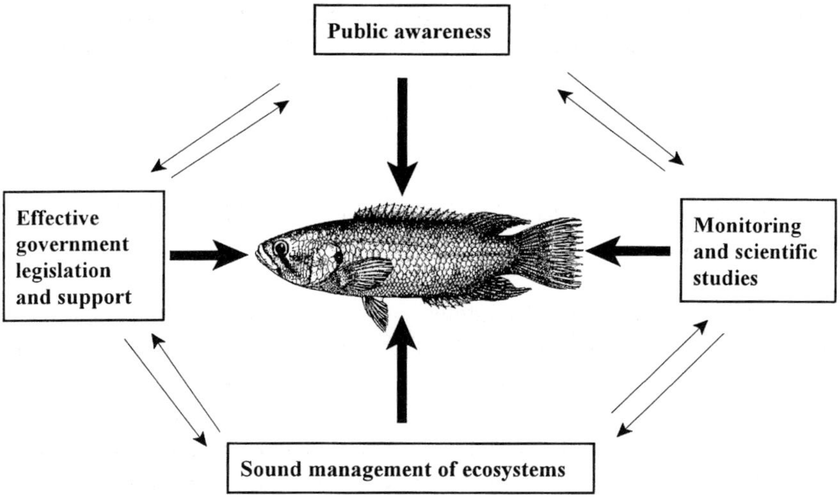

Figure 37.1 Conservation of a fish species using the holistic approach which demonstrates some of the key elements and their linkages. Ichthyologists must play a dominant role in linking these boxes

complicated by the political system where less than 10 years ago the majority did not vote. Conservation is a low political priority amongst people struggling to mobilise support for issues like housing, education and health.

Not withstanding, underlying the issue of conservation is the need to mobilise public awareness. It is not enough to produce scientific information, it must be disseminated to all sections of the community (Fig. 37.1). Good scientific research, clear government policies and policing will only succeed when supported by public sentiment. Furthermore, the public are known to appreciate hearing the opinion of scientists when backed by facts, even on controversial issues (e.g. Cambray 2000c).

Scientists must therefore not limit themselves to scientific publications, the facts and knowledge must also be actively disseminated to the public and policy makers.

Acknowledgements

The National Research Foundation of South Africa is thanked for financial support and a special travel grant which made it possible for J.A. Cambray to participate at the Freshwater Fish Conservation conference in Portugal, October 2000. WWF-SA supported the *Sandelia* project through the Ubuntu Welfare and Development Trust. Companies such as Total SA, Mazda, Eheim and Tetra are also thanked for their contributions to the *Sandelia* project. We would also like to thank various aquarist groups, such as the Anabantoid Association of Great Britain and the International Labyrinth Society for their positive support of the *S. bainsii* conservation campaign. Angling bodies such as the Aguabonita angling club are thanked for their efforts and contribution to the conservation of the golden trout. Ms Eve Cambray is thanked for her logic, editorial skills and encouragement.

References

Bruton M.N. & van As J.G. (1986) Faunal invasions of aquatic ecosystems in southern Africa, with suggestions for their management. In I.A.W. MacDonald, F.J. Kruger & A.A. Ferrar (eds) *The Ecology and Management of Biological Invasions in Southern Africa*. Cape Town: Oxford University Press, pp. 47–61.

Cambray J.A. (1993) Ein labyrinthfisch ohne labyrinth braucht überlebenshilfe, *Sandelia bainsii*. *DATZ* No. **2**/93, pp. 118–119.

Cambray J.A. (2000a) 'Threatened fishes of the world' series, an update. *Environmental Biology of Fishes* **59**, 353–357.

Cambray J.A. (2000b) Community involvement and the Eastern Cape rocky (*Sandelia bainsii*, Pisces, Anabantidae) from the perspective of a researcher. *South African Journal of Wildlife Research* **30**, 114–117.

Cambray J.A. (2000c) The Crags and the alien trout issue – fish facts. *CXpress* 15–31 October 2000, **58**, p. 9.

Cambray J.A. & Bianco P.G. (1998) Freshwater fish crisis: a blue planet perspective. *Italian Journal of Zoology* **65**(Suppl.), 345–356.

Cambray J.A. & Jubb R. (1977) Dispersal of fishes via the Orange-Fish Tunnel, South Africa. *Journal of the Limnological Society of Southern Africa* **3**, 33–35.

CIFA (1985) Introduction of species and conservation of genetic resources. *Report of the Committee for Inland Fisheries of Africa*, Report No. 85/13, pp. 1–18.

de Moor I. & Bruton M.N. (1988) Atlas of alien and translocated indigenous aquatic animals in southern Africa. *South African National Scientific Programmes*, Report No. 144. Pretoria: CSIR, 310 pp.

Erman D.C. & Pister E.P. (1989) Ethics and the environmental biologist. *Fisheries* **14**(2), 4–7.

Groombridge B. (ed.) (1993) *1994 IUCN Red List of Threatened Animals*. Gland: IUCN, 286 pp.

Harrison I.J. & Stiassny M.L.J. (1999) The quiet crisis, a preliminary listing of the freshwater fishes of the world that are extinct or 'missing in action.' In R.D.E. MacPhee & H.-D. Sues (eds) *Extinctions in Near Time*. New York: Academic/Plenum Publishers, pp. 271–329.

Hey D. (1977) *Water is Life*. Cape Town: Oxford University Press, 138 pp.

Holcík J. (1990) Conservation of the huchen, *Hucho hucho* (L.), (Salmonidae) with special reference to Slovakian rivers. *Journal of Fish Biology* **37**(Suppl. A), 113–121.

Huntley R. (1996) The future of fishes. *Aquatic Survival* **5**, 1–6.

Kuhn T.S. (1970) *The Structure of Scientific Revolutions*, second edition, enlarged. International Encyclopaedia of Unified Science, Vol. 2, No. 2. Chicago: The University of Chicago Press, 210 pp.

MacDonald I.A.W., Kruger F.J. & Ferrar A.A. (1986) *The Ecology and Management of Biological Invasions in Southern Africa*. Cape Town: Oxford University Press, 324 pp.

Magin C.D., Johnson T.H., Groombridge B., Jenkins M. & Smith H. (1994) Species extinctions, endangerment and captive breeding. In P.J.S. Olney, G.M. Mace & A.T.C. Feistner (eds) *Creative Conservation: Interactive Management of Wild and Captive Animals*. London: Chapman & Hall, pp. 3–31.

Meffe G.K. (1990) Genetic approaches to conservation of rare fishes: examples from North American desert species. *Journal of Fish Biology* **37**(Suppl. A), 105–112.

Miller R., Williams J. & Williams J. (1989) Extinctions of North American fishes during the past century. *Fisheries* (Bethesda) **14**(6), 22–39.

Moureau J., Arrignon J. & Jubb R.A. (1988) Les introductions d'espèces étrangères dans les eaux continentales africaines. Intérêts et limites. In C. Lévêque, M.N. Bruton & G.W. Ssentongo (eds)

Biologie et Ecologie des Poissons d'eau douce Africains/Biology and Ecology of African Freshwater Fishes. Paris: ORSTOM, pp. 395–425.

Pister E. (1985) Desert pupfishes: reflections on reality, desirability, and conscience. *Environmental Biology of Fishes* **12**, 3–11.

Pister E. (1999) Professional obligations in the conservation of fishes. *Environmental Biology of Fishes* **55**, 13–20.

Pullin R. (1990) Down-to-earth thoughts on conserving aquatic genetic diversity. *NAGA* January 1990, 5–8.

Redford R. (1991) Message/endorsement. In J. Porritt (ed.), *Save the Earth*. London: Dorling Kindersley Ltd, p. 45.

Smith F.D.M., May R.M., Pellew R., Johnston T.H. & Walter K.R. (1993) How much do we know about the current extinction rate? *Trends in Ecology & Evolution* **8**, 375–378.

Stiassny M. (1998) The medium is the message: freshwater biodiversity in peril. In J. Cracraft & F. Grifo (eds) *The Living Planet in Crisis: Biodiversity Science and Policy*, New York: Columbia University Press, pp. 53–71.

Welcomme R.L. (1984) International transfers of inland fish species. In W.R. Courtenay Jr & J.R. Stauffer Jr (eds) *Distribution, Biology, and Management of Exotic Fishes*, Baltimore: John Hopkins University Press, pp. 22–40.

Williams J.E., Johnson J.E., Hendrickson D.A., Contreras-Balderas S., Williams J.D., Navarro-Mendoza M., McAllister D.E. & Deacon J.J. (1989) Fishes of North America endangered, threatened, or of special concern: 1989. *Fisheries* (Bethesda) **14**(6), 2–20.

Chapter 38
How far could a monitoring programme contribute to species conservation? An example of the nase, *Chondrostoma nasus*, in Switzerland

D. HEFTI*

Fisheries Section, Swiss Agency for the Environment, Forests and Landscape, Bern, Switzerland

Abstract

A nation-wide monitoring programme targeting the nase, *Chondrostoma nasus*, was started by the Federal Fisheries Authority in Switzerland in 1995. Most attention was focused on the breeding sites as these represent indicators for recruitment. Between 1995 and 1999, 26 breeding sites, containing 72 spawning grounds were recorded in the Swiss catchment area of the Rhine. The preferred spawning substrate was stones ranging from 3 to 8 cm in diameter. The number of breeding sites recorded was low compared with the 57 reproduction sites known historically, which confirms the decline of the species in Switzerland. Spawning activity was not confirmed at each site throughout the whole period investigated. Precise mapping of the spawning grounds at the regional level identified hot spots where protection measures are needed as a matter of priority at the local level.

Keywords: conservation, monitoring, nase, reproduction.

38.1 Introduction

The nase, *Chondrostoma nasus* (L.), is a litho-rheophilic cyprinid. In Switzerland, the species inhabits the grayling, *Thymallus thymallus* (L.), and barbel, *Barbus barbus* (L.) zones of lowland streams in the Rhine catchment. This zone has been severely altered by systematic damming of rivers and interruption of the natural transport of sediment, which has a strong impact on the availability of the ecological requirements of the nase. The species has little or no relevance to recreational fishing, so it is not regularly restocked. In the Swiss Red Data Book, *C. nasus* is considered highly endangered (Kirchhofer, Zaugg & Pedroli 1994).

In 1995, the Federal Fisheries Authority in Switzerland started a nation-wide monitoring programme for nase, based on the poor status of the wild populations in national waters (Maier, Zeh, Ortlepp, Zbinden & Hefti 1995). Monitoring focused on the breeding sites of the species, as they represent indicators for recruitment, to establish a list

*Correspondence: Fisheries Section, Swiss Agency for the Environment, Forests and Landscape, CH-3003 Bern, Switzerland (email: daniel.hefti@buwal.admin.ch).

of the current reproduction areas. This chapter presents the results of the monitoring programme and assesses their relevance for the conservation of the species.

38.2 Materials and methods

Between 1995 and 1999, local fisheries authorities in each of the Swiss cantons were asked to record any spawning activities of *C. nasus* on its territory. The following information was recorded each year: name of the river; co-ordinates of the spawning grounds; date of the observation; and number of spawners (estimate). Water temperature was also recorded.

During the spawning season, April to May, stretches of rivers where the reproduction of *C. nasus* might take place were visited regularly. Direct observations of spawning behaviour (quite unambiguous in the case of the nase) or the presence of eggs on the substrate were considered proof that a site was a spawning area. The information gathered was centralised and mapped at the national and local levels for reproduction sites and spawning grounds respectively.

A reproduction site was defined as a stretch of river within which fish can migrate freely, and where the reproduction of *C. nasus* was observed at least once during the monitoring period. The lengths of the reproduction sites ranged from a few hundred metres to about 3 km. A spawning ground was defined as a locality on the riverbed where reproduction of the nase took place.

The substrate of 10 spawning grounds was determined using the method of Fehr (1987), modified to take into account 1-cm particle size classes. Only the surface layer was investigated.

38.3 Results

The spawning period ranged from mid-April to mid-May (≈90% of the cases), with water temperatures from 10°C to 14°C. No clear pattern in the distribution of spawning time and water temperatures over the years was observed. The latest spawning activity was recorded in the River Aare (site no. 15) between 19 and 21 June 1995.

Altogether, 26 reproduction sites of *C. nasus* were identified throughout the whole Swiss Rhine catchment area (Fig. 38.1). Each site contains at least one spawning ground. At only six sites (sites nos 1, 17, 18, 20, 21 and 25) was the reproduction of *C. nasus* observed every year. These sites were considered the most important reproductive areas for the species. At six other sites (site nos 5, 6, 8, 22, 23, and 26) observation of spawning activity was not confirmed over the whole period investigated because of high water discharges during the spawning season. At the other sites, reproduction did not take place regularly, as no positive identification of spawning activity was recorded in all years, despite good hydrological conditions.

A total of 72 spawning grounds of *C. nasus* were identified during the monitoring period. To visualise their location, each reproduction site was mapped individually at a large scale (1 : 20 000). For example, Figure 38.2 shows a map of reproduction site

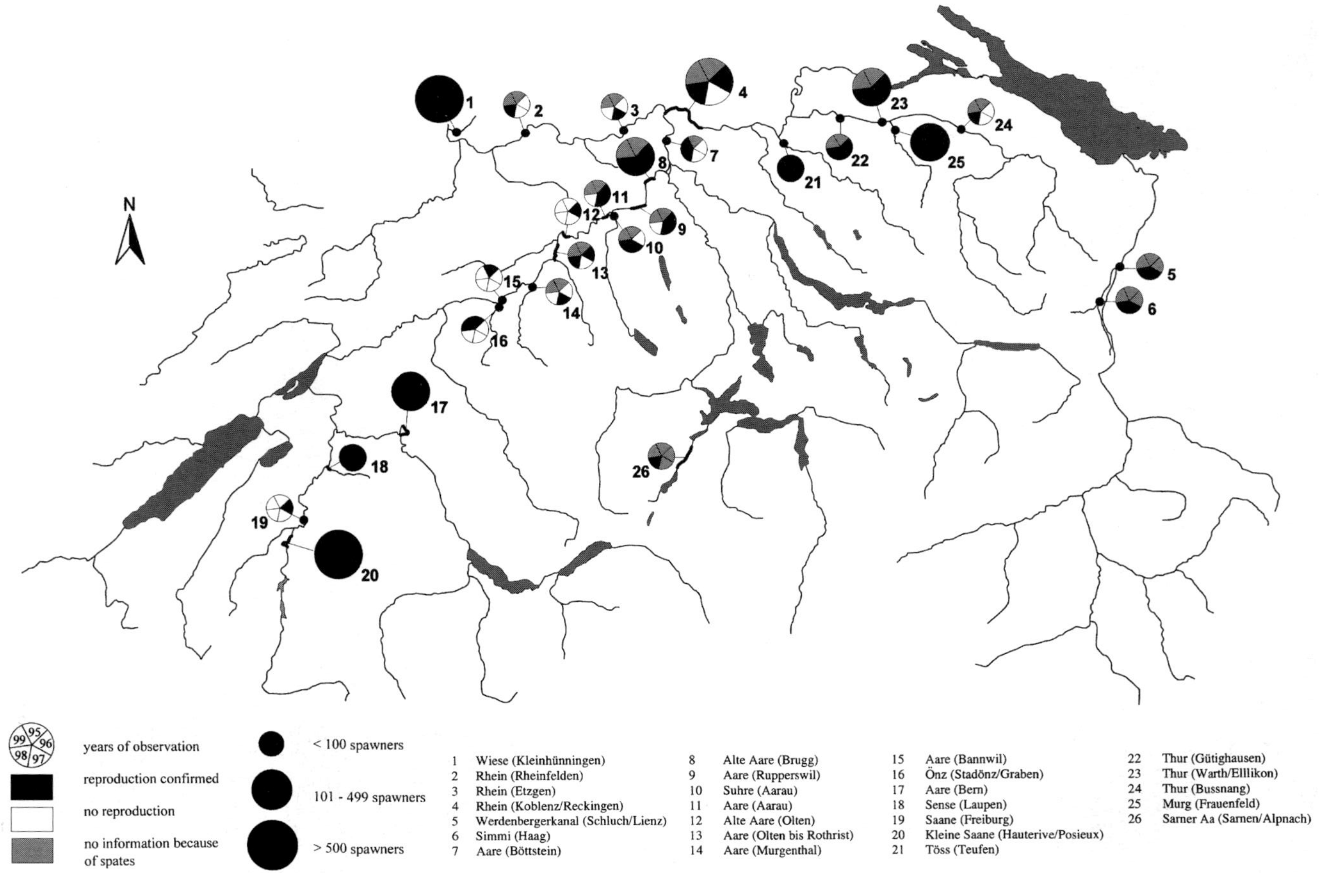

Figure 38.1 Location of the 26 reproduction sites with gross estimate of the number of spawners involved yearly (indicated by the size of the circle) and occurrence of reproduction in each of the 5 years (slices in the circle)

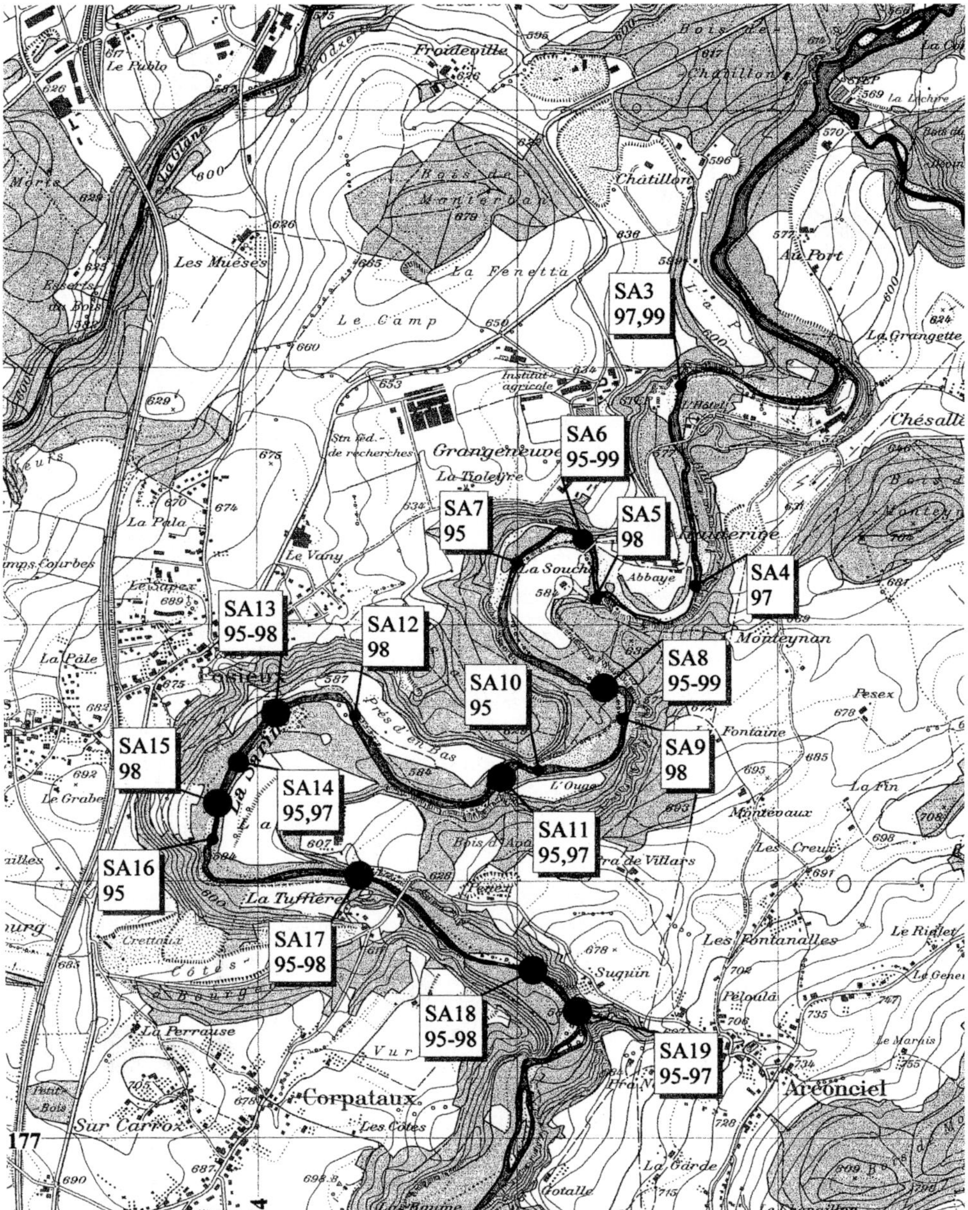

Figure 38.2 Map of the site number 20 at regional scale (1 : 20 000) showing the location of the 17 spawning grounds. 95–99 – spawning activity recorded from 1995 to 1999; 95, 97 – spawning activity recorded in 1995 and 1997

number 20 containing 17 spawning grounds. The location of the spawning grounds may change slightly from year to year because of the natural dynamics of the system. Spawning grounds of *C. nasus* were situated in shallow water with high water velocity. In rivers with altered morphology, spawning grounds were always located in stretches of rivers that remained more or less natural. The composition of the sediment in the upper layer of 10 spawning grounds was analysed (Fig. 38.3) indicated the mean diameter of the substrate ranged from 3.4 to 8 cm, and the dominant fraction size was 2–8 cm. Some spawning grounds exhibited a quite narrow distribution of substrate size (site nos 4, 17 and 21); others had a broader distribution with stones up to 20 cm in diameter (site nos 8 and 20). The variability observed among the 10 spawning grounds can be explained by differences in the sizes of the rivers, and hence discharge.

38.4 Discussion

Penàz (1996) reported a marked decline of nase populations in Central European rivers where the species was formerly one of the most abundant fishes. Comparing the present and historical distributions of *C. nasus* in Switzerland, Zbinden & Maier (1996) came to the same conclusion. Although 57 reproduction sites were known historically, Zbinden and Maier (1996) recorded only 14 sites with 25 spawning grounds, and one site where reproduction was considered as probable. Of the 26 reproduction sites identified during the present monitoring, Zbinden and Maier (1996) mentioned 11 as functional. By contrast, three reproduction sites mentioned by Zbinden and Maier (1996) have not been confirmed since. The differences between Zbinden and Maier (1996) and the present monitoring are probably because the former study only investigated one spawning year (1993). The present monitoring, carried out over 5 years, allowed 35 additional spawning grounds to be identified, belonging to 15 reproduction sites. The present number of reproduction sites was less than half of the historical average.

The monitoring programme gives only a broad estimate of the reproduction potential of *C. nasus* in Switzerland, and has to be interpreted with caution. The collection of data was probably not homogeneous throughout the country because it is dependent on the commitment of the various local fisheries authorities and societies. Another problem was the difficulty following each identified reproduction site over the whole period, because of the highly variable water discharges ordinarily occurring in the spring. Finally, the relation between the number of spawning grounds and the effective recruitment of the species is only partial, because no information was obtained about the viability of the eggs in the gravel (siltation of the substrate was not investigated) and the post-embryonic survival of the species.

The morphological parameters of the spawning grounds of *C. nasus* found in the present study were similar to those observed by Lelek and Penàz (1962), Dedual (1990), Maier *et al.* (1995), and Hofer and Kirchhofer (1996). Two other *Chondrostoma* species live in distinct catchment areas of Switzerland: *Chondrostoma toxostoma* (Vallot) in the Rhône basin and *Chondrostoma soetta* Bonaparte in the Pô basin. No difference was found between the spawning grounds of *C. nasus* and *C. toxostoma*. By contrast, slight differences of substratum composition were observed for *C. soetta*,

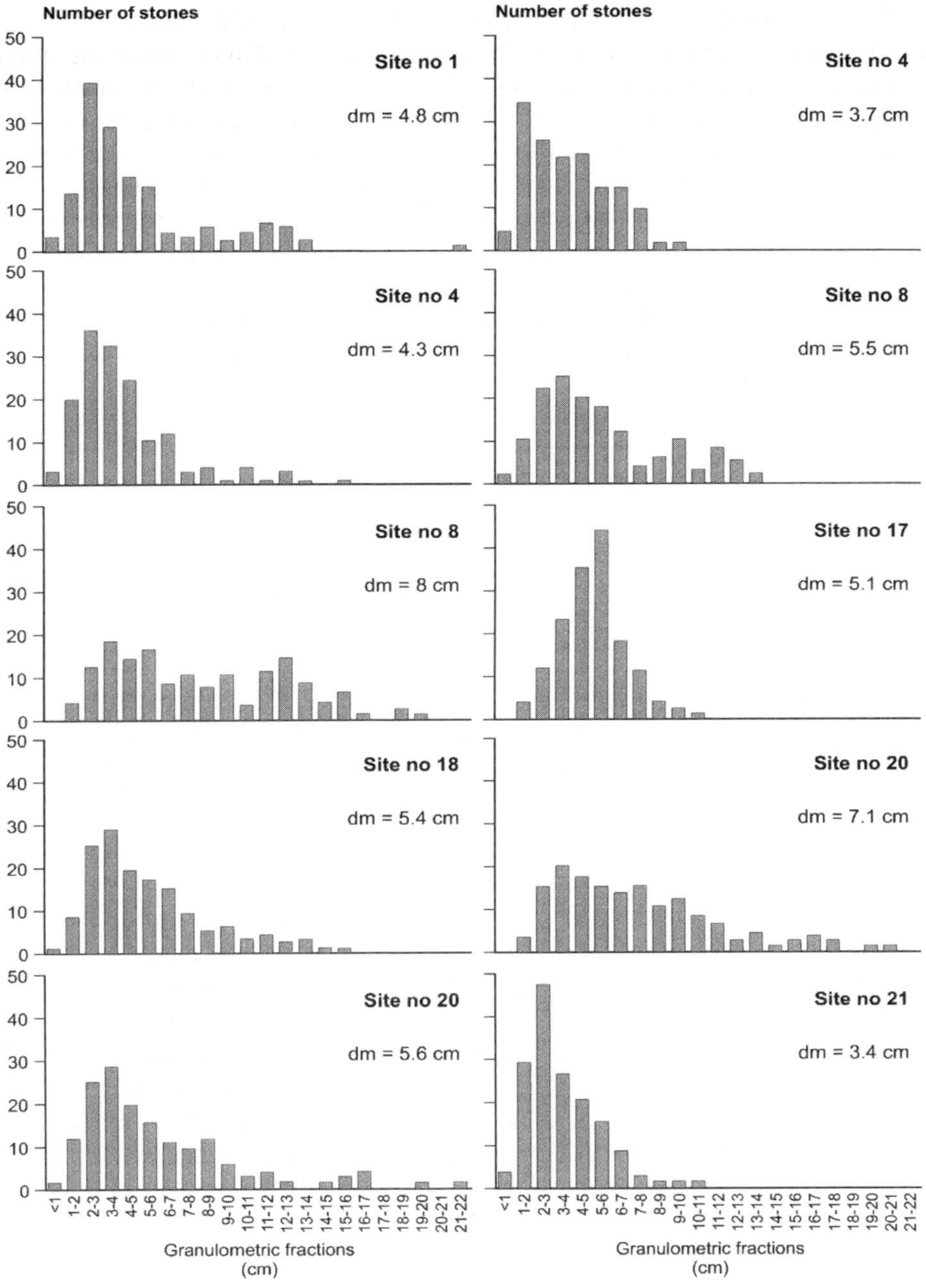

Figure 38.3 Granulometric fractions of 10 spawning grounds of *C. nasus* belonging to different sites. dm – median diameter of stones

which spawned not only on gravel, but also on submerged vegetation, sand, rocks and even concrete slabs (Zeh & Ortlepp 1996). The preservation of stretches of river with adequate substrate type and transport capacity is an important factor to ensure the survival of the nase in Switzerland. The conservation of *C. nasus* also needs further efforts to improve stretches of river that have been channelised or impounded.

The present monitoring programme is a first attempt to map the reproduction sites of an endangered species at the national level. However, one may question how far this contributes to the conservation of the species. Mapping a species does not prevent it from becoming extinct. The main goal of the study was to provide a useful tool for local managers. This needs to be built into a conservation strategy for *C. nasus* to re-establish the connection between the main streams and their tributaries, to protect the spawning areas and to enhance habitat diversity (Maier *et al.* 1995). In this sense, the mapping gives precious information to set priorities when planning environmental restoration measures. The mapping of each spawning ground at the local level precisely identifies hot spots, allowing the manager to take adequate protective measures, for example, during technical operations in the riverbed.

Acknowledgements

I am very grateful to all the authorities, technicians, volunteers and other people involved in the field work during the monitoring period. The assistance provided by Stephane Zbinden for the synthesis of the present monitoring is gratefully acknowledged.

References

Dedual M. (1990) *Biologie et problème de dynamique du nase* (Chondrostoma nasus nasus) *dans la Petite Sarine*. PhD Thesis, University of Fribourg, Switzerland, 168 pp. (in French).

Fehr R. (1987) Einfache Bestimmung der Korngrössenverteilung von Geschiebematerial mit Hilfe der Linienzahlanalyse. *Schweizer Ingenieur und Architekt* **38**, 1104–1109 (in German).

Hofer K. & Kirchhofer A. (1996) Drift, habitat choice and growth of the nase (*Chondrostoma nasus*, Cyprinidae) during early life stages. In A. Kirchhofer & D. Hefti (eds) *Conservation of Endangered Freshwater Fish in Europe*. Basel: Birkhäuser Verlag, pp. 269–278.

Kirchhofer A., Zaugg B. & Pedroli J.C. (1994) Rote Liste der Fische und Rundmäuler der Schweiz. In P. Duelli (ed.) *Rote Listen der gefährdeten Tierarten der Schweiz*. Bern: EDMZ/BUWAL, pp. 35–37 (in German).

Lelek A. & Penàz M. (1962) Spawning of *Chondrostoma nasus* (L.) in the Brumovka river. *Zoologické Listy* **12**, 121–134.

Maier K.J., Zeh M., Ortlepp J., Zbinden S. & Hefti D. (1995) Distribution et reproduction des espèces du genre *Chondrostoma* en Suisse: le nase (*C. nasus*), la sofie (*C. toxostoma*) et la savetta (*C. soetta*). *Informations concernant la pêche* **53**, OFEFP, 62 pp. (in French).

Penàz M. (1996) *Chondrostoma nasus* – its reproduction strategy and possible reasons for a widely observed population decline – a review. In A. Kirchhofer & D. Hefti (eds) *Conservation of Endangered Freshwater Fish in Europe*. Basel: Birkhäuser Verlag, pp. 279–285.

Zbinden S. & Maier K.J. (1996) Contribution to the knowledge of the distribution and spawning grounds of *Chondrostoma nasus* and *Chondrostoma toxostoma* (Pisces, Cyprinidae) in Switzerland. In A. Kirchhofer & D. Hefti (eds) *Conservation of Endangered Freshwater Fish in Europe*. Basel: Birkhäuser Verlag, pp. 287–297.
Zeh M. & Ortlepp J. (1996) Spawning grounds of *Chondrostoma soetta* Bonaparte, 1840 (Cyprinidae) in Southern Switzerland. In A. Kirchhofer & D. Hefti (eds) *Conservation of Endangered Freshwater Fish in Europe*. Basel: Birkhäuser Verlag, pp. 299–304.

Chapter 39
A conservation plan for the unique and highly threatened freshwater fishes of the Cape Floral Kingdom

N.D. IMPSON*
Western Cape Nature Conservation Board, Stellenbosch, South Africa

I.R. BILLS
JLB Smith Institute of Ichthyology, Grahamstown, South Africa

J.A. CAMBRAY
Albany Museum, Grahamstown, South Africa

Abstract

Areas of critical importance to freshwater fish conservation in the Cape Floral Kingdom (CFK) were identified in the Cape Action Plan for the Environment (CAPE), a recently completed strategic biodiversity conservation planning exercise for the CFK. The CFK is the richest (species per unit area) and most threatened of the world's six plant kingdoms. Indigenous freshwater fishes are a priority group for conservation within the CFK because 16 of the 19 species (84%) are endemic with 15 species (79%) listed as threatened. The majority of species are not effectively conserved in CFK national parks and nature reserves as either most of their distribution range is outside these areas or indigenous fishes species share protected area habitat with invasive alien fishes such as the predatory smallmouth blackbass *Micropterus dolomieu*. A project to more effectively conserve rivers and indigenous fishes was developed as part of the CAPE. This project includes land acquisition, addressing the shortage of manpower and funding at conservation authorities, eradicating invasive fishes and plants from key areas and increasing funding for research and monitoring.

Keywords: Cape Floral Kingdom, conservation, threatened endemic fishes.

39.1 Introduction

The Cape Floral Kingdom (CFK) (Fig. 39.1) is recognised as a centre for a distinct 'Cape' component of the ichthyofauna of Africa (Skelton 1983). The CFK's ichthyofauna is dominated by cyprinids (15 species or 79% of the fauna) of the genera *Pseudobarbus*, *Barbus* and *Labeo*. Two austroglanidids, an anabantid and a galaxiid, are also found here.

The ichthyofauna of this mainly Mediterranean-type climate region is characterised by isolated and geographically-restricted ranges, high levels of endemicity, inflexible

*Correspondence: Western Cape Nature Conservation Board, Private Bag X5014, Stellenbosch, 7599, South Africa (email: impsond@cncjnk.wcape.gov.za).

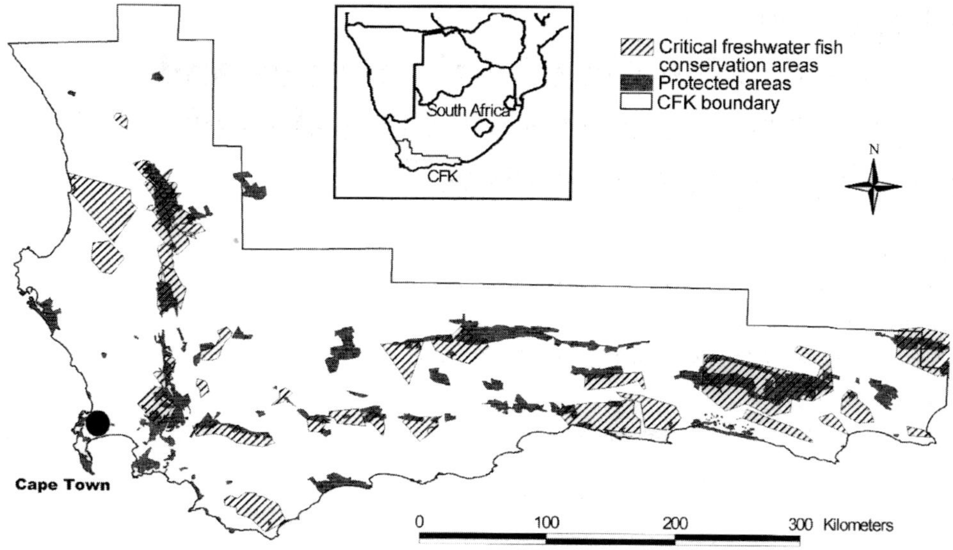

Figure 39.1 Critical areas for fish conservation and existing protected areas in the CFK. The insert shows the location of the CFK within South Africa

life history strategies and a low resilience to disturbance (Skelton 1987). This is reflected by the threatened status of the fishes. Of its 19 primary freshwater fish species, 15 (79%) are listed as threatened, including nine (47%) which are critically endangered or endangered (Baillie & Groombridge 1996). The most notable endemic freshwater fish hotspot in southern Africa is the Olifants River system of the CFK (Skelton, Cambray, Lombard & Benn 1995) which has 10 indigenous fish species, of which eight are endemic and threatened.

The major threats to these fishes are predation by, and competition, with invasive alien fishes, and habitat degradation and destruction by inappropriate agricultural development (Impson & Hamman 2000). The longitudinal nature of rivers makes them extremely difficult to conserve (Davies, O' Keeffe & Snaddon 1993), as ideally whole systems including their catchments require protection (Skelton *et al.* 1995). Rivers, especially their middle and lower reaches, are poorly conserved within the CFK, resulting in their biota encountering ever-increasing levels of anthropogenic disturbance.

This chapter describes a conservation plan to improve the management and conservation status of the rivers and freshwater fishes of the CFK, as part of the recently completed Cape Action Plan for the Environment (CAPE). The aim of this internationally funded and collaborative conservation project was to assess the biodiversity and identify biodiversity hotspots in the CFK, the most threatened of the world's six floral kingdoms. The primary purpose of the conservation plan was to review the distribution records of CFK freshwater fishes and identify river areas of critical importance for their future survival. In addition, key constraints to their conservation were identified. The chapter offers recommendations for the effective conservation of these fishes.

39.2 Materials and methods

Freshwater fish distribution records from the Western Cape Nature Conservation Board (WCNCB), the J.L.B. Smith Institute of Ichthyology and Albany Museum (including South African Museum records) were amalgamated. All records (apart from some WCNCB records) are based on voucher samples. Fish distribution maps were developed for the CFK and adjacent areas using a geographical information system (GIS) package (Arc View 3.1) which included overlays of the rivers and protected areas.

Distribution maps were evaluated for fish diversity (indigenous and alien), endemicity and viable breeding populations of threatened species. From these, critical areas for fish conservation were identified including the associated river catchment. A conservative approach was taken when identifying critical areas since the ichthyofauna is highly threatened and evolutionarily significant units (ESUs) (term after Waples 1991, in Meffe & Carroll 1994) have already been identified, for species such as *G. zebratus* Castelnau (Waters & Cambray 1997) and Berg River redfin, *Pseudobarbus burgi* (Boulenger) (Bloomer & Impson 2000). These areas, although insufficient to conserve the entire genetic diversity of a species, are aimed at conserving the extreme ranges of diversity within each species. This approach enhances future opportunities for formally conserving indigenous species and provides each species a greater capability of withstanding varied environmental impacts than a taxon with limited genetic diversity.

The guidelines of Moyle and Sato (1991) and Skelton *et al.* (1995) were incorporated for designing aquatic protected areas. These areas should:

(1) encompass the entire catchment if possible or alternatively be higher up in the catchment where protection and management is more effective;
(2) maintain natural hydrological cycles to secure the water quantity and quality requirements of the aquatic biota; and
(3) eradicate high-impact alien predators such as largemouth bass, *Micropterus salmoides* (Lacépède) and rainbow trout, *Oncorhynchus mykiss* (Walbaum).

The effectiveness of the CFK's protected area network for freshwater fish conservation was determined by how many protected areas each species had been recorded from. Finally, existing knowledge was summarised on key constraints to the conservation of CFK fishes and their associated ecosystems.

39.3 Results

39.3.1 *Fish diversity, endemicity and conservation status*

Records showed 35 freshwater fish species had been recorded in inland waters of the CFK (Table 39.1). The CFK is characterised by relatively few indigenous fishes (19 species) for an area of its size, but has a richness of endemic species (16 species, 84%). Disturbingly, 16 fish species alien to the CFK were recorded, including 13 species that have established invasive populations. The latter group comprises mainly North

Table 39.1 Freshwater fishes of the CFK

Indigenous fishes	IUCN status[a]	Alien fishes	Estimated effect on biota or habitat (high, moderate or low)
Austroglanis barnardi (Skelton)[b]	CR	*Barbus aeneus* (Burchell)[c]	In Gourits System – M
Austroglanis gilli (Barnard)[b]	VU	*Carassius auratus* (L.)	Urban – L
Barbus andrewi Barnard[b]	VU	*Clarias gariepinus* (Burchell)[c]	Widespread – H
Barbus anoplus Weber	NL	*Cyprinus carpio* L.[c]	Widespread – H
Barbus calidus (Barnard)[b]	EN	*Gambusia affinis* (Baird & Girard)[c]	Localised – M
Barbus capensis (A. Smith)[b]	VU	*Lepomis macrochirus* Rafinesque[c]	Widespread – M to H
Barbus erubescens Skelton[b]	CR	*Micropterus dolomieu* (Lacépède)[c]	Widespread – H
Barbus pallidus A. Smith	NL	*Micropterus punctulatus* (Rafinesque)[c]	Localised – H
Barbus serra Peters[b]	EN	*Micropterus salmoides* (Lacépède)[c]	Widespread – H
Labeo seeberi Gilchrist & Thompson[b]	CR	*Oncorhynchus mykiss* (Walbaum)[c]	In some mountain streams – M
Labeo umbratus (A. Smith)	NL	*Oreochromis mossambicus* (Peters)[c]	Widespread – M
Pseudobarbus afer (Peters)[b]	LR (nt)	*Oreochromis aureus* (Steindachner)	Very localised – L
Pseudobarbus asper (Boulenger)[b]	VU	*Tilapia sparrmanii* A. Smith[c]	Widespread – M
Pseudobarbus burchelli A. Smith[b]	EN	*Perca fluviatilis* L.	Very localised – L
Pseudobarbus burgi (Boulenger)[b]	CR	*Salmo trutta* L.[c]	Localised – M
Pseudobarbus phlegethon (Barnard)[b]	EN	*Tinca tinca* (L.)[c]	In Breede System – L to M
Pseudobarbus tenuis (Barnard)[b]	EN		
Galaxias zebratus Castelnau[b]	LR (nt)		
Sandelia capensis (Cuvier)[b]	NL		

[a]From Baillie and Groombridge (1996) where CR: critically endangered, EN: endangered, VU: vulnerable, LR (nt): lower risk, near threatened, and NL: not listed.
[b]Denotes endemic species.
[c]Denotes invasive species.

American (six species), Eurasian (three species) and southern African natives (four species).

The conservation status of CFK freshwater fishes is a particular cause of concern because 15 of its 19 species (79%) are threatened, including nine that are critically endangered or endangered (Table 39.1). Only four species, three of which are indigenous (*Barbus anoplus* Weber, *Barbus pallidus* Smith, *Labeo umbratus* (Smith)) and one endemic, *Sandelia capensis* (Cuvier), can be regarded as reasonably safe, although sub-populations of each of these species are under threat.

39.3.2 *Critical areas for conservation*

Forty-one river areas were mapped as being critically important to conserving the CFK's indigenous fishes. These areas and the existing network of protected areas (national parks, provincial nature reserves and local authority nature reserves) are shown in Figure 39.1. The critical areas are located across the CFK with the majority in the Olifants (15 areas) and Breede (6 areas) River systems respectively. The emphasis on the Olifants system relates its national importance as an endemic and threatened fish hotspot (Skelton *et al.* 1995). These areas are described in detail by Impson, Bills, Cambray and Le Roux (1999).

39.3.3 *Contribution of protected areas*

Freshwater fishes of the CFK, including threatened species, would appear to be well conserved as only two of the 19 indigenous species have not been recorded from protected areas (Table 39.2). Some species seem to be adequately conserved, for example,

Table 39.2 Contribution of protected areas in the CFK towards freshwater fish conservation at the species level

Species	Number of national parks in which recorded	Number of provincial nature reserves in which recorded	Number of local authority reserves in which recorded
Austroglanis barnardi			
A. gilli		2	
Barbus andrewi		4	1
B. anoplus		6	
B. calidus		2	
B. capensis		3	
B. erubescens			
B. pallidus		3	2
B. serra		2	
Pseudobarbus afer	1	6	
P. asper		7	
P. burchelli		5	3
P. burgi		2	
P. phlegethon		1	
P. tenuis		2	
Galaxias zebratus	3	12	3
Labeo seeberi		2	
L. umbratus		4	
Sandelia capensis		16	5

the widespread and relatively common *G. zebratus* and *S. capensis*. Contrastingly, *Barbus erubescens* Skelton, arguably South Africa's most threatened freshwater fish, has not been recorded from protected areas. Table 39.2 shows that South African National Parks within the CFK make a negligible contribution to the conservation of indigenous CFK freshwater fishes (two species recorded) compared to the moderate contribution of local authority nature reserves (five species) and substantial contribution of provincial nature reserves (17 of the 19 species recorded). There is also considerable overlap in some areas (e.g. Cederberg Mountains) between protected areas and the critical areas identified to conserve CFK fishes.

39.3.4 *Constraints to conservation*

Four constraints play a major role in restricting the effectiveness of the conservation agencies. These are the negative effects of invasive alien fishes, insufficient budgets and manpower for conservation work, unsustainable land use and poor public awareness.

The effects of invasive alien fishes in South Africa include competition for space and food, predation, hybridisation, the introduction of diseases and parasites, the disruption of ecological processes and a reduction in environmental quality (Skelton 1987; De Moor & Bruton 1988). In particular, the smallmouth bass *Micropterus dolomieu* Lacépède has had a severe effect on the highly endemic fauna of Cape rivers (Hamman & Jordaan 1988) and, like *M. salmoides*, is now widespread in the CFK, including several protected areas (Fig. 39.2). Detailed studies on most of these impacts are lacking,

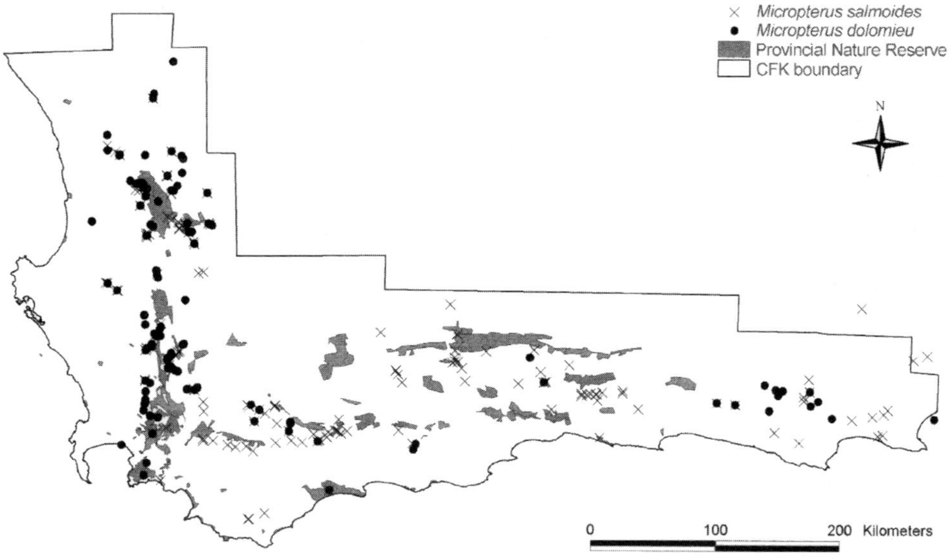

Figure 39.2 Distribution of the predatory and highly invasive smallmouth bass *Micropterus dolomieu* and largemouth bass *M. salmoides* in the CFK. These species are frequently recorded in provincial nature reserves

however, particularly the impact of predatory invasive alien fishes (Impson & Hamman 2000). Eradication of invasive alien fishes from rivers is possible using a variety of piscicides (Marking 1992), but this has not been undertaken in the CFK to date.

Critical capacity shortages have emerged at many South African conservation authorities since 1990 because of severance packages being offered by the state to reduce the size of the public service and posts not being filled due to budgetary constraints. For example, the Western Cape conservation authority had nine scientists and technicians in their aquatic section in 1992 compared with two at present. Continuing funding constraints have restricted monitoring and research on fishes at a time when these needs are at their most important.

Until 1994, environmental legislation in South Africa was largely ineffective in regulating and managing potentially harmful land-use practices such as intensive agriculture, afforestation and mining and their associated impacts. The channelisation of riverbeds, river diversion and water abstraction are important factors responsible for the decline of rivers and their fishes (Skelton 1987). The fall of apartheid in 1994 allowed outdated and environmentally harmful laws to be scrapped in favour of environmentally progressive national legislation such as the Impact Assessment Regulations of 1997 and Water Act of 1998. These Acts now provide South Africa's environmental custodians with the regulatory and enforcement capability to ensure sustainable utilisation of freshwater aquatic resources.

Lack of public awareness is another major problem. Few landowners have an understanding of river functioning and the ecological needs of rivers and regularly bulldoze rivers for flood control purposes and to create weirs. Similarly, some highly sensitive tributaries have abstraction points that remove the entire surface flow of the river during dry months. The attitude and activities of many freshwater anglers is also of concern. The CFK has few indigenous freshwater fishes of angling value and the three species that are so regarded are not easily accessible to the general public. Hence alien species are targeted and, being popular and accessible, are the ones that are stocked into new waters. Few stockings are legal (i.e. undertaken with the approval of conservation authorities) resulting in the continual spread of harmful invasive species.

39.4 Discussion

The indigenous freshwater fishes of the CFK are probably the most threatened and endemic component of its biota, yet they have been neglected to date in strategies and decisions to conserve the region. This is because most of the species are small, have little in the way of charismatic features and have no immediate economic value. In South Africa, protected areas have primarily been chosen because of their value in terms of terrestrial fauna and flora or for their contribution to water supply. The critical shortage of aquatic conservation staff has also prevented effective management of the CFK's freshwater and estuarine ecosystems when this has been most needed.

Fortunately, a conservation action plan has now been developed to more effectively conserve the inland waters of the CFK. The key components of this project, an integral part of the CAPE, which has been submitted to the Global Environment Facility for funding, are as follows.

- Develop a network of conservation areas that includes key aquatic systems for fish conservation (see Fig. 39.1). The urgent need for sanctuaries to protect freshwater fishes in the region and elsewhere in South Africa has been highlighted by Skelton (1987, 1993).
- Capacity needs within organisations (especially nature conservation authorities) involved in conserving freshwater fishes of the CFK have been determined and a further eight conservation staff are required at the WCNCB.
- Significantly increased funding was identified to allow urgent research projects, detailed field surveys and related management actions to proceed. Little is known about the biology, ecology and genetic diversity of several highly threatened species. These studies are essential to develop and implement effective species and habitat recovery plans (Bruton 1995). The conservation biology of the critically endangered *B. erubescens* has been determined (Marriot 1997), yet its recommendations have not been implemented due to a lack of capacity at the WCNCB. However, funding has now increased due to CAPE; three postgraduate studies are currently being undertaken on the conservation genetics of *G. zebratus*, *Pseudobarbus* and *Sandelia*.
- River rehabilitation projects, specifically removing alien fishes from critical fish conservation areas, are urgently required (Bills 1999) and are an important component of the conservation plan. Many threatened species recorded within protected areas actually receive little or no protection as they share protected area habitat with predatory invasive alien fishes such as *M. dolomieu* and *O. mykiss* and competitors such as *Cyprinus carpio* L. Permanent eradication of these species within protected areas is necessary for these areas to be effective. The incorporation of instream barriers to exclude invasive alien fishes is a key guideline in the design of aquatic reserves (Moyle & Sato 1991). Fortunately, the Working for Water Programme, a South African government poverty relief project has achieved significant success in clearing catchments and riparian areas of alien plants across South Africa (Impson & Hamman 2000).
- Environmental education and public awareness needs specific attention to change the way in which land-owners, anglers and the broader public view rivers and their indigenous fishes. A sense of custodianship of natural resources must be developed together with innovative methods of promoting the resource value of indigenous fishes.

Acknowledgements

The World Bank and Global Environment Facility (GEF) are especially thanked for funding the CAPE study. The GIS section of the WCNCB, in particular Annelise le Roux, are thanked for supplying IT and database design support. Freshwater fish databases were made available by the Albany Museum, J.L.B. Smith Institute of

Ichthyology, South African Museum and WCNCB. Peter Lloyd of the WCNCB is thanked for editorial comments. The Chief Executive Officer of the WCNCB is thanked for permission to publish this paper.

References

Baillie J. & Groombridge B. (1996) *1996 IUCN Red List of Threatened Animals*. Gland: IUCN, 368 pp.+annexes.

Bills R. (1999) Biology and conservation status of the Clanwilliam rock catfish and spotted rock catfish (Final Report). WWF-SA contract project. J.L.B. Smith Institute Investigational Report 59. 53 pp.

Bloomer P. & Impson N.D. (2000) Mitochondrial DNA differentiation in the critically endangered Berg River redfin *Pseudobarbus burgi*. *Journal of Heredity* **91**, 122–127.

Bruton M.N. (1995) Have fishes had their chips? The dilemma of threatened fishes. *Environmental Biology of Fishes* **43**, 1–27.

Davies B.R., O'Keeffe J.H. & Snaddon C.D. (1993) A synthesis of the ecological functioning, conservation and management of South African river ecosystems. *WRC Report No. TT 62/93*, Pretoria, 232 pp.

De Moor I.J. & Bruton M.N. (1988) Atlas of alien and translocated indigenous aquatic animals in southern Africa. *South African National Scientific Programmes Report* 144, 310 pp.

Hamman K. & Jordaan A. (1988) Invasive aquatic animals can harm ecosystems. *Custos* **18**, 38–45.

Impson N.D., Bills I.R., Cambray J.A. & Le Roux A. (1999) The primary freshwater fishes of the Cape Floral Kingdom: conservation needs for a unique and highly threatened fauna. Cape Nature Conservation Report, PBag X5014, Stellenbosch, 26 pp.

Impson N.D. & Hamman K.C.D. (2000) Conservation status, threats and future prospects for the survival of the freshwater fishes of the Western Cape Province, South Africa. In I.G. Cowx (ed.) *Management and Ecology of River Fisheries*. Oxford: Fishing News Books, Blackwell Science, pp. 418–427.

Marking L.L. (1992) Evaluation of toxicants for the control of carp and other nuisance fishes. *Fisheries* **17** (6), 6–13.

Marriot M.S. (1997) *Conservation biology and management of the Twee River redfin* Barbus erubescens *(Pisces: Cyprinidae)*. Unpublished MSc Thesis. Department of Ichthyology, Rhodes University, Grahamstown, 104 pp.

Meffe G.K. & Carroll C.R. (1994) *Principles of Conservation Biology*. Sunderland, Massachusetts: Sinaeur Associates, 600 pp.

Moyle P.B. & Sato G.M. (1991) On the design of preserves to protect native fishes. In W.L. Minckley and J.E. Deacon (eds) *Battle against Extinction: Native Fish Management in the American West*. Tuscon: The University of Arizona Press, pp. 155–173.

Skelton P.H. (1983) Perspectives on the conservation of threatened fishes in southern Africa. *The Naturalist* **27**, 3–12.

Skelton P.H. (1987) South African Red Data Book – Fishes. *South African National Scientific Programmes Report* 137, 199 pp.

Skelton P.H. (1993) *A Complete Guide to the Freshwater Fishes of Southern Africa*. Halfway House: Southern Book Publishers, 388 pp.

Skelton P.H., Cambray J.A., Lombard A. & Benn G.A. (1995) Patterns of distribution and conservation status of freshwater fishes in South Africa. *South African Journal of Zoology* **30**, 71–81.

Waters J.M. & Cambray J.A. (1997) Intraspecific phylogeography of the Cape galaxias from South Africa: evidence from mitochondrial DNA sequences. *Journal of Fish Biology* **50**, 1329–1338.

Section VI
Conclusions and recommendations

Chapter 40
Freshwater fish conservation: options for the future

I.G. COWX*
Hull International Fisheries Institute, University of Hull, UK

M.J. COLLARES-PEREIRA
Centro de Biologia Ambiental, Faculdade de Ciências, Universidade de Lisboa, Portugal

Abstract

Fresh waters have suffered the most intense intervention of all ecosystems over the past 100 years and many fish species are now extinct, rare or endangered and the need for conservation action is paramount. This chapter reviews our conclusions from an international symposium on freshwater fish conservation held in late 2000. It reviews the strengths, weaknesses, opportunities and threats relating to the conservation of freshwater fish as perceived by participants at the symposium. It concludes with options for action to ensure our freshwater fish natural heritage for future generations.

Keywords: conservation genetics, endangered species, habitat degradation, management, public awareness, stock enhancement.

40.1 Introduction

The conservation of freshwater fishes has been recognised as an important action throughout the world, since the number of species that are extinct, endangered or becoming rare is increasing dramatically as a result of a numerous anthropogenic perturbations. Unfortunately much of the knowledge on freshwater fish conservation is dispersed and fragmented and the need to collate the information was recognised during the tenure of the European Union (EU) LIFE-Saramugo project (Collares-Pereira, Cowx, Rodrigues, Rogado, Ribeiro, Mendes, Picciochi, Salgueiro, Alves & Coelho 2000). The symposium on *Freshwater Fish Conservation: Options for the Future* was convened in October/November 2000 to address this issue. The objectives of the symposium were to bring together experts in fish conservation to:

- exchange and disseminate scientific, technical and practical information about rare and endangered fish, and their conservation;
- highlight successes and failures of existing projects and their implementation; and
- identify current and future practices in the conservation of freshwater fish.

*Correspondence: Hull International Fisheries Institute, University of Hull, Hull HU6 7RX, UK (email: i.g.cowx@hull.ac.uk).

The symposium was, and these proceedings are, broken down into five themes to achieve these objectives.

- *Geographical overview*: review of the present status of threatened fish species and the conservation in Europe and the rest of the world.
- *Biology of rare and endangered species*: analysis of the biology of threatened species (e.g. case-studies from EU LIFE-NATURE programmes), methods of data collection and how the data are used for formulating conservation measures.
- *Analysis of constraints and opportunities*: analysis of the threats to fish populations from water resource development schemes, pollution, habitat degradation, introduction and translocation of fishes, climatic change; problems of river and lake conservation: ecosystem use and sustainability.
- *Conservation genetics*: use of genetic information in conservation: methodologies, units of conservation, other applications and main limitations.
- *Methods of managing freshwater fish conservation*: use of breeding endangered species in captivity for stock enhancement; role of legislation, institutions and policy making in fish conservation: past and present challenges; meeting freshwater fish conservation goals in an uncertain future.

To draw conclusions from such a diverse and wide coverage of information is complex. Many aspects need to be brought together to highlight limitations in our knowledge on freshwater fish conservation and promote the positive aspects for the benefit of future generations. In an attempt to achieve this, a SWOT (Strengths, Weaknesses, Opportunities and Threats) analysis was performed to draw key conclusions from the symposium and identify the way forward for freshwater fisheries conservation in the future. This chapter is broken down into the four components of a SWOT analysis, plus a concluding section of options for future action.

40.2 Strengths

It is widely accepted by the international community that there is a need to protect biodiversity, including freshwater fish. This is evident from the numerous Conventions and Directives (e.g. Bern and Washington Conventions, EU Habitats Directive 92/43/EEC, 2000 IUCN Red List) which underpin biodiversity protection and were promulgated through Agenda 21 of the Rio Convention. This international recognition, which has filtered into the political arena, should be used to drive future conservation management.

Despite the global recognition of conserve fish biodiversity being only a recent phenomenon, good legal and institutional frameworks (Skelton, Chapter 21; Crivelli, Chapter 34; Kirchhofer, Chapter 35) exist in many countries of the world. However, many of the institutions lack the resources to implement sound conservation plans (e.g. Impson, Bills & Cambray, Chapter 39) and representation of freshwater fishes is generally weak. Similarly, the resources to enforce legislation are lacking, thus freshwater fishes are not always afforded the protection they are designated under law.

Perhaps the greatest strength recognised in the conservation of freshwater fishes is the network of dedicated, enthusiastic persons working towards a common goal. This was epitomised by the symposium on *Freshwater Fish Conservation* on which these proceedings are based, but also the desire of all concerned to protect fish biodiversity for the benefit of future generations. This is reflected in the diverse range of projects, undertaken throughout the world, described throughout these proceedings. However, despite strong representations, few conservation projects can be considered truly successful. Those that appear successful generally involve the general public or have wider environmental benefits, e.g. restoration of Atlantic salmon, *Salmo salar* L., into the rivers Thames and Rhine (Mills 1990). These successes must be reviewed in greater detail, in conjunction with abject failures, to elucidate the factors that contributed to the outcome of the project and positive elements should be transferred to existing and future actions for the conservation of freshwater fishes.

Finally, molecular markers have provided a powerful new tool to determine historical and extant genetic diversity between populations and to define strategies for species recovery (Coelho, Salgueiro, Alves & Collares-Pereira, Chapter 30) and stock enhancement programmes (Carvalho & Cross 1998). The information can be used to identify the donor populations in such programmes to avoid genetic changes and/or erosion. These markers have also highlighted the negative impact of extensive stocking programmes which have diluted gene pools (Weiss, Uiblein & Jagsch, Chapter 7; Almodóvar, Nicola & Suárez, Chapter 31) and possibly contributed to the extinction of many fish populations (Persat & Keith, Chapter 32). Furthermore, the use of genetic discrimination to identify evolutionary diversity within species, linked to the concept of evolutionarily significant units (ESU) (Moritz, McGuigan & Bernatchez, Chapter 27) offers an opportunity to define conservation units within species. A problem lies, however, in the definition of species or conservation units for a wider audience, to justify specific or local conservation status and appropriate management strategies (see Section 40.3).

40.3 Weaknesses

When reviewing the preceding chapters, it becomes evident that despite fish being the taxonomic groups with the high number of endangered species (http://www.redlist.org/tables/table1a.html) there are comparatively few fish species conservation projects (e.g. between 1996 and 1999 only eight EU LIFE projects on fish were supported whereas 20 projects targetted mammals and 54 targetted birds; http://europa.eu.int/comm/environment/nature/home.htm). Furthermore, there are few examples of truly successful fish conservation projects, and those that exist, have involved considerable investment, in financial terms or human resources, from central and national agencies. Questions should be asked to why this is the case. Part of the reason is lack of public awareness and poor/uninformed management, which is discussed later. However, there is undoubtedly a link to weak information on which to formulate conservation action plans. The paucity of information of freshwater fishes is universal, partly because of the intrinsic difficulties in studying this species group (Cowx 1996), but also because of lack of investment in freshwater fish research *per se*. This is again epitomised by the apparent lack of recognition of the importance of freshwater fish and fisheries by the

EU and national institutions when formulating research priorities (e.g. see EU research programmes on http://www.cordis.lu/en/src/f_001_en.htm). If we are to improve the success of conservation projects, there is an urgent need to carry out research on the basic ecology of species, including habitat preferences and environmental tolerances (e.g. Economidis, Chapter 8; Cambray, Chapter 9; Almeida & Quintella, Chapter 12; Hefti, Chapter, 38), and on the historical and contemporary evolutionary process (Moritz *et al.*, Chapter 27), to underpin rehabilitation and recovery programmes. There is, however, a note of caution with respect to research; it should not be used to delay taking action to conserve endangered fish species or communities. Too many species are now in need of immediate intervention, and decisions need to be derived on best available information to save them from pending extinction.

A major problem that faces the credibility of conservationists is linked to species identification. There are many inherent problems with the recognition of species, especially from different geographical regions. The advent of molecular techniques has exacerbated this problem because genetic diversity is now being used to discriminate populations (Lusk, Lusková, Halacka, Šlechta & Šlechtová, Chapter 2; Persat & Keith, Chapter 32). The discovery of hybridogenetic complexes (Bohlen, Ráb, Šlechtová, Rábová, Ditterbusch & Freyhof, Chapter 28) is further complicating the issue. Consequently, the separation of species in sub-units, each linked to an ESU (i.e. historically isolated and thus independently evolving sets of populations) and the desire to designate local management units within an ESU for conservation purposes (Persat & Keith, Chapter 32) will alienate development enterprises and policy makers alike. Such a breakdown of species into distinct conservation units, will effectively designate most aquatic systems as sites of conservation interest and prevent any further development opportunities. At some stage more equitable criteria for defining the lowest taxonomic unit for conservation purposes will have to be developed to overcome this problem, and priorities for conservation will have to be established based on costs and resources. If this is not forthcoming, it is envisaged that the value of defining traditional conservation status (e.g. IUCN categories, Baillie & Groombridge 1996) will be diluted and become ineffective when lobbying for conservation of biodiversity.

Intrinsic within this issue are the procedures for designating species under the various Conventions (e.g. IUCN or Bern, EU Habitats Directive). This is mostly based on species abundance, whereas it is more often the habitat that is at risk, and indirectly the species. The criteria for designating status should thus be widened to take on board a species' vulnerability across its entire range, and threats to its habitat (Cowx, Chapter 20), both now and in the future.

It should be recognised that fish conservation in a multiple user environment is fraught with problems. Fish and fisheries are often considered of marginal importance because the value of the resource is usually ill-defined and poorly represented from an economic and social perspective (Cowx, Chapter 20). Fisheries are traditionally managed based on the quality of the fishing experience, and few are managed from an economic perspective (Cowx 2002), an issue born out by the paucity of information on the economic value of such fisheries (e.g. Kennedy & Crozier 1997; Peirson, Tingley, Spurgeon & Radford 2001). This problem spills over into fish conservation because the value of the species, particularly endangered species, has rarely been assessed. Consequently, fish conservation is given low priority in any consultation process and it is difficult to argue for conservation of the resource. If fish conservation is to be promoted in the future, there is an urgent need to provide robust, defensible, social and economic valuation of fish populations and species diversity (Cowx, Chapter 20). Once this information is available, value will be a powerful tool for arguing the case of fish conservation. However, it must be recognised that it is not the only tool to be used because the economic value of a major water resource scheme will far outweigh fish conservation value. This is primarily because the methods used are often fisheries specific and do not consider the upstream economic value in terms of aesthetic and conservation value or the downstream value associated with the service sectors.

Although an enthusiastic network of persons was considered a strength (Section 40.2), it can equally be considered a weakness. This contradiction arises because the freshwater fish fraternity frequently work in isolation of other resource practitioners, i.e. they live in a 'piscicentric' world where consideration of the needs of other resource users are often ignored or given little respect. Inevitably this leads to conflict which generally favours the strongest economic argument, often to the detriment of other user or ecological entities, especially freshwater fish and fisheries. Freshwater fisheries have suffered heavily because, as pointed out earlier, they are poorly valued in economic terms and in a market-driven environment will be largely overlooked.

Some of the blame must fall on conservation practitioners because scientists, managers and policy makers each have their own interests and modes of operation, and they respond to different motivations, constituencies and reward systems (Meffe, Chapter 33). However, as Meffe (*op. cit.*) pointed out, if conservation is to succeed, each must be engaged and work effectively with the others as a unit. Conservation often fails because these three components do not work together well, and even work at cross purposes.

Linked to this scenario is the apparent lack of appeal of fish species in general, i.e. they are not considered 'charismatic megafauna', such as elephants or whales. There is also a general lack of public and political awareness of the plight of many fish species, probably because they cannot easily be seen. Consequently their role in influencing any decision-making process is marginalised. Lessons can, however, be learnt from conservation programmes focussing on more charismatic species or the few fish species that have captured the public attention. There are several well-documented conservation/restoration projects in Europe, e.g. recovery of Atlantic salmon, *Salmo salar* L., into the rivers Thames and Rhine (Mills 1990) or conservation of saramugo, *Anaecypris hispanica* (Steindachner), in the River Guadiana, Portugal (Collares-Pereira, Cowx, Rodrigues & Rogado, Chapter 19). In these cases the public have been used to drive both the initial activities but also maintain pressure in the political arena for continued support of the conservation/restoration actions. The role of the public in influencing policy makers cannot be over emphasised, and due attention should focus on gaining public support through education and extension programmes in schools and the local communities, despite the general philosophy of scientists that this is not their role (Cambray & Pister, Chapter 37).

40.4 Opportunities

The concept of ecological indicators is well developed in the management of biological resources. For example, the Index of Biological Integrity promoted by

Angermeier and Karr (1986) is now being adopted as a mechanism for predicting ecosystem disturbance and health (Schmutz & Jungwirth 1999), and the presence of threatened fish species can be used in the same vane. This has great advantages because conservation actions focussing on endangered species will invariably benefit other species, especially where the actions involve rehabilitation of degraded habitats or protection of fragile environments. This concept was advanced by Filipe, Seabra, Marques, Silva, Tiago, Ribeiro, Cowx and Collares-Pereira (2000) who developed a simple index to define priority areas for the conservation of fish based on species rarity and endemicity. The use of such indices is valuable and should be used by conservation lobbies to show the importance of threatened habitats.

The principle of prioritising conservation areas using indices can be used to promote nature reserves or sanctuaries (Cambray, Chapter 9; Crivelli, Chapter 34). To date, few have been established specifically for freshwater fishes (Crivelli, Chapter 34), but this should change as knowledge of the status and distribution of fish species improves. Appropriate authorities and institutions should extend the existing network of nature reserves (or NATURA 2000 sites in Europe) to offer more protection for freshwater fishes. Care must, however, be taken not to designate small sections of rivers where the species persists but to identify the reasons for the reduction in population size and distribution and protect the extant populations from further deterioration. In essence the designation of watersheds, including all the catchment (land and water) upstream (Collares-Pereira *et al.*, Chapter 19; Persat & Keith, Chapter 32; Crivelli, Chapter 34) are necessary and must be negotiated.

Molecular studies have demonstrated the impact of stocking on the loss of genetic diversity. This information should be used to both restrict and direct stock enhancement programmes more effectively. Strong arguments should be formulated, based on the deleterious impacts of stocking on extant populations and loss of fitness (Carvalho & Cross 1998), to drive managers towards rehabilitation actions, which remove bottlenecks in the fish's life cycle. This would not only be a more acceptable long-term solution to the problem because the environment will be improved and natural recruitment procssses will be re-instated, but probably also benefit other stressed and threatened species in the target water body.

Information on genetic variability can also be used to define which populations are under threat and which populations are most appropriate in correctly planned restocking programmes (Coelho *et al.*, Chapter 30). The tool should also be used to show when mixing of stocks should be avoided because of loss of genetic and phenotypic adaptations (Moritz *et* al., Chapter 27). This philosophy should be built into all aquaculture programmes established for fish conservation purposes. Indiscriminant mixing of stocks in aquaculture facilities should be avoided unless the purpose is for reintroduction when a wide gene pool could potentially benefit the colonisation process (Cowx 1994).

One of the factors common to 'successful' conservation projects was involvement of people. The general public are excellent ambassadors to promote the fish cause (Cambray & Pister, Chapter 37). The biggest problem identified, however, was that the general populace have poor awareness of the issues and problems facing freshwater fish, thus greater opportunity should be made of their willingness to support conservation campaigns by promoting education and extension programmes

(Cambray & Pister, Chapter 37). Similarly, NGOs should be encouraged to support conservation of freshwater fish and front environmental lobbying of potentially damaging development projects.

40.5 Threats

Freshwater fishes are threatened by a wide array of factors, but anthropogenic disturbance seems to underlie the decline and extinction of many fish species (see Cowx, Chapter 20 for review). The main perturbations can be broken down into five key problems, viz: species introductions and translocations (e.g. Winfield, Fletcher & James, Chapter 25; Persat & Keith, Chapter 32), impoundment of rivers (dams and weirs, water abstraction and water transfer schemes) (e.g. Barbieri, Economou, Stoumboudi & Economidis, Chapter 5; Rodríguez-Ruiz, Encina & Granado-Lorencio, Chapter 26), water quality deterioration (pollution, eutrophication, acidification) (e.g. Bervoets, Lodts, Van Campenhout & Blust, Chapter 23; Van Campenhout, De Coen, Vandenbergh, Denayer & Janssen, Chapter 24), habitat degradation and fragmentation (channelisation and land use change, mineral extraction) (e.g. Chapman & Chapman, Chapter 22) and overexploitation (Freyhof, Chapter 1; Pustelnik, Guerri & Lepage, Chapter 14; Almodóvar *et al.*, Chapter 31). These problems seem to be universal (as exemplified throughout these proceedings). Although many of the issues are being addressed in developed countries through environmental legislation, the rate of progress in reversing the impacts is pathetically slow. Furthermore the cost of implementing rehabilitation programmes or seeking alternative solutions to the demands on water resources, which underlies many of the issues, is prohibitive and at best only a *status quo* is being achieved with respect to habitat quality, and at worst, as is still commonly found throughout the developing world where financial resources are limited, progressive deterioration is rife.

Of these key threats, water resource development schemes are a particular problem because the economic value of such schemes outweighs the conservation arguments, as mentioned earlier. Similarly, stock enhancement programmes are a much used and frequently abused management activity. This is because the social value of recreational and commercial fishing is high and conservation interests are largely ignored (Cowx 2002). To reverse these philosophies is going to be a major challenge to conservation managers, but neither will be achieved if the true economic value of conserving fish species is not enunciated or the fishing fraternity are not educated in deleterious effects of introductions and translocations of fish species (Cowx 1998, 2002).

Climatic change is a problem that seems to have received little attention with respect to fish conservation. However, Cowx (Chapter 20) emphasised the need to recognise that this phenomenon goes beyond the increased prevalence to floods and droughts and their direct effects, to the influence of environment change on ecological processes, including physiological tolerances and ecosystem dynamics. This could lead to the natural spread and proliferation of alien species with impacts similar to those observed for species introductions.

Notwithstanding the key issues identified above, the underlying cause for continued environmental degradation and overexploitation of aquatic resources is undoubtedly the ever expanding human population. Demands on water resources are continuously expanding, especially in arid and semi-arid regions, and schemes to increase supply for human consumption, agriculture, industry and power generation will inevitably increase pressure on, what must be considered, finite resources. The consequences are obvious with more species becoming threatened or extinct. Coupled with the demand for water is the increased exploitation of the fisheries and terrestrial resources (especially land and forestry; Chapman & Chapman, Chapter 22 and more recently tourism, see Cowx & Collares-Pereira 2000; Cowx, Chapter 20), which are exacerbating the problems. Regulation and control of access to the resources, and their utilisation in an environmentally sustainable manner are fundamental if fish diversity is not to be compromised further.

Skelton (Chapter 21) indicated that environmental conservation depends upon skills and knowledge of the people involved and the general decline of taxonomy and of the number of taxonomists itself is a threat to the conservation of fishes. The demise of tertiary level (university) education in taxonomy must be recognised as a threat to the future of species because it will strangle the flow of knowledge on biodiversity to society. Consequently, more emphasis must be placed on taxonomic studies in tertiary level training institutions to ameliorate this problem.

40.6 Options for the future

In view of the critical status of many freshwater habitats and their biota, there is a pressing need to take action and not fall back on the old addage that more research is required to ensure the decisions being made are appropriate. The urgency for direct management intervention is so great in relation to many species that decisions should be based on the best available science and existing experience to support management options. Where information on which to make a decision remains inadequate, the precautionary approach (FAO 1996, 1997) must be adopted. This is particularly important where development schemes are likely to impact on fish communities about which little is known. However, a networking mechanism for reporting the successes and failures in conservation management needs to be developed so lessons can be transferred to all. To achieve this, the network of fish conservationists, many of whom participated in this symposium (on *Freshwater Fish Conservation*) from which these proceedings were collated, must be expanded and become more transparent in their activities and more prolific in their information exchange.

Although information on the distribution and abundance of many species will always remain a limitation, existing information should be collated to reformulate the species status under international conventions such as the Bern Convention and the Habitats Directive (see, e.g. the wealth of information available in Chapters 1–19). The criteria by which species are designated should be widened to take account of environmental degradation or stress and how this could contribute towards the potential demise of a species. In addition, the EU Habitats Directive needs urgently updating to account for the changes in species status and the environment since it was first formulated pre-1992.

Increasing pressures on aquatic resources dictate that fish conservation can no longer be treated in isolation and an integrated approach to aquatic resource management is required (Cowx 1998). As shown throughout this volume, fish biodiversity is being constantly eroded, not only by exploitation of fish directly but mainly through degradation of their habitat. However, the demands for sustainability that grew out of the Rio conference of 1992 have put emphasis on the need to not only manage exploited resources but also promote biodiversity. Consequently, conflicts between these various interests must be resolved by involving all stakeholders in the management process. This can be achieved through integrated aquatic resource planning and management. River basin management plans, at both the national and multi-national scale, which will be obligatory under the new EU Water Framework Directive, will support this process but the profile of fish conservation needs to be raised and be better integrated into the planning process. Without this involvement the future of freshwater fishes remains bleak.

Similarly, there is a need to develop partnerships with stakeholders in affected ecosystems to strengthen and implement fish conservation activities, and develop mechanisms to influence other players. To achieve this, scientists must expand their range of activities from monitoring and reporting the status of endangered species to more influential and preventative work. They must use the best available data to educate other stakeholders and the wider public. They need to be involved in accurate environmental impact assessments and rehabilitation programmes to argue the case for fish conservation, i.e. there is need to develop a risk based approach to fish conservation. There is also a need to develop fiscal measures, such as the 'polluter-pays principle', and enforce legislation through the appropriate channels and institutions. This will only be achieved through valuation of threatened resources, an issue that is acting against the fish conservation lobby, and will be essential for integration into river basin management plans. There is an urgent need to adapt environmental economic evaluation tools to value the social and economic importance of freshwater fish biodiversity. Until this is undertaken fish conservation will continue to be given low priority in any consultation process and it will remain difficult to attract investment or credit for protection of the fishes.

Finally, the science and management of freshwater fish conservation must be considered in its infancy, and protocols and procedures utilised by many institutions need collation in a format that is understandable by all practitioners and interested parties. It is recommended that a manual of best fish species conservation practices is collated, and is written in a sympathetic manner, which can be referred to by managers, planners, engineers and assist in their decision-making processes.

References

Angermeier P.L. & Karr J.R. (1986) Applying an index of biological integrity based on stream fish communities: considerations in sampling and interpretation. *North American Journal of Fisheries Management* **6**, 418–429.

Baillie J. & Groombridge B. (eds) (1996) *1996 IUCN Red List of Threatened Animals.* Gland, Switzerland and Cambridge: IUCN, 368 pp.+annexes.

Carvalho G.R. & Cross T.F. (1998) Enhancing fish production through introductions and stocking: genetic perspectives. In I.G. Cowx (ed.) *Stocking and Introduction of Fish*. Oxford: Fishing News Books, Blackwell Science, pp. 329–337.

Collares-Pereira M.J., Cowx I.G., Rodrigues J.A., Rogado L., Ribeiro F., Mendes A., Picciochi P., Salgueiro P., Alves M.J. & Coelho M.M. (2000) Uma estratégia de conservação para o saramugo (*Anaecypris hispanica*), um endemismo piscícola em extinção. *Final Technical Report, LIFE-NATURE Programme, contract B4-3200/97/280,* Vol. I (121 pp.) and Vol. II (13 annexes) (in Portuguese with an English summary).

Cowx I.G. (1994) Stocking strategies. *Fisheries Management and Ecology* **1**, 15–30.

Cowx I.G. (1996) The integration of fish stock assessment into fisheries management. In I.G. Cowx (ed.) *Stock Assessment in Inland Fisheries*. Oxford: Fishing News Books, Blackwell Science, pp. 495–506.

Cowx I.G. (1998) Stocking strategies: issues and options for future enhancement programmes. In I.G. Cowx (ed.) *Stocking and Introduction of Fish*. Oxford: Fishing News Books, Blackwell Science, pp. 3–13.

Cowx I.G. (2002) Recreational fisheries. In P.J.B. Hart & J Reynolds (eds) *The Fisheries Handbook*. Oxford: Blackwell Science (in press).

Cowx I.G. & Collares-Pereira M.J. (2000) Conservation of endangered fish species in the face of water resource development schemes in the Guadiana River, Portugal: harmony of the incompatible. In I.G. Cowx (ed.) *Management and Ecology of River Fisheries* Oxford: Fishing News Books, Blackwell Science, pp. 428–438.

FAO (1996) *Precautionary approach to capture fisheries and species introductions*. Rome: FAO Fisheries Department Technical Guidelines for Responsible Fisheries No. 2, 54 pp.

FAO (1997) *Inland fisheries*. Rome: FAO Fisheries Department Technical Guidelines for Responsible Fisheries No. 6, 36 pp.

Filipe A.F., Seabra S., Marques T., Silva R., Tiago P., Ribeiro F., Cowx I.G. & Collares-Pereira M.J. (2000) A methodology for the definition of priority areas for conservation. International Symposium on Freshwater Fish Conservation: Options for the Future, Algarve (Portugal), 30 October–4 November, 64 pp. (Abstract).

Kennedy G.J.A. & Crozier W.W. (1997) What is the value of a wild salmon, *Salmo salar* L., smolt? *Fisheries Management and Ecology* **4**, 103–110.

Mills D.H. (ed.) (1990) *Strategies for the Rehabilitation of Salmonid Rivers*. London: Linnean Society, 208 pp.

Peirson G., Tingley D., Spurgeon J. & Radford A. (2001) Economic valuation of inland fisheries in England and Wales. *Fisheries Management and Ecology* **8**, 415–424.

Schmutz S. & Jungwirth M. (1999) Fish as indicators for large river connectivity: The Danube and its tributaries. *Archiv für Hydrobiologie* **115**(Suppl.), 329–348.

Subject index

Species index